STUDENT'S STUDY GUIDE AND SOLUTIONS MANUAL

Emmett M. Larson
Brevard Community College

Linda R. Beller
Brevard Community College

to accompany

MATHEMATICAL IDEAS
NINTH EDITION AND EXPANDED NINTH EDITION

Miller • Heeren • Hornsby

Addison
Wesley

Boston San Francisco New York
London Toronto Sydney Tokyo Singapore Madrid
Mexico City Munich Paris Cape Town Hong Kong Montreal

Reproduced by Addison Wesley Longman from camera-ready copy supplied by the authors.

Copyright © 2001 Addison Wesley Longman.

ISBN   0-321-07606-0

4 5 6 7 8 9 10   03 02

Contents

Contents

Contents

PREFACE

The *Study Guide and Solutions Manual* is intended to accompany *Mathematical Ideas*, Ninth Edition. The solutions to all odd-numbered exercises are included for Chapters 1 through 15. Solutions to the exercises in chapter tests are also included for the first time. The general style of the solutions is to explain all the "little steps" and to provide at least one problem of each type worked in detail. Chapter summaries provide, in varying detail, a review of key points, extra examples, and an enumeration of major topic objectives.

The *Study Guide and Solutions Manual* supports several styles of use by students: those who learn by exploring many problems and their solutions; those who need additional help on particular problems; and finally those who have studied the material previously and need only to review.

We would like to thank Dr. Al Koller, President of Brevard Community College, Titusville Campus, for the support provided to us during this project. Many thanks also go to several people who assisted us in various ways: Fayek Abdel Al, Lena Frame, William Maneker, Virginia Monroe, Angela Pletcher, Bill and Lois Queen, John Ryan Quinn, Lakshmi Ravindran, and Chris Stankiewicz for their thoughtful and careful chapter reviews. We would also like to thank Matthew Beller for his work on the Metric Appendix and assistance on quality check, and Kara Larson for her work in creating the logic in the Logic Puzzles solutions. Emmett would like to thank his family for "hanging in there" those long nights and weekends and especially his wife, Andrea, who did such a great job so often in the role of both mother and father during this period.

We would like to give our special thanks to Arthur Beller, Linda's husband, for managing this project, for keeping us focused and sane, for updating and creating our graphic art, for scheduling assistance, for his needed interfacing with AWL and for his continuous technical support. His long hours, thoughtful perspective, and strong management skills made him a critical asset to our writing team. Thank you Art!

Emmett Larson and Linda Beller
October, 2000
Titusville, Florida

1   THE ART OF PROBLEM SOLVING

1.1 Solving Problems by Inductive Reasoning
1.2 An Application of Inductive Reasoning: Number Patterns
1.3 Strategies for Problem Solving
1.4 Calculating, Estimating, and Reading Graphs
Chapter 1 Test

Chapter Goals

The general goal of this chapter is to provide an understanding of problem solving techniques and strategies.   Upon completion of this chapter, the student should be able to

- Distinguish between inductive and deductive reasoning.
- Identify number patterns.
- Use successive differences to determine terms in a number sequence.
- Adapt problem solving strategies to the solution of application problems.
- Do calculations using a calculator, estimate answers without the use of a calculator, and interpret graphs.

Chapter Summary

In this section, we examine two types of reasoning. Reasoning forms are known as arguments.

An **argument** is a claim that a statement, called the conclusion, follows from a set of statements called the premises.

A **premise** can be an assumption, law, rule, widely held idea, or observation.   From the premises, we then reason inductively or deductively to obtain a conclusion.   The premises and conclusion make up a logical argument. People, including students, philosophers, and scientists, use two basic kinds of reasoning processes.   It is important that one has fixed clearly in mind the distinction and usefulness of each.

The first type of reasoning is called inductive reasoning. This is the process of drawing conclusions on the basis of many observations.   Scientists use inductive reasoning when they make observations and discover definite patterns from which they propose a general law.   In science this procedure is often called the experimental or scientific method.

Inductive reasoning is the process of reasoning to a general conclusion (making a conjecture) through observations of specific cases or examples.

Inductive reasoning is the type of reasoning that is often used in our everyday lives. We frequently make decisions based on the results of similar experiences that we have had or observed.   Everyday we are exposed to arguments from the media and individuals, which are intended to persuade us to accept certain ideas or products. Many of these arguments are examples of inductive reasoning.

Although inductive reasoning provides a powerful tool to obtain conclusions, it is important to realize that there is no assurance that the conjecture reached will always be true. The difficulty with inductive reasoning is the conclusion or conjecture reached may well be true for the specific cases studied but may not be true for all cases.   In testing a conjecture reached by inductive reasoning, it takes only one example, referred to as a counterexample, to prove the conjecture false. Of the two forms of reasoning discussed here, however, inductive reasoning is the most creative.

On the other hand, the second type of reasoning, called deductive reasoning, is free of this problem associated with inductive reasoning.   If the premises are true in a (valid) deductive argument, one can be sure that the conclusion will be true.   That is, one is guaranteed a true conclusion given a deductive argument which is valid and has true premises. For this reason, deductive reasoning is used most often in mathematical proofs.

Deductive reasoning is the process of reasoning to a specific conclusion through general observations and facts.

EXAMPLE A   Inductive or Deductive?

Identify the following arguments as inductive or deductive reasoning.

(a) Every Friday we serve fish for dinner. Today is Friday. Therefore, today we will serve fish for dinner.

Solution
Since we are moving from a general rule to a specific conclusion, this argument is deductive.

(b) Fish was served for last Friday's dinner. Fish was served this Friday for dinner.
Therefore, fish will be served for all Friday dinners.

Solution
Since one is moving from specific pattern to a general rule, this argument is inductive.

Inductive reasoning is the process of arriving at a general conclusion from a specific pattern.

EXAMPLE B   Conjecture–Pattern

Determine the most probable next term in the list of numbers: 1, 4, 9, 16, 25

Solution
The next term is probably 36, which is 6^2. Notice that the other numbers can be computed as $1^2, 2^2, 3^2, 4^2, 5^2$.

Inductive reasoning can sometimes be used to predict an answer in a list of similarly constructed computation exercises, as shown in the next example.

EXAMPLE C   Conjecture–Computation

Use the list and inductive reasoning to predict the next equation, and then verify your conjecture.

$$\frac{1}{1\cdot2} = \frac{1}{2}$$
$$\frac{1}{1\cdot2} + \frac{1}{2\cdot3} = \frac{2}{3}$$
$$\frac{1}{1\cdot2} + \frac{1}{2\cdot3} + \frac{1}{3\cdot4} = \frac{3}{4}$$
$$\frac{1}{1\cdot2} + \frac{1}{2\cdot3} + \frac{1}{3\cdot4} + \frac{1}{4\cdot5} = \frac{4}{5}$$

Solution
Reasoning inductively from the above pattern we may arrive at
$$\frac{1}{1\cdot2} + \frac{1}{2\cdot3} + \frac{1}{3\cdot4} + \frac{1}{4\cdot5} + \frac{1}{5\cdot6} = \frac{5}{6}.$$

To verify,
$$\frac{1}{1\cdot2} + \frac{1}{2\cdot3} + \frac{1}{3\cdot4} + \frac{1}{4\cdot5} + \frac{1}{5\cdot6} =$$
$$\frac{1}{2} + \frac{1}{6} + \frac{1}{12} + \frac{1}{20} + \frac{1}{30} =$$
$$\frac{30}{60} + \frac{10}{60} + \frac{5}{60} + \frac{3}{60} + \frac{2}{60} =$$
$$\frac{30+10+5+3+2}{60} =$$
$$\frac{50}{60} = \frac{5}{6}.$$

When a number sequence is too complicated to make an obvious conjecture about the next term, the method of successive differences may be applied as in the next example.

EXAMPLE D   Method of Successive Differences

Consider the sequence 21, 34, 51, 72, 97 ... Use the successive differences to find the next term.

Solution
Find the differences between each term. Repeat the process until the differences are constant.

$$\begin{array}{ccccccccccc} 21 && 34 && 51 && 72 && 97 && 126 \\ & 13 && 17 && 21 && 25 && 29 \\ && 4 && 4 && 4 && (4) \end{array}$$

Observe that 29 comes from the addition of (4) to 25 and that 126 comes from the addition of 29 to 97. Thus, our next term in the sequence is found by $29 + 97 = 126$.

EXAMPLE E   Guessing and Checking

Several equations are given illustrating a suspected number pattern.

$$1^2 + 1 = 2^2 - 2$$
$$2^2 + 2 = 3^2 - 3$$
$$3^2 + 3 = 4^2 - 4$$

Determine the next equation and verify.

Solution
Consider the pattern. Using inductive reasoning, our next guess might be

$$4^2 + 4 = 5^2 - 5.$$

To verify, evaluate each side of the equation to get

$$16 + 4 = 25 - 5$$
$$20 = 20.$$

The Greek mathematicians of the Pythagorean brotherhood are given credit for studying numbers of geometric arrangements of points, such as triangular numbers (T), square numbers (S), and pentagonal numbers (P). (See Figure 5 in Section 1.2 of the text.) Formulas to generate these numbers as well as hexagonal numbers (H), heptagonal numbers (Hp), and octagonal numbers (O) are given below. Two "sum formulas" are also given.

Two Special Sum Formulas

For any counting numbers n,

$$(1 + 2 + 3 + \cdots + n)^2 = 1^3 + 2^3 + 3^3 + \cdots + n^3, \text{ and}$$

$$1 + 2 + 3 + \cdots + n = \frac{n(n + 1)}{2}.$$

Figurate Numbers Formulas

$$T_n = \frac{n(n + 1)}{2}, \quad S_n = n^2, \quad P_n = \frac{n(3n - 1)}{2},$$

$$H_n = \frac{n(4n - 2)}{2}, \quad Hp_n = \frac{n(5n - 3)}{2}, \quad O_n = \frac{n(6n - 4)}{2}.$$

EXAMPLE F   Summation Formula

Use the formula for the sum of the first n counting numbers to find $1 + 2 + 3 + \cdots + 675$.

Solution

Using $1 + 2 + 3 + \cdots + n = \dfrac{n(n + 1)}{2}$, we have

$$1 + 2 + 3 + \cdots + 675 = \frac{675(675 + 1)}{2}$$

$$= \frac{675(676)}{2}$$

$$= \frac{456,300}{2}$$

$$= 228,150.$$

EXAMPLE G   Figurate Numbers Formulas

Use the proper formulas to find each of the following.
(a) the fifth triangular number
(b) the fourth pentagonal number
(c) the sixth heptagonal number
(d) the third octagonal number

Solution

(a) $T_5 = \dfrac{5(5 + 1)}{2} = \dfrac{5(6)}{2} = \dfrac{30}{2} = 15.$

(b) $P_4 = \dfrac{4[3(4) - 1]}{2} = \dfrac{4(12 - 1)}{2} = \dfrac{4(11)}{2} = \dfrac{44}{2} = 22.$

(c) $H_6 = \dfrac{6[4(6) - 2]}{2} = \dfrac{6(24 - 2)}{2} = \dfrac{6(22)}{2} = \dfrac{132}{2} = 66.$

(d) $O_3 = \dfrac{3[6(3) - 4]}{2} = \dfrac{3(18 - 4)}{2} = \dfrac{3(14)}{2} = \dfrac{42}{2} = 21.$

As we have seen, two important approaches to solving problems include the use of deductive reasoning and inductive reasoning, or pattern recognition.

The Hungarian mathematician, George Polya, stated a four-step process that is often cited as a basic set of guidelines or problem solving strategy:

1. Understand the problem.
2. Devise a plan.
3. Carry out the plan.
4. Look back and check.

The following is a general summary of suggested problem solving strategies used in the text. See the text and EXERCISES (1.3) in the *Study Guide and Solutions Manual* for examples of the use of these strategies.

Suggested Problem Solving Strategies

Make a table or a chart.
Look for a pattern.
Solve a similar simpler problem.
Draw a sketch.
Use common sense.
Use inductive reasoning.
Write an equation and solve it.
If a formula applies, use it.
Work backward.
Guess and check.
Use trial and error.
Look for a "catch" if an answer seems obvious.

1.1 EXERCISES

1. This is an example of a deductive argument because a specific conclusion, "you can expect it to be ready in four days" is drawn from the two given premises.

3. This is an example of inductive reasoning because you are reasoning from a specific pattern to the conclusion that "It will also rain tomorrow".

5. This represents deductive reasoning, since you are moving from a general rule (addition) to a specific result (the sum of 95 and 20).

7. This is a deductive argument where you are reasoning from the two given premises. The first, "If you build it, they will come" is a general statement and the conclusion, "they will come" is specific.

9. This is an example of inductive reasoning, because you are reasoning from a specific pattern of all previous attendees to a conclusion that the next one "I" will also be accepted into medical school.

11. This is an example of inductive reasoning, because you are reasoning from a specific pattern to a generalization as to what is the next element in the sequence.

13. Writing exercise

15. Each number on the list is obtained by adding 3 to the previous number. The most probable next term is $18 + 3 = 21$.

17. Each number in the list is obtained by multiplying the previous number by 4. The most probable next term is $4 \times 768 = 3072$.

19. Each number is a multiple of 3. To find each term after the first two, multiply 3 by the sum of the factors which were multiplied by 3 in the two preceding terms.

$$3 = 3 \times 1$$
$$6 = 3 \times 2$$
$$9 = 3 \times 3 = 3 \times (1 + 2)$$
$$15 = 3 \times 5 = 3 \times (2 + 3)$$
$$24 = 3 \times 8 = 3 \times (3 + 5)$$
$$39 = 3 \times 13 = 3 \times (5 + 8)$$

The most probable next term is

$$3(8 + 13) = 3(21) = 63.$$

21. The numerators and denominators are consecutive counting numbers. The probable next term is 11/12.

23. The most probable next term is $6^3 = 216$. Observe the sequence:

$$1 = 1^3$$
$$8 = 2^3$$
$$27 = 3^3$$
$$64 = 4^3$$
$$125 = 5.^3$$

This sequence is made up of the cubes of each counting number.

25. The probable next term is 52. Note that each term (after the first) may be computed by adding successively 5, 7, 9, and 11 to each preceding term. Thus, it follows that a probable next term would come from

$$39 + 13 = 52.$$

27. The probable next term is 2 since the sequence of numbers seems to add one more 2 each time the 2's follow the number 8.

29. There are many possibilities. One such list is 10, 20, 30, 40, 50, ...

31.
$$(9 \times 9) + 7 = 88$$
$$(98 \times 9) + 6 = 888$$
$$(987 \times 9) + 5 = 8888$$
$$(9876 \times 9) + 4 = 88,888$$

Observe that on the left, the pattern suggests that the digit 5 will appended to the first number. Thus, we get $(98,765 \times 9)$ which is added to 3. On the right, the pattern suggests appending another digit 8 to obtain 888,888. Therefore,

$$(98,765 \times 9) + 3 = 888,885 + 3 = 888,888$$

By computation, the conjecture is verified.

33.
$$3367 \times 3 = 10,101$$
$$3367 \times 6 = 20,202$$
$$3367 \times 9 = 30,303$$
$$3367 \times 12 = 40,404$$

Observe that on the left, the pattern suggests that 3367 will be multiplied by the next multiple of 3, which is 15. On the right, the pattern suggests the result $50,505$. The pattern suggests the following equation:

$$3367 \times 15 = 50,505.$$

Multiply 3367×15 to verify the conjecture.

35.
$$34 \times 34 = 1156$$
$$334 \times 334 = 111,556$$
$$3334 \times 3334 = 11,115,556$$

The pattern suggests the following equation:

$$33,334 \times 33,334 = 1,111,155,556.$$

Multiply $33,334 \times 33,334$ to verify the conjecture.

37.

$$3 = \frac{3(2)}{2}$$

$$3 + 6 = \frac{6(3)}{2}$$

$$3 + 6 + 9 = \frac{9(4)}{2}$$

$$3 + 6 + 9 + 12 = \frac{12(5)}{2}$$

The pattern suggests the following equation:

$$3 + 6 + 9 + 12 + 15 = \frac{15(6)}{2}.$$

Since both the left and right sides equal 45, the conjecture is verified.

39.

$$5(6) = 6(6-1)$$
$$5(6) + 5(36) = 6(36-1)$$
$$5(6) + 5(36) + 5(216) = 6(216-1)$$
$$5(6) + 5(36) + 5(216) + 5(1296) = 6(1296-1)$$

Observe that the last equation may be written as:

$$5(6^2) + 5(6^2) + 5(6^3) + 5(6^4) = 6(6^4 - 1).$$

Thus, the next equation would likely be:

$$5(6) + 5(36) + 5(216) + 5(1296) + 5(6^5) = 6(6^5 - 1)$$

or

$$5(6) + 5(36) + 5(216) + 5(1296) + 5(7776) = 6(7776 - 1).$$

41.

$$\frac{1}{2} = 1 - \frac{1}{2}$$

$$\frac{1}{2} + \frac{1}{4} = 1 - \frac{1}{4}$$

$$\frac{1}{2} + \frac{1}{4} + \frac{1}{8} = 1 - \frac{1}{8}$$

$$\frac{1}{2} + \frac{1}{4} + \frac{1}{8} + \frac{1}{16} = 1 - \frac{1}{16}$$

Observe that the last equation may be written as

$$\frac{1}{2^1} + \frac{1}{2^2} + \frac{1}{2^3} + \frac{1}{2^4} = 1 - \frac{1}{2^4}.$$

The next equation would be

$$\frac{1}{2^1} + \frac{1}{2^2} + \frac{1}{2^3} + \frac{1}{2^4} + \frac{1}{2^5} = 1 - \frac{1}{2^5}, \text{ or}$$

$$\frac{1}{2} + \frac{1}{4} + \frac{1}{8} + \frac{1}{16} + \frac{1}{32} = 1 - \frac{1}{32}.$$

Using the common denominator 32 for each fraction, the left and right side add (in each case) to 31/32. The conjecture is therefore, verified.

43. $1 + 2 + 3 \ldots + 200$

Pairing and adjoining the first term to the last term, the second term to the second to last term, etc. we have:

$$1 + 200 = 201, 2 + 199 = 201, 3 + 198 = 201, \ldots.$$

There are 100 of these sums. Therefore,

$$100 \times 201 = 20,100.$$

45. $1 + 2 + 3 + \ldots + 800$

Pairing and adjoining the first term to the last term, the second term to the second to last term, etc. we have:

$$1 + 800 = 801, 2 + 799 = 801, 4 + 798 = 801, \ldots.$$

There are 400 of these sums. Therefore,

$$400 \times 801 = 320,400.$$

47. $1 + 2 + 3 + \ldots + 175$.

Note that there are an odd number of terms. So consider omitting for the moment, the last term, and take $1 + 174 = 175, 2 + 173 = 175, 3 + 172 = 175$, etc. There are $(174/2) = 87$ of these pairs in addition to the last term. Thus,

$$(87 \times 175) + 175 = 15,400.$$

49.

$$2 + 4 + 6 + \ldots + 100 = 2(1 + 2 + 3 + \ldots + 50)$$
$$= 2[25(1 + 50)]$$
$$= 2(1275)$$
$$= 2550$$

51. The pattern in Figures (a) − (d) shows a clockwise shading of the middle squares. In Figures (e) and (f), the pattern of the shading of the outer square corners appears in the clockwise directions. By inductive reasoning, we conclude that the lower right-hand corner should be shaded next. The next figure in the sequence is shown below.

53. These are the number of chimes a clock rings, starting with 12 o'clock, if the clock rings the number of hours on the hour, and 1 chime on the half-hour. The next most probable number is the number of chimes at 3:30, which is 1.

55. (a) Here are three examples.

$$
\begin{array}{rrr}
623 & 841 & 584 \\
-326 & -148 & -485 \\
\hline
297 & 693 & 99
\end{array}
$$

In each result, the middle digit is always 9, and the sum of the first and third digits is always 9 (considering 0 as the first digit if the difference has only two digits).

(b) Writing exercise

57.
$$142,857 \times 1 = 142,857$$
$$142,857 \times 2 = 285,714$$
$$142,857 \times 3 = 428,714$$
$$142,857 \times 4 = 571,428$$
$$142,857 \times 5 = 714,285$$
$$142,857 \times 6 = 857,142$$

Each result consists of the same six digits but in a different order.

$$142,857 \times 7 = 999,999$$

Thus, the pattern doesn't continue.

59. Count the chords and record the results.

No. of points	No. of chords	No. of chords added
2	1	1
3	3	2
4	6	3
5	10	4
6	15	5

By the pattern 6 more chords would be added for a total of 21 chords.

61. Writing exercise

1.2 EXERCISES

1.
$$
\begin{array}{ccccccc}
1 & 4 & 11 & 22 & 37 & 56 & \underline{79} \\
& 3 & 7 & 11 & 15 & 19 & \underline{23} \\
& & 4 & 4 & 4 & 4 & (4)
\end{array}
$$

Each line represents the difference of the two numbers above it. The number 23 is found from adding the predicted difference, (4), in line three, to 19 in line 2. And 79 , is found by adding 23 , in line two, to 56 in line one. Thus, our next term in the sequence is 79.

3.
$$
\begin{array}{ccccccc}
6 & 20 & 50 & 102 & 182 & 296 & 450 \\
14 & 30 & 52 & 80 & 114 & 154 & \\
16 & 22 & 28 & 34 & 40 & & \\
6 & 6 & 6 & (6) & & &
\end{array}
$$

Thus, our next term in the sequence is $154 + 296 = 450$.

5.
$$
\begin{array}{cccccccc}
0 & 12 & 72 & 240 & 600 & 1260 & 2352 & 4032 \\
12 & 60 & 168 & 360 & 660 & 1092 & 1680 & \\
48 & 108 & 192 & 300 & 432 & 588 & & \\
60 & 84 & 108 & 132 & 156 & & & \\
24 & 24 & 24 & (24) & & & &
\end{array}
$$

Thus, our next term in the sequence is

$$1680 + 2352 = 4032.$$

7.
$$
\begin{array}{cccccccc}
5 & 34 & 243 & 1022 & 3121 & 7770 & 16799 & 32758 \\
29 & 209 & 779 & 2099 & 4649 & 9029 & 15959 & \\
180 & 570 & 1320 & 2550 & 4380 & 6930 & & \\
390 & 750 & 1230 & 1830 & 2550 & & & \\
360 & 480 & 600 & 720 & & & & \\
120 & 120 & (120) & & & & &
\end{array}
$$

Thus, our next term in the sequence is

$$15959 + 16799 = 32,758.$$

9.
$$
\begin{array}{ccccccc}
1 & 2 & 4 & 8 & 16 & 31 & (57) & 99 \\
1 & 2 & 4 & 8 & 15 & 26 & 42 \\
1 & 2 & 4 & 7 & 11 & 16 \\
1 & 2 & 3 & 4 & 5 \\
1 & 1 & 1 & (1)
\end{array}
$$

The next term of the sequence is 57. Following this pattern, we predict that the number of regions determined by 8 point is 99. Use $n = 8$ in the formula

$$\frac{n^2 - 6n^3 + 23n^2 - 18n + 24}{24}.$$

$$\frac{8^2 - 6 \times 8^3 + 23 \times 8^2 - 18 \times 8 + 24}{24}$$
$$= \frac{4096 - 3,072 + 1472 - 144 + 24}{24}$$
$$= \frac{2376}{24}$$
$$= 99.$$

Thus, the result agrees with our prediction.

11.
$$(1 \times 9) - 1 = 8$$
$$(21 \times 9) - 1 = 188$$
$$(321 \times 9) - 1 = 2888$$

By the pattern, the next equation is

$$(4321 \times 9) - 1 = 38,888.$$

To verify calculate left side and compare,

$$(38,889) - 1 = 38,888.$$

13.
$$999,999 \times 2 = 1,999,998$$
$$999,999 \times 3 = 2,999,997$$

By the pattern, the next equation is

$$999,999 \times 4 = 3,999,996.$$

To verify, multiply left side to get,

$$3,999,996 = 3,999,996.$$

15.
$$3^2 - 1^2 = 2^3$$
$$6^2 - 3^2 = 3^2$$
$$10^2 - 6^2 = 4^3$$
$$15^2 - 10^2 = 5^3$$

Following this pattern, we see that the next equation will start with 21^2 since $15 + 6 = 21$. This equation will be $21^2 - 15^2 = 6^3$. The left side is $441 - 225 = 216$. The right side is also equivalent to 216.

17.
$$2^2 - 1^2 = 2 + 1$$
$$3^2 - 2^2 = 3 + 2$$
$$4^2 - 3^2 = 4 + 3$$

Following this pattern, we see that the next equation will be

$$5^2 - 4^2 = 5 + 4.$$

To verify, the left side is $25 - 16 = 9$.
The right side is also equivalent to 9.

19.
$$1 = 1 \times 1$$
$$1 + 5 = 2 \times 3$$
$$1 + 5 + 9 = 3 \times 5$$

The last term on the left side is 4 more than the previous last term. The first factor on the right side is the next counting number; the second factor is the next odd number. Thus the probable next equation is

$$1 + 5 + 9 + 13 = 4 \times 7.$$

To verify, calculate both sides to arrive at $28 = 28$.

21. $1 + 2 + 3 + \ldots + 300$

$$S = \frac{300(300 + 1)}{2} = \frac{90300}{2} = 45,150$$

23. $1 + 2 + 3 + \ldots + 675$

$$S = \frac{675(675 + 1)}{2} = \frac{456300}{2} = 228,150$$

25. $1 + 3 + 5 + 7 + \ldots + 101$
Note that

$$n = \frac{1 + 101}{2} = 51 \text{ terms so that}$$

$$S = 51^2 = 2601.$$

27. $1 + 3 + 5 + \ldots + 999$
Observe that

$$n = \frac{1 + 999}{2} = 500 \text{ terms, so that}$$

$$S = 500^2 = 250,000.$$

29. Since each term in the second series is twice that of the first series, we might expect the sum to be twice as large or

$$S = 2 \times \frac{n(n + 1)}{2} = n(n + 1).$$

31. Writing exercise

33. Figurate

Number	1st	2nd	3rd	4th	5th	6th	7th	8th
Triangular	1	3	6	10	15	21	28	36
Square	1	4	9	16	25	36	49	64
Pentagonal	1	5	12	22	35	51	70	92
Hexagonal	1	6	15	28	45	66	91	120
Heptagonal	1	7	18	34	55	81	112	148
Octagonal	1	8	21	40	65	96	133	176

35. $8(1) + 1 = 9 = 3^2$; $8(3) + 1 = 25 = 5^2$;
$9(6) + 1 = 49 = 7^2$; $8(10) + 1 = 81 = 9^2$.

37. The square numbers are $1, 4, 9, 25, 36, \ldots$.

$$1 \div 4 = 0, \text{ remainder } 1$$
$$4 \div 4 = 1, \text{ remainder } 0$$
$$9 \div 4 = 2, \text{ remainder } 1$$
$$16 \div 4 = 4, \text{ remainder } 0$$
$$25 \div 4 = 6, \text{ remainder } 1$$
$$36 \div 4 = 9, \text{ remainder } 0$$

The pattern of remainders is $1, 0, 1, 0, 1, 0, \ldots$.

39. The square number 25 may be represented by the sum of the two triangular numbers 10 and 15. The square number 36 may be represented by the sum of the two triangular numbers 15 and 21.

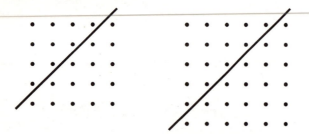

41.

n	2	3	4	5	6	7	8
A Square of n	4	9	16	25	36	49	64
B (Square of n) $+ n$	6	12	20	30	42	56	72
C One-half of Row B entry	3	6	10	15	21	28	36
D (Row A entry) $- n$	2	6	12	20	30	42	56
E One-half of Row D entry	1	3	6	10	15	21	28

(a) The results are 3 (e.g., $\frac{4+2}{2}$), 6, 10, 21, 28, 36; or triangular numbers.

(b) 1 (e.g., $\frac{2^2-2}{2}$), 3, 6, 10, 21, 28, 36; or triangular numbers.

43. To find the sixteenth square number, use

$$S_n = n^2 \text{ with } n = 16.$$
$$S_n = 16^2 = 256$$

45. To find the ninth pentagonal number, use

$$P_n = \frac{n(3n-1)}{2} \text{ with } n = 9.$$
$$P_9 = \frac{9(26)}{2} = 117$$

47. To find the tenth heptagonal number, use

$$H_{p_n} = \frac{n(5n-3)}{2} \text{ with } n = 10.$$
$$H_{p_{10}} = \frac{10(47)}{2} = 235$$

49. Since each coefficient in parentheses appears to step up by 1 we would predict:

$$N_n = \frac{n(7n-5)}{2}$$
$$N_6 = \frac{6(37)}{2} = 111.$$

verifies our prediction for $n = 6$.

51. The triangular numbers are
$$1, 3, 6, 10, 15, 21, 28, 36, 45, \ldots .$$
Adding consecutive triangular numbers, for example,
$$1 + 3 = 4, \ 3 + 6 = 9, \ 6 + 10 = 16, \ldots,$$
will give square numbers.

53. In each case, you get a perfect cube number. That is, if we take the 2nd and 3rd triangular numbers 3 and 6;

$$6^2 - 3^2 = 36 - 9 = 27$$

which is the perfect cube number, 3^3.

1.3 EXERCISES

1. The answer is 0. For example,

$$2 \times 3 \times 4 \times 5 \times 6 \times 7 = 5040.$$

Try several. You might observe that all such products will contain a tens number, hundreds number, or the two factors of 2 and 5 or other factors which end in 2 or 5. So in each case, the product will contain a factor ending in 0 or two of the factor will multiply to end in 0.

3. Since each box now has a different colored ball and the green ball is in the red box , the blue box must not have a blue ball (or green) so it must be red.

5. From the first to the third floor is a distance of 2 floors. From the first to the sixth floor is a distance of 5 floors. So $5 \div 2 = 2\frac{1}{2}$ times as long. (This, of course, assumes that you don't run out of breath or slow down as you run up the stairs.)

7. Subject to the following conditions

$R + E + E + E = N$ (or $R + 3E = N$),
$U + N + E + E = E$ (with possible carry),
$O + O + R + R = V$ (with possible carry),
$F + H + H = E = E$ (with possible carry),
$T + T = EL$ (*with possible carry*), use trial and error to arrive at $F = 9$, $O = 8$, $U = 2$, $R = 4$, $N = 7$, $E = 1$, $T = 6$, $H = 0$, $L = 3$ and, $V = 5$.

9. Use trial and error. One possible solution is as follows.

11. Use trial and error. Start with multiplication of two largest numbers ($8 \times 9 = 72$) and add others to see how close to 100 is the result. We see that

$$1 + 2 + 3 + 4 + 5 + 6 + 7 = 28.$$

This works since $28 + 72 = 100$. That is,

$$1 + 2 + 3 + 4 + 5 + 6 + 7 + 8 \times 9 = 100.$$

13. Examine the following pattern.

$34^2 = 1156$ Observe 1-"3", 2-"1"s, and 1- "5"

$334^2 = 111556$ Observe 2-"3"s, 3-"1"s, and 2- "5"s

$3334^2 = 11115556$ Observe 3-"3"s, 4-"1"s, and 3-"5"s

$33334^2 = 1111155556$ Observe 4-"3"s, 5-"1"s, and 4-"5"

Continuing this same pattern we would expect

$$3333333334^2 = 11111111115555555556 \text{ or } 10\text{-"1"s}$$

(one more than the number of "3"s) and 9-"5"s (the same number as the number of "3"s). Thus, the sum of the digits would be given by

$$10 \times 1 + 9 \times 5 + 6 = 61.$$

15. This exercise can be solved algebraically. If we let

$D =$ the total distance of the trip, and

$x =$ the distance traveled while asleep,

then

$$x + \frac{1}{2}x = \frac{1}{2}D$$
$$2x + x = D$$
$$3x = D$$
$$x = \frac{1}{3}D.$$

Thus, the distance traveled while asleep is $\frac{1}{3}$ of the total distance traveled.

17. Fill the big bucket. Pour into the small bucket. This leaves 4 gallons in the larger bucket. Empty the small bucket. Pour from the big bucket to fill up the small bucket. This leaves 1 gallon in the big bucket. Empty the small bucket. Pour 1 gallon from the big bucket to the small bucket. Fill up the big bucket. Pour into the small bucket. This leaves 5 gallons in the big bucket. Pour out the small bucket. This leaves exactly 5 gallons in the big bucket to take home. The above sequence is indicated by the following table.

Big bucket	7	4	4	1	1	0	7	5	5
Small bucket	0	3	0	3	0	1	1	3	0

19. Counting systematically we have:

$3 \times 5 = 15$ (one unit rectangles)

$2 \times 5 = 10$ (vertical two unit rectangles)

$3 \times 4 = 12$ (horizontal two unit rectangles)

$3 \times 3 = 9$ (horizontal three unit rectangles)

$3 \times 2 = 6$ (horizontal four, 4×1, unit rectangles)

$2 \times 4 = 8$ (horizontal four , 2×2, unit rectangles)

$2 \times 3 = 6$ (horizontal six unit rectangles)

$2 \times 2 = 4$ (horizontal eight, 4×2, unit rectangles)

3 (horizontal five unit rectangles)

2 (horizontal ten unit rectangles)

5 (vertical three unit rectangles)

4 (vertical six unit rectangles)

3 (vertical nine unit rectangles)

2 (vertical twelve unit rectangles)

1 (fifteen unit rectangles).

This gives a total of 90 rectangles.

21. One strategy is to assume the car was driving near usual highway speed limits ($55 - 75$ mph). We begin by trying 55 mph. In two hours the car would have traveled 110 miles. Adding 110 miles to the odometer reading, 15051. We get $110 + 15051 = 16061$ miles, which is palindromic. Thus, the speed of the car was 55 mile per hour.

23. Similar to Example 5 in the text , we might examine the units place and tens place for repetitive powers of 7 in order to explore possible patterns.

$7^1 =$	07	$7^5 =$	16, 807
$7^2 =$	49	$7^6 =$	117, 649
$7^3 =$	343	$7^7 =$	823, 543
$7^4 =$	2401	$7^8 =$	5, 764, 801

Since the final two digits cycle over four values we might consider dividing the successive exponents by 4 and examine their remainders . (Note: We are using inductive reasoning when we assume that this pattern will continue and will apply when the exponent is 1997.) Dividing the exponent 1997 by 4, we get a remainder of 1. This is the same remainder we get when dividing the exponent 1 (on 7^1) and 5 (on 7^5). Thus, we expect that the last two digits for 7^{1997} would be 07 as well.

25. A kilogram of $10 gold pieces is worth twice as much as a half a kilogram of $20 gold pieces. (The denomination has nothing to do with the value, only the weight does!)

27. Similar to Example 5 in the text (and Exercise 26 above), we might examine the units place for repetitive powers of 7 in order to explore possible patterns.

$$7^1 = 7 \qquad 7^5 = 16,807$$
$$7^2 = 49 \qquad 7^6 = 117,649$$
$$7^3 = 343 \qquad 7^7 = 823,543$$
$$7^4 = 2,401 \qquad 7^8 = 5,764,801$$

Since the units digit cycles over four values we might consider dividing the successive exponents by 4 and examine their remainders . Divide the exponent 491 by 4 to get a quotient of 122 and a remainder of 3. Reasoning inductively, the units digit would be the same as that of 7^3 and 7^7, which is 3.

29. The final number is 37, which represents the sum of 12 and the previous result, so subtract 12 from 37 to get 25. Since this result was half the previous result, multiply 25 by 2 to get 50, which is double the previous result. Now find half of 50, or 25. This is the square of the positive number so find the positive square root of 25, or 5, which is the original number.

31. To find the minimum number of socks to pull out, guess and check. There are two colors of socks. If you pull out 2 socks, you could have 2 of one color or 1 of each color. You must pull out more than 2 socks If you pull out 3 socks, you might have 3 of one color or 1 of one color and 2 of the other   In either case you have a matching pair, so 3 is the minimum number of socks to pull out.

33. To count the triangles, it helps to draw sketches of the figure several times. There are 5 triangles formed by two sides of the pentagon and a diagonal.
There are 4 triangles formed with each side of the pentagon as a base, so there are $4 \times 5 = 20$ triangles formed in this way. Each point of the star forms a small triangle, so there are 5 of these. Finally, there are 5 triangles formed with a diagonal as a base. In each, the other two sides are inside the pentagon. (None of these triangles has a side common to the pentagon.) Thus, the total number of triangles in the figure is $5 + 20 + 5 + 5 = 35$.

35. Use trial and error to find the smallest perfect number. Try making a chart such as the following one.

Number	Divisors other than itself	Sum
1	None	
2	1	1
3	1	1
4	1, 2	3
5	1	1
6	1, 2, 3	6

Six is the smallest perfect number.

37. Working backward, we see that if the lily pad doubles its size so that it completely overs the pond on the twentieth day, the pond was half covered on the previous (or nineteenth) day.

39. Solve by drawing a sketch. The following figure satisfies the description. Only three birds are needed.

41. From condition (2), we can figure that since the author is living now, the year must be 196, since $9 - 3 = 6$ (or perhaps, 230 which isn't possible, yet). Then, from condition (1),

$$23 - (1 + 9 + 6) = 7,$$

so the year is 1967.

43. By Eve's statement, Adam must have $2 more than Eve. But according to Adam, a loss of $1 from Eve to Adam gives Adam twice the amount that Eve has. By trial and error, the counting numbers 5 and 7 are the first to satisfy both conditions. Thus Eve has $5, and Adam has $7.

45. The first digit in the answer cannot be 0, 2, or 3, since these digits have already been used. It cannot be more than 3 since one of the factors is a number in the 30's, 50 it would be impossible to get a product over 45,000. Thus, the first digit of the answer must be 1. To find the first digit in the 3-digit factor, use estimation. Dividing a number between 15,000 and 16,000 by a number between 30 and 40 could give a result with a first digit of 3, 4, or 5. Since 3 and 5 have already been used, this first digit must be 4. Thus, the 3-digit factor is 402. We now have the following.

$$\begin{array}{r} 4\ 0\ 2 \\ \times \qquad 3 \\ \hline \underline{1}\ \ 5, \end{array}$$

To find the units digit of the 2-digit factor, use trial and error with the digits that have not yet been used: 6, 7, 8, and 9.

$36 \times 402 = 14,472$ (Too small and reuses 2 and 4)
$37 \times 402 = 14,874$ (Too small and reuses 4)
$38 \times 402 = 15,276$ (Reuses 2)
$39 \times 402 = 15,678$ (Correct)

The correct problem is as follows.

$$
\begin{array}{r}
4\ 0\ 2 \\
\times\ \ \ \ 3\ 9 \\
\hline
1\ 5,\ 6\ 7\ 8
\end{array}
$$

Notice that a combination of strategies was used to solve this problem.

47. Notice that the first column has three given numbers. Thus,

$$34 - (6 + 11 + 16) = 1$$

is the first number in the second row. (Note you could use the diagonal to solve for missing number in the same manner.) Then,

$$34 - (1 + 15 + 14) = 4$$

is in the second row, third column. The diagonal from upper left to lower right has three given numbers. Therefore,

$$34 - (6 + 15 + 10) = 3$$

is in the fourth row, fourth column. Continue filling in the missing numbers until the magic square is completed.

6	12	7	9
1	15	4	14
11	5	10	8
16	2	13	3

49. 25 pitches: Game tied 0 to 0 going into the 9th inning. Each pitcher has pitched a minimum of 24 pitches (three per inning). The winning pitcher pitches 3 more (fly ball/out) pitches for a total of 27. The losing (visiting team) pitcher pitches 1 more (for a total of 25) which happens to be a home run, thus, losing the game by a score of 1–0.

51. The two children row across. One stays on the opposite bank, and the other returns. One soldier rows across, and the child on the opposite bank then rows back. The two children row across. One stays, and the other returns. Now another soldier rows across. This process continues until all the soldiers are across.

53. A sketch may be helpful in solving this problem. The person takes the goat across and returns alone. On the second trip, the person takes the wolf across and returns with the goat. On the third trip, the goat is left on the first side while the person takes the cabbage across. Then the person returns alone and brings the goat back across.

55. Draw a sketch, visualize, or cut a piece of paper to build the cube. The cube may be folded with Z on the front.

Then, E is on top and M is on the left face. This places Q opposite the face marked Z. (D is on the bottom and X is on the right face.)

57. This may be worked algebraically or in reverse as Example 2 in text. Multiplying 2 by 10 gives 20. Subtract 8 to give 12 and then square to get 144. Add 52 to get 196. This represents a number times itself. The number is 14 coming from the fact that $14 \times 14 = 196$ (or the square root of $196 = 14$). The quotient must be 21 since $21 - \frac{1}{3} \times 21 = 14$. Multiplying 21 by 7, we get 147 which represents 3 times the original number plus 3/4 of that same product. The original number must be 28 since $3 \times 28 = 84$ and 3/4 of 84 is 63. And $84 + 63 = 147$.

59. A solution, found by trial and error, is shown here.

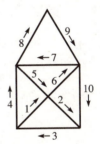

61. Solve this problem by looking for a pattern. 1/7 has a decimal representation of .142857..., where the group of 6 digits, 142857, is repeated indefinitely. When 100 is divided by 6, the remainder is 4, so the fourth digit of the repeating group, which is 8, is the 100th digit of the decimal representation.

63. Common sense tells you that the CEO is a woman.

65. Jessica is married to James or Dan. Since Jessica is married to the oldest person in the group, she is not married to James, who is younger than Cathy. So Jessica is married to Dan, and Cathy is married to James. Since Jessica is married to the oldest person, we know that Dan is 36. Since James is older than Jessica but younger than Cathy, we conclude that Cathy is 31, James is 30, and Jessica is 29.

67. This is a problem with a "catch." The obvious answer is that only one month, February, has 28 days. However, the problem does not specify exactly 28 days, so any month with at least 28 days qualifies. All 12 months have 28 days.

69. This is a problem with a "catch." Someone reading this problem might go ahead and calculate the volume of a cube 6 feet on each side, to get the answer 216 cubic feet. However, common sense tells us that since holes are by definition empty, there is no dirt in a hole.

71. "Madam, I'm Adam."

73. 6   The following represents one of several possible arrays.

	X	X
X		X
X	X	

1.4 EXERCISES

Using a graphing calculator, such as the TI-83, we would enter the expressions as indicated on the left side of the equality then push [Enter] to arrive at the answer. When using scientific or other types of calculators some adjustments will have to be made. See observations related to the solutions for Exercise 8 and Exercise 13 below. It is a good idea to review your calculator handbook for related examples.

1. $39.7 + (8.2 - 4.1) = 43.8$

3. $\sqrt{5.56440921} = 2.3589$

5. $\sqrt[3]{418.508992} = 7.48$

7. $2.67^2 = 7.1289$

9. $5.76^5 = 6340.338097$

Observe that when using a calculator, the numerator must be grouped in parenthesis as must the denominator. This will make the last operation (the indicated) division.

11. $\dfrac{14.32 - 8.1}{2 \times 3.11} = 1$

13. $\sqrt[5]{1.35} = 1.061858759$ Observe that many scientific calculators have only the $\sqrt[2]{}$ function built into the calculator. For an index larger than 2 you may want to think of the nth root of a number b, as equivalent to the exponential expression $b^{\frac{1}{n}}$. For example, $\sqrt[5]{1.35} = (1.35)^{\frac{1}{5}}$. Then use your exponentiation function (button) to calculate the $5th$ root of 1.35.

15. $\dfrac{\pi}{\sqrt{2}} = 2.221441469$

17. $\sqrt[4]{\dfrac{2143}{22}} = 3.141592653$

19. Choose a five-digit number such as 73,468.

$$73468 \times 9 = 661212$$
$$6 + 6 + 1 + 2 + 1 + 2 = 18$$
$$1 + 8 = 9$$

Choose a six-digit number such as 739,216.

$$739216 \times 9 = 6652944$$
$$6 + 6 + 5 + 2 + 9 + 4 + 4 = 36$$
$$3 + 6 = 9$$

Yes, the same result holds.

21.
$$(-3) \times (-8) = 24$$
$$(-5) \times (-4) = 20$$
$$(-2.7) \times (-4.3) = 11.61$$

Multiplying a negative number and a positive number gives a <u>positive</u> product.

23. $5.6^0 = 1; \; \pi^0 = 1; \; 2^0 = 1; \; 120^0 = 1; \; .5^0 = 1$

Raising a non zero number to the power 0 gives a result of <u>1</u>.

25.
$$\frac{1}{7} \approx .1428714$$
$$\frac{1}{(-9)} \approx -.1111111$$
$$\frac{1}{3} \approx .3333333$$
$$\frac{1}{(-8)} \approx -.1250000$$

The sign of the reciprocal of a number is the <u>same as the sign</u> of the number.

27. $(0/8) = 0; (0/2) = 0; (0/(-3)) = 0; (0/\pi) = 0$

Zero divided by a non zero number gives a quotient of 0.

29.
$$(-3) \times (-4) = 12;$$
$$(-3) \times (-4) \times (-5) \times (-6) = 360;$$
$$(-3) \times (-4) \times (-5) \times (-6) \times (-7) \times (-8) = 20160$$

Multiplying an even number of negative numbers gives a <u>positive</u> product.

31. Writing exercise

33. The result of multiplying any digit except 0 by 429 and then multiplying the result by 259 is a six-digit number consisting of only the digit you started with. Multiplying by 429 and 259 is the same as multiplying by 111,111 ($429 \times 259 = 111111$), which gives the same digits.

35. $(100 \div 20) \times 14,215,469 = \underline{71077345}$. One of the biggest petroleum companies in the world is Sh<u>ELLOIL</u>.

37.

$$60^2 - \frac{368}{4} = \underline{3508}$$ This electronic manufacturer produces the Wave Radio. <u>BOSE</u>.

39. Writing exercise

41. $431 \div 9 \approx 47.888$. Since more than 47 are needed, we require 48 cases.

43. $400 \div 30 \approx 13.333$. Since more than 13 are needed, we require 14 containers.

45. $140,000 \div 80 \approx 160,000 \div 80 = \$2000$; option B.

47. $34,671 \div 1005 \approx (35,000/1000) = 35$; option A.

49. $299 \div 19 \approx (300/20) = 15$; option C.

51. Using the two odometer readings, we have $010240.5 - 010140.5 = 100$ miles. This represents the one-way distance. For the round trip we have $2 \times 100 = 200$ miles.

53. The gas used was $1 - \frac{3}{4} = \frac{4}{4} - \frac{3}{4} = \frac{1}{4}$ of a tank or, $\frac{1}{4}$ of 20 gallons $= \frac{1}{4} \times 20$ gallons $= 5$ gallons. Since we want the miles per gallon (mileage) we will divide the number of miles traveled by the number of gallons used. Thus, 200 miles $\div$ 5 gallons $= 20$ miles per gallon.

55. From the graph, it is clear that there was 4.0 billion dollars invested in 1992.

57. From the graph, it is clear that there was $10.0 - 4.5 = 5.5$ billion dollars more spent in 1996 than in 1993.

59. There was a decrease between 1994 and 1955. The greatest increase occurred between 1993 and 1994 (line segment with greatest slope or steepest graph).

61. From 1993 to 1994 the increase is about $22,000 - 17,200 = 4800$ products. From 1995 to 1996 the increase is about $24,300 - 21,000 = 3300$ products. <u>Yes</u>, this answer does confirm the answer to Exercise 59.

63. The largest category is the technical/sales/administrative support; the smallest is farming/forestry/fishing.

65. Since the graph indicates $\frac{3}{100}$ of the total are employed in this category, we would expect $\frac{3}{100} \times 7500 = 225$ of the sample to be employed.

67. We would expect Detroit to be in the 70s.

69. It will be hot and stormy in Miami.

71. There appears to be a trough over Texas.

CHAPTER 1 TEST

1. This is an example of inductive reasoning since you are reasoning from a specific pattern to the general conclusion that he will again exceed his annual sales goal.

2. This is a deductive argument because you are reasoning from the stated general property, to the specific result, 101^2 is a natural number.

3. Observing a pattern such as $1 = 1^1, 4 = 2^2, 27 = 3^3$, leads us to the possibility that the fourth term may be $4^4 = 256$, the fifth term $5^5 = 3125$ and so on. Since the next term (6th term) is given as 46656, we have an added check for our pattern as $6^6 = 46656$. (The nth term of the sequence is n^n.)

4. The specific pattern seems to indicate that the second factor in the product is a multiple of 17 and the digits on the right side of the equation increase by 1. If this pattern is correct than the next term in the sequence would be

$$65,359,477,124,183 \times 68 = 4,444,444,444,444,444$$

since, $4 \times 17 = 68$. This can be verified by multiplying $65,359,477,124,183 \times 68$ on your calculator.

5. 3   11   31   69   131   223   <u>351</u>
       8   20   38   62   92   <u>128</u>
         12   18   24   30   36
            6   6   6   (6)

6. Using the method of Gauss, we have $1 + 250 = 251$, $2 + 259 = 251 \ldots$ etc. There are $\frac{250}{2} = 125$ such pairs, so the sum can be calculated as $125 \times 251 = 31,375$.

7. The next predicted octagonal number is 65. Since the next equation on the list would be $65 = 1 + 7 + 13 + 19 + 25$, where $25 = 19 + 6$.

8. Beginning with the first five octagonal numbers and applying the method of successive differences we get
    1   8   21   40   65   96   133   176
      7   13   19   25   31   37   43
        6   6   6   6   (6)   (6) . Dividing each octagonal number by 4 we get the following pattern of remainders: $1, 0, 1, 0, 1, 0, 1, 0 \ldots$.

9. After the first two terms, we can find the next by adding the two proceeding terms. That is, to get the $3rd$, term add $1 + 1 = 2$; the $4th, 1 + 2 = 3$; the $5th$ term, $2 + 3 = 5$; and so forth.

10. Examine the units place for repetitive powers of 9 in order to explore possible patterns.

$$9^1 = 9 \qquad 9^3 = 729 \qquad 9^5 = 59049$$
$$9^2 = 81 \qquad 9^4 = 6561 \qquad 9^6 = 531441$$

If we divide the exponent 1997 by 2 (since the pattern of the units digit cycles after every $2nd$ power), we get a reminder of 1. Noting that in the line of 9^1 where each exponent, when we divide by 2, yields a remainder of one has a units digit of 9, we reason inductively that 9^{1997} has the same units digit, 9.

11. The third error is that there are only two errors.

12. The pattern indicates an additional side for both the internal and external regular polygon as one moves from figure to figure. In addition, the "antennae" alternate between one and two.

13. One method of solution is to create an algebraic equation and solve. Let $n =$ number of bees. If $\frac{8}{9}n$ are left in the hive then $\frac{1}{9}n$ are away from the hive then $\sqrt{\frac{n}{2}} + 2 = \left(\frac{1}{9}\right)n$. Note the "2" represents the female and male bee at the lotus flower.

$$\sqrt{\frac{n}{2}} + 2 = \left(\frac{1}{9}\right)n$$
$$\sqrt{\frac{n}{2}} = \frac{n}{9} - \frac{18}{9}$$
$$\left(\sqrt{\frac{n}{2}}\right)^2 = \left(\frac{n-18}{9}\right)^2$$
$$\frac{n}{2} = \frac{n^2 - 36n + 324}{81}$$
$$81n = 2n^2 - 72n + 648$$
$$0 = 2n^2 - 153n + 648$$
$$0 = (2n - 9)(n - 72)$$

Thus, $n = \frac{9}{2}$ (extraneous)

or, $n = 72$ (answer).

14. We might group pages the following way
$(1,2)(3,4)(5,6)(7,8) \ldots (53,54)(55, 56)(57,58)(59,60)$
Observe that pages (1,2) will be printed on the same sheet as pages (59,60). In the same manner, pages (3,4) will be printed on the same sheet as (57,58), etc. Thus, pages (7,8) will be printed on the same sheet of paper as pages (53,4). So, if the sheet with page 7 is missing, then 8, 53, and 54 will also be missing.

15. Observe the following patterns on success powers of 11, 14, and 16 in order to determine the units value of each term in the sum $11^{11} + 14^{14} + 16^{16}$.

$$11^1 = 11 \qquad 14^1 = 14 \qquad 16^1 = 16$$
$$11^2 = 121 \qquad 14^2 = 196 \qquad 16^2 = 256$$
$$11^3 = 1331 \qquad 14^3 = 2744 \qquad 16^3 = 4096$$
$$14^4 = 38146$$

Thus, we would expect 11^{11} to have the same unit digit value of 1. Since powers of 14 cycle have units digits which cycle between 4 and 6, we observer that division of the exponents by 2 yield remainders of 1 or 0. We might expect the same pattern to continue to 14^{14}. Division of the exponent by 2 gives a remainder of 0. We get the same remainder, 0, for all even powers on 14 and each of these numbers has a units digit of 6. The powers of 16 seem to all have the same unit value of 6. Thus, if we add the units digits $1 + 6 + 6 = 13$, we see that the units digit of this sum is 3.

16. Making the following observations

$$9 \times 1 = 9$$
$$9 \times 2 = 18 \ (1 + 8 = 9)$$
$$9 \times 3 = 27 \ (2 + 7 = 9)$$
$$9 \times 3 = 27 \ (2 + 7 = 9)$$
$$9 \times 5 = 45 \ (4 + 5 = 9)$$
$$\cdots$$

suggests that the sum of the digits in the product will always be nine.

17. $\sqrt{98.16} = 9.907572861$ But answers may vary depending upon what calculator you are using.

18. $3.25^3 = 34.328125$

19. Estimate by $400 \div 100 = 4$ as an approximation. Thus, B is the correct option.

20. (a) Reading the graph gives about 850 (thousands of dollars) = $850,000. Thus B ($848,000) is the best estimate for the answer.

(b) The year 1993 showed the smallest increase from the proceeding year 1992. Note that the rate of increase (slope of the line segment) for that year is the smallest (most horizontal).

2   THE BASIC CONCEPTS OF SET THEORY

Chapter Goals

The general goal of this chapter is to provide a basic understanding of the language of sets. Upon completion of this chapter, the student should be able to:

- Identify and describe sets using standard set notation.
- Identify and describe set relationships.
- Identify and use set operations.
- Interpret and use Venn diagrams in the solution of problems.
- Identify cardinalities of finite and infinite sets.

Chapter Summary

The concept of a **set** is a generalization of an idea that is familiar in everyday life. For example, a collection of tools is often called a set of tools and a collection of bedroom furniture is called a bedroom set. Sets need not consist of physical objects as they may be a set of abstract ideas. State statutes, for instance, represent a set of laws. In mathematics sets consisting of numbers are of special interest. Because mathematical ideas can often be stated most simply using the language of sets, set theory and language form a common thread through many diverse mathematical topics.

A **set** can be thought of as a collection of objects. Each object that belongs to a set is called a **member** or an **element** of the set. Upper-case letters are used to represent the names of sets while lower-case letters are often used to represent elements. Thus, a $\in A$ suggests symbolically that a is a member or element of set A while b $\notin A$ means that b is not an element of set A. In writing a set, braces are used to enclose the objects making up the set. The set of whole numbers is defined as $\{0, 1, 2, 3, \ldots\}$ which is an example of an infinite set. The set used in the following discussion about notation forms is an example of a finite set.

There are three basic notations or ways to describe sets: A word description such as "the set of counting numbers between 1 and 100" may sometimes be used; a **listing method** suggesting the same set of numbers would be $\{2, 3, 4, \ldots 98, 99\}$ is also used; or alternatively we might use a **set-builder** (sometimes called set-generator) method, that is, $\{x \mid 1 < x < 100, x \text{ is a counting number}\}$. In general, set-builder notation is used when the set has many elements and a pattern of representation is difficult or impossible to find. **Set-builder** notation takes the form of $\{x \mid x \text{ has some property}\}$ and is read as "the set of elements x such that x has some property." The set $\{x \mid x \text{ is a book in the Library of Congress}\}$ is an example of a large set which is difficult to describe by the listing method since a pattern would be difficult to identify.

The number of elements in a set is called the cardinal number or cardinality of the set. The symbol $n(A)$ represents the cardinal number of set A and is read as "the number of elements in set A." If the cardinality of a set is 0 (that is, there are no elements in the set), we call that set the **empty set** or **null set** and indicate it symbolically as $\{\ \}$ or $\emptyset$. In other words, $n(\emptyset) = 0$.

In every problem, there is either a stated or implied universe of discourse. The universe of discourse includes all things under discussion at a given time and is referred to as the **universal set** which is symbolized as U. In measuring the length of a table, for example, the implied universe of discourse would be the set of positive rational numbers (integers and fractions greater than 0). When using sets, the universe of discourse is called the universal set. Once the universal set is stated or identified, then all subsequent sets or elements must be contained in U.

SET RELATIONSHIPS

Equal sets vs. Equivalent sets   If two sets contain the exact same elements, we say that the two sets are equal. If two sets have the same number of elements (same cardinality), then they are said to be equivalent sets where $A \sim B$ is read set A is equivalent to set B.   Consider, for example, the sets $A = \{1, 2, 3\}$, $B = \{3, 1, 2\}$ and $C = \{a, b, c\}$.   Here $A = B$, $A \neq C$, and A, B, and C are equivalent sets; that is, $A \sim B \sim C$. Observe that if two sets are equal, then they are also equivalent. However, if two sets are equivalent, they are not necessarily equal.

One-to-one correspondence   If two sets are equivalent (they contain the same number of elements), then there exists a one-to-one correspondence among their members. That is, each element of one set can be paired with a unique element of the second set. For example, if $A = \{1, 2, 3\}$ and $B = \{x, y, z\}$, we might establish a one-to-one correspondence by pairing the first elements in each set, the second elements of each set, and the third elements of each set as follows:

$$\{1, \quad 2, \quad 3\}$$
$$\updownarrow \quad \updownarrow \quad \updownarrow$$
$$\{x, \quad y, \quad z\}.$$

If each set has more than one element, then there is more than one correspondence that can be established. That is, there is more than one way in which a one-to-one correspondence can be established.

The establishing of a one-to-one correspondence between two sets is a way to show that two sets are equivalent, thus, one-to-one correspondence is often used to define equivalent sets. That is, two sets that may be put into a one-to-one correspondence are said to be equivalent.

Subsets   In our work with sets, we frequently encounter two sets that are neither equal nor equivalent however, do share common elements.   For example, if $A = \{3, 5, 9\}$ and $B = \{1, 3, 5, 9, 11\}$, we see that each element in set A is also contained in set B. We also see that A and B are not equal; thus, there are elements in B which are not found in A. In such cases, we say that set A is a **proper subset** of B or symbolically as $A \subset B$.

If it is possible, however, that $A = B$, then we say that A is a **subset** of B ($A \subseteq B$). That is, we are indicating that each element in set A is also an element of set B, but B doesn't have to contain elements which are not in A.

EXAMPLE   A   Set relationships

Indicate whether each statement is true or false.
(a) $\{5, 4\} \subset \{1, 3, 4, 5, 6\}$

True, since the elements 5 and 4 are contained in the second set and the two sets are not equal.

(b) $\{1, 2, 3\} \subset \{3, 2, 1\}$

False, the first set is a subset of the second but is not a proper subset since they are equal
(c) $\{0\} \subseteq \emptyset$

False, since 0 is not a member of the second set.

(d) $0 \in \emptyset$

False, since the empty set contains no elements.

(e) $0 \subset \{0, 1\}$

False, since the symbol $\subset$ means proper subset, the relationship must be between sets, not elements and sets.

These relations stated in more formal mathematical terms are as summarized as follows:

SET RELATIONS

Equal sets Two sets are equal if and only if every element of the first set is an element of the second set and vice versa.

Equivalent sets Two sets are equivalent if and only if they can be paired in a one-to-one correspondence.

Subset If every element in set A is contained in set B, then A is a subset of B, denoted as $A \subseteq B$.

Proper subset If every element in set A is also an element in set B, and $A \neq B$, then A is a proper subset of B, denoted as $A \subset B$.

Observe that proper subset is the stronger or more restrictive of the two set relations.

Reminders About Subsets

- When using either symbol for subset, $\subset$ or $\subseteq$, the item on either side of the symbol must be a set. This should not be confused with the symbol used to show set membership, $\in$. When using $\in$, only the item on the right will be a set; the item on the left must be an element of the set.
- The empty set is considered a subset of every set.
- The number of subsets contained within a given set is equal to 2^n; the number of proper subsets contained within a given set is equal to $2^n - 1$, where n is the number of elements in the given set.
- If $A \subset B$, then $n(A) < n(B)$ and if $A \subseteq B$, then $n(A) \leq n(B)$, where $n(A)$ and $n(B)$ represent the number of elements in A and in B.

SET OPERATIONS

Intersection   To help in defining the intersection of two sets, let us define the intersection of two streets. The intersection is where the streets cross or the section of pavement that may be considered to be on both streets. In the same way, given two sets A and B, the intersection of A and B, denoted $A \cap B$, is a new set that contains all the elements common to both A and B.

For example, if $A = \{2, 3, 4, 5\}$ and $B = \{1, 3, 5, 7\}$, then $A \cap B = \{3, 5\}$ since the elements 3 and 5 are the only elements contained in both A and B. Two sets that have no elements in common are called disjoint sets (their intersection is empty) or $A \cap B = \emptyset$.

Union   Given the two sets A and B, the union of A and B, denoted $A \cup B$, is a new set that contains all the elements in either A or B, or both. The union simply unites the two sets without repeating the values that are in both.

For example, if $A = \{1, 2, 4, 5\}$ and $B = \{1, 4, 5, 6, 7\}$, then $A \cup B = \{1, 2, 4, 5, 6, 7\}$.

Complement   Given the universal set, U, and any set A, then the complement of set A, denoted A', is the set of all elements in U which are not in A.

For example, if $U = \{1, 2, 3, 4, 5, 6, 7, 8, 9, 10\}$ and $A = \{4, 5, 7, 9, 10\}$, then we get A' by crossing out all elements in U which are contained in A; that is, $A' = \{1, 2, 3, 6, 8\}$.

Difference   The difference between two sets A and B, denoted $A - B$, is a new set consisting of those elements in set A which are not in set B.

For example, if $A = \{1, 2, 4, 5, 8\}$ and $B = \{1, 4, 5, 6, 7\}$, then $A - B = \{2, 8\}$.

Observe that $A - B$ and $A \cap B'$ indicate the same set.

Cartesian Product   A set may contain ordered pairs (pairs of numbers where the order of the elements is important versus sets themselves where the order of elements is unimportant) as elements. Such a set might look like $\{(1, 2), (3, 5), (7, 6)\}$. We can generate such sets by finding the Cartesian product of two original sets. If A and B are sets, then each element of A can be paired with each element of B, and the results can be written as ordered pairs. The set of all such ordered pairs is called the Cartesian product of A and B, written $A \times B$, and read "A cross B."

For example, if $A = \{1, 3\}$ and $B = \{2, 3, 4\}$, then $A \times B = \{(1, 2), (1, 3), (1, 4), (3, 2), (3, 3), (3, 4)\}$ where the first element in each ordered pair comes from set A and the second element comes from set B and $A \times B$ is the set of all such pairings.

EXAMPLE B   Set operations

Let $U = \{1, 2, 3, 4, 5, 6, 7, 8, 9, 10\}$,
$A = \{1, 2, 3, 4\}$, $B = \{1, 3, 5, 7, 9\}$,
$C = \{2, 4, 6, 8, 10\}$, $D = \{3, 5\}$.

(a) Find $A \cap B$.

Solution

$$A \cap B = \{1, 2, 3, 4\} \cap \{1, 3, 5, 7, 9\}$$
$$= \{1, 3\};$$

since the elements 1 and 3 are common to both.

(b) Find $A \cup B$.

Solution

$$A \cup B = \{1, 2, 3, 4\} \cup \{1, 3, 5, 7, 9\}$$
$$= \{1, 2, 3, 4, 5, 7, 9\};$$

since the union of A with B contains all of the elements in A along with those in B.

(c) Find C'.

Solution

$$C' = \{2,4,6,8,10\}'$$
$$= \{1,3,5,7,9\}$$
$$= B;$$

since B represents the set of all elements which are not in C.

(d) Find $B - A$.

Solution

$$B - A = \{1,3,5,7,9\} - \{1,2,3,4\}$$
$$= \{5,7,9\};$$

representing those elements in B which are not in A

(e) Find $D \times A$.

Solution

$$D \times A = \{3,5\} \times \{1,2,3,4\}$$
$$= \{(3,1),(3,2),(3,3),(3,4),(5,1),$$
$$(5,2),(5,3),(5,4)\};$$

representing all possible ordered pairs with the first element coming from set D and the second from set A.

EXAMPLE C   Set operations

Let $U = \{1,2,3,4,5,6,7,8,9,10\}$,
$A = \{1,2,3,4\}$, $B = \{1,3,5,7,9\}$,
$C = \{2,4,6,8,10\}$, $D = \{3,5\}$.

(a) Find $D \cup C'$.

Solution

$$D \cup C' = \{3,5\} \cup \{2,4,6,8,10\}'$$
$$= \{3,5\} \cup \{1,3,5,7,9\}$$
$$= \{1,3,5,7,9\}$$

(b) Find $C' \cap A$.

Solution

$$C' \cap A = \{2,4,6,8,10\}' \cap \{1,2,3,4\}$$
$$= \{1,3,5,7,9\} \cap \{1,2,3,4\}$$
$$= \{1,3\}$$

Observe that the set $C' \cap A$ is the same as $A - C$.

(c) Find $B \cup (C' \cap A)$.

Solution

$$B \cup (C' \cap A) = \{1,3,5,7,9\} \cup \{1,3\}$$
$$\text{[from (b) above]}$$
$$= \{1,3,5,7,9\}$$

These operations, using formal mathematical notation, are stated as follows:

SET OPERATIONS

Intersection
$A \cap B = \{x \mid x \in A \text{ and } x \in B\}$.
Union
$A \cup B = \{x \mid x \in A \text{ or } x \in B\}$.
Complement
$A' = \{x \mid x \in U \text{ and } x \notin A\}$.
Difference
$A - B = \{x \mid x \in A \text{ and } x \notin B\}$.
Cartesian Product
$A \times B = \{(a, b) \mid a \in A \text{ and } b \in B\}$.

PROPERTIES AND LAWS OF SETS

The following includes several useful laws and properties associated with the above set operations which will allow for simplification and the use of equivalent forms. The laws are especially useful in one's study of symbolic logic (Chapter 3). The properties apply (in part or in total) to other mathematical systems and operations as well, the most familiar system being the algebra of real numbers.

Commutative Properties
For all sets A and B,
$A \cup B = B \cup A$, and
$A \cap B = B \cap A$.

Associative Properties
For all sets A, B, and C,
$(A \cup B) \cup C = A \cup (B \cup C)$, and
$(A \cap B) \cap C = A \cap (B \cap C)$.

Identity Properties
For all sets A,
$A \cup \emptyset = A$, and
$A \cap U = A$.

Distributive Properties
For all sets A, B, and C,
$A \cap (B \cup C) = (A \cap B) \cup (A \cap C)$, and
$A \cup (B \cap C) = (A \cup B) \cap (A \cup C)$.

De Morgan's Laws
For all sets A and B,
$$(A \cap B)' = A' \cup B', \text{ and}$$
$$(A \cup B)' = A' \cap B'$$

VENN DIAGRAMS (Pictures of sets)

A Venn diagram is a pictorial or geometric representation of sets and their relationships. The universal set, U, is diagrammed as a rectangle, and any sets that are contained in U are diagrammed or shown as circles, or sometimes ovals, within the rectangle.

The following Venn diagram, for example, shows that $B \subset A$ and that sets A and C are disjoint, as are sets B and C.

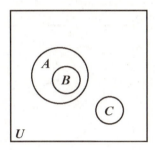

Shading or numbering of regions in a Venn diagram is commonly used to identify particular sets or combinations of sets.

EXAMPLE D   Shading regions of interest

Use a Venn diagram to shade the set
$B' \cup (A' \cap B')$.

Solution
The indicated set operations mean those elements not in B or those not in A as long as they are also not in B. It is a help to shade the region representing "not in A" first, then that region representing "not in B." Identify the intersection of these regions (covered by both shadings). As in algebra, the general strategy when deciding which order to do operations is to begin inside parentheses and work out.

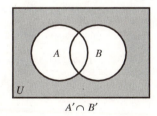

$A' \cap B'$

Finally, the region of interest will be that "not in B" along with (union of) the above intersection—$(A' \cap B')$.

That is, the final region of interest is given by

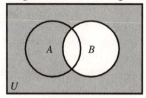

$B' \cup (A' \cap B')$

Venn Diagram Applications

One application of Venn diagrams is to indicate that two sets are equal as shown by the following example.

EXAMPLE E   Equal sets
Show that $A \cap (B \cup C) = (A \cap B) \cup (A \cap C)$
(Distributive property of intersection over union).

Solution
Draw a Venn diagram for the set on the left of the equality and a Venn diagram for the set on the right of the equality. If both regions are the same, then equal sets are indicated.

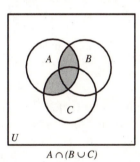 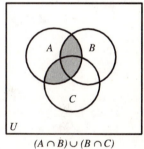

$A \cap (B \cup C)$ $(A \cap B) \cup (B \cap C)$

Another common application is the solution of survey problems as in the following example.

EXAMPLE F   Survey problem

A survey of 80 sophomores at a certain western college showed the following:
36 take English
32 take history
32 take political science
16 take history and English
16 take political science and history
14 take political science and English
 6 take all three.

How many students
(a) take English and neither of the other two?
(b) take none of the three courses?
(c) take history but neither of the other two?
(d) take political science and history but not English?
(e) do not take political science?

Solution

Construct a Venn diagram showing the universal set, U, representing all of the 80 sophomores surveyed; set E, representing those taking English; set H, representing those taking history; and set P, representing those taking political science. Always draw the Venn diagrams to represent the maximum number of intersections (unless it is clearly known that some of the sets are mutually exclusive or disjoint sets).

Indicate the cardinalities (number of elements) in each region by starting with the intersection of all three sets—in this exercise 6. See the completed Venn diagram below. Then work up the list by examining the cardinalities associated with the intersection of two sets at a time. For example, since there is a total of 14 students taking political science and English and we already have 6 taking all three, we can deduce that there are only 8 in the region represented by the intersection of P and E and also outside of the H set.

Similarly, we conclude that there are 10 students in the region representing P and H but not E (i.e., 10 students taking political science and history but not English) and 10 students in E and H but not in P. Since there are 32 taking political science altogether and we have already accounted for 24 ($8 + 6 + 10$) of them in the intersections of P with E and H, there must be only 8 students ($32 - 24$) in P and at the same time not in E or H.

We can conclude in a similar manner that there are 6 students ($32 - 26$) in H but not in E or P and that there are 12 students ($36 - 14$) in E but not in P or H. If we add all of the cardinalities of each subregion within the

three circles representing the three courses, we get $12 + 8 + 8 + 10 + 6 + 10 + 6 = 60$ students taking at least one course. Since there were 80 students surveyed, there must be $80 - 60 = 20$ in the region outside the circles (i.e., not taking one of these courses).

Note that the general strategy is to complete the cardinalities of the regions innermost to the diagram and work one's way out using deduction until the cardinalities of all subregions are known.

Once these cardinalities are indicated on the Venn diagram, most all questions about the survey can easily be answered.

Completed Venn diagram:

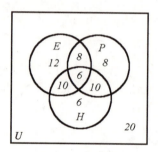

Thus, by inspecting the Venn diagram:

(a) 12 take English and neither of the other two,
(b) 20 take none of the three courses,
(c) 6 take history but neither of the other two,
(d) 10 take political science and history but not English,
(e) 48 are not taking political science ($80 - 32$ which are taking political science).

2.1 EXERCISES

1. $\{2, 4, 6, 8\}$ matches C, the set of even positive integers less than 10.

3. $\{\ldots, -4, -3, -2, -1\}$ matches E, the set of all negative integers.

5. $\{2, 4, 8, 16, 32\}$ matches B, the set of the five least positive integer powers of 2, since each element represents a successive power of 2 beginning with 2^1.

7. $\{2, 4, 6, 8, 10\}$ matches H, the set of the five least positive integer multiples of 2, since this set represents the first five positive even integers. Remember that all even numbers are multiples of 2.

9. The set of all counting numbers less than or equal to 6 can be expressed by listing as $\{1, 2, 3, 4, 5, 6\}$.

11. The set of all whole numbers not greater than 4 can be expressed by listing as $\{0, 1, 2, 3, 4\}$.

13. In the set $\{6, 7, 8 \ldots, 14\}$ the ellipsis (three dots) indicates a continuation of the pattern. A complete listing of this set is $\{6, 7, 8, 9, 10, 11, 12, 13, 14\}$.

15. The set $\{-15, -13, -11, \ldots, -1\}$ contains all integers from -15 to -1 inclusive. Each member is two larger than its predecessor. A complete listing of this set is $\{-15, -13, -11, -9. -7, -5, -3, -1\}$.

17. The set $\{2, 4, 8, \ldots, 256\}$ contains all powers of two from 2 to 256 inclusive. A complete listing of this set is $\{2, 4, 8, 16, 32, 64, 128, 256\}$.

19. A complete listing of the set $\{x \mid x$ is an even whole number less than 11$\}$ is $\{0, 2, 4, 6, 8, 10\}$. Remember that 0 is the first whole number.

21. The set of all counting numbers greater than 20 is represented by the listing $\{21, 22, 23, \ldots\}$.

23. The set of Great Lakes is represented by {Lake Erie, Lake Huron, Lake Michigan, Lake Ontario, Lake Superior}.

25. The set, $\{x \mid x$ is a positive multiple of 5$\}$, is represented by the listing $\{5, 10, 15, 20, \ldots\}$.

27. The set, $\{x \mid x$ is the reciprocal of a natural number$\}$, is represented by the listing $\{1, 1/2, 1/3, 1/4, 1/5, \ldots\}$.

29. The set of all rational numbers may be represented using set-builder notation as $\{x \mid x$ is a rational number$\}$.

31. The set, $\{1, 3, 5, \ldots, 75\}$, may be represented using set-builder notation as $\{x \mid x$ is an odd natural number less than 76$\}$.

33. The set, $\{2, 4, 6, \ldots, 32\}$ is finite since there is a cardinal number associated with this set. Hence, there are a countable number of elements in the set.

35. The set $\{1/2, 2/3, 3/4, \ldots\}$ is infinite since there is no last element, and we would be unable to count all of the elements.

37. The set $\{x \mid x$ is a natural number greater than 50$\}$, is infinite since there is no last number and hence is uncountable.

39. The set $\{x \mid x$ is a rational number$\}$ is infinite, since there is no last number and therefore is uncountable.

41. For any set A, $n(A)$ represents the cardinal number of the set, that is, the number of elements in the set. The set, $A = \{0, 1, 2, 3, 4, 5, 6, 7\}$, contains 8 elements. Thus, $n(A) = 8$.

43. The set $A = \{2, 4, 6, \ldots, 1000\}$ contains 500 elements. Thus, $n(A) = 500$.

45. The set $A = \{a, b, c, \ldots, z\}$ has 26 elements (letters of the alphabet). Thus, $n(A) = 26$.

47. The set $A = $ the set of integers between -20 and 20 has 39 members. The set can be indicated as $\{-19, -18, \ldots, 18, 19\}$, or 19 negative integers, 19 positive integers, and 0. Thus, $n(A) = 39$.

49. The set $A = \{1/3, 2/4, 3/5, 4/6, \ldots, 27/29, 28/30\}$ has 28 elements. Thus, $n(A) = 28$.

51. Writing exercise

53. The set $\{x \mid x$ is a real number$\}$ is well defined since we can always tell if a number is real and belongs to this set.

55. The set $\{x \mid x$ is good athlete$\}$ is not well defined, since set membership, in this case, is a value judgment, and there is no clear-cut way to determine whether a particular athlete is "good."

57. The set $\{x \mid x$ is a difficult course$\}$ is not well defined since set membership is a value judgment and there is no clear-cut way to determine whether a particular course is "difficult."

59. $5 \boxed{\in} \{2, 4, 5, 6, 7\}$ since 5 is a member of the set.

61. $-4 \boxed{\notin} \{4, 7, 8, 12\}$ since -4 in not contained in the set.

63. $0 \boxed{\in} \{-2, 0, 5, 9\}$ since 0 is a member of the set.

65. $\{3\} \boxed{\notin} \{2, 3, 4, 6\}$ since the elements are not sets themselves.

67. The statement $3 \in \{2, 5, 6, 8\}$ is false since the element 3 is not in the set.

69. The statement $b \in \{h, c, d, a, b\}$ is true since b is contained in the set.

71. The statement $9 \notin \{6, 3, 4, 8\}$ is true since 9 is not a member of the set.

73. The statement $\{k, c, r, a\} = \{k, c, a, r\}$ is true since both sets contain exactly the same elements.

75. The statement $\{5, 8, 9\} = \{5, 8, 9, 0\}$ is false because the second set contains a different element from the first set, 0.

77. The statement $\{x \mid x$ is a natural number less than $3\} = \{1, 2\}$ is true since both represent sets with exactly the same elements.

79. That statement $4 \in A$ is true since 4 is a member of set A.

81. The statement $4 \notin C$ is false since 4 is a member of the set C.

83. Every element of C is also an element of A is true since the members, 4, 10, and 12 of set C, are also members of set A.

85. Writing exercise

87. An example of two sets that are not equivalent and not equal would be $\{3\}$ and $\{c, f\}$. Other examples are possible.

89. An example of two sets that are equivalent but not equal would be $\{a, b\}$ and $\{a, c\}$. Other examples are possible.

91. (a) The stocks with share volumes ≥ 188.7 million are those the listed in the set {Viacom (Class B), Trans World Airlines, Harken Energy, Echo Bay Mines}.

    (b) The stocks with share volumes ≤ 188.7 million are those listed in the set {Echo Bay Mines, JTS, Nabors Industries, Hasbro, Royal Oak Mines, Grey Wolf Industries, IVAX}.

2.2 EXERCISES

1. $\{p\}, \{q\}, \{p, q\}, \emptyset$ matches F, the subsets of $\{p, q\}$.

3. $\{a, b\}$ matches C, the complement of $\{c, d\}$, if $U = a, b, c, d$.

5. U matches A, the complement of $\emptyset$.

7. $\{-2, 0, 2\} \not\subseteq \{-2, -1, 1, 2\}$

9. $\{2, 5\} \subseteq \{0, 1, 5, 3, 4, 2\}$

11. $\emptyset \subseteq \{a, b, c, d, e\}$, since the empty set is considered a subset of any given set.

13. $\{-7, 4, 9\} \not\subseteq \{x \mid x$ is an odd integer$\}$ since the element "4" is not an element of the second set.

15. $\{B, C, D\} \subseteq \{B, C, D, F\}$ and $\{B, C, D\} \subseteq \{B, C, D, F\}$, i.e., both.

17. $\{9, 1, 7, 3, 5\} \subseteq \{1, 3, 5, 7, 9\}$

19. $\emptyset \subseteq \{0\}$ or $\emptyset \subseteq \{0\}$, i.e., both.

21. $\{-1, 0, 1, 2, 3\} \not\subseteq \{0, 1, 2, 3, 4\}$; therefore, neither. Note that if a set is not a subset of another set, it can not be a proper subset either.

23. $A \subset U$ is true, since all sets must be subsets of the Universal set by definition, and U contains at least one more element than A.

25. $D \subseteq B$ is false, since the element "d" in set D is not also a member of set B.

27. $A \subset B$ is true. All members of A are also members of B, and there are elements in set B not contained in set A.

29. $\emptyset \subset A$ is true since the empty set, $\emptyset$, is considered a subset of all sets. It is a proper subset since there are elements in A not contained in $\emptyset$.

31. $\emptyset \subseteq \emptyset$ is true since the empty set, $\emptyset$, is considered a subset of all sets including itself. Note that all sets are subsets of themselves.

33. $D \not\subseteq B$ is true. Set D is not a subset of B because the element "d", though a member of set D, is not also a member of set B.

35. There are exactly 6 subsets of C is false. Since there are 3 elements in set C, there are $2^3 = 8$ subsets.

37. There are exactly 3 subsets of A is false. Since there are 2 elements in set A, there are $2^2 = 4$ subsets.

39. There is exactly one subset of $\emptyset$ is true. The only subset of $\emptyset$ is $\emptyset$ itself.

41. The Venn diagram does not represent the correct relationships among the sets since C is not a subset of A. Thus, the answer is false.

43. Since the given set has 3 elements, there are $2^3 = 8$ subsets and $2^3 - 1 = 7$ proper subsets.

45. Since the given set has 6 elements, there are $2^6 = 64$ subsets and $2^6 - 1 = 63$ proper subsets.

47. The set $\{x \mid x$ is an odd integer between -6 and $4\} = \{-5, -3, -1, 1, 3\}$. Since the set contains 5 elements, there are $2^5 = 32$ subsets and $2^5 - 1 = 32 - 1 = 31$ proper subsets.

49. The complement of $\{1, 4, 6, 8\}$ is $\{2, 3, 5, 7, 9, 10\}$, that is, all of the elements in U not also in the given set.

51. The complement of $\{1, 3, 4, 5, 6, 7, 8, 9, 10\}$ is $\{2\}$.

53. The complement of $\emptyset$, the empty set, is $\{1, 2, 3, 4, 5, 6, 7, 8, 9, 10\}$, the universal set.

55. In order to contain all of the indicated characteristics, the universal set, $U = \{$Higher cost, Lower cost, Educational, More time to see the sights, Less time to see the sights, Cannot visit relatives along the way, Can visit relatives along the way$\}$.

57. Since D contains the set of characteristics of the driving option, $D' = \{$Higher cost, More time to see the sights, Cannot visit relatives along the way$\}$.

59. The set of element(s) common to F' and D' is $\emptyset$, the empty set, since there are no common elements.

61. The only possible set is $\{A, B, C, D, E\}$. (All are present.)

63. The possible subsets of three people would include $\{A, B, C\}, \{A, B, D\}, \{A, B, E\}, \{A, C, D\},$ $\{A, C, E\}, \{A, D, E\}, \{B, C, D\}, \{B, C, E\},$ $\{B, D, E\},$ and $\{C, D, E\}$.

65. The possible subsets consisting of one person would include $\{A\}, \{B\}, \{C\}, \{D\}$ and $\{E\}$.

67. Adding the number of subsets in Exercises $61 - 66$, we have $1 + 5 + 10 + 10 + 5 + 1 = 32$ ways that the group can gather.

69. (a) There are s subsets of B that do not contain e. These are the subsets of the original set A.

   (b) There is one subset of B for each of the original subsets of set A, which is formed by including e as the element of that subset of A. Thus, B has s subsets which do contain e.

   (c) The total number of subsets of B is the sum of the numbers of subsets containing e and of those not containing e. This number is $s + s$ or $2s$.

   (d) Adding one more element will always double the number of subsets, so we conclude that the formula 2^n is true in general.

71. (a) Consider all possible subsets of a set with five elements (the number of coins). The number of proper subsets would be $2^5 = 32$. But the 32 includes also the empty set and we must select at least one coin. Thus there are $32 - 1 = 31$ sums of money.

   (b) Removing the condition says, in effect, that we may also choose no coins. Thus, there are $2^4 = 16$ subsets or possible sums of coins including - no coins.

2.3 EXERCISES

1. The intersection of A and B, $A \cap B$, matches B, the set of elements common to both A and B.

3. The difference of A and B, $A - B$, matches A, the set of elements in A that are not in B.

5. The Cartesian product of A and B, $A \times B$, matches E, the set of ordered pairs such that each first element is from A and each second element is from B, with every element of A paired with every element of B.

7. $X \cap Y = \{$a, c$\}$ since these are the elements that are common to both X and Y.

9. $Y \cup Z = \{$a, b, c, d e, f$\}$ since these are the elements that are contained in X or Z (or both).

11. $X \cup U = \{$a, b, c, d, e, f, g$\} = U$. Observe that any set union with the universal set will give the universal set.

13. $X' = \{$b, d, f$\}$ since these are the only elements in U not contained in X.

15. $X' \cap Y' = \{$b, d, f$\} \cap \{$d, e, f, g$\} = \{$d, f$\}$

17. $X \cup (Y \cap Z) = \{$a, c, e, g$\} \cup \{$b, c$\} = \{$a, b, c, e, g$\}$ Observe that the intersection must be done first.

19. $(Y \cap Z') \cup X = (\{$a, b, c$\} \cap \{$a, g$\}) \cup \{$a, c, e, g$\} = \{$a$\} \cup \{$a, c, e, g$\} = \{$a, c, e, g$\} = X$

21. $(Z \cup X')' \cap Y = (\{b, c, d, e, f\} \cup \{b, d, f\})' \cap \{a, b, c\} = \{b, c, d, e, f\}' \cap \{a, b, c\} = \{a, g\} \cap \{a, b, c\} = \{a\}$

23. $X - Y = \{$e, g$\}$ Since these are the only two elements that belong to X and not to Y.

25. $X' - Y = \{$b, d, f$\} - \{$a, b, c$\} = \{$d, f$\}$. Observe that we must find X' first.

27. $X \cap (X - Y) = \{$a, c, e, g$\} \cap \{$e, g$\} = \{$e, g$\}$. Observe that we must find $X - Y$ first.

29. $A \cup (B' \cap C')$ is the set of all elements that either are in A or are not in B and not in C.

31. $(C - B) \cup A$ is the set of all elements that are in C but not in B, or they are in A.

33. $(A - C) \cup (B - C)$ is the set of all elements that are in A but not C, or are in B but not in C.

35. The smallest set representing the universal set U is $\{e, h, c, l, b\}$.

37. T', the complement of T, is the set of effects in U that are not adverse effects of tobacco use: $T' = \{l, b\}$.

39. $T \cup A$ is the set of all adverse effects that are either tobacco related or alcohol related: $T \cup A = \{$e, h, c, l, b$\} = U$.

41. $B \cup C$ is the set of all tax returns showing business income or filed in 1999.

43. $C - A$ is the set of all tax returns filed in 1999 without itemized deductions.

45. $(A \cup B) - D$ is the set of all tax returns with itemized deductions or showing business income, but not selected for audit.

47. $A \subseteq (A \cup B)$ is always true since $A \cup B$ will contain all of the elements of A.

49. $(A \cap B) \subseteq A$ is always true since the elements of $A \cap B$ must be in A.

51. $n(A \cup B) = n(A) + n(B)$ is not always true. If there are any common elements to A and B, they will be counted twice.

53. $n(A \cup B) = n(A) + n(B) - n(A \cap B)$ is always true, since any elements common to sets A and B which are counted twice by $n(A) + n(B)$ are returned to a single count by the subtraction of $n(A \cap B)$.

55. (a) $X \cup Y = \{1, 2, 3, 5\}$

    (b) $Y \cup X = \{1, 2, 3, 5\}$

    (c) For any sets X and Y,
$$X \cup Y = Y \cup X.$$

    This conjecture indicates that set union is a commutative operation.

57. (a) $X \cup (Y \cup Z) = \{1, 3, 5\} \cup (\{1, 2, 3\} \cup \{3, 4, 5\})$
$$= \{1, 3, 5\} \cup \{1, 2, 3, 4, 5\}$$
$$= \{1, 3, 5, 2, 4\}$$

    (b) $(X \cup Y) \cup Z = (\{1, 3, 5\} \cup \{1, 2, 3\}) \cup \{3, 4, 5\}$
$$= \{1, 3, 5, 2\} \cup \{3, 4, 5\}$$
$$= \{1, 3, 5, 2, 4\}$$

    (c) For any sets X, Y and Z,
$$X \cup (Y \cup Z) = (X \cup Y) \cup Z.$$

    This conjecture indicates that set union is a associative operation.

59. (a) $(X \cup Y)' = \{1, 3, 5, 2\}' = \{4\}$

    (b) $X' \cap Y' = \{2, 4\} \cap \{4, 5\} = \{4\}$

    (c) For any sets X and Y,
$$(X \cup Y)' = X' \cap Y'.$$

    Observe that this conjecture is one form of DeMorgan's Laws.

61. (a) $X \cup \emptyset = \{1, 3, 5\} \cup \emptyset = X$

    (b) For any set X,
$$X \cup \emptyset = X.$$

63. The statement $(3, 2) = (5 - 2, 1 + 1)$ is true.

65. The statement $(6, 3) = (3, 6)$ is false. The parentheses indicate an ordered pair (where order is important) and corresponding elements in the ordered pairs must be equivalent.

67. The statement $\{6, 3\} = \{3, 6\}$ is true since order is not important when listing elements in sets.

69. The statement $\{(1, 2), (3, 4)\} = \{(3, 4), (1, 2)\}$ is true. Each set contains the same two elements, the order of which is unimportant.

71. To form the Cartesian product $A \times B$, list all ordered pairs in which the first element belongs to A and the second element belongs to B:
With $A = \{2, 8, 12\}$ and $B = \{4, 9\}$,
$A \times B = \{(2, 4), (2, 9), (8, 4), (8, 9), (12, 4), (12, 9)\}$.
To form the Cartesian product $B \times A$, list all ordered pairs in which the first element belongs to B and the second element belongs to A:
$$B \times A = \{(4, 2), (4, 8), (4, 12), \; (9, 2), (9, 8), (9, 12)\}.$$

73. For $A = \{d, o, g\}$ and $B = \{p, i, g\}$,
$A \times B = \{(d, p), (d, i), (d, g), (o, p), (o, i), (o, g),$
$(g, p), (g, i), (g, g)\}$;
$B \times A = \{(p, d), (p, o), (p, g), (i, d), (i, o), (i, g),$
$(g, d), (g, o), (g, g)\}$.

75. For $A = \{2, 8, 12\}$ and $B = \{4, 9\}$,
$n(A \times B) = n(A) \times n(B) = 3 \times 2 = 6$, or by counting the generated values in Exercise 71 we also arrive at 6. In the same manner $n(B \times A) = 2 \times 3 = 6$.

77. For $n(A) = 35$ and $n(B) = 6$,
$n(A \times B) = n(A) \times n(B) = 35 \times 6 = 210$.
$n(B \times A) = n(B) \times n(A) = 6 \times 35 = 210$.

79. To find $n(B)$ when $n(A \times B) = 36$ and $n(A) = 12$, we have:
$$n(A \times B) = n(A) \times n(B)$$
$$36 = 12 \times n(B)$$
$$3 = n(B).$$

81. Let $U = \{a, b, c, d, e, f, g\}$,
$A = \{b, d, f, g\}$, and $B = \{a, b, d, e, g\}$.

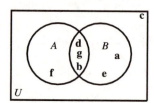

83. The set operations for $B \cap A'$ indicate those elements in B and not in A or

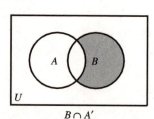

$B \cap A'$

85. The set operations for $A' \cup B$ indicate those elements not in A or in B.

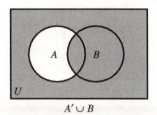

$A' \cup B$

87. The set operations for $B' \cup A$ indicate those elements not in B or in A.

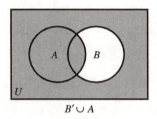

$B' \cup A$

89. The set operations $B' \cap B$ indicate those elements not in B and in B at the same time, and since there are no elements that can satisfy both conditions, we get the null set (empty set), $\emptyset$.

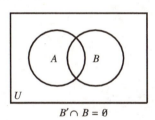

$B' \cap B = \emptyset$

91. The indicated set operations mean those elements not in B or those not in A as long as they are also not in B. It is a help to shade the region representing "not in A" first, then that region representing "not in B." Identify the intersection of these regions (covered by both shadings). As in algebra, the general strategy when deciding which order to do operations is to begin inside parentheses and work out.

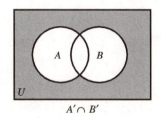

$A' \cap B'$

Finally, the region of interest will be that "not in B" along with (union of) the above intersection—$(A' \cap B')$. That is, the final region of interest is given by

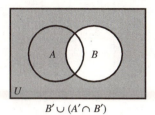

$B' \cup (A' \cap B')$

93. The complement of U, U', is the set of all elements not in U. But by definition, there can be no elements outside the universal set. Thus, we get the null (or empty) set, $\emptyset$, when we complement U.

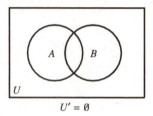

$U' = \emptyset$

95. Let $U = \{m, n, o, p, q, r, s, t, u, v, w\}$,
$A = \{m, n, p, q, r, t\}$,
$B = \{m, o, p, q, s, u\}$, and
$C = \{m, o, p, r, s, t, u, v\}$.

Placing the elements of these sets in the proper location on a Venn diagram will yield the following diagram.

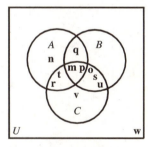

It helps to identify those elements in the intersect of A, B, and C first, then those elements not in this intersection but in each of the two set intersections (e.g., $A \cap B$, etc.), next, followed by elements that lie in only one set etc.

97. The set operations $(A \cap B) \cap C$ indicate those elements common to all three sets.

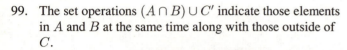

$(A \cap B) \cap C$

99. The set operations $(A \cap B) \cup C'$ indicate those elements in A and B at the same time along with those outside of C.

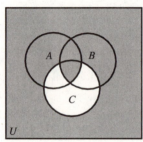

$(A \cap B) \cup C'$

101. The set operations $(A' \cap B') \cap C$ indicate those elements that are in C while simultaneously outside of both A and B.

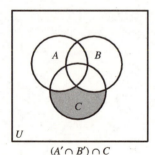

$(A' \cap B') \cap C$

103. The set operations $(A \cap B') \cup C$ indicate those elements that are in A and at the same time outside of B along with those in C.

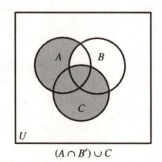

$(A \cap B') \cup C$

105. The set operations $(A \cap B') \cap C'$ indicate the region in A and outside B and at the same time outside C.

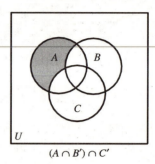

$(A \cap B') \cap C'$

107. The set operations $(A' \cap B') \cup C'$ indicate the region that is both outside A and at the same time outside B along with the region outside C.

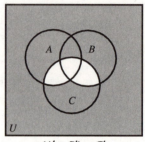

$(A' \cap B') \cup C'$

109. The shaded area indicates the region $(A \cup B)'$ or $A' \cap B'$.

111. Since this is the region in A or in B but, at the same time, outside of A and B, we have the set $(A \cup B) \cap [(A \cap B)']$ or $(A \cup B) - (A \cap B)$.

113. The shaded area may be represented by the set $(A \cap B) \cup (A \cap C)$; that is, the region in the intersection of A and B along with the region in the intersection of A and C or, by the distributive property, $A \cap (B \cup C)$.

115. The region is represented by the set $(A \cap B) \cap C'$, that is, the region outside of C but inside both A and B, or $(A \cap B) - C$.

117. If $A = A - B$, then A and B must not have any common elements, or $A \cap B = \emptyset$.

119. $A = A - \emptyset$ is true for any set A.

121. $A \cup \emptyset = \emptyset$ is true only if A has no elements, or $A = \emptyset$.

123. $A \cap \emptyset = A$ is true only if A has no elements, or $A = \emptyset$.

125. $A \cup A = \emptyset$ is true only if A has no elements, or $A = \emptyset$.

127. $A \cap A' = \emptyset$.

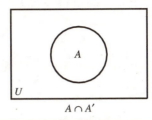

Thus, by the Venn diagrams, the statement is always true.

129. $(A \cap B) \subseteq A$.

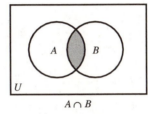

Thus, by the Venn diagrams, the shaded region is in A; therefore, the statement is always true.

131. If $A \subseteq B$, then $A \cup B = A$.

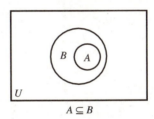

Thus, the statement is not always true.

133. $(A \cup B)' = A' \cap B'$ (De Morgan's second law).

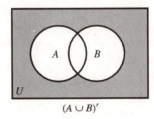

Thus, by the Venn diagrams, the statement is always true.

135. Writing exercise

2.4 EXERCISES

1. (a) $n(A \cap B) = 1$ since A and B share only one element.

   (b) $n(A \cup B) = 9$, since there are a total of 9 elements in A or in B.

(c) $n(A \cap B') = 6$ since there are 6 elements which are in A and, at the same time, outside B.

(d) $n(A' \cap B) = 2$ since there are 2 elements which are in B and, at the same time, outside A.

(e) $n(A' \cap B') = 5$ since there are only 5 elements which are outside of A and, at the same time, outside of B.

3. (a) $n(A \cap B \cap C) = 1$ since there is only one element shared by all three sets.

   (b) $n(A \cap B \cap C') = 2$ since there are 2 elements in A and B while at the same time, outside of C.

   (c) $n(A \cap B' \cap C) = 6$ since there are 6 elements in A and C while, at the same time, outside of B.

   (d) $n(A' \cap B \cap C) = 7$ since there are 7 elements which are outside of A, while at the same time, are in B and C.

   (e) $n(A' \cap B' \cap C) = 8$ since there are 8 elements outside of A and outside of B while at the same time, are inside of C.

   (f) $n(A \cap B' \cap C') = 3$ since there are 3 elements in A which, at the same time, are outside of B and outside of C.

   (g) $n(A' \cap B \cap C') = 4$ since there are 4 elements outside of A and, at the same time, outside of C but inside of B.

   (h) $n(A' \cap B' \cap C') = 5$ since there are 5 elements which are outside all three sets at the same time.

5. Using the Cardinal Number Formula, $n(A \cup B) = n(A) + n(B) - n(A \cap B)$, we have $n(A \cup B) = 8 + 14 - 5 = 17$.

7. Using the Cardinal Number Formula, $n(A \cup B) = n(A) + n(B) - n(A \cap B)$, we have $30 = n(A) + 20 - 6$. Solving for $n(A)$, we get $n(A) = 16$.

9. Use deduction to complete the cardinalities of the unknown regions. For example since $n(B') = 30$, there is a total of 13 elements in B $[n(U) - n(B') = 43 - 30]$; therefore, 8 elements in B that are not in A $[n(B) - n(A \cap B) = 13 - 5]$. Since there is a total of 25 elements in A and 5 are accounted for in $A \cap B$, there must be 20 elements in A that are not in B. This leaves a total of 33 elements in the regions formed by A along with B. Thus, there are 10 elements left in U that are not in A or in B. Completing the cardinalities for each region we arrive at the following Venn diagram.

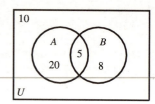

11. Use deduction to complete the cardinalities of each region outside of $A \cap B$. Since there is a total of 13 elements in A and 8 accounted for in the intersect of A and B, there must be 5 elements in A outside of B.

Since there is a total of 15 elements in the union of A and B with 13 accounted for in A, there must be 2 elements in B which are not in A. The region $A' \cup B'$ is equivalent (by De Morgan's law) to $(A \cap B)'$, or the elements outside the intersection. Since this totals 11, we must have 4 $[11 - 7]$ elements outside the union but inside U. Completing the cardinalities for each region we arrive at the following Venn diagram.

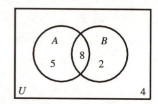

13. Fill in the cardinal numbers of the regions, beginning with the $(A \cap B \cap C)$. Since there is a total of 15 elements in $A \cap C$ of which 6 are accounted for in $A \cap B \cap C$, we conclude that there are 9 elements in $A \cap C$ but outside of B. Similarly, there must be 2 elements in $B \cap C$ but outside A and 4 elements in $A \cap B$ but outside of C. Since there are 26 elements in C of which we have accounted for 17 $[9 + 6 + 2]$, there must be 9 elements in C but outside of A or B.

Similarly, there are 12 elements in B outside of A or C and 5 elements in A outside of B or C. And finally, adding the elements in the regions of A, B, and C gives a total of 47. Thus, the number of elements outside of A, B, and C is $n(U) - 47$ or $50 - 47 = 3$. Completing the cardinalities for each region we arrive at the following Venn diagram.

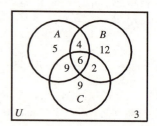

15. Fill in the cardinal numbers of the regions, beginning with $n(A \cap B \cap C) = 6$. Since there are 21 elements in A and B, $n(A \cap B) = 21$ and there must be 6 elements in all three sets, that is, $n(A \cap B \cap C) = 6$.

At the same time, the elements in A and B but not in C are given by $n(A \cap B \cap C') = 21 - 6 = 15$. Since $n(A \cap C) = 26$, we have those elements in A and C but not in B as $n(A \cap C \cap B') = 26 - 6 = 20$. Since $n(B \cap C) = 7$ then, $n(B \cap C \cap A') = 7 - 6 = 1$.

Since $n(A \cap C') = 20$, we have those elements in A but not in either B or C as $n(A \cap (B \cup C)') = 20 - 15 = 5$. Since $n(B \cap C') = 25$, we have those elements in B but not in A or C as $n(B \cap (A \cup C)') = 25 - 15 = 10$. Since $n(C) = 40$, we have those elements in C but not in A or B as $n(C \cap (A \cup B)') = 40 - (20 + 6 + 1) = 13$.

Observe that $n(A' \cap B' \cap C') = n(A \cap B \cap C)'$ (by De Morgan's). That is, there are 2 elements outside the union of the three sets.

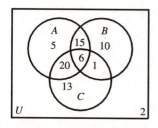

17. Complete a Venn diagram showing the cardinality for each region. Let $S =$ the set of CDs featuring Paul Simon, $G =$ the set of CDs featuring Art Garfunkel.

Beginning with $n(S \cap G) = 5$ and $n(S) = 8$, we conclude that $n(S \cap G') = 8 - 5 = 3$. Since $n(G) = 7$, we conclude that $n(S' \cap G) = 7 - 5 = 2$.

There are 12 CDs on which neither sing, so $n(S \cup G)' = 12$

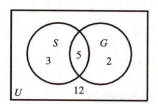

Interpreting the resulting cardinalities we see that:
(a) There are 3 CDs which feature only Paul Simon.

(b) There are 2 CDs which feature only Art Garfunkel.

(c) There are 10 CDs which feature at least one of these two artists.

19. Let U be the set of mathematics majors receiving federal aid. Since half of the 48 mathematics majors receive federal aid, $n(U) = \frac{1}{2}(48) = 24$. Let $P, W,$ and T be the sets of students receiving Pell Grants, participating in Work Study, and receiving TOPS scholarships, respectively.

Construct a Venn diagram and label the number of elements in each region. Since the 5 with Pell Grants had no other federal aid, 5 goes in set P, which does not intersect the other regions. The most manageable data remaining are the 2 who had TOPS scholarships and participated in Work Study. Place 1 in the intersection of W and T. Then since 14 altogether participated in Work Study, $14 - 2 = 12$ is the number who participated in Work Study but did not have TOPS scholarships or Pell Grants; in symbols, $n(W \cap T' \cap P')$. Also, $4 - 2 = 2$ is the number who had Stafford Loans, but did not participate in Work Study nor had Pell Grants, $n(T \cap W' \cap P')$. Finally,

$$\begin{aligned} n(P' \cap W' \cap T') &= n(P \cup W \cup T)' \\ &= 24 - (5 + 12 + 2 + 2) \\ &= 3. \end{aligned}$$

is the number in the region not included in the sets $P, W,$ and T. The completed Venn diagram is as follows.

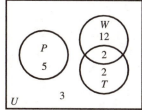

(a) Since half of the 48 mathematics majors received federal aid, the other half, 24 math majors, received no federal aid.

(b) There were only 2 students who received more than one of these three forms of aid. This is shown by the number in the region $T \cap W$.

(c) Since 24 students received federal aid, but only 21 are accounted for in the given information (the sum of the numbers in the three circles), there were 3 math majors who received other kinds of federal aid.

(d) The number of students receiving a TOPS scholarship or participating in Work Study is $12 + 2 + 2 = 16$.

21. Let U be the set of people interviewed, and let $M, E,$ and G represent the sets of people using microwave ovens, electric ranges, and gas ranges respectively.

Construct a Venn diagram and label the cardinal number of each region, beginning with the region $(M \cap E \cap G)$. $n(M \cap E \cap G) = 1$. Since $n(M \cap E) = 19, n(M \cap E \cap G') = 19 - 1 = 18$. Since $n(M \cap G) = 17, n(M \cap G \cap E') = 17 - 1 = 16$. Since $n(G \cap E) = 4, n(G \cap E \cap M') = 4 - 1 = 3$. Since $n(M) = 58, n(M \cap G' \cap E') = n(M \cap (G \cup E)') = 58 - (18 + 16 + 1) = 23$. Since $n(E) = 63$, $n(E \cap M' \cap G') = 63 - (18 + 3 + 1) = 41$. Since $n(G) = 58$, $n(G \cap M' \cap E') = n(G \cap (M \cup E)') = 58 - (16 + 3 + 1) = 38$. Also, $n(M \cup E \cup G)' = 2$ (these are the people who cook with only solar energy).

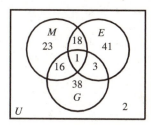

The sum of the numbers in all the regions is 142, while only 140 people were interviewed. Therefore, Jim's data is incorrect and he should be reassigned again.

23. Let $L, P,$ and T represent the sets of songs about love, prison, and trucks, respectively.

Construct a Venn diagram and label the cardinal number of each region.

$n(T \cap L \cap P) = 12$. Since $n(P \cap L) = 13, n(P \cap L \cap T') = 13 - 12 = 1$. Since $n(T \cap L) = 18, n(T \cap L \cap P') = 18 - 12 = 6$.

We have $n(T \cap P \cap L') = 3, n(P \cap L' \cap T') = 2$, and $n(P' \cap L' \cap T') = 8$. Since $n(L) = 28$, $n(L \cap P' \cap T') = 28 - (1 + 12 + 6) = 9$. Since $n(T \cap P') = 16, n(T \cap P' \cap L') = 16 - 6 = 10$.

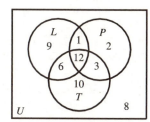

(a) The total in all eight regions is 51, the total number of songs.

(b) $n(T) = 6 + 12 + 3 + 10 = 31$.

(c) $n(P) = 1 + 2 + 12 + 3 = 18$.

(d) $n(T \cap P) = 12 + 3 = 15$.

(e) $n(P') = n(U) - n(P) = 51 - 18 = 33$.

(f) $n(L') = n(U) - n(L) = 51 - 28 = 23$.

25. Construct a Venn diagram and label the cardinal number of each region beginning with the intersection of all three sets, $n(W \cap F \cap E) = 80$.

Since $n(E \cap F) = 90$,
$n(E \cap F \cap W') = 90 - 80 = 10$.
Since $n(W \cap F) = 95$,
$n(W \cap F \cap E') = 95 - 80 = 15$.
Since $n(E \cap F) = 90$,
$n(E \cap F \cap W') = 90 - 80 = 10$.
Since $n(F) = 140$,
$n(F \cap W' \cap E') = 140 - (15 + 10 + 80) = 35$.
$n(W' \cap F' \cap E') = n(W \cup F \cup E)' = 10$.

Since $n(W \cap E)$ is not given, it is not obvious how to label the three remaining regions. We need to use the information that $n(E') = 95$. The only region not yet labeled that is outside of E is $(W \cap E' \cap F')$. Since $n(E') = 95$, $n(W \cap E' \cap F') = 95 - (10 + 35 + 15) = 35$. Since $n(W) = 160$, $n(W \cap E \cap F') = 160 - (35 + 15 + 80) = 30$. Since $n(E) = 130$, $n(E \cap W' \cap F') = 130 - (30 + 10 + 80) = 10$.

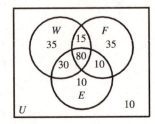

Add the cardinal numbers of all the regions to find that the total number of students interviewed was 225.

27. (a) The set $A \cap B \cap C \cap D$ is region 1.

(b) The set $A \cup B \cup C \cup D$ includes the regions 1, 2, 3, 4, 5, 6, 7, 8, 9, 10, 11, 12, 13, 14, and 15.

(c) The set $(A \cap B) \cup (C \cap D)$ includes the set of regions $\{1, 3, 9, 11\} \cup \{1, 2, 4, 5\}$ or the regions 1, 2, 3, 4, 5, 9, and 11.

(d) The set $(A' \cap B') \cap (C \cup D)$ includes the set of regions $\{5, 13, 8, 16\} \cap \{1, 2, 3, 4, 5, 6, 7, 8, 9, 10, 12, 13\}$ which is represented by regions 4, 8, and 13.

29. (a) $n(W \cap O) = 6$, coming from the intersection of second row with the third column.

(b) $n(C \cup B) = 473$, coming from the union of the first row along with the first column is given by $95 + 390 - 12$. Observe that the 12 are counted twice when adding the totals for the respective row and column and hence must be subtracted (once).

(c) $n(R' \cup W') = n(R \cap W)' = 835$, which is the number of army personnel outside of the intersection of the second column (R) with the second row (W) or $840 - 5$.

(d) $n((C \cup W) \cap (B \cup R)) = 12 + 29 + 4 + 5 = 50$. This comes from adding the numbers in the first two rows (representing $C \cup W$) that are also in the first two columns (representing $B \cup R$).

(e) Using the cardinal number formula,

$n((C \cap B) \cup (E \cap O))$
$\quad = n(C \cap B) + n(E \cap O) - n(C \cap B \cap E \cap O)$
$\quad = 12 + 285 - 0 = 297$.

(f) $n(B \cap (W \cup R)') = n(B \cap (W' \cap R'))$
$\qquad\qquad = n(B \cap W' \cap R')$
$\qquad\qquad = n(B \cap W')$
$\qquad\qquad = n(B) - n(B \cap W)$
$\qquad\qquad = 390 - 4$
$\qquad\qquad = 386$

31. Writing exercise

2.5 EXERCISES

1. The set $\{3\}$ has the same cardinality as B, $\{26\}$. The cardinal number is 1.

3. The set $\{x \mid x$ is a natural number$\}$ has the same cardinality as A, $\aleph_0$.

5. The set $\{x \mid x$ is an integer between 0 and 1$\}$ has the same cardinality, 0, as F since there are no members in either set.

7. One correspondence is:

$$\{\text{I}, \quad \text{II}, \quad \text{III}\}$$
$$\updownarrow \quad \updownarrow \quad \updownarrow$$
$$\{x, \quad y, \quad z\}.$$

Other correspondences are possible.

9. One correspondence is:

$$\{a, \quad d, \quad i, \quad t, \quad o, \quad n\}$$
$$\updownarrow \quad \updownarrow \quad \updownarrow \quad \updownarrow \quad \updownarrow \quad \updownarrow$$
$$\{a, \quad n, \quad s, \quad w, \quad e, \quad r\}.$$

Other correspondences are possible.

11. $n(\{a, b, c, d, \dots, k\}) = 11$. By counting the number of letters a through k we establish the cardinality to be 11.

13. $n(\phi) = 0$ since there are no members.

15. $n(\{300, 400, 500, \dots\}) = \aleph_0$ since this set can be placed in a one-to-one correspondence with the counting numbers (i.e., is a countable infinite set}.

17. $n(\{-1/4, -1/8, -1/12, \dots\}) = \aleph_0$ since this set can be placed in a one-to-one correspondence with the counting numbers.

19. $n(\{x \mid x \text{ is an odd counting number}\}) = \aleph_0$ since this set can be placed in a one-to-one correspondence with the counting numbers.

21. $n(\{\text{Jan, Feb, Mar}, \dots, \text{Dec}\}) = 12$ since there are twelve months indicated in the set.

23. "$\aleph_0$ bottles of beer on the wall, $\aleph_0$ bottles of beer, take one down and pass it around, $\boxed{\aleph_0}$ bottles of beer on the wall." This is true because $\aleph_0 - 1 = \aleph_0$.

25. The answer is both. Since the sets $\{u, v, w\}$ and $\{v, u, w\}$ are equal sets (same elements), they must then have the same number of elements and thus are equivalent.

27. The sets $\{X, Y, Z\}$ and $\{x, y, z\}$ are equivalent because they contain the same number of elements (same cardinality) but not the same elements.

29. The sets $\{x \mid x \text{ is a positive real number}\}$ and $\{x \mid x \text{ is a negative real number}\}$ are equivalent because they have the same cardinality, c. They are not the equal since they contain different elements.

Note that each of the following answers shows only one possible correspondence.

31.
$$\{2, \quad 4, \quad 6, \quad 8, \quad 10, \quad 12, \quad \dots, \quad 2n, \quad \dots\}$$
$$\updownarrow \quad \updownarrow \quad \updownarrow \quad \updownarrow \quad \updownarrow \quad \updownarrow \qquad \updownarrow$$
$$\{1, \quad 2, \quad 3, \quad 4, \quad 5, \quad 6, \quad \dots, \quad n, \quad \dots\}$$

33.
$$\{1{,}000{,}000 \quad 2{,}000{,}000 \quad 3{,}000{,}000, \dots, 1{,}000{,}000n, \dots\}$$
$$\updownarrow \qquad\qquad \updownarrow \qquad\qquad \updownarrow \qquad\qquad \updownarrow$$
$$\{1, \qquad\qquad 2, \qquad\qquad 3, \quad \dots, \quad n, \qquad \dots\}$$

35.
$$\{2, \quad 4, \quad 8, \quad 16, \quad 32, \quad \dots, \quad 2^n, \quad \dots\}$$
$$\updownarrow \quad \updownarrow \quad \updownarrow \quad \updownarrow \quad \updownarrow \qquad \updownarrow$$
$$\{1, \quad 2, \quad 3, \quad 4, \quad 5, \quad \dots, \quad n, \quad \dots\}$$

37. The statement "If A and B are infinite sets, then A is equivalent to B" is not always true. For example, let $A = $ the set of counting numbers and $B = $ the set of real numbers. Each has a different cardinality.

39. The statement "If set A is an infinite set and A is not equivalent to the set of counting numbers, then $n(A) = c$" is not always true. For example, A could be the set of all subsets of the set of real numbers. Then, $n(A)$ would be an infinite number greater than c.

41. (a) Use the figure (in the text), where the line segment between 0 and 1 has been bent into a semicircle and positioned above the line, to prove that $\{x \mid x \text{ is a real number between 0 and 1}\}$ is equivalent to $\{x \mid x \text{ is a real number}\}$.

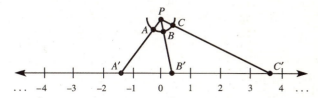

Rays emanating from point P will establish a geometric pairing of the points on the semicircle with the points on the line.

(b) The fact part (a) establishes about the set of real numbers is that the set of real numbers is infinite, having been placed in a one-to-one correspondence with a proper subset of itself.

43.
$$\{3, \quad 6, \quad 9, \quad 12, \quad \dots, \quad 3n, \quad \dots\}$$
$$\updownarrow \quad \updownarrow \quad \updownarrow \quad \updownarrow \qquad \updownarrow$$
$$\{6, \quad 9, \quad 12, \quad 15, \quad \dots, \quad 3n+3, \quad \dots\}$$

45.
$$\{3/4, \quad 3/8, \quad 3/12, 3/16, \dots, 3/(4n), \quad \dots\}$$
$$\updownarrow \quad \updownarrow \quad \updownarrow \quad \updownarrow \qquad \updownarrow$$
$$\{3/8, 3/12, \ 3/16, 3/20, \dots, 3/(4n+4), \dots\}$$

47.
$$\{1/9, \quad 1/18, \quad 1/27, \quad \dots, \quad 1/(9n), \quad \dots\}$$
$$\updownarrow \quad \updownarrow \quad \updownarrow \qquad \updownarrow$$
$$\{1/18, \ 1/27, \ 1/36, \quad \dots, \quad 1/(9n+9), \ \dots\}$$

49. Writing exercise

51. Writing exercise

CHAPTER 2 TEST

1. $A \cup C = \{a, b, c, d\} \cup \{a, e\} = \{a, b, c, d, e\}$

2. $B \cap A = \{b, e, a, d\} \cap \{a, b, c, d\} = \{a, b, d\}$

3. $B' = \{b, e, a, d\}' = \{c, f, g, h\}$

4. $A - (B \cap C') = A - (\{b, e, a, d\} \cap \{b, c, d, f, g, h\})$
   $= \{a, b, c, d\} - \{b, d\}$
   $= \{a, c\}$

5. $b \in A$ is true since b is member of set A.

6. $C \subseteq A$ is false since the element e, which is a member of set C, is not also a member of set A.

7. $B \subset (A \cup C)$ is true since all members of set B are also members of $A \cup C$.

8. $c \notin C$ is true because c is not a member of set C.

9. $n[(A \cup B) - C] = 4$ is false. Because,

$$n[(A \cup B) - C] = n[\{a, b, c, d, e\} - \{a, e\}]$$
$$= n(\{b, c, d\})$$
$$= 3.$$

10. $\emptyset \subset C$ is true. The empty set is considered a subset of any set. C has more elements then $\emptyset$ which makes $\emptyset$ a proper subset of C.

11. $(A \cap B')$ is equivalent to $(B \cap A')$ is true. Because,

$$n(A \cap B') = n(\{c\}) = 1$$
$$n(B \cap A') = n(\{e\}) = 1.$$

12. $(A \cup B)' = A' \cap B'$ is true by one of De Morgan's laws.

13. $n(A \times C) = n(A) \times n(C) = 4 \times 2 = 8$.

14. The number of proper subsets of A is

$$2^4 - 1 = 16 - 1 = 15.$$

Answers may vary for Exercises 15–18.

15. A word description for $\{-3, -1, 1, 3, 5, 7, 9\}$ is the set of all odd integers between -4 and 10.

16. A word description for $\{$January, February, March, $\ldots,$ December$\}$ is the set of months of the year.

17. Set-builder notation for $\{-1, -2, -3, -4, \ldots\}$ would be $\{x \mid x$ is a negative integer$\}$.

18. Set-builder notation for $\{24, 32, 40, 48, \ldots, 88\}$ would be $\{x \mid x$ is a multiple of 8 between 20 and 90$\}$.

19. $\emptyset \boxed{\subseteq} \{x \mid x$ is a counting number between 17 and 18$\}$ since the empty set is a subset of any set.

20. $\{4, 9, 16\} \boxed{\text{neither}} \{4, 5, 6, 7, 8, 9, 10\}$ since the element 16 is not a member of the second set.

21. $X \cup Y'$

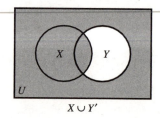

$X \cup Y'$

22. $X' \cap Y'$

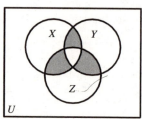

$X' \cap Y'$

23. $(X \cup Y) - Z$

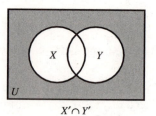

$(X \cup Y) - Z$

24. $[(X \cap Y) \cup (Y \cap Z) \cup (X \cap Z)] - (X \cap Y \cap Z)$

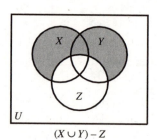

$[(X \cap Y) \cup (Y \cap Z) \cup (X \cap Z)] - (X \cap Y \cap Z)$

25. $A \cap T =$
$\{$Electric razor, Telegraph, Zipper$\} \cap \{$Electric razor, Fiber optics, Geiger counter, Radar$\} = \{$Electric Razor$\}$.

26. $(A \cup T)' =$
$(\{$Electric razor, Telegraph, Zipper$\} \cup \{$Electric razor, Fiber optics, Geiger counter, Radar$\})' = \{$Electric razor, Fiber optics, Geiger counter, Radar, Telegraph, Zipper$\}' = \{$Adding machine, Barometer, Pendulum clock, Thermometer$\}$.

27. $A - T' = \{$Electric razor, Telegraph, Zipper$\} -$
$\{$Electric razor, Fiber optics, Geiger counter, Radar$\}'$
$= \{$Electric razor, Telegraph, Zipper$\} -$
$\{$Adding machine, Barometer, Pendulum clock,
Telegraph, Thermometer, Zipper$\}$
$= \{$Electric razor$\}$.

28. Writing exercise

29. (a) $n(A \cup B) = 12 + 3 + 7 = 22$.

(b) $n(A \cap B') = n(A - B) = 12$. These are the elements in A but outside of B.

(c) $n(A \cap B)' = n(\{3\}') = 12 + 7 + 9 = 28$.

30. Let $G = $ set of students who are receiving government grants. Let $S = $ set of students who are receiving private scholarships. Let $A = $ set of students who are receiving aid from the college. Complete a Venn diagram by inserting the appropriate cardinal number for each region in the diagram. Begin with the intersection of all three sets: $n(G \cap S \cap A) = 8$. Since $n(S \cap A) = 28$, $n(S \cap A \cap G') = 28 - 8 = 20$. Since $n(G \cap A) = 18$, $n(G \cap A \cap S') = 18 - 8 = 10$. Since $n(G \cap S) = 23$, $n(G \cap S \cap A') = 23 - 8 = 15$.

Since
$n(A) = 43, n\big(A \cap (G \cup S)'\big) = 43 - (10 + 8 + 20)$
$= 44 - 38 = 5$. Since $n(S) = 55, n\big(S \cap (G \cup A)'\big)$
$= 55 - (15 + 8 + 20) = 55 - 43 = 12$.

Since
$n(G) = 49, n\big(G \cap (S \cup A)'\big) = 49 - (10 + 8 + 15)$
$= 49 - 33 = 16$.

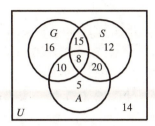

Thus,
(a) $n\big(G \cap (S \cup A)'\big) = 16$ have a government grant only.

(b) $n(S \cap G') = 32$ have a private scholarship but not a government grant.

(c) $16 + 12 + 5 = 33$ receive financial aid from only one of these sources.

(d) $10 + 15 + 20 = 45$ receive aid from exactly two of these sources.

(e) $n(G \cup S \cup A)' = 14$ receive no financial aid from any of these sources.

(f) $n\big(S \cap (A \cup G)'\big) + n(A \cup G \cup S)' = 12 + 14 = 26$ received private scholarships or no aid at all.

3   INTRODUCTION TO LOGIC

Chapter Goals

The general goal of this chapter is to provide an overview of basic symbolic logic and some of its applications to the science of thinking and reasoning. Upon completion of this chapter, the student should be able to

- Understand the (logical) character of the most commonly used logical connectives.
- Determine when compound statements are true or false.
- Determine negations of simple, compound, and quantified statements.
- Determine equivalency of statements.
- Recognize common equivalent statements.
- Create truth tables for compound statements.
- Determine the validity of quantified arguments (using Euler diagrams).
- Determine the validity of non-quantified arguments (using truth tables).

Chapter Summary

Logic is the study of correct reasoning. Application and uses of logic abound in such diverse fields as the sciences, computer technologies, mathematics, rhetoric, law, and psychology. This chapter contains some of the basic principles of logic including the language, the symbols, and some applications.

Naturally, a brief exposure will not turn someone into a logician. But the need for clear thinking will become apparent and, with practice and effort, proficiency with the English language and the ability to reason abstractly should improve.

One of the goals of this chapter is to determine whether complicated (compound) statements are true or false, based on the truth or falsity of their individual parts (simple statements). By a simple statement, we mean the following:

A **simple statement** is a single assertion that has the property of being true or false.

Examine carefully the types of English sentences that satisfy this definition of a simple statement. Many assertions do not have the indicated property. For example, opinions, questions, paradoxes and commands, do not have the property of being true or false and thus are excluded from our meaning of simple statements. Remember that to be considered a statement from the point of view of applying logic, it <u>must have the property of being true or false</u>.

An efficient way to decide the truth value of compound statements is to symbolize the simple statements using letters– p, q, and r, for example.

LOGICAL CONNECTIVES

Compound statements are formed by combining simple statements using connecting words. These connecting words are called logical connectives, and they have their own symbolization different from the simple statements. The commonly used connectives and their symbolizations are summarized in the following table.

Logical connective	Symbol	Compound statement
not	~	$\sim p$
and	$\wedge$	$p \wedge q$
or	$\vee$	$p \vee q$
if ..., then	$\rightarrow$	$p \rightarrow q$
if and only if	$\leftrightarrow$	$p \leftrightarrow q$

Each logical connective is defined by means of a truth table. A truth table gives all possible combinations of truth value for each of the simple statements and the truth value of the compound statement based upon the logical nature of the connectives. Our first use of the truth table is to define the logical character (truth values) of the individual connectives.

The following is a review of the logical connectives and the associated truth values. Learn these carefully as they are the foundation for achieving the remaining goals of this chapter.

Negation   A simple statement, p, that has been negated (its truth value has been reversed), is known as a **negation** and symbolized, as we saw above, by **$\sim p$**. The expression $\sim p$ is typically read as "not p," but other phrases such as "it is false that," "it is not true that," and "it is not the case that" are commonly used as well. The logical definition (character) of this connective is given in the truth table:

Negation

p	$\sim p$
T	F
F	T

The sentence "Mike loves mathematics" is a statement; the negation of this statement is "Mike does not love mathematics." If it is true that "Mike loves mathematics," then it must be false that "Mike does not love mathematics"

and vice versa. The negation must always have the opposite truth value of the original statement.

Quantified Statements   The words "all," "each," "every," and "no(ne)" are called **universal quantifiers**, while words and phrases like "some," "there exists," and "(for) at least one" are called **existential quantifiers**. Quantifiers are commonly used in mathematics, logic, and the sciences to indicate how many cases of a particular situation exist. One must be very careful when forming the negation of a statement involving quantifiers. Consider the statement "All dogs growl." This is a false statement (I know because my dog doesn't growl). Thus, its negation must be a true statement. Some people would say that the negation is "No dogs growl." However, this is also false, so we have not negated the original statement. One may negate the original statement by "It is not the case that all dogs growl." A more typical negation would be "Some dogs do not growl." Both negations are equivalent.

The key to correctly negating a quantified statement is to change the original quantifier (universal to existential or vice versa) and negate the internal statement. The following table, found in the text, can be used to generalize this method of finding the negation of a statement involving quantifiers.

EXAMPLE A   Negation of quantified statements

Write the negation of the following statements.

Original statement	*Negation*
All boys love girls.	*Some boys do not love girls.*
Some students deserve an A.	*All students do not deserve an A. (meaning no students deserve an A)*
All teachers are poor.	*Some teachers are not poor.*
Everyone is happy.	*Some are not happy.*

The negation of compound statements is reviewed later (see "Negations of Some Compound Statements" in the *Study Guide and Solutions Manual*).

Conjunction   Any two statements, for example p and q, connected by the word "and" forms an expression known as a **conjunction**, denoted **$p \wedge q$**. As indicated in the table above, the expression is read "p and q." In everyday language, the connective *and* implies the idea of "both." The logical definition given by the following truth table summarizes this idea.

Conjunction

p	q	$p \wedge q$
T	T	T
T	F	F
F	T	F
F	F	F

Symbolically ($p \rightarrow q$)	English translations
if p, then q	If it is a rainy day, then it is a day that I stay indoors.
if p, q	If it is a rainy day, it is a day that I stay indoors.
q, if p	It is a day that I stay indoors, if it is a rainy day.
p implies q	It is a rainy day implies that it is a day that I stay indoors.
p only if q	It is a rainy day only if it is a day that I stay indoors.
p is sufficient for q	Being a rainy day is sufficient to being a day that I stay indoors.
q is necessary for p	Being a day that I stay indoors is necessary for it to be a rainy day.
all p's are q's	All rainy days are days that I stay indoor.
not p unless q	It is not a rainy day unless it is a day that I stay indoors.

The following observation is worth remembering:
$p \wedge q$ is true only when both conjuncts (component simple statements) are true.

Disjunction   Any two statements, for example p and q, connected by the word "or" form an expression known as **disjunction**. There are two types of "or" statements: inclusive and exclusive. **Inclusive** disjunction or just disjunction, denoted $\boldsymbol{p \vee q}$, means "either one or the other, or both." **Exclusive** disjunction, denoted $\boldsymbol{p \veebar q}$, means "either one or the other, but not both." Inclusive is commonly used in mathematics and logic. Exclusive is commonly used in everyday language. The logical definition of each is given by the following truth tables.

Disjunction

		Disjunction	Exclusive disjunction
p	q	$p \vee q$	$p \veebar q$
T	T	T	F
T	F	T	T
F	T	T	T
F	F	F	F

The following observation is worth remembering:
- $p \vee q$ is false only when both disjuncts (component simple statements) are false.

Conditional   If two statements, for example p and q, are connected in such a manner that the first statement is preceded by the word "if" and the second is preceded by the word "then," the resulting statement is referred to as a **conditional statement**, denoted $\boldsymbol{p \rightarrow q}$. The "if" part of the statement (p in this case) is called the **antecedent**, while the "then" part of the statement (q in this case) is called the **consequent**. As indicated above, this expression can be read as "if p, then q" or alternatively "p implies q". There are also other ways of stating or translating "$p \rightarrow q$". These are commonly used in mathematics and the sciences. The following table summarizes some of the usual ways in which "$p \rightarrow q$" can be translated into words.

For example, let p represent the simple statement "It is a rainy day" and q represent the simple statement "It is a day that I stay indoors." We can then make the following translations:

Sometimes the conditional connective is "hidden" in everyday expressions. For example, "all basketball players are tall" can be written or thought of as a conditional statement in the following way(s): "If you are a basketball player, then you are tall," or "being tall is a necessary condition for being a basketball player," and so on, as suggested by the above table.

The logical definition of the conditional connective is given by the following truth table.

Conditional

p	q	$p \rightarrow q$
T	T	T
T	F	F
F	T	T
F	F	T

The following observations are worth remembering:
- $p \rightarrow q$ is false only when the antecedent is true and the consequent is false.
- If the antecedent is false, then $p \rightarrow q$ is automatically true.
- If the consequent is true, then $p \rightarrow q$ is automatically true.

Forms of Conditional Statements   The direct statement "$p \rightarrow q$" can be altered by negating and/or interchanging the antecedent and consequent. In some cases, the results are equivalent. The following table summarizes these results.

Name	Symbolic form	Read as
Direct Statement	$p \rightarrow q$	If p, then q
Converse	$q \rightarrow p$	If q, then p
Inverse	$\sim p \rightarrow \sim q$	If not p, then not q
Contrapositive	$\sim q \rightarrow \sim p$	If not q, then not p

The following observations are worth remembering:
- The direct statement (original conditional) is equivalent to the contrapositive.
- The converse is equivalent to the inverse.
- The direct statement and the converse are not equivalent.

See the section "Equivalent Statements" in the *Study Guide and Solutions Manual* for review.

Biconditional A **biconditional** statement is the conjunction of a conditional statement and its converse. A biconditional is denoted $p \leftrightarrow q$. This expression can be read any one of the following ways as summarized in the table.

Let p, for example, represent the simple statement "It is an equilateral triangle" and q represent "It is a triangle with three equal sides."

Symbolically ($p \leftrightarrow q$)	English translation
If p, then q, and if q, then p	If it is an equilateral triangle, then it has three equal sides, and if it has three equal sides, then it is an equilateral triangle.
p if and only if q (or p iff q)	It is an equilateral triangle if and only if it has three equal sides.
p is necessary and sufficient for q	Being an equilateral triangle is necessary and sufficient for it to have three equal sides.

The logical definition of the biconditional connective is given by the following truth table:

Biconditional

p	q	$p \leftrightarrow q$
T	T	T
T	F	F
F	T	F
F	F	T

The following observation is worth remembering:
- $p \leftrightarrow q$ is true only if both parts have the same truth value.

We can now analyze the truth value of more complicated compound statements using the above definitions. If we know, or can determine, the truth values of the simple statements which make up the compound statement, we are able to determine the truth value of the complete compound statement. The following example shows how to do this.

EXAMPLE B **Compound statement truth value given simple statements' truth values**

Given that the simple statements p and q are true and that r is false, what is the truth value for the compound statement $(p \vee \sim q) \rightarrow r$?

Solution
Replace (or write under) each simple statement with its known truth value.

$$(p \vee \sim q) \rightarrow r$$
$$(T \vee \sim T) \rightarrow F$$

Then simplify by deciding the truth value of each component connective for the compound statement starting inside the parentheses first.

$$(T \vee F) \rightarrow F,$$

since the negation of a true statement is false.

$$T \rightarrow F,$$

since disjunction is true if either disjunct is true.

$$F,$$

since a conditional statement is false when the antecedent is true and the consequent is false. Thus, the given compound statement is false.

EXAMPLE C **Truth value with recognizable truth values of the simple statements**

Determine the truth value of the following statements.

(a) Four is an even number and five is an even number.

Solution
Determine the truth value of each simple statement. "Four is an even number" is true while the statement "five is an even number" is false. Symbolically, this is

$$p \wedge q,$$

where p represents "Four is an even number" and q represents "five is an even number." Replacing the truth values for the simple statements in symbolic form gives

$$T \wedge F = F,$$

since a conjunction (and) is true only if both parts are true. Thus, the compound statement is false.

(b) It is not the case that the moon is made of green cheese or Disney World is in Florida.

Solution
Symbolically the statement looks like

$$\sim(p \vee q)$$

where p represents "the moon is made of green cheese" and q represents "Disney World is in Florida." NASA has indicated that p is a false statement, and we know that q is a true statement. Thus, we have

$$\sim(F \vee T) = \sim(T) = F,$$

since disjunction (or) is true if either disjunct is true, and the negation of true is false. We see that the compound statement is false.

Often we do not know the truth values of the individual simple statements, or, we are interested in all possible combinations of truth values and the corresponding truth values for the compound statement. In either case, the construction of a truth table is an efficient way to organize the information and results.

COMPOUND STATEMENTS AND TRUTH TABLES

Truth tables are used whenever we are interested in the truth values of a compound statement. The truth values of such a statement are typically based upon all combinations of truth values for the (component) simple statements and the definitions of the logical connectives. Truth tables are useful tools to characterize (i.e., define) the logical connectives as we did above. Truth tables have other important uses. We can compare statements, for example, to see if they are equivalent (or opposite) in meaning. Finally, we will be able to decide in many cases whether arguments are valid or invalid (i.e., fallacious) by means of truth tables. Thus, the ability to create truth tables based upon the definitions of the individual logical connectives is vital.

In summary, truth tables are useful tools to:

- Define (or characterize) the logical connectives.
- Establish truth values of more complicated compound statements made up of several connectives and simple statements.
- Establish relationships between statements (e.g., equivalent, opposite, etc.).
- Establish whether arguments are logically valid or invalid in form.

There are two basic methods that can be used to construct truth tables. These methods are demonstrated in the following examples.

EXAMPLE D   First method

Construct a truth table for $\sim(p \vee q)$.

Solution
Step 1 As with all truth tables, first list all possible truth values for the letters representing all simple statements (indexing columns). In this case, with two simple statements, there are four combinations of truth values. It is not strictly necessary to list the possibilities in this order, but it is standard and makes for consistency of presentation and answers.

p	q
T	T
T	F
F	T
F	F

Step 2 Make a column for the expression within the parentheses using the truth values from the definition of the connective disjunction (or).

p	q	$p \vee q$
T	T	T
T	F	T
F	T	T
F	F	F

Remember that disjunction is only false when both disjuncts are themselves false.

Step 3 Construct one more column for $\sim(p \vee q)$. Use the definition of the negation connective.

p	q	$p \vee q$	$\sim(p \vee q)$
T	T	T	F
T	F	T	F
F	T	T	F
F	F	F	T

This completes the truth table for $\sim(p \vee q)$.

Many students, especially after some practice, prefer the following shorter method to create a truth table. It is important to keep track of which order to complete the connectives and which columns one is comparing, to arrive at the truth values for a particular connective.

EXAMPLE E   Alternative method

Construct a truth table for $p \rightarrow \sim q$.

Solution
Step 1 As with all truth tables, first list all possible truth values for the letters representing the simple statements (indexing columns).

p	q
T	T
T	F
F	T
F	F

Step 2 Complete a column under "p" (same as index column) and a column under "$\sim$" which represents the negation of q.

p	q	p	$\rightarrow$	$\sim q$
T	T	T		F
T	F	T		T
F	T	F		F
F	F	F		T

Step 3 Complete the column under " $\rightarrow$ " which is the major (final) connective for the complete compound statement.   We are working with the two previous columns (3 and 4) and using the definition of the conditional.   Remember that the conditional is always true except when the antecedent (if part) is true and the consequent (then part) is false.

p	q	p	$\rightarrow$	$\sim q$
T	T	T	F	F
T	F	T	T	T
F	T	F	T	F
F	F	F	T	T

This final column represents the truth values for the compound statement given the various truth values of the component simple statements.   Thus, this particular compound statement is only false when the simple statements p and q are both true.

EQUIVALENT STATEMENTS

One important application of truth tables is to decide if two statements are equivalent in meaning. Equivalent statements are used often in mathematics, logic and elsewhere to simplify complicated expressions. Two statements are equivalent if they have the same truth value in every possible situation. That is, they must have the same truth value for each possible combination of truth values of the component simple statements. In other words, if we create a truth table for each

statement that we are comparing and the resulting tables are the same, then the statements must be equivalent.

EXAMPLE F   Equivalent statements (DeMorgan's Law)

Are the statements $\sim(p \wedge q)$ and $\sim p \vee \sim q$ equivalent?

Solution
Construct a truth table for each statement.

p	q	$\sim$	$(p \wedge q)$	$\sim p$	$\vee$	$\sim q$
T	T	F	T	F	F	F
T	F	T	F	F	T	T
F	T	T	F	T	T	F
F	F	T	F	T	T	T
		2	1	1	2	1

The numbers, found underneath the table, indicate the order in which the columns were created. Observe that the first (1) column is completed by the definition of conjunction applied to the two indexing columns for p and q.   The second and third (1) columns come from the definition of negation as applied to the indexing column for p and then for q. The first (2) column comes from negating the first (1) column (i.e., we are negating the "and" statement). The second (2) column comes from combining the corresponding (1) columns using the definition of disjunction.

Since the (2) columns represent the truth values of each of the compound statements and they have the same values, we know that the two statements are equivalent.

There are many examples of equivalent statements, and it is useful to remember some of these. EXAMPLE F above and EXAMPLE 11 (3.2) in the text exemplify a particularly useful set of equivalencies called De Morgan's laws. The symbol " $\equiv$ " is often used to identify statements which are equivalent.

De Morgan's laws   For any statements p and q,

$$\sim(p \vee q) \equiv \sim p \wedge \sim q \text{ and}$$
$$\sim(p \wedge q) \equiv \sim p \vee \sim q.$$

Thus, if we negate a disjunctive statement, we get the conjunction of the individual negations. If we negate a conjunctive statement, we get the disjunction of the individual negations.

Other useful equivalencies include the following:

Double negation equivalence

$$\sim(\sim p) \equiv p$$

Conditional equivalence

$$p \rightarrow q \equiv \sim p \vee q$$

Application of De Morgan's law to the conditional equivalence

$$p \rightarrow q \equiv \sim(p \wedge \sim q)$$

Contrapositive equivalence

$$p \rightarrow q \equiv \sim q \rightarrow \sim p$$

Biconditional equivalence

$$p \leftrightarrow q \equiv (p \rightarrow q) \wedge (q \rightarrow p)$$

Distributive equivalencies

$$p \wedge (q \vee r) \equiv (p \wedge q) \vee (p \wedge r)$$
$$p \vee (q \wedge r) \equiv (p \vee q) \wedge (p \vee r)$$

Two statements may have opposite truth values in their respective truth tables. That is, when one statement is true, the other is false and vice versa. This, of course, is the logical character (definition) of negation and means that the one statement is the negation of the other statement.

Negations of Some Compound Statements The De Morgan equivalencies give us a direct method to negate conjunctive and disjunctive statements. For example, if we wanted to negate the statement "$p \wedge q$," an obvious negation would be "$\sim(p \wedge q)$," but this is equivalent, according to De Morgan, to "$\sim p \vee \sim q$" which is typically used as a standard negation. In the same manner, if we wanted the negation of "$p \vee q$" which is directly "$\sim(p \vee q)$," we can apply a form of De Morgan's law to arrive at the equivalent negation "$\sim p \wedge \sim q$."

EXAMPLE G Negation of disjunction and conjunction

Find the negation of each statement.

(a) I will go to great lengths to succeed, or I will quit.

Solution
Let p represent "I will go to great lengths to succeed" and q represent "I will quit." The symbolization of the statement then becomes

$$p \vee q.$$

The negation is

$$\sim(p \vee q) \equiv \sim p \wedge \sim q,$$

by De Morgan's law or, using words, "I will not go to great lengths to succeed, and I will not quit."

(b) It is afternoon, and I am interested in going.

Solution
Let p represent "It is afternoon" and q represent "I am interested in going."

The symbolic form of the statement then becomes

$$p \wedge q.$$

The negation is

$$\sim(p \wedge q) \equiv \sim p \vee \sim q,$$

by De Morgan's law, or, using words, "It is not afternoon, or I am not interested in going."

We are often called upon to negate or recognize the negation of conditional statements. The conditional equivalence along with De Morgan's laws allows us to do this easily. For example, the negation of $p \rightarrow q$ is directly $\sim(p \rightarrow q)$, but since $p \rightarrow q$ is equivalent to $\sim p \vee q$, we have

$$\sim(p \rightarrow q) \equiv \sim(\sim p \vee q)$$
$$\equiv p \wedge \sim q$$

by the conditional equivalence and by De Morgan's law.

EXAMPLE H Negation of the conditional

Write the negation of "If I am elected, all of your problems are solved."

Solution
Let p represent the statement "I am elected" and q represent the statement "all of your problems are solved." The symbolic form then becomes

$$p \rightarrow q.$$

The negation is

$$\sim(p \rightarrow q) \equiv p \wedge \sim q,$$

or, using words, "I am elected, and your problems are not solved."

With a little practice, it is not difficult to write (or recognize) negations like those above without actually symbolizing the statements. When in doubt, it is a good idea to symbolize. If you do not remember the corresponding negation, check by truth tables.

The following table summarizes the above negations:

Statement	Negation
$p \wedge q$	$\sim p \vee \sim q$
$p \vee q$	$\sim p \wedge \sim q$
$p \rightarrow q$	$p \wedge \sim q$

No study of logic can be complete without a discussion of the principles involved in correct reasoning. In Chapter 1 two types of reasoning were introduced: **inductive** and **deductive**. Chapter 1 concentrated on the use of inductive reasoning to observe patterns. Here we look at how deductive reasoning may be used to determine whether logical arguments are valid or invalid. From Chapter 1 we remember that:

An **argument** is a claim that a statement, called the conclusion, follows from a set of statements, called the premises.

A premise can be an assumption, law, rule, widely held idea, or observation. We then reason deductively from the premises to obtain a conclusion. The premises and conclusion make up a logical argument.

EULER DIAGRAMS AND VALID ARGUMENTS

When reasoning from the premises of an argument to obtain a conclusion, we want the structure or form of the argument to be valid.

An argument is **valid** if the fact that all the premises are true forces the conclusion to be true. An argument that is not valid is **invalid** (a fallacy).

Arguments that contain quantified statements (using words like "all," "some," "no(ne)," etc.) can often be analyzed using a visual approach known as Euler diagrams. In these diagrams, regions (e.g., circles or ellipses, etc.) are used to indicate relationships of premises to conclusions just as circles are used in Venn diagrams to visualize set relationships.

Consider the following argument: All dogs are faithful animals. Sheena is a dog. Therefore, Sheena is a faithful animal.

EXAMPLE I   The quantifier "all"

Use an Euler diagram to determine the validity of the above argument.

Solution
The above argument can be stated in the following commonly used form where we place one premise over another, with the conclusion below a line.

All dogs are faithful animals.	*premise*
Sheena is a dog.	*premise*
Therefore, Sheena is a faithful animal.	*conclusion*

Draw an Euler diagram that represents the situation in which all of the premises are true.

The first premise, All dogs are faithful animals involves the quantifier "all." This statement claims that the set of dogs is included in the set of faithful animals. We diagram this below with the "dog" region (set of all dogs) inside the "faithful animal" region (set of all faithful animals).

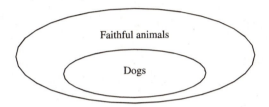

A <u>helpful hint</u> is that whatever word immediately follows the word "all" will represent the set of elements contained in the smaller region. Observe also that sometimes the word "all" is implied by the premise but not stated explicitly. Thus, the first premise, "All dogs are faithful animals," may also have been stated in the following way: Dogs are faithful animals.

The second premise, "Sheena is a dog," suggests that Sheena must be represented by some point or value inside the dog region. Thus, if we represent Sheena by say x, then x must lie inside the dog region. The following diagram, therefore, represents both premises as being true statements.

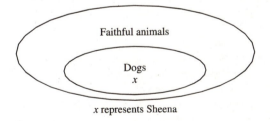

x represents Sheena

Notice that there is no other way to draw the regions that would support the given premises, that is, show the premises as true statements. If there were other ways to diagram the premises as true statements, we would do so.

Finally, the conclusion, "Sheena is a faithful animal," can be clearly seen in the diagram to be a true statement since it must lie within the region of faithful animals. Since the true premises force a true conclusion, we know that the argument is <u>valid</u>. Observe that there is no way to diagram the premises true and get a false conclusion.

EXAMPLE J   The quantifier "all"

Is the following argument valid?

All oak trees have green leaves.
That plant has green leaves.
Therefore, that plant is an oak tree.

Solution

Draw the region for "oak trees" entirely inside the region for "plants having green leaves." Let x represent "that plant"; then we have a choice for locating the x.

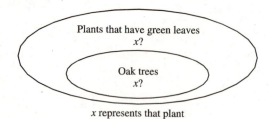

x represents that plant

The x must go inside the region for "plants having green leaves" but can go either inside or outside the region for "oak trees." Either way the figure is diagrammed so that the premises are true, but in the second case the diagram yields a conclusion that is false. Thus, it is possible to have true premises and a false conclusion. Since true premises do not force a true conclusion, the argument is <u>invalid</u> or fallacious.

EXAMPLE K   The quantifier "some"

Is the following argument valid?

Some roommates make a lot of noise.
Robert is my roommate.
Therefore, Robert makes a lot of noise.

Solution

Diagram a region representing roommates to intersect a region representing those who make a lot of noise. Let x = Robert. Thus, "x" must lie in the region representing roommates but may or may not lie in the region representing those who make a lot of noise.

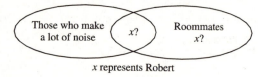

x represents Robert

Thus, the argument is <u>invalid</u> since the conclusion may be false.

It is important to observe that validity of an argument is not the same as the truth of a statement (such as the conclusion). In fact, one may have a valid argument and a false conclusion. For this to happen, one or more of the premises must be false.

One may also have an argument which is invalid with a true or a false conclusion.

The validity of an argument describes only the structure or form of an argument. Arguments are not true or false since this is a property of statements only; rather, they are valid or invalid in form. The important thing to remember is that an argument which is valid in form, along with premises known to be true, forces or guarantees that the conclusion must be true. The following example allows us to concentrate only on the structure, or form, of two related arguments.

EXAMPLE L   Validity is a characteristic only of the form of an argument

Consider the following set of premises. Two conclusions are indicated. Use an Euler diagram to determine if either (or both) of the arguments is (are) valid.

All zoys are glachos.
All glachos are oosters.
No shfnexi is an ooster.
Therefore, all zoys are oosters.   *conclusion 1*
Therefore, no zoy is a shfnexi.   *conclusion 2*

Solution

Certainly, we know nothing about zoys, glachos, oosters, and shfnexis and cannot, therefore, judge whether any of these five statements is true or false. Nevertheless, the two arguments made of the premises and each conclusion individually sound reasonable. That is, each conclusion may seem true if we accept the premises as being true. What we must catch is, that in some way, the premises and conclusions are related by the structure of the argument. To determine if the argument (or structure) is valid, we will diagram the premises as true statements and decide if this forces either or both of the conclusions to be true.

We will draw a region representing zoys. Since all "zoys are glachos," we will draw a region representing glachos so that it contains the region for zoys. Also since "all glachos are oosters," we will draw a region for oosters so that all of the glachos are in that region. Now, since "no shfnexi is an ooster," we must draw a region representing shfnexi completely outside the oosters' region. The drawing might appear as

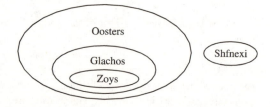

It is now easy to see that both conclusions can be made (i.e., each argument is valid) since the assumption that the premises were true forces, in each case, a conclusion which is true.

In more complex arguments, Euler diagrams do not work so well. This is because Euler diagrams require a sketch showing every possible case. In complex arguments it is hard to be sure that all cases have been considered. Typically arguments that do not involve quantified statements are more readily determined by a method using truth tables. In order to check an argument for validity using truth tables, we will need the following definition.

A statement that is true in all cases is called a **tautology**.

That is, if a compound statement is true for any combination of truth values for the corresponding simple statements, then we call the compound statement a tautology. Examples of tautologies include

(a) $p \vee \sim p$,
(b) $p \rightarrow p$, and
(c) $\sim(p \wedge q) \rightarrow (\sim p \vee \sim q)$.

These can be demonstrated by examining a truth table to see if for each combination of truth values for the simple statements, the compound statement is true. You might remember that in mathematics the equivalent idea is called an identity. Examples of identities include $x + 1 > x$ and $x^2 - 9 = (x - 3)(x + 3)$, since each is a true statement for any replacement of the variable x.

We are interested in determining whether a particular compound statement is a **tautology** when deciding if an argument is valid.

Return for a moment to the definition of validity. An argument is **valid** if the fact that all the premises are true forces the conclusion to be true, that is, if the conclusion follows directly from the premises. Stated in other words, if (the conjunction of) the premises imply the conclusion (in all cases), then the argument is valid. More formally we restate the definition of validity in the following way.

An argument is said to be **valid** when the conjunction of the premises implying the conclusion forms a tautology.

TRUTH TABLES AND VALID ARGUMENTS

As an example of this method, consider the following argument.

If my room needs cleaning, then I will clean it.
I will not clean my room.
Therefore, my room does not need cleaning.

In order to test the validity of this argument, we begin by identifying the simple statements found in the argument. They are "my room needs cleaning" and "I will clean it." Let us assign the letters p and q to represent these statements: Let p represent "my room needs cleaning" and q represent "I will clean it."

We may symbolize the argument as

$$p \rightarrow q \quad \textit{premise}$$
$$\underline{\sim q} \qquad \textit{premise}$$
$$\sim p. \qquad \textit{conclusion}$$

To decide if this argument is valid, we must determine whether the conjunction of both premises implies the conclusion for all possible cases of truth values for p and q. We will therefore, write the conjunction of the premises as the antecedent of a conditional statement, and the conclusion as the consequent.

$$[(p \rightarrow q) \quad \wedge \quad \sim q] \quad \rightarrow \quad \sim p$$
$$\uparrow \qquad \uparrow \qquad \uparrow \qquad \uparrow \qquad \uparrow$$
$$\textit{premise} \quad \text{and} \quad \textit{premise} \quad \text{implies} \quad \textit{conclusion}$$

Finally, construct the truth table for this conditional statement, as shown below.

First Method

p q	$(p \rightarrow q)$	$\sim q$	$(p \rightarrow q) \wedge \sim q$	$\sim p$	$[(p \rightarrow q) \wedge \sim q] \rightarrow \sim p$
T T	T	F	F	F	T
T F	F	T	F	F	T
F T	T	F	F	T	T
F F	T	T	T	T	T

Observe that in using this method, you are building the truth values for the components of the compound statement and combining these components using the corresponding connectives. The last connective completed indicates the truth values of the compound statement.

With a little practice, the following alternative method is found by most students to be quicker and seems to provide less room for error.

Alternative Method

p q	$[(p \rightarrow q)$	$\wedge$	$\sim q]$	$\rightarrow$	$\sim p$
T T	T	F	F	T	F
T F	F	F	T	T	F
F T	T	F	F	T	T
F F	T	T	T	T	T
	1	2	1	3	2

In this method you are beginning with the entire conditional statement, formed by the conjunction of the premises implying the conclusion. Complete the truth table columns in the order indicated by the connectives. Note that the order is indicated by the numbers at the bottom of the table. Columns with the same number are combined by the corresponding connective to yield the next numbered column. (Note that this is slightly different from the numbering process in the text. Here we are combining the "1s" columns to create the "2s" columns, etc., rather than specifying the columns to complete in numbered order.) This is similar to the "order of operations" you learn in an algebra or arithmetic course. Be sure to keep the truth values directly under the connective that you are working with when using this method.

Since the final column, using either method, indicates the truth values of the conditional statement is always true, we can conclude that the argument itself is valid. In other words, if the conditional statement formed from the conjunction of the premises is a tautology, then the argument is valid.

The form (structure) of the above argument

$$p \rightarrow q$$
$$\frac{\sim q}{\sim p}$$

is a common one and is called **modus tollens** or **law of contraposition**. Modus tollens is derived from Latin and means "a manner of denying." The conclusion of the conditional statement in the first premise is being denied.

A table of common valid and invalid arguments follows Example O.

An important observation is that when setting up a truth table, a statement with two different simple statements requires a truth table with $2^2 = 4$ lines (combinations of truth values). A compound statement made up of three simple statements would require a truth table with $2^3 = 8$ lines. This may be generalized in the same manner as the number of subsets of a set, as discussed in Chapter 1 (and again in Chapter 10).

> A statement with n different letters requires a truth table with 2^n lines.

In the following examples, we will use the shorter alternate method for creating the truth tables. See the textbook for more examples where the truth table is built from the component parts as in the preceding example. In addition, see textbook and/or corresponding EXERCISES (in the *Study Guide and Solutions Manual*) for examples of arguments with more than two simple statements.

EXAMPLE M   Validity by truth table

Use truth tables to decide if the following argument is valid. If Erek passes this course, then he will graduate. Erek graduated. Therefore, he passed this course.

Solution

Let p represent "Erek passes (or passed) this course," and q represents "Erek will (or has) graduate(d)." Thus, the symbolic form of the argument is given by

$$p \rightarrow q$$
$$\frac{q}{p.}$$

Joining the premises together by conjunction to imply the conclusion gives

$$[(p \rightarrow q) \wedge q] \rightarrow p.$$

Setting up a truth table for this compound statements gives

p	q	$[(p \rightarrow q)$	$\wedge$	$q]$	$\rightarrow$	p
T	T	T	T	T	T	T
T	F	F	F	F	T	T
F	T	T	T	T	F	F
F	F	T	F	F	T	F
		1	2	1	3	2

We see from the third line of the truth table that the conditional statement is not a tautology, thus, the argument is <u>invalid</u>. The structure (pattern) of the above argument is a common fallacy called the **fallacy of the converse**.

Notice that if we were completing the table row by row instead of column by column, we could have stopped once we arrived at an "F" for the conditional statement (the third line in this case). This approach to completing a truth table can often shorten the process of deciding that an argument is invalid.

EXAMPLE N   Validity by truth table

Use truth tables to decide if the following argument is valid. If you are infected with a virus, then it can be transmitted. The consequences are serious, and it cannot be transmitted. Therefore, if the consequences are not serious, then you are not infected with a virus.

Solution

Let p represent "You are infected with a virus," q represent "It can be transmitted," and r represent "The consequences are serious." Thus, the symbolic form of the argument is given by

$$p \to q$$
$$\underline{r \land \sim q}$$
$$\sim r \to \sim p.$$

Joining the premises together by conjunction to imply the conclusion gives

$$[(p \to q) \land (r \land \sim q)] \to (\sim r \to \sim p).$$

Complete the corresponding truth table.

p	q	r	$[(p$	$\to$	$q)$	$\land$	$(r$	$\land$	$\sim q)]$	$\to$	$(\sim r$	$\to$	$\sim p)$
T	T	T	T	T	T	F	T	F	F	T	F	T	F
T	T	F	T	T	T	F	F	F	F	T	T	F	F
T	F	T	T	F	F	F	T	T	T	T	F	T	F
T	F	F	T	F	F	F	F	F	T	T	T	F	F
F	T	T	F	T	T	T	T	F	F	T	F	T	T
F	T	F	F	T	T	F	F	F	F	T	T	T	T
F	F	T	F	T	F	T	T	T	T	T	F	T	T
F	F	F	F	T	F	F	F	F	T	T	T	T	T
			1	2	1	3	1	2	1	4	2	3	2

Since the conjunction of the premises implying the conclusion forms a tautology, the argument is <u>valid</u>.

Let us now consider the following argument that does not contain any conditional statements in the premises or the conclusion.

EXAMPLE O   Validity by truth table

Use a truth table to decide if the following argument is valid or invalid. The water pump is broken. I know this because the radiator is leaking or the water pump is broken, and the radiator definitely is not leaking.

Solution
Observe that, in this example, the first statement is the conclusion of the argument. The premises follow the words "I know this because." Remember that the conclusion must be derived or follow from the premises.

The argument can be arranged as follows:

The radiator is leaking or the water pump is broke
The radiator is not leaking.
Therefore, the water pump is broken.

If we let p represent "The radiator is leaking" and q represent "the water pump is broken," then the symbolic form of the argument is

$$p \lor q$$
$$\underline{\sim p}$$
$$q.$$

Joining the premises together by conjunction to imply the conclusion gives

$$[(p \lor q) \land \sim p] \to q.$$

Setting up a truth table for this compound statement gives

p	q	$[(p$	$\lor$	$q)$	$\land$	$\sim p]$	$\to$	q
T	T	T	T	T	F	F	T	T
T	F	T	T	F	F	F	T	F
F	T	F	T	T	T	T	T	T
F	F	F	F	F	F	T	T	F
		1	2	1	3	2	4	3

Since the conditional statement is a tautology, we see that the argument is valid.

The form (structure or pattern) of the above argument

$$p \lor q$$
$$\underline{\sim p}$$
$$q.$$

is a common one and is called a **disjunctive syllogism**. Observe that in such an argument, if either of the original disjuncts are negated, then we can conclude the other disjunct.

Below is a summary found in the text of commonly used valid and invalid argument forms. It can be very helpful to memorize these various arguments.

Valid Arguments		Invalid Arguments	
Modus ponens	$p \to q$ $\underline{p}$ q	Fallacy of the converse	$p \to q$ $\underline{q}$ p
Modus tollens	$p \to q$ $\underline{\sim q}$ $\sim p$	Fallacy of the inverse	$p \to q$ $\underline{\sim p}$ $\sim q$
Disjunctive syllogism	$p \lor q$ $\underline{\sim p}$ q		
Reasoning by transitivity	$p \to q$ $\underline{q \to r}$ $p \to r$		

See textbook and/or (3.6) "Exercises" in the *Study Guide and Solutions Manual* for examples of the remaining valid and invalid argument forms.

One reason that it is useful to have the above forms memorized is that one may decide if an argument is valid or invalid by recognizing the form of the argument without having to use a truth table, as in the following example.

EXAMPLE P   Validity by recognizing the form of the argument

Decide if the following arguments are valid or invalid.

(a)
If you like mathematics, then you'll like logic.
You don't like mathematics.
Therefore, you don't like logic.

Solution
Let p represent "you like mathematics" and q represent "you (will) like logic."   We can then symbolize the argument as follows:

$$p \to q$$
$$\underline{\sim p}$$
$$\sim q.$$

We could test the validity by a truth table as we have been doing, or we might recognize this argument form as the commonly misused invalid argument form—**fallacy of the inverse**.

(b)
Kara will receive an A for this course, if she works hard.
If Kara receives an A for this course, she will feel satisfied.
So, if Kara works hard, then she will feel satisfied.

Solution
Let p represent "Kara works hard," q represent "she (will) receive(s) an A for this course," and r represent "she will feel satisfied."   We can then symbolize the argument as follows:

$$p \to q$$
$$\underline{q \to r}$$
$$p \to r.$$

Notice that for the first premise, the antecedent came at the end of the statement.

We could again test the validity by a truth table, or we might recognize this argument form as **reasoning by transitivity** (also called **hypothetical syllogism**), a valid argument.

3.1 EXERCISES

1. Because the declarative sentence "December 7, 1941, was a Sunday" has the property of being true or false, it is considered a statement.

3. "Listen my children and you shall hear of the midnight ride of Paul Revere" is not a declarative sentence and does not have the property of being true or false. Hence, it is not considered a statement.

5. "$5 + 8 = 13$ and $4 - 3 = 1$" is a declarative sentence that is true and therefore, is considered a statement.

7. "Some numbers are negative" is a declarative sentence that is true and therefore, is a statement.

9. "Accidents are the main cause of deaths of children under the age of 8" is a declarative sentence that has the property of being true or false and therefore, is considered to be a statement.

11. "Where are you going today?" is a question, not a declarative sentence, and therefore, is not considered a statement.

13. "Kevin 'Catfish' McCarthy once took a prolonged continuous shower for 340 hours, 40 minutes" is a declarative sentence that has the property of being either true or false and therefore, is considered to be a statement.

15. "I read the Chicago Tribune and I read the New York Times" is a compound statement because it consists of two simple statements combined by the connective "and."

17. "Tomorrow is Sunday" is a simple statement because only one assertion is being made.

19. "Jay Beckenstein's wife loves Ben and Jerry's ice cream" is not compound because only one assertion is being made.

21. "If Julie Ward sells her quota, then Bill Leonard will be happy" is a compound statement because it consists of two simple statements combined by the connective "if … then."

23. The negation of "Her aunt's name is Lucia" is "Her aunt's name is not Lucia."

25. A negation of "Every dog has its day" is "At least one dog does not have its day."

27. A negation of "Some books are longer than this book" is "No book is longer than this book."

29. A negation of "No computer repairman can play blackjack" is "At least one computer repairman can play blackjack."

31. A negation of "Everybody loves somebody sometime" is "Someone does not love somebody sometime."

33. A negation for "$y > 12$" (without using a slash sign) would be "$y \leq 12$."

35. A negation for "$q \geq 5$" would be "$q < 5$."

37. Writing exercise

Let p represent the statement "She has green eyes," and let q represent "He is 48 years old." Translate each symbolic compound statement into words.

39. A translation for "$\sim p$" is "She does not have blue eyes."

41. A translation for "$p \wedge q$" is "She has green eyes and he is 48 years old."

43. A translation for "$\sim p \vee q$" is "She does not have green eyes or he is 48 years old."

45. A translation for "$\sim p \vee \sim q$" is "She does not have green eyes or he is not 48 years old."

47. A translation for "$\sim (\sim p \wedge q)$" is "It is not the case that she does not have green eyes and he is 48 years old."

49. "Chris collects videotapes and Jack does not play the tuba" may be symbolized as $p \vee \sim q$.

51. "Chris does not collect videotapes or Jack plays the tuba" may be symbolized as $\sim p \vee q$.

53. "Neither Chris collects videotapes nor Jack plays the tuba" may be symbolized as $\sim (p \vee q)$ or equivalently, $\sim p \wedge \sim q$.

55. Writing exercise

57. Since all whole numbers are integers, the statement "Every whole number is an integer" is true.

59. Since 1/2 is a rational number but is not an integer, the statement "There exists a rational number that is not an integer" is true.

61. Since rational numbers are real numbers, the statement "All rational numbers are real numbers" is true.

63. Since 1/2 is a rational number but is not an integer, the statement "Some rational numbers are not integers" is true.

65. The number, 0, is a whole number but is not positive. Thus, the statement "Each whole number is a positive number" is false.

Refer to the sketches labeled A, B, and C, in the text, and identify the sketch (or sketches) that is (are) satisfied by the given statement involving a quantifier.

67. The condition that "all pictures have frames" is satisfied by group C.

69. The condition that "At least one picture does not have a frame" is met by groups A and B.

71. The condition that "At least one picture has a frame" is satisfied by groups A and C.

73. The condition that "all pictures do not have frames" is satisfied by group B. Observe that this statement is equivalent to "No pictures have a frame."

75. Writing exercise

77. We might write the statement "There is no one here who has not done that at one time or another" using the word "every" as "Everyone here has done that at one time or another."

3.2 EXERCISES

1. If q is false, then $(p \wedge \sim q) \wedge q$, since both conjuncts (parts of the conjunction) must be true for the compound statement to be true.

3. If $p \wedge q$ is true, and p is true, then q must be also in order for the conjunctive statement to be true. Observe that both conjuncts must be true for a conjunctive statement to be true.

5. If $\sim(p \vee q)$ is true, both components (disjuncts) must be false. Thus, the disjunction itself is false making its negation true.

In exercises 7–17, p represents a false statement and q represents a true statement.

7. Since $p = F$,
$$\sim p = \sim F$$
$$= T.$$

That is, replace p by F and determine the truth of ~F.

9. Since p is false and q is true, we may consider the "or" statement as
$$F \vee T$$
$$T,$$

by the logical definition of an "or" statement. That is, $p \vee q$ is true.

11. With the given truth values for p and q we may consider $p \vee \sim q$ as
$$F \vee \sim T$$
$$F \vee F$$
$$F,$$

by the logical definition of " $\vee$ ".

13. With the given truth values for p and q we may consider $\sim p \vee \sim q$ as
$$\sim F \vee \sim T$$
$$T \vee F$$
$$T.$$

Thus, the compound statement is true.

15. Replacing p and q with the given truth values, we have
$$\sim(F \wedge \sim T)$$
$$\sim(F \wedge F)$$
$$\sim F$$
$$T.$$

Thus, the compound statement $\sim(p \wedge \sim q)$ is true.

17. Replacing p and q with the given truth values, we have
$$\sim[\sim F \wedge (\sim T \vee F)]$$
$$\sim[T \wedge (F \vee F)]$$
$$\sim[T \wedge F]$$
$$\sim F$$
$$T.$$

Thus, the compound statement $\sim[\sim p \wedge (\sim q \vee p)]$ is true.

19. The statement $3 \geq 1$ is a disjunction since it means "3 > 1" or "3 = 1."

In exercises 21–27, p represents a true statement, and q and r represent false statements.

21. Replacing p, q and r with the given truth values, we have
$$(T \wedge F) \vee \sim F$$
$$F \vee T$$
$$T.$$

Thus, the compound statement $(p \wedge r)) \vee \sim q$ is true.

23. Replacing p, q and r with the given truth values, we have
$$T \wedge (F \vee F)$$
$$T \wedge F$$
$$F.$$

Thus, the compound statement $p \wedge (q \vee r)$ is false.

25. Replacing p, q and r with the given truth values, we have
$$\sim(T \wedge F) \wedge (F \vee \sim F)$$
$$\sim F \wedge (F \vee T)$$
$$T \wedge T$$
$$T.$$

Thus, the compound statement $\sim(p \wedge q) \wedge (r \vee \sim q)$ is true.

27. Replacing p, q and r with the given truth values, we have

$$\sim[(\sim T \wedge F) \vee F]$$
$$\sim[(F \wedge F) \vee F]$$
$$\sim[F \vee F]$$
$$\sim F$$
$$T.$$

Thus, the compound statement $\sim[(\sim p \wedge q) \vee r]$ is true.

Let p represent the statement "$2 > 7$," which is false, let q represent "$8 \not> 6$," which is false and let r represent "$19 \geq 19$," which is true. [E.g. $p = F$, $q = F$ and $r = T$.]

29. Replacing p, r with the given truth values, we have

$$F \wedge T$$
$$F.$$

The compound statement $p \wedge r$ is false.

31. Replacing q and r with the observed truth values, we have
$$\sim F \vee \sim T$$
$$T \vee F$$
$$T.$$

The compound statement $\sim q \vee \sim r$ is true.

33. Replacing p, q and r with the observed truth values, we have
$$(F \wedge F) \vee T$$
$$F \vee T$$
$$T.$$

The compound statement $(p \wedge q) \vee r$ is true.

35. Replacing p, q and r with the observed truth values, we have
$$(\sim T \wedge F) \vee \sim F$$
$$(F \wedge F) \vee T$$
$$F \vee T$$
$$T.$$

The compound statement $(\sim r \wedge q) \vee \sim p$ is true.

37. Since there are two simple statements (p and r), we have $2^2 = 4$ combinations or truth value, or rows in the truth table, to examine.

39. Since there are four simple statements (p, q, r, and s), we have $2^4 = 32$ combinations or truth value, or rows in the truth table, to examine.

41. Since there are seven simple statements (p, q, r, s, t, u, and v), we have $2^7 = 128$ combinations or truth value, or rows in the truth table, to examine.

43. If the truth table for a certain compound statement has 64 rows, then there must be six distinct component statements ($2^6 = 64$).

45. $\sim p \wedge q$

p	q	$\sim p$	$\sim p \wedge q$
T	T	F	F
T	F	F	F
F	T	T	T
F	F	T	F

47. $\sim(p \wedge q)$

p	q	$p \wedge q$	$\sim(p \wedge q)$
T	T	T	F
T	F	F	T
F	T	F	T
F	F	F	T

49. $(q \vee \sim p) \vee \sim q$

p	q	$\sim p$	$\sim q$	$q \vee \sim p$	$(q \vee \sim p) \vee \sim q$
T	T	F	F	T	T
T	F	F	T	F	T
F	T	T	F	T	T
F	F	T	T	T	T

In Exercises 51–59 we are using the alternative method, filling in columns in the order indicated by the numbers. Observe that columns with the same number are combined (by the logical definition of the connective) to get the next numbered column. Note that <u>this is different</u> then the way the numbered columns are used in the textbook. Remember that the last column (highest numbered column) completed yields the truth values for the complete compound statement. Be sure to align truth values under the appropriate logical connective or simple statement.

51. $(\sim p \ \vee q)$

p	q	$\sim q$	$\wedge$	$(\sim p \ \vee q)$		
T	T	F	F	F	T	T
T	F	T	F	F	F	F
F	T	F	F	T	T	T
F	F	T	T	T	T	F
		2	3	1	2	1

53. $(p \lor \sim q) \land (p \land q)$

p	q	(p ∨ ~q)	∧	(p ∧ q)
T	T	T T F	T	T T T
T	F	T T T	F	T F F
F	T	F F F	F	F F T
F	F	F T T	F	F F F
		1 2 1	3	1 2 1

55. $(\sim p \land q) \land r$

p	q	r	(~p ∧ q)	∧	r
T	T	T	F F T	F	T
T	T	F	F F T	F	F
T	F	T	F F F	F	T
T	F	F	F F F	F	F
F	T	T	T T T	T	T
F	T	F	T T T	F	F
F	F	T	T F F	F	T
F	F	F	T F F	F	F
			1 2 1	3	2

57. $(\sim p \land \sim q) \lor (\sim r \lor \sim p)$

p	q	r	(~p ∧ ~q)	∨	(~r ∨ ~p)
T	T	T	F F F	F	F F F
T	T	F	F F F	T	T T F
T	F	T	F F T	F	F F F
T	F	F	F F T	T	T T F
F	T	T	T F F	T	F T T
F	T	F	T F F	T	T T T
F	F	T	T T T	T	F T T
F	F	F	T T T	T	T T T
			1 2 1	3	1 2 1

59. $\sim(\sim p \land \sim q) \lor (\sim r \lor \sim s)$

p	q	r	s	~(~p ∧ ~q)	∨	(~r ∨ ~s)
T	T	T	T	T F F F	T	F F F
T	T	T	F	T F F F	T	F T T
T	T	F	T	T F F F	T	T T F
T	T	F	F	T F F F	T	T T T
T	F	T	T	T F F T	T	F F F
T	F	T	F	T F F T	T	F T T
T	F	F	T	T F F T	T	T T F
T	F	F	F	T F F T	T	T T T
F	T	T	T	T T F F	T	F F F
F	T	T	F	T T F F	T	F T T
F	T	F	T	T T F F	T	T T F
F	T	F	F	T T F F	T	T T T
F	F	T	T	F T T T	F	F F F
F	F	T	F	F T T T	T	F T T
F	F	F	T	F T T T	T	T T F
F	F	F	F	F T T T	T	T T T
				3 1 2 1	4	2 3 2

61. "You can pay me now or you can pay me later" has the symbolic form $(p \lor q)$. The negation, $\sim(p \lor q)$, is equivalent, by one of De Morgan's laws, to $(\sim p \land \sim q)$. The corresponding word statement is "You can't pay me now and you can't pay me later."

63. "It is summer and there is no snow" has the symbolic form $p \land \sim q$. The negation, $\sim(p \land \sim q)$, is equivalent by De Morgan's, to $\sim p \lor q$. The word translation for the negation is "It is not summer or there is snow."

65. "I said yes but she said no" is of the form $p \land q$. The negation, $\sim(p \land q)$, equivalent, by De Morgan's, to $\sim p \lor \sim q$. The word translation for the negation is "I did not say yes or she did not say no." (Note that the connective "but" is equivalent to that of "and.")

67. "$5 - 1 \neq 4$ and $9 + 12 \neq 7$" is of the form $\sim p \land q$. The negation, $\sim(\sim p \land q)$, equivalent, by De Morgan's, to $p \lor \sim q$. The translation for the negation is "$5 - 1 \neq 4$ or $9 + 12 = 7$."

69. "Cupid or Vixen will lead Santa's sleigh next Christmas" is of the form $p \lor q$. The negation, $\sim(p \lor q)$, equivalent, by De Morgan's, to $\sim p \land \sim q$. A translation for the negation is "Neither Cupid nor Vixen will lead Santa's sleigh next Christmas."

71. "For every real number $y, y < 13$ or $y > 6$" is <u>true</u> since for any real number at least one of the component statements is true.

73. "For some integer $p, p \geq 4$ and $p \leq 4$" is <u>true</u> since both component statements are true for the integer $p = 4$.

75. $p \underline{\vee} q$

p	q	$p \underline{\vee} q$
T	T	F
T	F	T
F	T	T
F	F	F

Observe that it is only the first line in the truth table that changes for "exclusive disjunction" since the component statements can not both be true at the same time.

77. "$3 + 1 = 4 \underline{\vee} 2 + 5 = 7$" is <u>false</u> since both component statements are true.

79. "$3 + 1 = 7 \underline{\vee} 2 + 5 = 7$" is <u>true</u> since the first component statement is false and the second is true.

3.3 EXERCISES

1. The statement "It must be alive if it is breathing" becomes "If it is breathing, then it must be alive."

3. The statement "Faith Sherlock visits Ireland every summer" becomes "If it is summer, then Faith Sherlock visits Ireland."

5. The statement "Every picture tells a story" becomes "If it is a picture, then it tells a story."

7. The statement "No guinea pigs are scholars" becomes "If it is a guinea pig, then it is not a scholar."

9. The statement "Running Bear loves Little White Dove" becomes "If he is Running Bear, then he loves Little White Dove."

11. The statement "If the antecedent of a conditional statement is false, the conditional statement is true" is <u>true</u>, since a false antecedent will always yield a true conditional statement."

13. The statement "If q is true, then $(p \wedge q) \to q$ is true" is <u>true</u>, since with a true consequent the conditional statement is always true (even though the antecedent may be false)."

15. The negation of "If pigs fly, I'll believe it" is "If pigs don't fly, I won't believe it." This statement is <u>false</u>. The negation is "pigs fly and I don't believe it."

17. "Given that $\sim p$ is true and q is false, the conditional $p \to q$ is true" is a <u>true</u> statement since the antecedent, p, must be false.

19. Writing exercise

21. "$F \to (4 \neq 7)$" is a <u>true</u> statement, since a false antecedent always yields a conditional statement which is true.

23. "$(6 \geq 6) \to F$" is a <u>false</u> statement, since the antecedent is true and the consequent is false.

25. "$(4 = 11 - 7) \to (8 > 0)$" is <u>true</u>, since the antecedent and the consequent are both true.

Let s represent the statement "She has a snake for a pet," let p represent the statement "he trains ponies," and let m represent "they raise monkeys."

27. "$\sim m \to p$" expressed in words, becomes "If they do not raise monkeys, then he trains ponies."

29. "$s \to (m \wedge p)$" expressed in words, becomes "If she has a snake for a pet, then they raise monkeys and he trains ponies."

31. "$\sim p \to (\sim m \vee s)$" expressed in words, becomes "If he does not train ponies, then they do not raise monkeys or she has a snake for a pet."

Let b represent the statement "I ride my bike," let r represent the statement "it rains" and let p represent "the play is canceled."

33. The statement "If it rains, then I ride my bike" can be symbolized as "$r \to b$."

35. The statement "If I do not ride my bike, then it does not rain" can be symbolized as "$\sim b \to \sim r$."

37. The statement "If I ride my bike, or if the play is canceled, then it rains" can be symbolized as "$b \vee (p \to r)$."

39. The statement "I'll ride my bike if it doesn't rain" can be symbolized as "$\sim r \to b$."

Assume that p and r are false, and q is true.

41. Replacing r and q with the given truth values, we have

$$\sim F \to T$$
$$T \to T$$
$$T.$$

Thus, the compound statement $\sim r \to q$ is true.

43. Replacing p and q with the given truth values, we have

$$T \to F$$
$$F.$$

Thus, the compound statement $q \to p$ is false.

45. Replacing p and q with the given truth values, we have

$$F \to T$$
$$T.$$

Thus, the compound statement $p \to q$ is true.

47. Replacing p, r and q with the given truth values, we have

$$\sim F \rightarrow (T \wedge F)$$
$$T \rightarrow F$$
$$F.$$

Thus, the compound statement $\sim p \rightarrow (q \wedge r)$ is false.

49. Replacing p, r and q with the given truth values, we have

$$\sim T \rightarrow (F \wedge F)$$
$$F \rightarrow F$$
$$T.$$

Thus, the compound statement $\sim q \rightarrow (p \wedge r)$ is true.

51. Replacing p, r and q with the given truth values, we have

$$(\sim F \rightarrow \sim T) \rightarrow (\sim F \wedge \sim F)$$
$$(T \rightarrow F) \rightarrow (T \wedge T)$$
$$F \rightarrow T$$
$$T.$$

Thus, the compound statement $(p \rightarrow \sim q) \rightarrow (\sim p \wedge \sim r)$ is true.

53. Writing exercise

55. $\sim q \rightarrow p$

p	q	$\sim q$	$\rightarrow$	p
T	T	F	T	T
T	F	T	T	T
F	T	F	T	F
F	F	T	F	F
		1	2	1

57. $(\sim p \rightarrow q) \rightarrow p$

p	q	($\sim p$	$\rightarrow$	q)	$\rightarrow$	p
T	T	F	T	T	T	T
T	F	F	T	F	T	T
F	T	T	T	T	F	F
F	F	T	F	F	T	F
		1	2	1	3	2

59. $(p \vee q) \rightarrow (q \vee p)$

p	q	(p	$\vee$	q)	$\rightarrow$	(q	$\vee$	p)
T	T	T	T	T	T	T	T	T
T	F	T	T	F	T	F	T	T
F	T	F	T	T	T	T	T	F
F	F	F	F	F	T	F	F	F
		1	2	1	3	1	2	1

Since this statement is always true (column 3), it is a tautology.

61. $(\sim p \rightarrow \sim q) \rightarrow (p \wedge q)$

p	q	($\sim p$	$\rightarrow$	$\sim q$)	$\rightarrow$	(p	$\wedge$	q)
T	T	F	T	F	T	T	T	T
T	F	F	T	T	F	T	F	F
F	T	T	F	F	T	F	F	T
F	F	T	T	T	F	F	F	F
		1	2	1	3	1	2	1

63. $[(r \vee p) \wedge \sim q] \rightarrow p$

p	q	r	[(r	$\vee$	p)	$\wedge$	$\sim q]$	$\rightarrow$	p
T	T	T	T	T	T	F	F	T	T
T	T	F	F	T	T	F	F	T	T
T	F	T	T	T	T	T	T	T	T
T	F	F	F	T	T	T	T	T	T
F	T	T	T	T	F	F	F	T	F
F	T	F	F	F	F	F	F	T	F
F	F	T	T	T	F	T	T	F	F
F	F	F	F	F	F	F	T	T	F
			1	2	1	3	2	4	3

65. $(\sim p \wedge \sim q) \rightarrow (\sim r \rightarrow \sim s)$

p	q	r	s	($\sim p$	$\wedge$	$\sim q$)	$\rightarrow$	($\sim r$	$\rightarrow$	$\sim s$)
T	T	T	T	F	F	F	T	F	T	F
T	T	T	F	F	F	F	T	F	T	T
T	T	F	T	F	F	F	T	T	F	F
T	T	F	F	F	F	F	T	T	T	T
T	F	T	T	F	F	T	T	F	T	F
T	F	T	F	F	F	T	T	F	T	T
T	F	F	T	F	F	T	T	T	F	F
T	F	F	F	F	F	T	T	T	T	T
F	T	T	T	T	F	F	T	F	T	F
F	T	T	F	T	F	F	T	F	T	T
F	T	F	T	T	F	F	T	T	F	F
F	T	F	F	T	F	F	T	T	T	T
F	F	T	T	T	T	T	T	F	T	F
F	F	T	F	T	T	T	T	F	T	T
F	F	F	T	T	T	T	F	T	F	F
F	F	F	F	T	T	T	T	T	T	T
				1	2	1	3	1	2	1

67. The negation of "If that is an authentic Persian rug, I'll be surprised" is "That is an authentic Persian rug and I won't be surprised."

69. The negation of "If the English measures are not converted to metric measures, then the spacecraft will crash on the surface of Mars" is "The English measures are not converted to metric measures and the spacecraft does not crash on the surface of Mars"

71. The negation of "If you want to be happy for the rest of your life, never make a pretty woman your wife" is "You want to be happy for the rest of your life and you make a pretty woman you wife."

73. An equivalent statement to "If you give your plants tender, loving care, they will flourish" is "you do not give your plants tender, loving care or they flourish."

75. An equivalent statement to "If she doesn't, he will" is "She does or he will."

77. An equivalent conditional statement to "All residents of Butte are residents of Montana" is "If you are a resident of Butte, then you are a resident of Montana." A negation of this statement would be "The person is a not a resident of Butte or is a resident of Montana."

79. The statements $p \rightarrow q$ and $\sim p \vee q$ equivalent if they have the same truth tables.

p	q	$p \rightarrow q$			$\sim p \vee q$		
T	T	T	T	T	F	T	T
T	F	T	F	F	F	F	F
F	T	F	T	T	T	T	T
F	F	F	T	F	T	T	F
		1	2	1	1	2	1

Since the truth values in the final columns for each statement are the same, the statements are equivalent.

81.

p	q	$p \rightarrow q$			$\sim q \rightarrow \sim p$		
T	T	T	T	T	F	T	F
T	F	T	F	F	T	F	F
F	T	F	T	T	F	T	T
F	F	F	T	F	T	T	T
		1	2	1	1	2	1

Since the truth values in the final columns for each statement are the same, the statements are equivalent.

83.

p	q	$p \rightarrow \sim q$			$\sim p \vee \sim q$		
T	T	T	F	F	F	F	F
T	F	T	T	T	F	T	T
F	T	F	T	F	T	T	F
F	F	F	T	T	T	T	T
		1	2	1	1	2	1

Since the truth values in the final columns for each statement are the same, the statements are equivalent.

85.

p	q	$p \wedge \sim q$			$\sim q \rightarrow \sim p$		
T	T	T	F	F	F	T	F
T	F	T	T	T	T	F	F
F	T	F	F	F	F	T	T
F	F	F	F	T	T	T	T
		1	2	1	1	2	1

Since the truth values in the final columns for each statement are not the same, the statements are not equivalent. Observe that since they have opposite truth values, each statement is the negation of the other.

87. In the diagram, two series circuits are shown, which correspond to $p \wedge q$ and $p \wedge \sim q$. These circuits, in turn, form a parallel circuit. Thus, the logical statement is

$$(p \wedge q) \vee (p \wedge \sim q).$$

One pair of equivalent statements listed in the text includes

$$(p \wedge q) \vee (p \wedge \sim q) \equiv p \wedge (q \vee \sim q).$$

Since $(q \vee \sim q)$ is always true, $p \wedge (q \vee \sim q)$ simplifies to

$$p \wedge T \equiv p.$$

89. In the diagram, a series circuit is shown, which corresponds to $\sim q \wedge r$. This circuit, in turn, forms a parallel circuit with p. Thus, the logical statement is

$$p \vee (\sim q \wedge r).$$

91. In the diagram, a parallel circuit corresponds to $p \vee q$. This circuit is parallel to $\sim p$. Thus, the total circuit corresponds to the logical statement

$$\sim p \vee (p \vee q).$$

This statement in turn, is equivalent to

$$(\sim p \vee p) \vee (\sim p \vee q).$$

Since $\sim p \vee p$ is always true, we have

$$T \vee (\sim p \vee q) \equiv T.$$

93. The logical statement, $p \wedge (q \vee \sim p)$, can be represented by the following circuit.

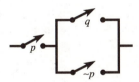

The statement, $p \wedge (q \vee \sim p)$, simplifies to $p \wedge q$ as follows:

$$p \wedge (q \vee \sim p) \equiv (p \wedge q) \vee (p \wedge \sim p)$$
$$\equiv (p \wedge q) \vee F$$
$$\equiv p \wedge q.$$

95. The logical statement, $(p \vee q) \wedge (\sim p \wedge \sim q)$, can be represented by the following circuit.

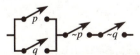

The statement, $(p \vee q) \wedge (\sim p \wedge \sim q)$, simplifies to F as follows:

$$(p \lor q) \land (\sim p \land \sim q) \equiv p \lor q \land \sim p \land \sim q$$
$$\equiv p \lor \sim p \land q \land \sim q$$
$$\equiv (p \lor \sim p) \land (q \land \sim q)$$
$$\equiv T \land F$$
$$\equiv F.$$

$$(\sim q \land \sim p) \lor (\sim p \lor q) \equiv [\sim q \lor (\sim p \lor q)] \land [\sim p \lor (\sim p \lor q)]$$
$$\equiv [\sim q \lor \sim p \lor q] \land [\sim p \lor \sim p \lor q]$$
$$\equiv [(\sim q \lor q) \lor \sim p] \land [(\sim p \lor \sim p) \lor q]$$
$$\equiv (T \lor \sim p) \land (\sim p \lor q)$$
$$\equiv T \land (\sim p \lor q)$$
$$\equiv \sim p \lor q.$$

97. The logical statement, $[(p \lor q) \land r] \land \sim p$, can be represented by the following circuit.

The statement, $[(p \lor q) \land r] \land \sim p$, simplifies to $(r \land \sim p) \land q$ as follows:

$$[(p \lor q) \land r] \land \sim p \equiv [(p \land r) \lor (q \land r)] \land \sim p$$
$$\equiv [(p \land r) \land \sim p] \lor [(q \land r) \land \sim p]$$
$$\equiv [p \land r \land \sim p] \lor [q \land r \land \sim p]$$
$$\equiv [(p \land \sim p) \land r] \lor [(r \land \sim p) \land q]$$
$$\equiv (F \land r) \lor [(r \land \sim p) \land q]$$
$$\equiv F \lor [(r \land \sim p) \land q]$$
$$\equiv [(r \land \sim p) \land q].$$

99. The logical statement, $\sim q \rightarrow (\sim p \rightarrow q)$, can be represented by the following circuit.

The statement, $\sim q \rightarrow (\sim p \rightarrow q)$, simplifies to $p \lor q$ as follows:

$$\sim q \rightarrow (\sim p \rightarrow q) \equiv \sim q \rightarrow (p \lor q)$$
$$\equiv q \lor (p \lor q)$$
$$\equiv q \lor p \lor q$$
$$\equiv p \lor q \lor q$$
$$\equiv p \lor (q \lor q)$$
$$\equiv p \lor q.$$

101. Writing exercise

3.4 EXERCISES

For each given direct statement (symbolically as $p \rightarrow q$), write (a) the converse ($q \rightarrow p$), (b) the inverse ($\sim p \rightarrow \sim q$), and (c) the contrapositive ($\sim q \rightarrow \sim p$) in if ... then forms. Wording may vary in the answers to Exercises 1–10.

1. The direct statement: If beauty were a minute, then you would be an hour.

    (a) *Converse*: If you were an hour, then beauty would be a minute.

    (b) *Inverse*: If beauty were not a minute, then you would not be an hour.

    (c) *Contrapositive*: If you were not an hour, then beauty would not be be a minute.

3. The direct statement: If it ain't broken, don't fix it.

    (a) *Converse*: If you don't fix it, then it ain't broke.

    (b) *Inverse*: If it's broke, then fix it.

    (c) *Contrapositive*: If you fix it, then it's broke.

It is helpful to restate the direct statement in an if ... then form for the exercises 5–9.

5. The direct statement: If you walk in front of a moving car, then it is dangerous to your health.

    (a) *Converse*: If it is dangerous to your health, then you walk in front of a moving car.

    (b) *Inverse*: If you do not walk in front of a moving car, then it is not dangerous to your health.

    (c) *Contrapositive*: If it is not dangerous to your health, then you not walk in front of a moving car.

7. The direct statement: If they are birds of a feather, then they flock together.

    (a) *Converse*: If they flock together, then they are birds of a feather.

    (b) *Inverse*: If they are not birds of a feather, then they do not flock together.

    (c) *Contrapositive*: If they do not flock together, then they are not birds of a feather.

9. The direct statement: If you build it, then he will come.

    (a) *Converse*: If he comes, then you built it.

    (b) *Inverse*: If you don't build it, then he won't come.

    (c) *Contrapositive*: If he doesn't come, then you didn't build it.

11. The direct statement: $p \rightarrow \sim q$.

   (a) *Converse*: $\sim q \rightarrow p$.

   (b) *Inverse*: $\sim p \rightarrow q$.

   (c) *Contrapositive*: $q \rightarrow \sim p$.

13. The direct statement: $\sim p \rightarrow \sim q$.

   (a) *Converse*: $\sim q \rightarrow \sim p$.

   (b) *Inverse*: $p \rightarrow q$.

   (c) *Contrapositive*: $q \rightarrow p$.

15. The direct statement: $p \rightarrow (q \vee r)$.

   (a) *Converse*: $(q \vee r) \rightarrow p$.

   (b) *Inverse*: $\sim p \rightarrow \sim (q \vee r)$ or $\sim p \rightarrow (\sim q \wedge \sim r)$.

   (c) *Contrapositive*: $(\sim q \wedge \sim r) \rightarrow \sim p$.

17. Writing exercise

Writing the statements, Exercises 19–40, in the form "if p, then, q" we arrive at the following results.

19. The statement "If it is muddy, I'll wear my galoshes" becomes "If it is muddy, then I'll wear my galoshes."

21. The statement "'17 is positive' implies that $17 + 1$ is positive" becomes "If '17 is positive,' then $17 + 1$ is positive"

23. The statement "All integers are rational numbers" becomes "If a number is an integer, then it is a rational number."

25. The statement "Doing crossword puzzles is sufficient for driving me crazy" becomes "If I do crossword puzzles, then I am Driven crazy."

27. "A day's growth of beard is necessary for Greg Tobin to shave" becomes "If Greg Tobin is to shave, then he must have a day's growth of beard."

29. The statement "I can go from Boardwalk to Connecticut Avenue only if I pass GO" becomes "If I go from Boardwalk to Connecticut, then I pass GO."

31. The statement "No whole numbers are not integers" becomes "If a number is a whole number, then it is an integer."

33. The statement "The Indians will win the pennant when their pitching improves" becomes "If their pitching improves, then the Indians will win the pennant."

35. The statement "A rectangle is a parallelogram with a right angle" becomes "If the figure is a rectangle, then it is a parallelogram with a right angle."

37. The statement "A triangle with two sides of the same length is isosceles" becomes "If a triangle has two sides of the same length, then it is isosceles."

39. The statement "The square of a two-digit number whose units digit is 5 will end in 25" becomes "If a two-digit number whose units digit is 5 is squared, then it will end in 25."

41. Option D is the answer since "r is necessary for s" represents the converse, $s \rightarrow r$, of all of the other statements.

43. Writing exercise

45. The statement "$5 = 9 - 4$ if and only if $8 + 2 = 10$" is true, since this is a biconditional composed of two true statements.

47. The statement "$8 + 7 \neq 15$ if and only if $3 \times 5 = 9$" is true, since this is a biconditional consisting of two false statements.

49. The statement "Bill Clinton was president if and only if Jimmy Carter was not president" is false, since this is a biconditional consisting of a true and a false statement.

51. The statements "Elvis is alive" and "Elvis is dead" are contrary, since both cannot be true at the same time.

53. The statements "That animal has four legs" and "That animal is a dog"54. The statements "The book is nonfiction" and "That book costs more than $70" are consistent, since both statements can be true.

55. The statements "This number is an integer" and "This number is irrational" are contrary, since both cannot be true at the same time.

57. Answers will vary. One example is: That man is Kent Merrill; That man sells books.

3.5 EXERCISES

1. Draw an Euler diagram where the region representing "boxers" must be inside the region representing "those who wear trunks" so that the first premise is true.

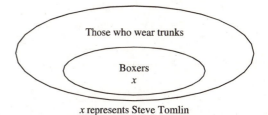

x represents Steve Tomlin

Let x represent Steve Tomlin. By premise 2, x must lie in the "boxers" region. Since this forces the conclusion to be true, the argument is valid.

3.   Draw an Euler diagram where the region representing "residents of New York" must be inside the region representing "those who love Coney Island hot dogs" so that the first premise is true.

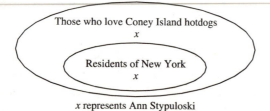

x represents Ann Stypuloski

Let x represent Ann Stypuloski. By the second premise, x must lie in the "those who love Coney Island hot dogs" region. Thus, she could be inside or outside the inner region. Since this allows for a false conclusion (she doesn't have to be in the "residents of New York" region for both premises to be true), the argument is <u>invalid</u>.

5.   Draw an Euler diagram where the region representing "contractors" must be inside the region representing "those who use cell phones" so that the first premise is true.

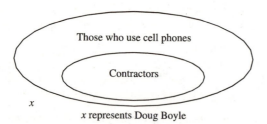

x represents Doug Boyle

Let x represent "Doug Boyle." By the second premise, x must lie outside the region representing "those who use cell phones." Since this forces the conclusion to be true, the argument is <u>valid</u>.

7.   Draw an Euler diagram where the region representing "people who apply for a loan" must be inside the region representing "people who pay for a title search" so that the first premise is true.

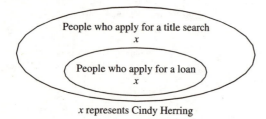

x represents Cindy Herring

Let x represent "Cindy Herring." By the second premise, x must lie in the "people who pay for a title search" region. Thus, she could be inside or outside the inner region. Since this allows for a false conclusion (she doesn't have to be in the "people who apply for a loan" region for both premises to be true), the argument is <u>invalid</u>.

9.   Draw an Euler diagram where the region representing "philosophers" intersects the region representing "those who are absent minded." This keeps the first premise true.

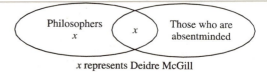

x represents Deidre McGill

Let x represent "Deidre McGill." By the second premise, x must lie in the region representing "philosophers." Thus, she could be inside or outside the region "representing people who are absent minded." Since this allows for a false conclusion, the argument is <u>invalid</u>.

11.   Draw an Euler diagram where the region representing "trucks" intersects the region representing "vehicles with sound systems." This keeps the first premise true. There are several ways to represent the 2nd premise as true. One way is as shown below. Examining the diagram, however, it is apparent that the conclusion is false.

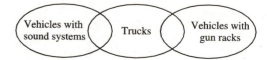

Since the diagram shows true premises but also a false conclusion, the argument is <u>invalid</u>. Note that all ways to draw an Euler diagram representing true premises must also yield a true conclusion for the argument to be valid.

13.   Interchanging the second premise and the conclusion of EXAMPLE 3 (in the text) yields the following argument,

>   All banana trees have green leaves.
>   <u>That plant is a banana tree.</u>
>   That plant has green leaves.

Draw an Euler diagram where the region representing "Banana trees" must be inside the region representing "Things that have green leaves" so that the first premise is true.

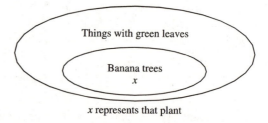

x represents that plant

Let x represent "That plant." By the second premise, x must lie inside the region representing "Banana trees." Since this forces the conclusion to be true, the argument is valid which makes the answer to the question <u>yes</u>.

15. The following is a valid argument which can be constructed from the given Euler diagram.

    People who have major surgery must go to the hospital.
    Andrea Sheehan is having major surgery.
    Andrea Sheehan must go to the hospital.

17. The following Euler diagram represents true premises.

    Since the diagram forces the conclusion to be true also, the argument is <u>valid</u>.

19. The following Euler diagram represents true premises. However, it is clear that, drawn this way, the conclusion is false.

    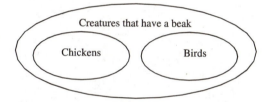

    The argument is <u>invalid</u> even though the conclusion is true.

21. The following Euler diagram yields true premises. It also forces the conclusion to be true.

    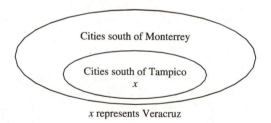

    Thus, the argument in <u>valid</u>. Observe that the diagram is the only way to show true premises.

23. The following Euler diagram represents true premises.

    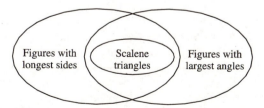

    No information, however is given regarding the relationship between the largest angle and the longest

side. The argument is <u>invalid</u> even though the conclusion is true.

The premises marked A, B, and C are followed by several possible conclusions (Exercises 25–30). Take each conclusion in turn, and check whether the resulting argument is valid or invalid.

A. All people who drive contribute to air pollution.
B. All people who contribute to air pollution make life a little worse.
C. Some people who live in a suburb make life a little worse.

Diagram the three premises to be true.

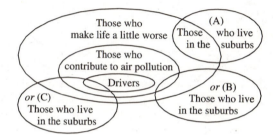

25. We are not forced into the conclusion, "Some people who live in a suburb drive" since diagrams (A) or (B) represent true premises where this conclusion is false. Thus, the argument is <u>invalid</u>.

27. We are not forced into the conclusion, "Some people who contribute to air pollution live in a suburb" since option (*A*) represents true premises and a false conclusion. Thus, the argument is <u>invalid</u>.

29. The conclusion, "All people who drive make life a little worse" yields a <u>valid</u> argument since all three options (*A–C*) represent true premises and force this conclusion to be true.

31. No answer

EXTENSION: LOGIC PUZZLES

1. Draw charts as indicated and complete using the initial information given. Use "•" for Yes and "x" for No. For any cell that is assigned "•", mark "x" in the remaining unmarked cells in that row and column.

   (1) *Neither Lauren nor Zach was the child accompanied by Ms. Reed. Tara was more interested in shopping at the country store than in picking pumpkins.* Mark (x)'s in the boxes for Lauren and Zach under Ms. Reed's column. Place (•) into the intersection of Tara and country store since that was her interest. Note: remember to always mark (x)'s in remaining cells of a row or column when marking a cell with (•)

(2) Xander and his father, Mr. Morgan, didn't go on the hay ride. Ms. Fedor's child (who isn't Zach or Lauren) was fascinated by the cider-making process. Place (•) into the cell representing the intersection of Xander and Mr. Morgan since you know that they are father and son. Place (x) into the cell representing the intersection of Xander and hay ride as well as the cell for his father, Mr. Morgan, and hay ride. Place (•) into the cell at the intersection of Ms. Fedor and cider-making. Place (x)'s in the cells representing Lauren and Zach in Ms. Fedor's column. Since neither Lauren nor Zach is Ms. Fedor's child neither can be interested in cider-making. Mark (x)'s in the cells for cider-making in Zach and Lauren's rows.

(3) Zach is neither the child who went on the hay ride nor the one who wanted to go apple picking. Mr. Hanson's child didn't go on the hay ride. Place (x) in the aspect cells for apple picking and hay ride in Zach's row. This leaves feeding animals as Zach's favorite aspect. Place (•) in the feeding animal cell for Zach. Mark (x) in the aspect cell for the hay ride in Mr. Hanson's column.

Tara and Raven are the only two children that could still belong to Ms. Fedor. Since Ms. Fedor's child liked cider-making check the aspect columns and notice Tara liked the country store. This leaves Raven as Ms. Fedor's child. Place (•) in the cell for Raven in Ms. Fedor's column and (•) in the cider-making cell in Raven's row. The only aspect now available to Xander is apple picking. Place (•) in that cell. The last aspect, the hay ride, is now Lauren's only possible aspect. Place (•) in the appropriate cell.

At this point *Chart 1* will have been completed.

Chart 1

		PARENT					ASPECT				
		MS. FEDOR	M. HANSON	MR. MAIER	MR. MORGAN	MS. REED	APPLE PICKING	CIDER MAKING	COUNTRY STR.	FEEDING ANM.	HAY RIDE
CHILD	LAUREN	X			X	X	X	X	X	X	•
	RAVEN	•	X	X	X	X	X	•	X	X	X
	TARA	X			X		X	X	•	X	X
	ZANDER	X	X	X	•	X	•	X	X	X	X
	ZACH	X			X	X	X	X	X	•	X
ASPECT	APPLE PICK.	X			•	X					
	CIDER MAK.	•	X	X	X	X					
	COUNTRY	X				•					
	FEEDING	X	•								
	HAY RIDE	X	X	•	X	X					

Tara is now the only child left in Ms. Reed's column. Place (•) at the intersection of Lauren and Ms. Reed. Notice that Xander's favorite aspect is apple picking and we know that Mr. Morgan is Xander's father. Thus we

can place (•) in Mr. Morgan's column in the apple picking cell. This leaves Lauren and Zach with the possibility of Mr. Hanson and Mr. Maier for parents. By comparing Lauren's favorite aspect, the hay ride, to Mr. Hanson and Mr. Maier's aspect columns we find that Mr. Hanson's child did NOT like the hay ride. Thus Mr. Maier must be Lauren's father. Mr. Hanson is left as Zach's father. Now it is a simple matter of matching each child's interest with the aspect cell for his or her parent. Zach likes feeding animals; mark (•) in the feeding animals cell for Mr. Hanson. Lauren likes the hay ride; mark (•) in the hay ride cell for Mr. Maier. The country store is now the only available interest in Ms. Reed's column. Check to see that Ms. Reed's child, Tara, is, in fact, the child that likes the country store. Place (•) in the appropriate cell and the logic puzzle is complete (*Chart 2*).

Chart 2

		PARENT					ASPECT				
		MS. FEDOR	M. HANSON	MR. MAIER	MR. MORGAN	MS. REED	APPLE PICKING	CIDER MAKING	COUNTRY STR.	FEEDING ANM.	HAY RIDE
CHILD	LAUREN	X	X	•	X	X	X	X	X	X	•
	RAVEN	•	X	X	X	X	X	•	X	X	X
	TARA	X	X	X	X	•	X	X	•	X	X
	ZANDER	X	X	X	•	X	•	X	X	X	X
	ZACH	X	•	X	X	X	X	X	X	•	X
ASPECT	APPLE PICK.	X	X	X	•	X					
	CIDER MAK.	•	X	X	X	X					
	COUNTRY	X	X	X	X	•					
	FEEDING	X	•	X	X	X					
	HAY RIDE	X	X	•	X	X					

Thus, Lauren, Mr. Mair, hay ride; Raven, Ms. Fedor, cider making; Tara, Ms. Reed, country store; Xander, Mr. Morgan, apple picking; Zach, Mr. Hanson, feeding animals.

3. Draw charts as indicated and complete using the initial information given. Use "•" for Yes and "x" for No. For any cell that is assigned "•", mark "x" in the remaining unmarked cells in that row and column.

(1) The tie with the grinning leprechauns wasn't a present from a daughter. Mark (x) in the cell at the intersection of daughter and grinning leprechaun.

(2) Mr. Crow's tie features neither the dancing reindeer nor the yellow happy faces. (x) the dancing reindeer and the yellow happy faces from Mr. Crow's row.

(3) Mr. Speigler's tie wasn't a present from his uncle. Place (x) in the uncle cell in Mr. Speigler's row.

(4) *The tie with the yellow happy faces wasn't a gift from a sister.* Mark (x) in the cell at the intersection of yellow happy faces and sister.

(5) *Mr. Evans and Mr. Speigler won the tie with the grinning leprechauns and the tie that was a present from a father-in-law, in some order.* The leprechaun tie could not have come from the father-in-law so mark (x) at that intersection. For either Mr. Evans or Mr. Speigler to receive the tie from his father-in-law no one else could have received a tie from his father-in-law so (x) the father-in-law cells for Mr. Crow and Mr. Hurley. The same logic applies to the leprechaun; neither Mr. Crow nor Mr. Hurley could have received the leprechaun tie. (x) the appropriate cells. Mr. Crow's only tie option is now the tie with cupids so place (●) in the cell at the intersection of cupids and Mr. Crow.

(6) *Mr. Hurley received his flamboyant tie from his sister.* Place (●) for the cell for sister in Mr. Hurley's row. Since Mr. Hurley received the tie from his sister and the sister did NOT give the happy faces tie (x) the happy faces cell in Mr. Hurley's row. This leaves reindeer as the only choice for Mr. Hurley's row. (●) the cell for reindeer in Mr. Hurley's row and (●) the cell for reindeer in the sister row because Mr. Hurley received his tie from his sister.

Since Mr. Crow did NOT receive his tie from his father-in-law or sister (x) those cells in the column under cupids. The father-in-law could now only have given the happy faces tie so (●) the appropriate cell. This leaves the cupids tie for the daughter and the leprechaun tie for the uncle. Since the cupid tie belongs to Mr. Crow, place (●) in the daughter cell in Mr. Crow's row.

Mr. Speigler's tie could now only have come from his father-in-law, leaving the uncle's tie for Mr. Evans. Now notice that Mr. Evan's tie came from his uncle and the uncle purchased the leprechaun tie so Mr. Evans received the tie with leprechauns and Mr. Speigler received the only remaining tie, the tie with happy faces.

Your completed chart should look like the following.

Chart

	Cupids	Happy faces	Leprechauns	Reindeer	Daughter	Father-in-law	Sister	Uncle
Mr. Crow	●	X	X	X	●	X	X	X
Mr. Evans	X	X	●	X	X	X	X	●
Mr. Hurley	X	X	X	●	X	X	●	X
Mr. Speigler	X	●	X	X	X	●	X	X
Daughter	●	X	X	X				
Father-in-law	X	●	X	X				
Sister	X	X	X	●				
Uncle	X	X	●	X				

Thus, Mr. Crow, cupids, daughter; Mr. Evans, leprechauns, uncle; Mr. Hurley, reindeer, sister; Mr. Speigler, happy faces, father-in-law.

3.6 EXERCISES

1. Let p represent "you use binoculars, "q represent "you get a glimpse of the comet, " and r represent "you will be amazed." The argument is then represented symbolically by:
$$p \rightarrow q$$
$$\underline{q \rightarrow r}$$
$$p \rightarrow r.$$

This is the <u>valid</u> argument form "reasoning by transitivity."

3. Let p represent "Kevin O'Brien sells his quota" and q represent "He will get a bonus." The argument is then represented symbolically by:
$$p \rightarrow q$$
$$\underline{p \quad\quad}$$
$$q.$$

This is the <u>valid</u> argument form "modus ponens."

5. Let p represent "She buys another pair of shoes" and q represent "her closet will overflow." The argument is then represented symbolically by:
$$p \rightarrow q$$
$$\underline{q \quad\quad}$$
$$p.$$

Since this is the form "fallacy of the converse," it is invalid and considered a <u>fallacy</u>.

7. Let p represent "Patrick Roy plays" and q represent "the opponent gets shut out." The argument is then represented symbolically by:
$$p \rightarrow q$$
$$\underline{\sim q \quad\quad}$$
$$\sim p.$$

This is the <u>valid</u> argument form "modus tollens."

9. Let p represent "we evolved a race of Isaac Newtons." and q represent "that would not be progress." The argument is then represented symbolically by:
$$p \rightarrow q$$
$$\underline{\sim p \quad\quad}$$
$$\sim q.$$

Note: that since we let q represent "that <u>would not</u> be progress," then $\sim q$ represents "that <u>is</u> progress."
Since this is the form "fallacy of the inverse," it is invalid and considered a <u>fallacy</u>.

11. Let p represent "Pat Quinlin jogs" and q represent "John Remington pumps iron" The argument is then represented symbolically by:

$$p \vee q \quad (\text{or } q \vee p)$$
$$\frac{\sim q}{p.}$$

Since this is the form "disjunctive syllogism," it is a <u>valid</u> argument.

To show validity for the arguments in the following exercises we must show that the conjunction of the premises implies the conclusion. That is, the conditional statement $[P_1 \wedge P_2 \wedge \ldots \wedge P_n] \rightarrow C$ must be a tautology. For exercises 13 and 14 we will use the standard (long format) to develop the corresponding truth tables. For the remainder of the exercises we will use the alternate (short format) to create the truth tables.

13. Form the conditional statement

$$[(p \vee q) \wedge p] \rightarrow \sim q$$

from the argument. Complete a truth table.

p	q	$p \vee q$	$(p \vee q) \wedge p$	$\sim q$	$[(p \vee q) \wedge p] \rightarrow \sim q$
T	T	T	T	F	F
T	F	T	T	T	T
F	T	T	F	F	T
F	F	F	F	T	T

Since the conditional, formed by the conjunction of premises implying the conclusion, is not a tautology, the argument is <u>invalid</u>.

15. Form the conditional statement

$$[(\sim p \rightarrow \sim q) \wedge q] \rightarrow p$$

from the argument. Complete a truth table.

p	q	$[(\sim p$	$\rightarrow$	$\sim q)$	$\wedge$	$q]$	$\rightarrow$	p
T	T	F	T	F	T	T	T	T
T	F	F	T	T	F	F	T	T
F	T	T	F	F	F	T	T	F
F	F	T	T	T	F	F	T	F
		1	2	1	3	2	4	3

Since the conditional, formed by the conjunction of premises implying the conclusion, is a tautology, the argument is <u>valid</u>.

17. Form the conditional statement

$$[(p \rightarrow q) \wedge (q \rightarrow p)] \rightarrow (p \wedge q)$$

from the argument. Complete a truth table.

p	q	$[(p \rightarrow q)$	$\wedge$	$(q \rightarrow p)]$	$\rightarrow$	$(p \wedge q)$
T	T	T	T	T	T	T
T	F	F	F	T	T	F
F	T	T	F	F	T	F
F	F	T	T	T	F	F
		1	3	2	4	3

Since the conditional, formed by the conjunction of premises implying the conclusion, is not a tautology, the argument is <u>invalid</u>.

19. Form the conditional statement

$$[(p \rightarrow \sim q) \wedge q] \rightarrow \sim p$$

from the argument. Complete a truth table.

p	q	$[(p \rightarrow \sim q)$	$\wedge$	$q]$	$\rightarrow$	$\sim p$
T	T	T F F	F	T	T	F
T	F	T T T	F	F	T	F
F	T	F T F	T	T	T	T
F	F	F T T	F	F	T	T
		1 2 1	3	2	4	3

Since the conditional, formed by the conjunction of premises implying the conclusion, is a tautology, the argument is <u>valid</u>.

21. Form the conditional statement

$$[(\sim p \vee q) \wedge (\sim p \rightarrow q) \wedge p] \rightarrow \sim q$$

from the argument. Complete a truth table.

p	q	$[(\sim p \vee q)$	$\wedge$	$(\sim p \rightarrow q)]$	$\wedge$	$p]$	$\rightarrow$	$\sim q$
T	T	F T T	T	F T T	T	T	F	F
T	F	F F F	F	F T F	F	T	T	T
F	T	T T T	T	T T T	F	F	T	F
F	F	T T F	F	T F F	F	F	T	T
		1 2 1	3	2	4	3	5	4

Since the conditional, formed by the conjunction of premises implying the conclusion, is not a tautology, the argument is <u>invalid</u>.

23. Form the conditional statement

$$\{[(\sim p \wedge r) \to (p \vee q)] \wedge (\sim r \to p)\} \to (q \to r)$$

from the argument.

p	q	r	$\{[(\sim p \wedge r) \to (p \vee q)] \wedge (\sim r \to p)\} \to (q \to r)$
T	T	T	F F T T T T F T T T T
T	T	F	F F F T T T T T T F F
T	F	T	F F T T T T F T T T T
T	F	F	F F F T T T T T T T T
F	T	T	T T T T T T F T F T T
F	T	F	T F F T T F T F F T F
F	F	T	T T T F F F F T F T T
F	F	F	T F F T F F T F F T T
			1 2 1 3 2 4 2 3 2 5 4

The F in the final column 5 shows us that the statement is not a tautology and hence, the argument is <u>invalid</u>.

25. Writing exercise

27. Let p represent "Jeff loves to play golf," q represent "Joan likes to sew," and r represent "Brad sings in the choir." The argument is then represented symbolically by:

$$p$$
$$q \to \sim p$$
$$\sim q \to r$$
$$r.$$

Construct the truth table for

$$[p \wedge (q \to \sim p) \wedge (\sim q \to r)] \to r.$$

p	q	r	$[p \wedge (q \to \sim p) \wedge (\sim q \to r)] \to r$
T	T	T	T F T F F F F T T T T
T	T	F	T F T F F F F T F T F
T	F	T	T T F T F T T T T T T
T	F	F	T T F T F F T F F T F
F	T	T	F F T T T F F T T T T
F	T	F	F F T T T F F T F T F
F	F	T	F F F T T F T T T T T
F	F	F	F F F T T F T F F T F
			2 3 1 2 1 4 2 3 2 5 4

Since the conditional, formed by the conjunction of premises implying the conclusion, is a tautology, the argument is <u>valid</u>.

29. Let p represent "the Pokémon craze continues," q represent "Beanie Babies will remain popular," and r represent "Barbie dolls continue to be favorites." The argument is then represented symbolically by:

$$p \to q$$
$$r \vee q$$
$$\sim r$$
$$\sim p.$$

Construct the truth table for

$$[(p \to q) \wedge (r \vee q) \wedge (\sim r)] \to \sim p.$$

Note: we do not have to complete a column under each simple statement $p, q,$ and r, (as we did in exercises above) since it is easy to compare the appropriate index columns to create the truth value for each connective.

p	q	r	$[(p \to q) \wedge (r \vee q) \wedge (\sim r)] \to \sim p$
T	T	T	T T T F F T F
T	T	F	T T T T T F F
T	F	T	F F T F F T F
T	F	F	F F F F T T F
F	T	T	T T T F F T T
F	T	F	T T T T T T T
F	F	T	T T T F F T T
F	F	F	T F F F T T T
			1 2 1 3 2 4 3

Since the conditional, formed by the conjunction of premises implying the conclusion, is not a tautology, the argument is <u>invalid</u>. Note: If you are completing the truth table along rows (rather than down columns), you could stop after completing the second row, knowing that with a false conditional the statement will not be a tautology.

31. Let p represent "I've got you under my skin," q represent "you are deep in the heart of me," and r represent "you are really a part of me." The argument is then represented symbolically by:

$$p \to q$$
$$q \to \sim r$$
$$q \vee r$$
$$p \to r.$$

Construct the truth table for

$$[(p \to q) \wedge (q \to \sim r) \wedge (q \vee r)] \to (p \to r).$$

p	q	r	$[(p \to q) \wedge (q \to \sim r) \wedge (q \vee r)] \to (p \to r)$
T	T	T	T F T F F F T T T
T	T	F	T T T T T T T F F
T	F	T	F F F T F F T T T
T	F	F	F F F T T F T F F
F	T	T	T F T F F F T T T
F	T	F	T T T T T T T T T
F	F	T	T F T F F F T T T
F	F	F	T T T T T F T T T
			2 3 1 2 1 4 3 5 4

Since the conditional, formed by the conjunction of premises implying the conclusion, is not a tautology, the argument is <u>invalid</u>. Note: If you are completing the truth table along rows (rather than down columns), you could stop after completing the first row, knowing that with a false conditional the statement will not be a tautology.

33. Let p represent "Otis is a disc jockey," q represent "he lives in Lexington," and r represent "he is a history buff." The argument is then represented symbolically by

$$p \rightarrow q$$
$$\underline{q \wedge r}$$
$$\sim r \rightarrow \sim p.$$

Construct the truth table for

$$[(p \rightarrow q) \wedge (q \wedge r)] \rightarrow (\sim r \rightarrow \sim p).$$

p	q	r	$[(p$	$\rightarrow$	$q)$	$\wedge$	$(q \wedge r)]$	$\rightarrow$	$(\sim r$	$\rightarrow$	$\sim p)$
T	T	T		T		T	T	T	F	T	F
T	T	F		T		F	F	T	T	F	F
T	F	T		F		F	F	T	F	T	F
T	F	F		F		F	F	T	T	F	F
F	T	T		T		T	T	T	F	T	F
F	T	F		T		F	F	T	T	T	T
F	F	T		T		F	F	T	F	T	T
F	F	F		T		F	F	T	T	T	T
				1		2	1	3	1	2	1

Since the conditional, formed by the conjunction of premises implying the conclusion, is a tautology, the argument is <u>valid</u>.

The following exercises involve Quantified arguments and can be analyzed, as such, by Euler diagrams. However, the quantified statements can be represented as conditional statements as well. This allows us to use a truth table – or recognize a valid argument form – to analyze the validity of the argument.

35. Let p represent "you are a man," q represent "you are created equal," and r represent "you are a women." The argument is then represented symbolically by:

$$p \rightarrow q$$
$$\underline{q \rightarrow r}$$
$$p \rightarrow r.$$

This is a "Reasoning by Transitivity" argument form and hence, is <u>valid.</u>

37. We apply reasoning by repeated transitivity to the six premises. A conclusion from this reasoning, which makes the argument valid, is reached by linking the first antecedent to the last consequent. This conclusion is "If I tell you the time, then my life will be miserable."

Answers in Exercises, 39 – 45 may be replaced by their contrapositives.

39. The statement "All my poultry are ducks" becomes "If it is my poultry, then it is a duck."

41. The statement "Guinea pigs are hopelessly ignorant of music" becomes "If it is a Guinea pig, then it is hopelessly ignorant of music."

43. The statement "No teachable kitten has green eyes" becomes "If it is a teachable kitten, then it does not have green eyes."

45. The statement "I have not filed any of them that I can read" becomes "If I can read it, then I have not filed it."

47. (a) "No ducks are willing to waltz" becomes "if it is a duck, then it is not willing to waltz."

(b) "No officers ever decline to waltz" becomes "if one is an officer, then one is willing to waltz."

(c) "All my poultry are ducks" becomes "if it is my poultry , then it is a duck."

(d) In symbols, the three premises are

$$p \rightarrow \sim s$$
$$r \rightarrow s$$
$$q \rightarrow p.$$

Begin with q, which only appears once. Replacing $r \rightarrow s$ with its contrapositive, $\sim s \rightarrow \sim r$, rearrange the three premises.

$$q \rightarrow p$$
$$p \rightarrow \sim s$$
$$\sim s \rightarrow \sim r$$

By repeated use of reasoning by transitivity, the conclusion which provides a valid argument is

$$q \rightarrow \sim r.$$

In words, "If it is my poultry, then it is not an officer," or "none of my poultry are officers."

49. (a) "Promise-breakers are untrustworthy" becomes "if one is a promise-breaker, then one is not trustworthy."

(b) "Wine-drinkers are very communicative" becomes "if one is a wine-drinker, then one is very communicative."

(c) "A person who keeps a promise is honest" becomes "if one is not a promise-breaker, then one is honest."

(d) "No teetotalers are pawnbrokers" becomes "if one is not a wine-drinker, then one is not a pawnbroker."

(e) "One can always trust a very communicative person" becomes "if one is very communicative, then one is trustworthy."

(f) In symbols, the statements are

$$r \to {\sim}s$$
$$u \to t$$
$${\sim}r \to p$$
$${\sim}u \to {\sim}q$$
$$t \to s.$$

Begin with q, which only appears once. Using the contrapositive of $\sim u \to \sim q$, $(q \to u)$, and $r \to \sim s$, $(s \to \sim r)$, rearrange the five premises as follows:

$$q \to u$$
$$u \to t$$
$$t \to s$$
$$s \to {\sim}r$$
$${\sim}r \to p.$$

By repeated use of reasoning by transitivity, the conclusion which provides a valid argument is

$$q \to p.$$

In words, this conclusion can be stated as "if one is a pawnbroker, then one is honest," or "all pawnbrokers are honest."

51. Begin by changing each quantified premise to a conditional statement:

(a) The statement "All the dated letters in this room are written on blue paper" becomes "If it is dated, then it is on blue-paper."

(b) The statement "None of them are in black ink, except those that are written in the third person" becomes "If is not in the third person, then it is not in black ink."

(c) The statement "I have not filed any of them that I can read" becomes "If I can read it, then it is not filed."

(d) The statement "None of them that are written on one sheet are undated" becomes "If it is on one sheet, then it is dated.

(e) The statement "All of them that are not crossed are in black ink" becomes "If it is not crossed, then it is in black ink.

(f) The statement "All of them written by Brown begin with 'Dear Doctor'" becomes "If it is written by Brown, then it begins with 'Dear Sir'."

(g) The statement "All of them written on blue paper are filed" becomes "If it is on blue paper, then it is filed."

(h) The statement "None of them written on more than one sheet are crossed" becomes "If it is not on more than one sheet, then it is not crossed."

(i) The statement "None of them that begin with "Dear Sir' are written in the third person" becomes "If it begins with 'Dear Sir,' then it is not written in the third person."

(j) In symbols, the statements are

(a) $\quad r \to w$
(b) $\quad {\sim}u \to {\sim}t$
(c) $\quad v \to {\sim}s$
(d) $\quad x \to r$
(e) $\quad {\sim}q \to t$
(f) $\quad y \to p$
(g) $\quad w \to s$
(h) $\quad {\sim}x \to {\sim}q$
(i) $\quad p \to {\sim}u.$

Begin with y, which appears only once. Using contrapositives of $v \to \sim s$ $(s \to \sim v)$, $\sim q \to t$ $(\sim t \to q)$, and $\sim x \to \sim q$ $(q \to x)$, rearrange the nine statements.

$$y \to p$$
$$p \to {\sim}u$$
$${\sim}u \to {\sim}t$$
$${\sim}t \to q$$
$$q \to x$$
$$x \to r$$
$$r \to w$$
$$w \to s$$
$$s \to {\sim}v.$$

By repeated use of reasoning by transitivity, the conclusion that makes the argument valid is

$$y \to {\sim}v.$$

In words, the conclusion can be stated as "if it is written by Brown, then I can't read it," or equivalently "I can't read any of Brown's letters."

CHAPTER 3 TEST

1. The negation of "$6 - 3 = 3$" is "$6 - 3 \neq 3$."

2. The negation of "All men are created equal" is "Some men are not created equal."

3. The negation of "Some members of the class went on the field trip" is "No members of the class went on the field trip." An equivalent answer would be "All members of the class did not go on the field trip."

4. The negation of "If that's the way you feel, then I will accept it" is "That's the way you feel and I won't accept it." Remember that $\sim(p \to q) \equiv (p \land \sim q)$.

5. The negation of "She passed GO and collected $200" is "She did not pass GO or did not collect $200."
    Remember that $\sim(p \wedge q) \equiv (\sim p \vee \sim q)$.

Let p represent "You will love me" and let q represent "I will love you."

6. The symbolic form of "If you won't love me, then I will love you" is "$\sim p \to q$."

7. The symbolic form of "I will love you if you will love me" (or equivalently, "if you will love me, then I will love you") is "$p \to q$."

8. The symbolic form of "I won't love you if and only if you won't love me" is "$\sim p \leftrightarrow \sim q$."

9. Writing the symbolic form "$\sim p \wedge q$" in words, we get "You won't love me and I will love you."

10. Writing the symbolic form "$\sim(p \vee \sim q)$" in words, we get "It is not the case that you will love me or I won't love you" (or equivalently, by DeMorgan's, "you won't love me and I will love you").

Assume that p is true and that q and r are false for Exercises 11–14.

11. Replacing q and r with the given truth values, we have

    $$\sim F \wedge \sim F$$
    $$T \wedge T$$
    $$T.$$

    The compound statement $\sim q \wedge \sim r$ is true.

12. Replacing p, q and r with the given truth values, we have
    $$F \vee (T \wedge \sim F)$$
    $$F \vee (T \wedge T)$$
    $$F \vee \quad T$$
    $$T.$$

    The compound statement $r \vee (p \wedge \sim q)$ is true.

13. Replacing r with the given truth value (s not known), we have
    $$F \to (s \wedge F)$$
    $$F \to \quad F$$
    $$T.$$

    The compound statement $r \to (s \wedge r)$ is true.

14. Replacing p and q with the given truth values, we have

    $$T \leftrightarrow (T \to F)$$
    $$T \leftrightarrow \quad (F)$$
    $$F.$$

    The compound statement $p \leftrightarrow (p \to q)$ is false.

15. Writing exercise

16. The necessary condition for

    (a) a conditional statement to be false is that the antecedent must be true and the consequent must be false.

    (b) a conjunction to be true is that both component statements must be true.

    (c) a disjunction to be false is that is that both component statements must be false.

17.

p	q	p	$\wedge$	$(\sim p$	$\vee$	$q)$
T	T	T	T	F	T	T
T	F	T	F	F	F	F
F	T	F	F	T	T	T
F	F	F	F	T	T	F
		2	3	1	2	1

18.

p	q	$\sim$	$(p \wedge q)$	$\to$	$(\sim p$	$\vee$	$\sim q)$
T	T	F	T	T	F	F	F
T	F	T	F	T	F	T	T
F	T	T	F	T	T	T	F
F	F	T	F	T	T	T	T
		2	1	3	1	2	1

    Since the last completed column (3) is all true, the conditional is a tautology.

19. The statement "Some negative integers are whole numbers" is <u>false</u>, since all whole numbers are non-negative.

20. The statement "All irrational numbers are real numbers" is true, because the real numbers are made up of both the rational and irrational numbers.

The wording may vary in the answers in Exercises 21–26.

21. "All integers are rational numbers" can be stated as "If the number is an integer, then it is a rational number."

22. "Being a rhombus is sufficient for a polygon to be a quadrilateral" can be stated as "If a polygon is a rhombus, then it is a quadrilateral."

23. "Being divisible by 3 is necessary for a number to be divisible by 9" can be stated as "If a number is divisible by 9, then it is divisible by 3." Remember that the "necessary" part of the statement becomes the consequent.

24. "She digs dinosaur bones only if she is a paleontologist" can be stated as "If she digs dinosaur bones, then she is a paleontologist." Remember that the "only if" part of the statement becomes the consequent.

25. The direct statement: If a picture paints a thousand words, the graph will help me understand it.

    (a) *Converse*: If the graph will help me understand it, then a picture paints a thousand words.

    (b) *Inverse*: If a picture doesn't paints a thousand words, the graph won't help me understand it.

    (c) *Contrapositive*: If the graph doesn't help me understand it, then a picture doesn't paint a thousand words.

26. The direct statement: $\sim p \to (q \wedge r)$.

    (a) Converse: $(q \wedge r) \to \sim p$.

    (b) Inverse: $p \to \sim(q \wedge r)$, or $p \to (\sim q \vee \sim r)$.

    (c) Contrapositive: $\sim(q \wedge r) \to p$, or $(\sim q \vee \sim r) \to p$.

27. Complete an Euler diagram as:

x represents Pat Pearson

    Since, when the premises are diagrammed as being true, we are forced into a true conclusion, the argument is valid.

28. (a) Let p represent "he eats liver" and q represent "he will eat anything." The argument is then represented symbolically by:

$$p \to q$$
$$\underline{p}$$
$$q.$$

    This is the valid argument form "modus ponens," hence the answer is A.

    (b) Let p represent "you use your seat belt" and q represent "you will be safer." The argument is then represented symbolically by:

$$p \to q$$
$$\underline{\sim p}$$
$$\sim q.$$

    The answer is F, a fallacy of the inverse.

    (c) Let p represent "I hear *Come Sunday Morning*," q represent "I think of her," and "I get depressed." The argument is then represented symbolically b

$$p \to q$$
$$\underline{q \to r}$$
$$p \to r.$$

This is the valid argument form "reasoning by transitivity," hence the answer is C.

(d) Let p represent "she sings" and q represent "she dances." Represented the argument symbolically:

$$p \vee q$$
$$\underline{\sim p}$$
$$q.$$

This is the valid argument form "disjunctive syllogism," hence the answer is D.

29. Let p represent "I write a check," q represent "It will bounce," and "The bank guarantees it." The argument is then represented symbolically by:

$$p \to q$$
$$r \to \sim q$$
$$\underline{r}$$
$$\sim p$$

Construct the truth table for

$$\{[(p \to q) \wedge (r \to \sim q)] \wedge r\} \to (\sim p).$$

p	q	r	$\{[(p \to q)$	$\wedge$	$(r \to \sim q)]$	$\wedge$	$r\}$	$\to$	$(\sim p)$
T	T	T	T	F	T F F	F	T	T	F
T	T	F	T	T	F T F	F	F	T	F
T	F	T	F	F	T T T	F	T	T	F
T	F	F	F	F	F T T	F	F	T	F
F	T	T	T	F	T F F	F	T	T	T
F	T	F	T	T	F T F	F	F	T	T
F	F	T	T	T	T T T	T	T	T	T
F	F	F	T	T	F T T	F	F	T	T
			2	3	1 2 1	4	3	5	4

Since the conditional, formed by the conjunction of premises implying the conclusion, is a tautology, the argument is valid.

30. Construct the truth table for

$$[(\sim p \to \sim q) \wedge (q \to p)] \to (p \vee q).$$

p	q	$[(\sim p \to \sim q)$	$\wedge$	$(q \to p)]$	$\to$	$(p \vee q)$
T	T	F T F	T	T T T	T	T T T
T	F	F T T	T	F T T	T	T T F
F	T	T F F	F	T F F	T	F T T
F	F	T T T	T	F T F	F	F F F
		1 2 1	3	1 2 1	4	2 3 2

Since the conditional, formed by the conjunction of premises implying the conclusion, is not a tautology, the argument is invalid.

4 | NUMERATION AND MATHEMATICAL SYSTEMS

4.1 Historical Numeration Systems
4.2 Arithmetic in the Hindu-Arabic System
4.3 Conversion Between Number Bases
4.4 Finite Mathematical Systems
4.5 Groups
Chapter 4 Test

Chapter Goals

The student should achieve an understanding of numbers and how they are symbolized. He should also gain a better understanding of how our number system is constructed. After completing the chapter the student should be able to

- Convert numerals from other number systems to our system.
- Convert between various number bases.
- Perform arithmetic operations in various number bases.
- Test a mathematical system for closure, associative, commutative, identity, inverse, and distributive properties.
- Test a mathematical system to see if it is a group.

Chapter Summary

A group of symbols used to represent numbers is called a numeration system. Three types of numeration systems are simple grouping, multiplicative grouping and positional. The idea of a number base is a central theme throughout the various systems. The Egyptian numeration system is a simple grouping system. To review this system consider the following example.

EXAMPLE A   Egyptian Symbols

The symbol ⌒ represents 10,000; ⌇ represents 1000; ⌒ represents 100; ∩ represents 10; and | represents 1. Notice that each represents a power of ten; this means that it is a base ten numeration system. Using these Egyptian symbols the number 82,653 is represented as follows:

⌒⌒⌒⌒⌒⌒⌒⌒ ⌇⌇ ⌒⌒⌒⌒⌒⌒ ∩∩∩∩∩ |||

These numerals are interpreted as

$$(8 \times 10,000) + (2 \times 1000) + (6 \times 100) + (5 \times 10)$$
$$+ (3 \times 1)$$
$$= 80,000 + 2000 + 600 + 50 + 3$$
$$= 82,653.$$

A multiplicative grouping system is used by the Chinese. The symbols are written vertically, and the number is read from top to bottom. The following example shows a number in Chinese numerals with the Hindu-Arabic (our system) written to the right.

EXAMPLE B   Chinese Symbols

八	8
百	100
七	7
十	10
二	2

These Chinese symbols mean

$$(8 \times 100) + (7 \times 10) + 2 = 872.$$

Today the Hindu-Arabic system is used worldwide. It is a positional system in which the face value indicates the inherent value of the symbol, and the place value indicates the power of the number base associated with its location or position. The meaning of the number is understood. It is a system that is based on powers of ten. Numbers can be written in expanded form to demonstrate this fact.

EXAMPLE C Expanded notation

1984 indicates

$$(1 \times 1000) + (9 \times 100) + (8 \times 10) + (4 \times 1)$$

or

$$(1 \times 10^3) + (9 \times 10^2) + (8 \times 10^1) + (4 \times 10^0)$$

Numbers can be added and subtracted in expanded form. This can make the rules that are used for "borrowing" or regrouping more understandable. Consider the next example.

EXAMPLE D Subtraction in Expanded Notation

Subtract these numbers in expanded notation: $645 - 439$

Solution
First write each number in expanded notation:

$$645 = (6 \times 10^2) + (4 \times 10^1) + (5 \times 10^0)$$

and

$$439 = (4 \times 10^2) + (3 \times 10^1) + (9 \times 10^0)$$

In order to subtract, the first number must be regrouped; i.e., one group of ten must be borrowed from (4×10^1), which becomes (3×10^1). The group of ten is added to (5×10^0), which becomes (15×10^0).

$$
\begin{aligned}
645 &= (6 \times 10^2) + (3 \times 10^1) + (15 \times 10^0) \\
-439 &= (4 \times 10^2) + (3 \times 10^1) + (\,9 \times 10^0) \\
\hline
&= (2 \times 10^2) + (0 \times 10^1) + (\,6 \times 10^0) \\
&= 200 + 0 + 6 \\
&= 206.
\end{aligned}
$$

Other number bases become important in work with computers. Consider some examples in converting between number bases.

EXAMPLE E Base Eight

Sometimes it is easier to evaluate the number beginning from the right:

456_{eight} indicates

$$(6 \times 8^0) + (5 \times 8^1) + (4 \times 8^2)$$

or

$$(6 \times 1) + (5 \times 8) + (4 \times 64) = 6 + 40 + 256$$
$$= 302_{\text{ten}}.$$

EXAMPLE F Conversion From Base Ten

To convert from base ten to another base such as eight, a shortcut method of repeated division by the base number can be used. To convert 85_{ten} to base eight:

$$
\begin{array}{r|ll}
8 & 85 & \text{Rem} \\
8 & 10 & \leftarrow \quad 5 \\
8 & 1 & \leftarrow \quad 2 \\
 & 0 & \leftarrow \quad 1
\end{array}
$$

Read the remainder column from the bottom to the top for the answer 125_{eight}.

EXAMPLE G Conversion to base ten

To change a number to base ten, the symbols must be interpreted appropriately. To change $3BC_{\text{sixteen}}$ to base ten, each symbol represents a different power of 16. Beginning at the right, the first position represents 16^0; the position to its left represents 16^1, etc. Also, because there are only 10 digits (0, 1, 2, 3, 4, 5, 6, 7, 8, and 9), in base sixteen it is necessary to have six more symbols to represent 10–15. The letters A–F are commonly used for these symbols.

Now look at the conversion of this number to base ten.

Solution

$$
\begin{aligned}
3BC_{\text{sixteen}} &= (3 \times 16^2) + (11 \times 16^1) + (12 \times 16^0) \\
&= 768 + 176 + 12 \\
&= 956_{\text{ten}}.
\end{aligned}
$$

A mathematical system is composed of three parts: a set of elements; one or more operations for combining the elements; and one or more relations between the elements. The formal properties of mathematical systems are closure, commutativity, associativity, identity, inverse, and the distributive property. In the following example, these properties are explored to see if the mathematical system forms a group.

EXAMPLE H   Testing a System to See if it a Group

The set of elements are $\{0, 1\}$ and the operation is addition.

Form the addition table for this system and check the necessary four properties.

+	0	1
0	0	1
1	1	2

1. Closure: The set is not closed. Closure indicates that whenever the operation is applied to any two elements of the set, the result is another element in the set. Because $1 + 1 = 2$, and 2 is not an element of the given set, the system is not closed.

2. Associative property: This property holds because the grouping of the elements can be changed without changing the result of the operation. For example,

$$(1 + 0) + 1 = 1 + (0 + 1)$$
$$(1) + 1 = 1 + (1)$$
$$2 = 2$$

3. Identity property: The identity element is 0 because if 0 is added to any element in the set, the result is the original element.

4. Inverse property: The inverse property uses the identity element in its explanation. If the inverse is added to an element, the result should be the identity element. The inverse property is not satisfied because although 0 has an inverse of 0, 1 has no inverse.

Therefore, $\{0, 1\}$ under addition does not form a group because closure is not satisfied and the inverse property is not satisfied.

Another property of a mathematical system is the distributive property of multiplication over addition. This property is demonstrated in the following example.

EXAMPLE I   Distributive Property

3×95 can be interpreted as

$$3(90 + 5) = 3 \times 90 + 3 \times 5$$
$$= 270 + 15$$
$$= 285.$$

A group is defined as a mathematical system that satisfies these properties: closure, associative, identity, and inverse. For example, the set of integers under the operation of addition is a group because it satisfies all four properties.

4.1 EXERCISES

For Reference:

EGYPTIAN

Number	Symbol	Description
1	│	Stroke
10	∩	Heel Bone
100	ꝯ	Scroll
1000	⚱	Lotus Flower
10,000	ℓ	Pointing Finger
100,000	ᗡ	Burbot Fish
1,000,000	⚐	Astonished Person

CHINESE

Number	Symbol
0	零
1	～
2	弌
3	三
4	田
5	上
6	六
7	七
8	八
9	九
10	十
100	百
1000	千

1. $(1 \times 10,000) + (3 \times 1000) + (0 \times 100) + (3 \times 10)$
    $+ (6 \times 1) = 13,036$

3. $(7 \times 1,000,000) + (6 \times 100,000) + (3 \times 10,000)$
    $= (0 \times 1000) + (7 \times 100) + (2 \times 10) + (9 \times 1)$
    $= 7,630,729$

5. ℓℓ ⚱⚱⚱ ꝯ ∩∩∩∩ │││││

7. ⚐⚐⚐⚐⚐⚐⚐ ᗡᗡᗡᗡᗡᗡ ℓℓℓℓ
    ⚱⚱⚱⚱⚱⚱⚱

9. ℓℓℓℓℓ ⚱⚱⚱⚱⚱⚱⚱⚱ ꝯꝯꝯ

11. ℓℓℓℓℓℓℓ ⚱⚱⚱⚱ ꝯꝯꝯꝯꝯꝯ

13. ℓℓℓℓℓ ⚱⚱ ꝯꝯꝯꝯꝯꝯꝯ

15. $(9 \times 100) + (3 \times 10) + (5 \times 1) = 935$

17. $(3 \times 1000) + (7 \times 1) = 3007$

19. 九
    百
    六
    十

21. 七
    十
    零
    十
    千

23. ～   ～
    十   to   十
    三   六
    百   百
    六   田
    十   十
    八   田

25. 六   to   九
    百   百
    十   零
    八   七

27. There is a total of one scroll, eleven heelbones, and six
    strokes. Group ten heelbones to create a second scroll.

    $$(2 \times 100) + (1 \times 10) + (6 \times 1) = 200 + 10 + 6$$
    $$= 216$$

29. There is a total of five pointing fingers, three lotus
    flowers, five scrolls, nine heelbones, and eleven strokes.
    Group ten strokes to create another heelbone. The ten
    heelbones then create another scroll.

    $$(5 \times 10,000) + (3 \times 1000) + (6 \times 100) + (0 \times 10)$$
    $$+ (1 \times 1) = 50000 + 3000 + 600 + 0 + 1$$
    $$= 53,601$$

31. After subtracting, there is one scroll, one heelbone, and
    three strokes.

    $$(1 \times 100) + (1 \times 10) + (3 \times 1) = 100 + 10 + 3$$
    $$= 113$$

33. Regroup the pointing finger to make ten lotus flowers for
    a total of eleven. Then one lotus flower must be
    regrouped to make ten scrolls for a total of twelve, and
    one scroll must be regrouped to make ten heelbones.
    Regroup one heelbone to make ten strokes. Then ten
    lotus flowers less three yields seven; eleven scrolls less
    six yields five; nine heelbones remain; fourteen strokes
    less six yields eight.

$$(7 \times 1000) + (5 \times 100) + (9 \times 10) + (8 \times 1)$$
$$= 7000 + 500 + 90 + 8$$
$$= 7598$$

35. Form two columns, headed by 1 and 53. Keep doubling each row until there are numbers in the first column that add up to 26.

	1	53	
→	2	106	←
	4	212	
→	8	424	←
→	16	848	←

$2 + 8 + 16 = 26$. Then add corresponding numbers from the second column:

$$106 + 424 + 848 = 1378.$$

37. Form two columns, headed by 1 and 103. Keep doubling each row until there are numbers in the first column that add up to 58:

	1	103	
→	2	206	←
	4	412	
→	8	824	←
→	16	1648	←
→	32	3296	←

$2 + 8 + 16 + 32 = 58$. Then add corresponding numbers from the second column:

$$206 + 824 + 1648 + 3296 = 5974.$$

39.

thirty golden basins	∩∩∩
a thousand silver basins	⌐
four hundred ten silver bowls	9999 ∩
thirty golden bowls	∩∩∩
3000 shekels	⌐⌐⌐
500 shekels	99999
50 shekels	∩∩∩∩∩
400 shekels	9999

30×3000

	1	3000	
→	2	6000	←
→	4	12,000	←
→	8	24,000	←
→	16	48,000	←

$$30 \times 3000 = 6000 + 12,000 + 24,000 + 48,000$$
$$= 90,000$$

500×1000

	1	1000	
	2	2000	
→	4	4000	←
	8	8000	
→	16	16,000	←
→	32	32,000	←
→	64	64,000	←
→	128	128,000	←
→	256	256,000	←

$$500 \times 1000 = 4000 + 16,000 + 32,000 + 64,000 +$$
$$128,000 + 256,000 = 500,000$$

50×410

	1	410	
→	2	820	←
	4	1640	
	8	3280	
→	16	6560	←
→	32	13,120	←

$$50 \times 410 = 820 + 6560 + 13,120 = 20,500$$

30×400

	1	400	
→	2	800	←
→	4	1600	←
→	8	3200	←
→	16	6400	←

$$30 \times 400 = 800 + 1600 + 3200 + 6400 = 12,000$$

Now add 90,000, 500,000, 20,500, and 12,000 using Egyptian symbols.

ℓℓℓℓℓℓℓℓ + ⌣⌣⌣⌣⌣ + ℓℓ
99999 + ℓ ⌐⌐ = ⌣⌣⌣⌣⌣
ℓℓℓℓℓℓℓℓℓℓ ⌐⌐ 99999

Regroup ten pointing fingers to one burbot fish.

⌣⌣⌣⌣⌣⌣ ℓℓ ⌐⌐ 99999

The total value of the treasure is

$$(6 \times 100,000) + (2 \times 10,000) + (2 \times 1000)$$
$$+ (5 \times 100) = 622,500 \text{ shekels.}$$

41. Writing exercise

43. Writing exercise

45. 99,999. Five distinct symbols allows only five positions.

47. The largest number is $44,444_{\text{five}}$, which is equivalent to 3124_{ten}.

49. $10^d - 1$. Examine Exercise 45 to see that
$10^5 - 1 = 100.000 - 1 = 99,999$.

51. $7^d - 1$.

53. Writing exercise

4.2 EXERCISES

1. $73 = (7 \times 10) + (3 \times 1) = (7 \times 10^1) + (3 \times 10^0)$

3. $3774 = (3 \times 1000) + (7 \times 100) + (7 \times 10)$
$+ (4 \times 1)$
$= (3 \times 10^3) + (7 \times 10^2) + (7 \times 10^1)$
$+ (4 \times 10^0)$

5. $4924 = (4 \times 1000) + (9 \times 100) + (2 \times 10)$
$+ (4 \times 1)$
$= (4 \times 10^3) + (9 \times 10^2) + (2 \times 10^1)$
$+ (4 \times 10^0)$.

7. $14,206,040 = (1 \times 10,000,000) + (4 \times 1,000,000)$
$+ (2 \times 100,000) + (0 \times 10,000)$
$+ (6 \times 1000) + (0 \times 100)$
$+ (4 \times 10) + (0 \times 1)$
$= (1 \times 10^7) + (4 \times 10^6) + (2 \times 10^5)$
$+ (0 \times 10^4) + (6 \times 10^3)$
$+ (0 \times 10^2) + (4 \times 10^1)$
$+ (0 \times 10^0)$

9. $(4 \times 10) + (2 \times 1) = 42$

11. $(6 \times 1000) + (2 \times 100) + (9 \times 1) = 6209$

13. $(7 \times 10,000,000) + (4 \times 100,000) + (1 \times 1000)$
$+ (9 \times 1) = 70,401,009$

15. $54 = (5 \times 10^1) + (4 \times 10^0)$
$+ 35 = (3 \times 10^1) + (5 \times 10^0)$
$= (8 \times 10^1) + (9 \times 10^0)$
$= 80 + 9$
$= 89$

17. $85 = (8 \times 10^1) + (5 \times 10^0)$
$- 53 = (5 \times 10^1) + (3 \times 10^0)$
$= (3 \times 10^1) + (2 \times 10^0)$
$= 30 + 2$
$= 32$

19. $75 = (7 \times 10^1) + (5 \times 10^0)$
$+ 34 = (3 \times 10^1) + (4 \times 10^0)$
$= (10 \times 10^1) + (9 \times 10^0)$
$= (1 \times 10^2) + (9 \times 10^0)$
$= 100 + 9$
$= 109$

21. $434 = (4 \times 10^2) + (3 \times 10^1) + (4 \times 10^0)$
$+ 299 = (2 \times 10^2) + (9 \times 10^1) + (9 \times 10^0)$
$= (6 \times 10^2) + (12 \times 10^1) + (13 \times 10^0)$
$= (6 \times 10^2) + (12 \times 10^1) + (1 \times 10^1)$
$+ (3 \times 10^0)$
$= (6 \times 10^2) + (13 \times 10^1) + (3 \times 10^0)$
$= (6 \times 10^2) + (1 \times 10^2) + (3 \times 10^1)$
$+ (3 \times 10^0)$
$= (7 \times 10^2) + (3 \times 10^1) + (3 \times 10^0)$
$= 700 + 30 + 3$
$= 733$.

23. $54 = (5 \times 10^1) + (4 \times 10^0)$
$- 48 = (4 \times 10^1) + (8 \times 10^0)$

Since, in the units position, we cannot subtract 8 from 4, we use the distributive property to modify the top expansion as follows:

$(4 \times 10^1) + (1 \times 10^1) + (4 \times 10^0)$
$(4 \times 10^1) + (10 \times 10^0) + (4 \times 10^0)$

$54 = (4 \times 10^1) + (14 \times 10^0)$
$- 48 = (4 \times 10^1) + (8 \times 10^0)$
$= (6 \times 10^0)$
$= 6$.

25. $645 = (6 \times 10^2) + (4 \times 10^1) + (5 \times 10^0)$
$- 439 = (4 \times 10^2) + (3 \times 10^1) + (9 \times 10^0)$

Since, in the units position, we cannot subtract 9 from 5, we use the distributive property to modify the top expansion as follows.

$(6 \times 10^2) + (3 \times 10^1) + (1 \times 10^1) + (5 \times 10^0)$
$(6 \times 10^2) + (3 \times 10^1) + (10 \times 10^0) + (5 \times 10^0)$

$645 = (6 \times 10^2) + (3 \times 10^1) + (15 \times 10^0)$
$- 439 = (4 \times 10)^2 + (3 \times 10^1) + (9 \times 10^0)$
$= (2 \times 10)^2 + (0 \times 10^1) + (6 \times 10^0)$
$= 200 + 6$
$= 206$

27. Reading the abacus from the right. The number represented by this abacus is
$$[(1 \times 5) + (1 \times 1)] + (1 \times 50) + (2 \times 100)$$
$$= 6 + 50 + 200 = 256.$$

29. The number represented by this abacus is
$$[(1 \times 5) + (4 \times 1)] + (1 \times 50) + (2 \times 100)$$
$$+ (3 \times 1000) + [(1 \times 50,000) + (1 \times 10,000)]$$
$$= 4 + 5 + 50 + 200 + 3000 + 50,000 + 10,000$$
$$= 63,259.$$

31. $38 = (3 \times 10) + [(1 \times 5) + (3 \times 1)]$

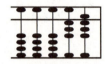

33. $2547 = (2 \times 1000) + (1 \times 500) + (4 \times 10)$
$$+ [(1 \times 5) + (2 \times 1)]$$

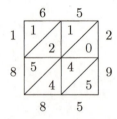

35. 65×29 is written around the top and right side.

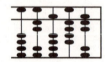

Obtain the numbers inside each box by finding the products of all the pairs of digits on the top and side: $6 \times 2 = 12$; $5 \times 2 = 10$ etc. Then add diagonally starting from the bottom right, placing the sums outside. For example, $0 + 4 + 4 = 8$. Finally, read the answer around the left side and the bottom as 1885.

37. 525×73 is written around the top and right side.

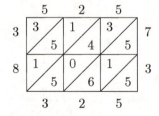

Find each number inside the boxes by finding the product of all the pairs of digits on the top and side. Then add diagonally beginning from the bottom right. For example, $5 + 1 + 6 = 12$. Write the 2 outside the box and carry the 1. Now add $1 + 3 + 4 + 5 = 13$.

Again carry to the next diagonal above. Read the answer around the left side and the bottom as 38,325.

39. 723×4198 is written around the top and right side.

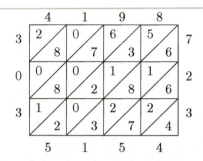

Find each number inside the boxes by finding the product of all the pairs of digits on the top and side. Then add diagonally beginning from the bottom right. For example, $6 + 2 + 7 = 15$. Write the 5 outside the box and carry the 1. Now add $1 + 6 + 1 + 8 + 2 + 3 = 21$. Again carry the 2 to the next diagonal above. Read the answer around the left side and the bottom as 3,035,154.

41. Select the rods for 6 and 2 and place them side by side. Use the index to locate the row or level for a multiplier of 8.

Index	6	2
1	0 ⋰ 6	0 ⋰ 2
2	1 ⋰ 2	0 ⋰ 4
3	1 ⋰ 8	0 ⋰ 6
4	2 ⋰ 4	0 ⋰ 8
5	3 ⋰ 0	1 ⋰ 0
6	3 ⋰ 6	1 ⋰ 2
7	4 ⋰ 2	1 ⋰ 4
→ 8	4 ⋰ 8	1 ⋰ 6
9	5 ⋰ 4	1 ⋰ 8

The resulting lattice is shown below.

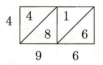

The product of 8 and 62 is 496.

43. Select the rods for 8, 3, 5, and 4 and place them side by side. Use the index to first locate the row or level for multipliers of 2 and 6.

Index	8	3	5	4
1	0 8	0 3	0 5	0 4
→ 2	1 6	0 6	1 0	0 8
3	2 4	0 9	1 5	1 2
4	3 2	1 2	2 0	1 6
5	4 0	1 5	2 5	2 0
→ 6	4 8	1 8	3 0	2 4
7	5 6	2 1	3 5	2 8
8	6 4	2 4	4 0	3 2
9	7 2	2 7	4 5	3 6

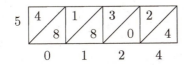

The product $6 \times 8354 = 50,124$.

Find the product of 2×8354 in a similar way, but using the index for a multiplier of 2.

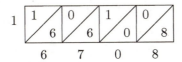

To create the table below write the multiplicand on the top row and the multiplier in the right hand column as shown. Insert the product of 6 and 8354 as the first entry. Insert the product of 2 and 8354 as the second entry, shifted one column to the left because it is actually 20×8354. The final answer is found by addition; $26 \times 8354 = 217,204$.

$$
\begin{array}{r}
8354 \\
\hline
50124 \,|\, 6 \\
16708 \,|\, 2 \\
\hline
217204
\end{array}
$$

45. Complete missing place value with 0.

$$
\begin{array}{r}
283 \\
- 041 \\
\end{array}
$$

Replace digits in subtrahend (041) with the nines complement of each and add.

$$
\begin{array}{r}
283 \\
+ 958 \\
\hline
1241
\end{array}
$$

Delete the first digit on left and add that 1 to the remaining part of the sum: $241 + 1 = 242$.

47. Complete missing place values with 0.

$$
\begin{array}{r}
50000 \\
- 00199 \\
\end{array}
$$

Replace digits in subtrahend (00199) with the nines complement of each and add.

$$
\begin{array}{r}
50000 \\
+ 99800 \\
\hline
149800
\end{array}
$$

Delete the first digit on left and add that 1 to the remaining part of the sum: $49,800 + 1 = 49,801$.

49. To multiply 5 and 92 using the Russian peasant method, write each number at the top of a column.

$$
\begin{array}{rrl}
\rightarrow & 5 & 92 \quad \leftarrow \\
 & 2 & 184 \\
\rightarrow & 1 & 368 \quad \leftarrow \\
\end{array}
$$

Divide the first column by 2 and double the second column until 1 is obtained in the first column. Ignore the remainders when dividing. Add the numbers in the second column that correspond to the odd numbers in the first: $92 + 368 = 460$.

51. To multiply 62 and 529 using the Russian peasant method, write each number at the top of a column.

$$
\begin{array}{rrl}
 & 62 & 529 \\
\rightarrow & 31 & 1058 \quad \leftarrow \\
\rightarrow & 15 & 2116 \quad \leftarrow \\
\rightarrow & 7 & 4232 \quad \leftarrow \\
\rightarrow & 3 & 8464 \quad \leftarrow \\
\rightarrow & 1 & 16,928 \quad \leftarrow \\
\end{array}
$$

Divide the first column by 2 and double the second column until 1 is obtained in the first column. Ignore the remainders when dividing. Add the numbers in the second column that correspond to the odd numbers in the first column.

$$1058 + 2116 + 4232 + 8464 + 16,928 = 32,798.$$

4.3 EXERCISES

1. 1, 2, 3, 4, 5, and 6 are the first six digits. To represent the number seven, 10 is used meaning $(1 \times 7^1) + (0 \times 7^0)$. The next six numbers would be 11, 12, 13, 14, 15, and 16. To express the number fourteen, 20 is used meaning $(2 \times 7^1) + (0 \times 7^0)$. Continue in this pattern: 21, 22, 23, 24, 25, 26.

3. 1, 2, 3, 4, 5, 6, 7, and 8 are the first eight digits. To represent the number nine, 10 is used which means $(1 \times 9^1) + (0 \times 9^0)$. The next eight numbers are 11, 12, 13, 14, 15, 16, 17, 18. To express the number eighteen, 20 is used which means $(2 \times 9^1) + (0 \times 9^0)$. Continue in this pattern: 21, 22.

5. 13_{five} is the number just before, and 20_{five} is the number just after the given number.

7. $B6E_{\text{sixteen}}$ is the number just before, and $B70_{\text{sixteen}}$ is the number just after the given number.

9. Three distinct symbols are needed.

11. Eleven distinct symbols are needed.

13. The smallest four-digit number in base three is 1000, which means $(1 \times 3^3) = 27$. The largest four-digit number is 2222, which means $(2 \times 3^3) + (2 \times 3^2) + (2 \times 3^1) + (2 \times 3^0)$. This is equivalent to $54 + 18 + 6 + 2 = 80$.

15. $(2 \times 5^1) + (4 \times 5^0) = 10 + 4 = 14$

Using the calculator shortcut: $(2 \times 5) + 4 = 14$.

17. $2^3 + 2^1 + 2^0 = 8 + 2 + 1 = 11$

Using the calculator shortcut:
$[(1 \times 2 + 0) \times 2 + 1] \times 2 + 1$
$= [5] \times 2 + 1$
$= 11$.

19. $(3 \times 16^2) + (11 \times 16^1) + (12 \times 16^0)$
$= 3 \times 256 + 11 \times 16 + 12$
$= 956$

Using the calculator shortcut:
$(3 \times 16 + 11) \times 16 + 12 = 956$.

21. $(2 \times 7^3) + (3 \times 7^2) + (6 \times 7^1) + (6 \times 7^0)$
$= 686 + 147 + 42 + 6$
$= 881$

Using the calculator shortcut:
$[(2 \times 7 + 3) \times 7 + 6] \times 7 + 6 = [125] \times 7 + 6 = 881$.

23. $(7 \times 8^4) + (0 \times 8^3) + (2 \times 8^2) + (6 \times 8^1)$
$\qquad + (6 \times 8^0)$
$= 28,672 + 128 + 48 + 6$
$= 28,854$

Using the calculator shortcut:
$\{[(7 \times 8 + 0) \times 8 + 2] \times 8 + 6\} \times 8 + 6 = 28,854$.

25. $(2 \times 4^3) + (0 \times 4^2) + (2 \times 4^1) + (3 \times 4^0)$
$= 128 + 8 + 3 = 139$

Using the calculator shortcut:
$[(2 \times 4 + 0) \times 4 + 2] \times 4 + 3 = 139$.

27. $(4 \times 6^4) + (1 \times 6^3) + (5 \times 6^2) + (3 \times 6^1)$
$\qquad + (3 \times 6^0)$
$= 5184 + 216 + 180 + 18 + 3$
$= 5601$

Using the calculator method:
$\{[(4 \times 6 + 1) \times 6 + 5] \times 6 + 3\} \times 6 + 3 = 5601$.

29. The base five place values, starting from the right, are 1, 5, 25, 125, and so on. Since 86 is between 25 and 125, we will need some 25's but no 125's. Begin by dividing 86 by 25; then divide the remainder obtained from this division by 5. Finally, divide the remainder obtained from the previous division by 1, giving a remainder of 0.

$$86 \div 25 = 3, \text{ remainder } 11$$
$$11 \div 5 = 2, \text{ remainder } 1$$
$$1 \div 1 = 1, \text{ remainder } 0$$

The digits of the answer are found by reading quotients from the top down.
$86 = 321_{\text{five}}$

Shortcut:

	5|86		Rem
	5|17	←	1
	5|3	←	2
	0	←	3

Read the answer from the remainder column, reading from the bottom up.

$$86_{\text{ten}} = 321_{\text{five}}.$$

31.

$$
\begin{array}{c|c c}
2 & 19 & \text{Rem} \\
2 & 9 & \leftarrow & 1 \\
2 & 4 & \leftarrow & 1 \\
2 & 2 & \leftarrow & 0 \\
2 & 1 & \leftarrow & 0 \\
& 0 & \leftarrow & 1 \\
\end{array}
$$

$19_{\text{ten}} = 10011_{\text{two}}$

33.

$$
\begin{array}{c|c c}
16 & 147 & \text{Rem} \\
16 & 9 & \leftarrow & 3 \\
& 0 & \leftarrow & 9 \\
\end{array}
$$

$147_{\text{ten}} = 93_{\text{sixteen}}$

35.

$$
\begin{array}{c|c c}
5 & 36401 & \text{Rem} \\
5 & 7280 & \leftarrow & 1 \\
5 & 1456 & \leftarrow & 0 \\
5 & 291 & \leftarrow & 1 \\
5 & 58 & \leftarrow & 1 \\
5 & 11 & \leftarrow & 3 \\
5 & 2 & \leftarrow & 1 \\
& 0 & \leftarrow & 2 \\
\end{array}
$$

$36,401_{\text{ten}} = 2131101_{\text{five}}$

37.

$$
\begin{array}{c|c c}
2 & 586 & \text{Rem} \\
2 & 293 & \leftarrow & 0 \\
2 & 146 & \leftarrow & 1 \\
2 & 73 & \leftarrow & 0 \\
2 & 36 & \leftarrow & 1 \\
2 & 18 & \leftarrow & 0 \\
2 & 9 & \leftarrow & 0 \\
2 & 4 & \leftarrow & 1 \\
2 & 2 & \leftarrow & 0 \\
2 & 1 & \leftarrow & 0 \\
& 0 & \leftarrow & 1 \\
\end{array}
$$

$586_{\text{ten}} = 1001001010_{\text{two}}$

39.

$$
\begin{array}{c|c c}
3 & 8407 & \text{Rem} \\
3 & 2802 & \leftarrow & 1 \\
3 & 934 & \leftarrow & 0 \\
3 & 311 & \leftarrow & 1 \\
3 & 103 & \leftarrow & 2 \\
3 & 34 & \leftarrow & 1 \\
3 & 11 & \leftarrow & 1 \\
3 & 3 & \leftarrow & 2 \\
3 & 1 & \leftarrow & 0 \\
& 0 & \leftarrow & 1 \\
\end{array}
$$

$8407_{\text{ten}} = 102112101_{\text{three}}$

41.

$$
\begin{array}{c|c c}
6 & 9346 & \text{Rem} \\
6 & 1557 & \leftarrow & 4 \\
6 & 259 & \leftarrow & 3 \\
6 & 43 & \leftarrow & 1 \\
6 & 7 & \leftarrow & 1 \\
6 & 1 & \leftarrow & 1 \\
& 0 & \leftarrow & 1 \\
\end{array}
$$

$9346_{\text{ten}} = 111134_{\text{six}}$

43. First convert 43_{five} to base ten.

$$(4 \times 5) + 3 = 23$$

Then convert 23 to base seven.

$$
\begin{array}{c|c c}
7 & 23 & \text{Rem} \\
7 & 3 & \leftarrow & 2 \\
& 0 & \leftarrow & 3 \\
\end{array}
$$

$43_{\text{five}} = 32_{\text{seven}}$

45. First convert 6748_{nine} to base ten.

$$(6 \times 9^3) + (7 \times 9^2) + (4 \times 9) + 8 = 4985$$

Then convert 4985 to base four.

$$
\begin{array}{c|c c}
4 & 4985 & \text{Rem} \\
4 & 1246 & \leftarrow & 1 \\
4 & 311 & \leftarrow & 2 \\
4 & 77 & \leftarrow & 3 \\
4 & 19 & \leftarrow & 1 \\
4 & 4 & \leftarrow & 3 \\
4 & 1 & \leftarrow & 0 \\
& 0 & \leftarrow & 1 \\
\end{array}
$$

$6748_{\text{nine}} = 1031321_{\text{four}}$

47. Replace each octal digit with its 3-digit binary equivalent. Then combine all the binary equivalents into a single binary numeral.

$$\begin{array}{ccc} 3 & 6 & 7 \\ \downarrow & \downarrow & \downarrow \\ 011 & 110 & 111 \end{array}$$

$367_{\text{eight}} = 11110111_{\text{two}}$

49. Starting at the right, break the digits into groups of three. Then convert the groups to their octal equivalents. (Refer to Table 7.)

$$\begin{array}{ccc} 100 & 110 & 111 \\ \downarrow & \downarrow & \downarrow \\ 4 & 6 & 7 \end{array}$$

$100110111_{\text{two}} = 467_{\text{eight}}$

51. Each hexadecimal digit yields a 4-digit binary equivalent. (See Table 8.)

$$\begin{array}{cc} D & C \\ \downarrow & \downarrow \\ 1101 & 1100 \end{array}$$

$DC_{\text{sixteen}} = 11011100_{\text{two}}$

53. Starting at the right, break the digits into groups of four. Then convert the groups to their hexadecimal equivalent. (Refer to Table 8.)

$$\begin{array}{cc} 10 & 1101 \\ \downarrow & \downarrow \\ 2 & D \end{array}$$

$101101_{\text{two}} = 2D_{\text{sixteen}}$

55. In order to compare these numbers, we need to write them in the same base. Convert each of them to decimal form (base ten).

$$\left(4 \times 7^1\right) + \left(2 \times 7^0\right) = 28 + 2 \text{ or } 30$$
$$\left(3 \times 8^1\right) + \left(7 \times 8^0\right) = 24 + 7 \text{ or } 31$$
$$\left(1 \times 16^1\right) + \left(13 \times 16^0\right) = 16 + 13 \text{ or } 29$$

The largest number is 37_{eight}.

57. $\left(9 \times 12^2\right) + \left(10 \times 12\right) + 11 = 1427$ copies

59. Since A is assigned the number 65, C is assigned the number 67. Change 67 from decimal form to binary form.

$$\begin{array}{rcl} 2\underline{|67} & & \text{Rem} \\ 2\underline{|33} & \leftarrow & 1 \\ 2\underline{|16} & \leftarrow & 1 \\ 2\underline{|8} & \leftarrow & 0 \\ 2\underline{|4} & \leftarrow & 0 \\ 2\underline{|2} & \leftarrow & 0 \\ 2\underline{|1} & \leftarrow & 0 \\ 0 & \leftarrow & 1 \end{array}$$

$C = 1000011_{\text{two}}$

61. Since a is assigned the number 97, k is assigned the number 107. (Since k is the eleventh letter of the alphabet, its corresponding number will be ten more than the number corresponding to a.) Change 107 to binary form.

$$\begin{array}{rcl} 2\underline{|107} & & \text{Rem} \\ 2\underline{|53} & \leftarrow & 1 \\ 2\underline{|26} & \leftarrow & 1 \\ 2\underline{|13} & \leftarrow & 0 \\ 2\underline{|6} & \leftarrow & 1 \\ 2\underline{|3} & \leftarrow & 0 \\ 2\underline{|1} & \leftarrow & 1 \\ 0 & \leftarrow & 1 \end{array}$$

$k = 1101011_{\text{two}}$

63. Convert each seven-digit binary number to decimal form; then find the corresponding letters.

Base Two	Base Ten	Letter
1000011	$64 + 2 + 1 = 67$	C
1001000	$64 + 8 = 72$	H
1010101	$64 + 16 + 4 + 1 = 85$	U
1000011	$64 + 2 + 1 = 67$	C
1001011	$64 + 8 + 2 + 1 = 75$	K

The given number represents CHUCK.

65.

Letter	Base Ten	Base Two
O	79	1001111
r	114	1110010
l	108	1101100
e	101	1100101
a	97	1100001
n	110	1101110
s	115	1110011

The word "Orleans" is represented by the following ASCII string:

O	r	l	e
1001111	1110010	1101100	1100101
a	n	s	
1100001	1101110	1110011.	

67. The largest base eight number that consists of two digits is 77_{eight}, which is equivalent to 63_{ten}. In base three this number is 2100_{three}. The smallest base eight number that consists of two nonzero digits is 11_{eight}, which is equivalent to 9_{ten}. However, in base three this number has only three digits, 100_{three}. In base three the smallest four-digit number is 1000_{three}, which equals 27_{ten}. In base eight, this number is 33_{eight}. The smallest number, then, is 27 and the largest, 63.

69. (a) The binary ones digit is 1.
    (b) The binary twos digit is 1.
    (c) The binary fours digit is 1.
    (d) The binary eights digit is 1.
    (e) The binary sixteens digit is 1.

71. In order to include all ages up to 63, we will need to add one more column to Table 9. This column will contain the numbers whose binary thirty-twos digit is 1. Thus, 6 columns would be needed.

73. In base two, every even number has 0 as its ones digit and every odd number has a 1 as its ones digit. Thus, we can distinguish odd and even numbers by looking at their ones digit. The criterion works.

75. In base four, every even number has 0 or 2 as its ones digit, while every odd number has 1 or 3 as its ones digit. Thus, we can distinguish odd and even numbers by looking at their ones digit. The criterion works.

77. In base six, every even number has 0, 2, or 4 as its ones digit, while every odd number has 1, 3, or 5 as its ones digit. The criterion works.

79. In base eight, every even number has 0, 2, 4, or 6 as its ones digit, while every odd number has 1, 3, 5, or 7 as its ones digit. The criterion works.

81. Writing exercise

83. A numeral in any base is divisible by that base only if the numeral ends in 0. Thus, a base five numeral is divisible by 5 if it ends in 0. Therefore, the only numerals on the list that are divisible by 5 are 200 and 2310.

4.4 EXERCISES

1. All properties are satisfied. 1 is the identity element; 1 is its own inverse, as is 2.

3. All properties are satisfied except the inverse. 1 is the identity element; 2, 4, and 6 have no inverses.

5. All properties are satisfied except the inverse. 1 is the identity element; 5 has no inverse.

7. All properties are satisfied. F is the identity element. A and B are inverses; F is its own inverse.

9. All properties are satisfied. t is the identity element. s and r are inverses; t and u are their own inverses.

11. The letter b represents a rotation of $90°$ and d represents a rotation of $270°$. If both rotations are performed, the square returns to its original position. The answer, then, is $b \square d = a$.

13. A rotation of $270°$ (represented by d) followed by a rotation of $90°$ (b), again returns the square to its original position. The answer, then, is $d \square b = a$.

15.

$\square$	a	b	c	d
a	a	b	c	d
b	b	c	$\underline{d}$	a
c	c	$\underline{d}$	a	$\underline{b}$
d	d	a	$\underline{b}$	$\underline{c}$

17. Call the new set $\{U, \phi, \{1\}, \{2\}, \dots\}$ U star.
    Closure: Yes. The set U^* is closed because the intersection of any two members in U^* is a member of U^*.
    Commutative: Yes. The order of the intersection does not change the outcome.
    Associative: Yes. The grouping can be changed without affecting the outcome.
    Identity: Yes. For the operation intersection, the identity is U.

19. Here is one possibility.

	a	b	c	d
a	a	b	c	d
b	b	a	d	c
c	c	d	a	b
d	d	c	b	a

21. Try $2 + (6 - 4)$. If the distributive property of addition over subtraction did hold, this expression would equal $2 + 6 + 2 - 4$. The first expression $2 + (6 - 4) = 4$. But $2 + 6 + 2 - 4 = 6$. The property does not hold.

23. (a) $a = 2, b = -5, c = 4$

$$2 + (-5 \times 4) = 2 + (-20) = -18$$
$$(2 + (-5)) \times (2 + 4) = -3 \times 6 = -18$$

The equation is true for these values.

(b) $a = -7, b = 5, c = 3$

$$-7 + (5 \times 3) = -7 + (15) = 8$$
$$(-7 + 5) \times (-7 + 3) = -2 \times -4 = 8$$

The equation is true for these values.

(c) $a = -8, b = 14, c = -5$

$$-8 + (14 \times (-5)) = -8 + (-70) = -78$$
$$(-8 + 14) \times (-8 + (-5)) = 6 \times -13 = -78$$

The equation is true for these values.

(d) $a = 1, b = 6, c = -6$

$$1 + (6 \times (-6)) = 1 + (-36) = -35$$
$$(1 + 6) \times (1 + (-6)) = 7 \times -5 = -35$$

The equation is true for these values.

25. The statement is true when $a + b + c = 1$ or $a = 0$.

27. (a) True only when $a = 0$.

(b) True only when $a = 0$.

29. Use the charts of Example 4 to see that $d \circ e = c$. Then $c \bigstar c = e$. Examine the right side of the equation to see that $c \bigstar d = b$ and $c \bigstar e = d$. Then $b \circ d = e$. Each side simplifies to e.

31. $d \bigstar (e \circ c) = d \bigstar b = d$ and $(d \bigstar e) \circ (d \bigstar c) = c \circ b = d$. The left side of the equation is equivalent to the right side.

33.

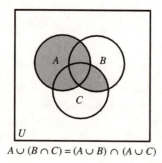

$A \cup (B \cap C) = (A \cup B) \cap (A \cup C)$

35.

p	q	r	$p \vee$	$(q \wedge r)$	$(p \vee q)$	$\wedge$	$(p \vee r)$
T	T	T	T	T	T	T	T
T	T	F	T	F	T	T	T
T	F	T	T	F	T	T	T
F	T	T	T	T	T	T	T
T	F	F	T	F	T	T	T
F	T	F	F	F	T	F	F
F	F	T	F	F	F	F	T
F	F	F	F	F	F	F	F
			2	1	1	2	1

Compare column 2 for each logic statement. Since they are equal row by row, the distributive property is proven to hold.

4.5 EXERCISES

1. The operation is not specified.

3. Form the multiplication table for this system and check the necessary four properties.

$\times$	0
0	0

The set $\{0\}$ under multiplication forms a group. All properties are satisfied: closure, associative, identity, and inverse.

5. Form the addition table for this system and check the necessary four properties.

$+$	0	1
0	0	1
1	1	2

The set $\{0, 1\}$ under addition does not form a group because closure is not satisfied, and the inverse property is not satisfied.

7.

$\div$	-1	1
-1	1	-1
1	-1	1

This system forms a group. All properties are satisfied: closure, associative, identity, and inverse.

9.

$\times$	-1	0	1
-1	1	0	-1
0	0	0	0
1	-1	0	1

This set does not form a group because the inverse property is not satisfied.

11. The set of integers under the operation of subtraction is not a group because the associative, identity, and inverse properties are not satisfied. Here are some examples to show why these properties are not satisfied:

Associative: $(2 - 3) - 1 = -1 - 1 = -2$, but
$$2 - (3 - 1) = 2 - 2 = 0.$$

Identity: $3 - 0 = 3$, but $0 - 3 \neq 3$.

If there is no identity element, the inverse property cannot be checked.

13. The closure, associative, and identity properties are satisfied, with 1 as the identity element. However, the inverse property is not satisfied. While -1 and 1 are their own inverses, 0 has no multiplicative inverse, and the multiplicative inverses of the other odd integers are not integers at all. For example, the multiplicative inverse of 3 is 1/3, which is not an element of the set of the system. Therefore, the set of odd integers under multiplication is not a group.

15. The closure, associative, and identity properties are all satisfied, with 0 as the identity element. Every rational number p/q has an additive inverse (or "opposite") number $-p/q$ such that

$$\frac{p}{q} + \left(\frac{-p}{q}\right) = \left(\frac{-p}{q}\right) + \frac{p}{q} = 0.$$

For example,

$$\frac{3}{5} + \left(\frac{-3}{5}\right) = \left(\frac{-3}{5}\right) + \frac{3}{5} = 0.$$

Since all of the required properties are satisfied, the set of rational numbers under multiplication is a group.

17. This system is not closed, since the sum of two prime numbers is not always a prime number. For example, $5 + 7 = 12$. The identity property is not satisfied since 0 is not a prime number. Since the identity property is not satisfied, the inverse property cannot be satisfied. Therefore, the set of prime numbers under addition is not a group.

19. Writing exercise

21. Read the chart in the text to see that $RN = S$.

23. Read the chart in the text to see that $TV = N$.

25. Use the chart in the text to see that $N(TR) = NP$ or M; also, $(NT)R = VR$ or M.

27. $T(VN) = TT = Q$; $(TV)N = NN = Q$.

29. The identity element is Q. Then the inverse of N is N, because $NN = Q$.

31. The inverse of R is R, because $RR = Q$.

33. The inverse of T is T, because $TT = Q$.

35. The symmetries of a square is not commutative. If the order in which the operation is done is changed, the answer changes. See the table in Example 4 to see that $R \, \square \, P = V$, but $P \, \square \, R = T$.

37. This group is commutative. For example, $4 + 2 = 6$ and $2 + 4 = 6$.

39. Writing exercise

41. Writing exercise

42. Writing exercise

CHAPTER 4 TEST

1. These are symbols from the Egyptian numeration system. The lotus flower represents 1000; the scroll represents 100; the heelbone represents 10; the stroke represents 1. Therefore, the number is 2536.

2. $(8 \times 1000) + (3 \times 100) + (6 \times 10) + (4 \times 1) = 8364$

3. $(6 \times 10,000) + (0 \times 1000) + (9 \times 100) + (2 \times 10) + (3 \times 1)$
$$= (6 \times 10^4) + (0 \times 10^3) + (9 \times 10^2) + (2 \times 10^1) + (3 \times 10^0)$$

4. Using the Egyptian method, form two columns, headed by 1 and 54. Keep doubling each row until there are numbers in the first column that add up to 37.

→	1	54 ←
	2	108
→	4	216 ←
	8	432
	16	864
→	32	1728 ←

$1 + 4 + 32 = 37$. Then add corresponding numbers from the second column: $54 + 216 + 1728 = 1998$.

5. 236×94

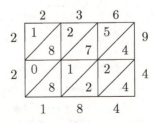

Read the answer around the left side and bottom: 22,184.

6. $21,325 - 8498$

    Complete missing place value with 0.

    $$21325$$
    $$- 08498$$

    Replace digits in subtrahend (08498) with the nines complement of each and add.

    $$21325$$
    $$+ 91501$$
    $$\overline{112826}$$

    Delete the first digit on left and add that 1 to the remaining part of the sum: $12,826 + 1 = 12,827$.

7. 424_{five} is equivalent to
    $(4 \times 5^2) + (2 \times 5^1) + (4 \times 5^0)$.

    Then
    $(4 \times 25) + (2 \times 5) + (4 \times 1) = 100 + 10 + 4 = 114$.

8. 100110_{two} is equivalent to
    $(1 \times 2^5) + (0 \times 2^4) + (0 \times 2^3) + (1 \times 2^2) + (1 \times 2^1) \times (0 \times 2^0)$.

    Then $(1 \times 32) + (1 \times 4) + (1 \times 2) = 38$.

9. $A80C_{\text{sixteen}}$ is equivalent to
    $(10 \times 16^3) + (8 \times 16^2) + (0 \times 16^1) \times (12 \times 16^0)$.

    Then $(10 \times 4096) + (8 \times 256) + (12 \times 1)$
    $= 40960 + 2048 + 12$
    $= 43,020$.

10. Using repeated division,

    $$
    \begin{array}{r r}
    2\,|\,58 & \text{Rem} \\
    2\,|\,29 \leftarrow & 0 \\
    2\,|\,14 \leftarrow & 1 \\
    2\,|\,7 \leftarrow & 0 \\
    2\,|\,3 \leftarrow & 1 \\
    2\,|\,1 \leftarrow & 1 \\
    0 \leftarrow & 1
    \end{array}
    $$

    $58_{\text{ten}} = 111010_{\text{two}}$.

11. Using repeated division,

    $$
    \begin{array}{r r}
    5\,|\,1846 & \text{Rem} \\
    5\,|\,369 \leftarrow & 1 \\
    5\,|\,73 \leftarrow & 4 \\
    5\,|\,14 \leftarrow & 3 \\
    5\,|\,2 \leftarrow & 4 \\
    0 \leftarrow & 2
    \end{array}
    $$

    $1846_{\text{ten}} = 24341_{\text{five}}$.

12. Starting at the right, break the digits into groups of three. Then convert the groups to their octal equivalents. (See Table 6.)

    $$
    \begin{array}{ccc}
    10 & 101 & 110 \\
    \downarrow & \downarrow & \downarrow \\
    2 & 5 & 6
    \end{array}
    $$

    $10101110_{\text{two}} = 256_{\text{eight}}$

13. Each hexadecimal digit yields a 4-digit binary equivalent. (See Table 6.)

    $$
    \begin{array}{ccc}
    B & 5 & 2 \\
    \downarrow & \downarrow & \downarrow \\
    1011 & 0101 & 0010
    \end{array}
    $$

    $B52_{\text{sixteen}} = 101101010010_{\text{two}}$

14. The advantage of multiplicative grouping over simple grouping is that there is less repetition of symbols.

15. The advantage, in a positional numeration system, of a smaller base over a larger base, is that there are fewer symbols to learn.

16. The advantage, in a positional numeration system, of a larger base over a smaller base, is that there are fewer digits in the numerals.

17. The identity element is zero for addition of whole numbers.

18. For multiplication of rational numbers, the inverse of 3 is 1/3.

19. $(a + b) + c = (b + a) + c$ illustrates the commutative property of addition. The order of the addends has changed, but not the grouping.

20. Writing exercise

21. (a) There is an identity element.

    (b) o

22. (a) Closure is satisfied.

    (b) Writing exercise

23. (a) The system is not commutative.

    (b) Writing exercise

24. (a) The distributive property is not satisfied.

    (b) Writing exercise

25. (a) The system is not a group.

    (b) Writing exercise

5 NUMBER THEORY

5.1 Prime and Composite Numbers
5.2 Selected Topics from Number Theory
5.3 Greatest Common Factor and Least Common Multiple
5.4 Modular Systems
5.5 The Fibonacci Sequence and the Golden Ratio
Extension: **Magic Squares**
Chapter 5 Test

Chapter Goals

The chapter goal is to expose the student to the standard properties of the natural numbers. Upon completion of this chapter the student should be able to

- Find the natural number factors of any number.
- Identify and use some tests for divisibility.
- Distinguish between prime and composite numbers.
- Find the prime factorization of a composite number.
- Find greatest common factor (GCF) and least common multiple (LCM).
- Distinguish between abundant and deficient numbers.
- Identify patterns related to the Fibonacci sequence and other sequences.

Chapter Summary

This chapter explains the properties of the natural numbers, which are also called the counting numbers. A natural number is said to be divisible by another number if the answer is a natural number and the remainder is zero when the numbers are divided. In the text Table 2 in Section 5.1 contains several tests to determine if one natural number is divisible by another. These divisors are also called factors.

EXAMPLE A   Divisors or factors of a natural number

The natural number factors of 20 are 1, 2, 4, 5, 10, and 20. That is, 20 is divisible by all of its factors.

EXAMPLE B   Divisibility tests

Here are several divisibility tests for 1092.
(a) It is divisible by 2 because it is even.

(b) It is divisible by 3 because the sum of the digits is 12, a multiple of 3.

(c) It is divisible by 4 because 92, the last two digits form a number that is divisible by 4.

(d) It is not divisible by 5 because it does not end in 5 or 0.

(e) It is divisible by 6 because it is divisible by both 2 and 3.

(f) It is not divisible by 8 because the last three digits form a number that is not divisible by 8.

(g) It is not divisible by 9 because the sum of the digits is 12, which is not divisible by 9.

(h) It is not divisible by 10 because the last digit is not 0.

(i) It is divisible by 12 because it is divisible by both 3 and 4.

Several different kinds of numbers are explored in Section 2 of Chapter 5: perfect numbers, deficient and abundant numbers, amicable numbers, as well as several others are defined and explained in this section. Abundant and deficient numbers are defined according to the sum of their factors. An abundant number is defined as a number whose proper divisors have a sum which is greater than the number; a deficient number has proper divisors that have a sum less than the number. To understand this more clearly, consider the following example.

EXAMPLE C   Abundant and deficient numbers

An example of a deficient number is 75. Its proper divisors are 1, 3, 5, 15, and 25 ; these are all the numbers except 75 that divide into 75 evenly. The sum of these proper divisors is $1 + 3 + 5 + 15 + 25 = 49$. Because 49 is less than 75, the number 75 is said to be a deficient number.

An example of an abundant number is 40. The proper divisors of 40 are 1, 2, 4, 5, 8, 10 and 20. The sum of these numbers is 50. Because 50 is greater than 40, the number is abundant.

In testing for divisibility prime numbers are encountered. The only divisors of a prime number are the number itself and one. If a number is not prime, it is said to be composite. For example, 15 is a composite number. It has divisors 15 and 1, of course, but also 3 and 5. It is often useful to find prime factors of a number. One method of finding the prime factors is by repeated division by prime numbers, usually beginning with the smallest prime number 2. The resulting factors are all prime numbers.

EXAMPLE D   Prime factorization of a natural number

To find the prime factors of 1850, do repeated division by prime numbers.

$$
\begin{array}{r|l}
2 & 1850 \\
5 & 925 \\
5 & 185 \\
  & 37
\end{array}
$$

The prime factorization is $2 \cdot 5 \cdot 5 \cdot 37 = 2 \cdot 5^2 \cdot 37$.

Two applications of prime numbers are finding the greatest common factor and the least common multiple. The greatest common factor (GCF) is the largest number that is a factor common to a set of numbers. The least common multiple (LCM) is the smallest number that is a multiple common to a set of numbers. The following examples demonstrate each of these.

EXAMPLE E   Least Common Multiple

The least common multiple is often used as a least common denominator (LCD) when adding or subtracting fractions. The LCM for 16 and 12 can be found by the following method.

$$
\begin{aligned}
16 &= 2^4 \\
12 &= 2^2 \cdot 3 \\
\text{LCM} &= 2^4 \cdot 3 = 48
\end{aligned}
$$

The LCM is made up of 2^4, the smallest multiple of 16, and $2^2 \cdot 3$, the smallest multiple of 12. Since 2^2 is already a part of 2^4, it is necessary to have only four factors of 2.

EXAMPLE F   Greatest Common Factor

The greatest common factor, which is often used to simplify fractions is found for 20 and 30 in this example. The prime factorization of $20 = 2^2 \cdot 5$; the prime factorization of $30 = 2 \cdot 3 \cdot 5$. The factors common to both numbers are 2 and 5; $2 \cdot 5 = 10$ is the GCF.

Number sequences are also considered in the study of number theory. The first several terms of the famous Fibonacci sequence are 1, 1, 2, 3, 5, 8, 13, 21, ... Notice that each term after the second is obtained by adding the two previous terms. The beauty of the Fibonacci sequence is more than the fact that there is a pattern to the numbers. Researchers continue to observe patterns in nature, such as the pattern of flower petals and the scales of a pineapple. These and other natural phenomena follow the Fibonacci pattern. It is a fascinating area of study.

If the quotients of successive Fibonacci numbers are examined, it can be shown that they approach a value that is known as the golden ratio. The value of this ratio is $\frac{1+\sqrt{5}}{2}$, and it is often symbolized by the Greek letter ϕ (phi). Like the Fibonacci sequence, this ratio can be found in many unlikely places such as architecture, music, and nature.

5.1 EXERCISES

1. True. Remember that the natural numbers are also called the counting numbers: 1, 2, 3, 4, . . .

3. False. The even number 2 is a prime number.

5. False. If $n = 5$ it is true that $5|5$, but it is not true that $10|5$.

7. True. Consider an example such as 7. It is a factor of itself because $7 \div 7 = 1$. It is also a multiple of itself because $7 \cdot 1 = 7$.

9. False. Every composite number has a unique prime factorization.

11. False. $2^{11} - 1 = 2047$. The factors of 2047 are 23 and 89.

13. Remember that all natural number factors are those that divide the given number evenly: 1, 2, 3, 4, 6, 12.

15. 1, 2, 4, 5, 10, 20.

17. 1, 2, 3, 4, 5, 6, 8, 10, 12, 15, 20, 24, 30, 40, 60, and 120.

19. (a) It is not divisible by 2 because it is an odd number.

(b) It is divisible by 3 because the sum of the digits is 9, a number divisible by 3.

(c) It is not divisible by 4 because 15, the number formed by the last two digits, is not divisible by 4.

(d) It is divisible by 5 because the last digit is 5.

(e) It is not divisible by 6 because although it is divisible by 3, it is not divisible by 2. It must be divisible by both.

(f) It is not divisible by 8 because the three digits form a number that is not divisible by 8.

(g) It is divisible by 9 because the sum of the digits is 9.

(h) It is not divisible by 10 because the last digit is not 0.

(i) It is not divisible by 12 because, although it is divisible by 3, it is not divisible by 4. It must be divisible by both.

21. (a) It is not divisible by 2 because it is an odd number.

(b) It is not divisible by 3 because the sum of the digits is 50, which is not divisible by 3.

(c) It is not divisible by 4 because 25, the number formed by the last two digits, is not divisible by 4.

(d) It is divisible by 5 because the last digit is 5.

(e) It is not divisible by 6 because it is not divisible by 3 or 2. It must be divisible by both.

(f) It is not divisible by 8 because the last three digits form a number that is not divisible by 8.

(g) It is not divisible by 9 because the sum of the digits is 50, which is not divisible by 9.

(h) It is not divisible by 10 because the last digit is not 0.

(i) It is not divisible by 12 because it is not divisible by 3 or 4. It must be divisible by both.

23. (a) It is not divisible by 2 because it is odd.

(b) It is divisible by 3 because the sum of the digits is 45, a number divisible by 3.

(c) It is not divisible by 4 because 89, the number formed by the last two digits, is not divisible by 4.

(d) It is not divisible by 5 because it does not end in 5 or 0.

(e) It is not divisible by 6 because, although it is divisible by 3, it is not divisible by 2. It must be divisible by both.

(f) It is not divisible by 8 because the last three digits form a number that is not divisible by 8.

(g) It is divisible by 9 because the sum of the digits is 45, a number divisible by 9.

(h) It is not divisible by 10 because it does not end in 0.

(i) It is not divisible by 12 because although it is divisible by 3, it is not divisible by 4. It must be divisible by both.

25. (a) Writing exercise

(b) The largest prime number whose multiples would have to be considered is 13; the square of 13 is 169, which is less than 200. The next prime number after 13 is 17, but the square of 17 is 289, a number greater than 200.

(c) square root; square root; square root

(d) prime

27. Two primes that are consecutive natural numbers are 2 and 3; there are no others because one of them would be an even number which is divisible by 2.

29. The last digit must be zero, because the number must be divisible by 10.

31.

$$2\underline{|240}$$
$$2\underline{|120}$$
$$2\underline{|60}$$
$$2\underline{|30}$$
$$3\underline{|15}$$
$$5$$

The prime factorization of 240 is $2^4 \cdot 3 \cdot 5$.

33.

$$2\underline{|360}$$
$$2\underline{|180}$$
$$2\underline{|90}$$
$$3\underline{|45}$$
$$3\underline{|15}$$
$$5$$

The prime factorization is $2^3 \cdot 3^2 \cdot 5$.

35.

$$3\underline{|663}$$
$$13\underline{|221}$$
$$17$$

The prime factorization is $3 \cdot 13 \cdot 17$.

37. To test 142,891 for divisibility by 7:
(a) and (b)
$$14,289 - 2 = 14,287$$
$$1428 - 14 = 1414$$
$$141 - 8 = 133$$
$$13 - 6 = 7.$$

(c) Because 7 is divisible by 7, the given number is also divisible by 7.

39. To test 458,485 for divisibility by 7:
(a) and (b)
$$45,848 - 10 = 45,838$$
$$4583 - 16 = 4567$$
$$456 - 14 = 442$$
$$44 - 4 = 40.$$

(c) Because 40 is not divisible by 7, the given number is not divisible by 7.

41. (a) $8 + 9 + 9 + 9 = 35$

(b) $4 + 3 + 6 = 13$

(c) $35 - 13 = 22$

(d) Because 22 is divisible by 11, the given number is also divisible by 11.

43. (a) $4 + 3 + 9 + 2 + 8 = 26$

(b) $5 + 8 + 6 + 4 = 23$

(c) $26 - 23 = 3$

(d) Because 3 is not divisible by 11, the given number is not divisible by 11

45. Based on the divisibility test for 6, which says that the number must be divisible by both 2 and 3, the divisibility test for 15 is that the number must be divisible by both 3 and 5. That is, the sum of the digits must be divisible by 3 and the last digit must be 5 or 0.

47. 0, 2, 4, 6, or 8 because they are all the even single digit numbers.

49. The last two digits must form a number that is divisible by 4. The possible values for x are 0, 4, and 8.

51. The number is divisible by 6 if it is divisible by both 2 and 3. Both 985,230 and 985,236 are divisible by 2 and 3. Then $x = 0$ or 6.

53. The sum of the existing digits is 30. The overall sum must be divisible by 9 so x must be 6.

55. Prime factorization: $36 = 3^2 \cdot 2^2$.

Add one to each exponent:

$$2 + 1 = 3$$
$$2 + 1 = 3.$$

Multiply: $3 \cdot 3 = 9$. There are 9 divisors of 36.

57. Prime factorization: $72 = 2^3 \cdot 3^2$.

Add one to each exponent:

$$3 + 1 = 4$$
$$2 + 1 = 3.$$

Multiply: $4 \cdot 3 = 12$. There are 12 divisors of 72.

59. Prime factorization: $2^8 \cdot 3^2$.

Add one to each exponent:

$$8 + 1 = 9$$
$$2 + 1 = 3.$$

Multiply: $9 \cdot 3 = 27$. There are 27 divisors of $2^8 \cdot 3^2$.

61. It is divisible by 4 and does not end in two zeros, so it is a leap year.

63. It is not divisible by 4, so it is not a leap year.

65. Because 2400 is divisible by 4 and 400, it is a leap year.

67. Writing exercise

69. Writing exercise

71. $41^2 - 41 + 41 = 41^2$, which is not a prime.

73. For $n > 41$, Euler's formula produces a prime sometimes. In Exercise 72, Euler's formula produces 1763 for $n = 42$. The square root of 1763 is approximately 41.99, so all of the prime numbers less than this number must be tested. If 41 is tested as a divisor of 1763, the quotient is 43. Therefore the number is not prime. On the other hand, Euler's formula produces 1847 for $n = 43$. The square root of 1847 is approximately 42.98. If all the prime numbers less than this number are tested, none divides 1847 evenly. Therefore, the number is prime.

75. (a) $2^{2^4} + 1 = 2^{16} + 1 = 65,536 + 1 = 65,537$

    (b) The square root of 65,537 is approximately 256. All primes less than 256 must be checked as possible factors of this number; 251 is then the largest potential prime factor.

77. Writing exercise

79. Writing exercise

81. $M_6 = 2^6 - 1 = 64 - 1 = 63$

83. $2^p - 1$

85. The two prime factors of 10 are 2 and 5.

$$2^2 - 1 = 4 - 1 = 3$$
$$2^5 - 1 = 32 - 1 = 31$$

Two distinct factors of M_{10} are 3 and 31.

5.2 EXERCISES

1. True

3. True

5. True. By definition, the only factors of a prime number are the number itself and 1. The proper divisors of a number are all the divisors except the number itself. For a prime number, then, the only proper divisor is 1. That means that any prime number must be deficient.

7. False. For every new Mersenne prime, there is a perfect number.

9. True. $2^6 = 64$ and $(2^7 - 1) = 128 - 1$ or 127. Then $64(127) = 8128$, which is even and ends in 28.

11. $1 + 2 + 4 + 8 + 16 + 31 + 62 + 124 + 248 = 496$

13. $2^{13} - 1 = 8191$, which is prime.
    $2^{13-1}(2^{13} - 1) = 4096(8192 - 1) = 33,550,336$
    This number is even and ends in 6.

15. The divisors of 6 are 1, 2, 3, and 6. The sum of the reciprocals of these numbers is

$$\frac{1}{1} + \frac{1}{2} + \frac{1}{3} + \frac{1}{6}$$
$$= \frac{6}{6} + \frac{3}{6} + \frac{2}{6} + \frac{1}{6}$$
$$= \frac{12}{6}$$
$$= 2.$$

17. The proper divisors of 36 are 1, 2, 3, 4, 6, 9 and 12. The sum of these numbers is 37. Because the sum is greater than 36, the number is abundant.

19. The proper divisors of 75 are 1, 3, 5, 15, and 25. The sum of these numbers is 49. Because the sum is less than 75, the number is deficient.

21. Examine the Sieve of Eratosthenes in Table 5.1 to see that the prime numbers 2, 3, 5, 7, 11, 13, 17, 19, and 23 can be deleted from the search. Examine the remaining numbers: 4 is deficient because $1 + 2 = 3$; 6 is perfect because $1 + 2 + 3 = 6$; 8 is deficient because $1 + 2 + 4 = 7$; 9 is deficient because $1 + 3 = 4$; 10 is deficient because $1 + 2 + 5 = 8$. The number 12 is abundant because the sum of its proper divisors is $1 + 2 + 3 + 4 + 6 = 16$. The numbers 14, 15, and 16 are deficient. Verify this by adding their proper divisors. The number 18 is the next abundant number; the sum of its proper divisors is 21. Then 20 is the third abundant number; the sum of its proper divisors is 22. The numbers 21 and 22 are both deficient. Verify this. Finally the fourth abundant number is 24; the sum of its proper divisors is 36.

23. The sum of the proper divisors is 975. You can verify this with a calculator. Therefore, the number is abundant.

25. The sum of the proper divisors of 1,184 is 1,210. The sum of the proper divisors of 1,210 is 1,184. Thus, they are amicable.

27. $3 + 11, 7 + 7$.

29. $3 + 23, 7 + 19, 13 + 13$

31. If $a = 5$ and $b = 3$, then $5 + 2 \cdot 3 = 11$.

33. Examine the Sieve of Eratosthenes, which is Table 1 in Section 5.1, to find 71 and 73. These numbers differ by 2.

35. Examine the Sieve of Eratosthenes to find 137 and 139. These numbers differ by 2.

37. (a) $5 = 9 - 4$ or $3^2 - 2^2$

    (b) $11 = 36 - 25$ or $6^2 - 5^2$

39. $5^2 + 2 = 3^3; 25 + 2 = 27$

41. Examine the numbers.

43. Writing exercise

45. Writing exercise

47. (b) Sometimes

	p	$2p+1$	Sophie Germaine prime?
48.	2	5	yes
49.	3	7	yes
50.	5	11	yes
51.	7	15	no
52.	11	23	yes
53.	13	27	no

Table for Exercises 48–53

49. For $p = 3, 2p + 1 = 7$, a Sophie Germaine prime .

51. For $p = 7, 2p + 1 = 15$, not a Sophie Germaine prime .

53. For $p = 13, 2p + 1 = 27$, not a Sophie Germaine prime.

54. – 56.

	n	$n!$	$n!-1$	$n!+1$	$n!-1$ prime?	$n!+1$ prime?
	2	2	1	3	no	yes
54.	3	6	5	7	yes	yes
55.	4	24	23	25	yes	no
56.	5	120	119	121	no	no

To calculate the underlined numerical values in the table, recall the meaning of n-factorial, $n!$.

$$n! = n(n-1)(n-2)\ldots(3)(2)(1).$$

n	$n!$	$n!-1$	$n!+1$
3	$3 \cdot 2 \cdot 1 = 6$	5	7
4	$4 \cdot 3 \cdot 2 \cdot 1 = 24$	23	25
5	$5 \cdot 4 \cdot 3 \cdot 2 \cdot 1 = 120$	119	121

The numbers 5 and 7 are both prime. Although 23 is prime, the number 25 is not as it can be written as $5 \cdot 5$. The number 119 can be written as $7 \cdot 17$, and 121 can be written as $11 \cdot 11$.

57. Writing exercise

59. (b) Sometimes

5.3 EXERCISES

1. True. They would have a common factor of at least 2. Two numbers that are relatively prime have a greatest common factor of 1.

3. True. Consider the prime number 5; $5^2 = 25$. The greatest common factor of 5 and 25 is 5.

5. False. The variable $p = 2$ proves this statement false.

7. True. All natural numbers have 1 as a common factor.

9. True. Consider some examples: 25 and 9, 14 and 33, 16 and 39, etc.

11. To find the greatest common factor by the prime factors method, first write the prime factorization of each number:

$$70 = 2 \cdot 5 \cdot 7$$
$$120 = 2^3 \cdot 3 \cdot 5$$

Now find the primes common to all factorizations, with each prime raised to the smallest exponent from either factorization: 2, 5.
Finally, form the product of these numbers:

$$2 \cdot 5 = 10.$$

13.
$$480 = 2^5 \cdot 3 \cdot 5$$
$$1800 = 2^3 \cdot 3^2 \cdot 5^2$$

Now find the primes common to all factorizations, with each prime raised to the smallest exponent from either factorization: $2^3, 3, 5$.

Finally, form the product of these numbers:

$$2^3 \cdot 3 \cdot 5 = 120.$$

15.
$$28 = 2^2 \cdot 7$$
$$35 = 5 \cdot 7$$
$$56 = 2^3 \cdot 7$$

Now find the primes common to all factorizations, with each prime raised to the smallest exponent from all factorizations: 7. The only factor common to all three numbers is 7.

17. To find the greatest common factor of 60 and 84 by the method of dividing by prime factors, first write the numbers in a row and divide by their smallest common prime factor, which is 2.

$$2 \,\underline{|\, 60 \quad 84\,}$$
$$\quad\;\; 30 \quad 42$$

Now divide 30 and 42 by their next largest prime factor, which is 2.

$$2 \,\underline{|\, 60 \quad 84\,}$$
$$2 \,\underline{|\, 30 \quad 42\,}$$
$$\quad\;\; 15 \quad 21$$

Divide again by the next largest prime factor, which is 3.

$$\begin{array}{c|cc} 2 & 60 & 84 \\ 2 & 30 & 42 \\ 3 & 15 & 21 \\ & 5 & 7 \end{array}$$

Since 5 and 7 have no common prime factors, there are no more divisions. To find the greatest common factor, find the product of the primes on the left, 2, 3 and 3.

$$2 \cdot 2 \cdot 3 = 12$$

19. To find the greatest common factor of 310 and 460 by the method of dividing by prime factors, first write the numbers in a row and divide by their smallest common prime factor, which is 2.

$$\begin{array}{c|cc} 2 & 310 & 460 \\ \hline & 155 & 230 \end{array}$$

Now divide 155 and 230 by their next largest prime factor, which is 5.

$$\begin{array}{c|cc} 2 & 310 & 460 \\ 5 & 155 & 230 \\ \hline & 31 & 46 \end{array}$$

Since 31 and 46 have no common prime factors, there are no more divisions. To find the greatest common factor, find the product of the primes on the left, 2 and 5.

$$2 \cdot 5 = 10$$

21. 12, 18, and 30

$$\begin{array}{c|ccc} 2 & 12 & 18 & 30 \\ 3 & 6 & 9 & 15 \\ \hline & 2 & 3 & 5 \end{array}$$

Since 2, 3, and 5 have no common prime factors, there are no more divisions. Find the product of the primes on the left.

$$2 \cdot 3 = 6$$

23.

	Remainder
$60 \div 36 = 1$	24
$36 \div 24 = 1$	12
$24 \div 12 = 2$	0

Then 12, the last positive remainder, is the greatest common factor.

25.

	Remainder
$180 \div 84 = 2$	12
$84 \div 12 = 7$	0

Then 12, the last positive remainder, is the greatest common factor.

27.

	Remainder
$560 \div 210 = 2$	140
$210 \div 140 = 1$	70
$140 \div 70 = 2$	0

Then 70, the last positive remainder, is the greatest common factor.

29. Writing exercise

31. To find the least common multiple by the prime factors method, first write the prime factorization of each number.

$$24 = 2^3 \cdot 3$$
$$30 = 2 \cdot 3 \cdot 5$$

Choose all primes belonging to any factorization, with each prime raised to the largest exponent that it has in any factorization: 2^3, 3, 5. Finally, form the product of these numbers:

$$2^3 \cdot 3 \cdot 5 = 120.$$

The least common multiple of 24 and 30 is 120.

33. Prime factorizations:

$$56 = 2^3 \cdot 7$$
$$96 = 2^5 \cdot 3.$$

$$\begin{aligned} \text{Least common multiple} &= 2^5 \cdot 3 \cdot 7 \\ &= 672 \end{aligned}$$

35. Prime factorizations:

$$30 = 2 \cdot 3 \cdot 5$$
$$40 = 2^3 \cdot 5$$
$$70 = 2 \cdot 5 \cdot 7.$$

$$\begin{aligned} \text{Least common multiple} &= 2^3 \cdot 3 \cdot 5 \cdot 7 \\ &= 840 \end{aligned}$$

37. Use the result of Exercise 23 and the following formula.

Least common multiple of m and n

$$= \frac{m \cdot n}{\text{Greatest common factor of } m \text{ and } n}$$

$$\text{Least common multiple of 36 and 60} = \frac{36 \cdot 60}{12}$$

$$= 180$$

39. Use the result of Exercise 25 and the following formula.

Least common multiple of m and n

$$= \frac{m \cdot n}{\text{Greatest common factor of } m \text{ and } n}$$

Least common multiple of 84 and 180 $= \dfrac{84 \cdot 180}{12}$

$$= 1260$$

41. Use the result of Exercise 27 and the following formula.

Least common multiple of m and n

$$= \frac{m \cdot n}{\text{Greatest common factor of } m \text{ and } n}$$

Least common multiple of 210 and 560 $= \dfrac{210 \cdot 560}{70}$

$$= 1680$$

43. Let $p = 3$, $q = 5$, and $r = 7$; $a = 4$, $b = 2$, and $c = 1$.

Then,
$$p^a q^c r^b = 3^4 \cdot 5^1 \cdot 7^2$$
$$p^b q^a r^c = 3^2 \cdot 5^4 \cdot 7^1.$$

The GCF is $3^2 \cdot 5^1 \cdot 7^1$, which is $p^b q^c r^c$.

(b) Use the same numerical values for $p, q, r, a, b,$ and c as in part (a).

Then

$$p^b q^a = 3^2 \cdot 5^4$$
$$q^b r^c = 5^2 \cdot 7^1,$$
$$p^a r^b = 3^4 \cdot 7^2.$$

The LCM is $3^4 \cdot 5^4 \cdot 7^2$, which is $p^a q^a r^b$.

45. 150, 210, and 240

1. $210 \div 150 = 1$ with a remainder of 60.
2. $150 \div 60 = 2$ with a remainder of 30.
3. $60 \div 30 = 2$ with remainder of 0.
4. Then 30, the last positive remainder, is the greatest common factor of 150 and 210.
5. $240 \div 30 = 8$ with a remainder of 0.
6. Then 30 is the greatest common factor of the three given numbers.

47. 90, 105, and 315

1. $105 \div 90 = 1$ with a remainder of 15.
2. $90 \div 15 = 6$ with a remainder of 0.
3. Then 15, the last positive remainder, is the greatest common factor of 90 and 105.

4. $315 \div 15 = 21$ with a remainder of 0.
5. Then 15 is the greatest common factor of the three given numbers.

49. 144, 180, and 192

1. $180 \div 144 = 1$ with a remainder of 36.
2. $144 \div 36 = 4$ with a remainder of 0.
3. Then 36, the last positive remainder, is the greatest common factor of 144 and 180.
4. $192 \div 36 = 5$ with a remainder of 12.
5. $36 \div 12 = 3$ with a remainder of 0.
6. Then 12 is the greatest common factor of the three given numbers.

51. (a) The GCF is found in the intersection: $2 \cdot 3 = 6$.

(b) The LCM is found in the union: $2^2 \cdot 3^2 = 36$.

53. (a) The GCF is found in the intersection: $2 \cdot 3^2 = 18$.

(b) The LCM is found in the union: $2^3 \cdot 3^3 = 216$.

55. The numbers p and q must be relatively prime.

57. Writing exercise

59. Find the least common multiple of 16 and 36. $16 = 2^4$; $36 = 2^2 \cdot 3^2$. The LCM is $2^4 \cdot 3^2 = 144$. The 144th calculator is the first that they will both inspect.

61. Find the greatest common factor of 240 and 288. $240 = 2^4 \cdot 3 \cdot 5$; $288 = 2^5 \cdot 3^2$. The GCF is $2^4 \cdot 3 = 48$.

63. To answer the first question find the least common multiple of the two dollar amounts.

$$24 = 2^3 \cdot 3 \text{ and } 50 = 2 \cdot 5^2.$$

The LCM is $2^3 \cdot 3 \cdot 5^2 = 600$. The answer is \$600. To answer the second question divide \$600 by \$24, the price per book to obtain 25. He would have sold 25 books at this price.

5.4 EXERCISES

1. $5 \equiv 19 \pmod 3$ is false. The difference $19 - 5 = 14$ is not divisible by 3. $5 \equiv 18 \pmod 3$.

3. $5445 \equiv 0 \pmod 3$ is true. The difference $5445 - 0 = 5445$ is divisible by 3.

5. $(12 + 7) \pmod 4$
First add 12 and 7 to get 19. Then divide 19 by 4. The remainder is 3, so $(12 + 7) \pmod 4 \equiv 3$.

7. $(35 - 22) \pmod 5$
First subtract 22 from 35 to get 13. Then divide 13 by 5. The remainder is 3, so $(35 - 22) \pmod 5 \equiv 3$.

9. $(5 \times 8) \pmod 3$
    $5 \times 8 = 40$. Then divide 40 by 3. The remainder is 1, so $(5 \times 8) \pmod 3 \equiv 1$.

11. $[4 \times (13 + 6)] \pmod{11}$
    $[4 \times (13 + 6)] = 4 \times 19 = 76$. Then divide 76 by 11. The remainder is 10, so $[4 \times (13 + 6)] \pmod{11} \equiv 10$.

13. $(3 - 27) \pmod 5$
    $3 - 27 = -24$. Since this number is negative, add as many multiples of 5 as necessary to obtain a positive number.
    $$-24 + (5 \times 5) = -24 + 25$$
    $$= 1$$
    $$\equiv 1 \pmod 5$$
    Therefore, $(3 - 27) \pmod 5 \equiv 1 \pmod 5$.

15. $[(-8) \times 11] \pmod 3$
    $(-8) \times 11 = -88$. Since this number is negative, add as many multiples of 3 as necessary to obtain a positive number.
    $$-88 + (30 \times 3) = -88 + 90$$
    $$= 2$$
    $$\equiv 2 \pmod 3$$
    Therefore, $[(-8) \times 11] \pmod 3 \equiv 2 \pmod 3$.

17. $\{3, 10, 17, 24, 31, 38, \dots\}$ is the solution set. To find these numbers, multiply 7 times the whole numbers and add 3 each time.
    $$(7 \times 0) + 3 = 3$$
    $$(7 \times 1) + 3 = 10$$
    $$(7 \times 2) + 3 = 17$$

19. $6x \equiv 2 \pmod 2$
    Try the integers 0 and 1. Both work so this is an identity.

21. The modulus is 100,000. When the odometer reaches 99,999.9, it "rolls over" to 00,000.0.

23. A value of x must be found that will satisfy the following equations:
    $$x \equiv 2 \pmod{10}$$
    $$x \equiv 2 \pmod{15}$$
    $$x \equiv 2 \pmod{20}$$
    The solution set for the first equation is:
    $\{12, 22, 32, 42, \dots, 62, \dots\}$.
    The solution set for the second equation is:
    $\{17, 32, 47, 62, \dots\}$.
    The solution set for the third equation is:
    $\{22, 42, 62, \dots\}$. The smallest number that satisfies all three equations is 62.

25. (a) The next year would contain 365 days.

    (b) Divide 365 days by 7 to obtain 52 with a remainder of 1. That means that it will be one day past Thursday, which is Friday.

27. Because there are 31 days in July and 31 days in August for a total of 62, it is unnecessary to find any integers larger than 62 for the modulo systems. Eva's 21-day schedule indicates modulo 21; Carrie's 30-day schedule indicates modulo 30.

    Chicago
    For Eva, $x = 1 \pmod{21}$ has solution set
    $\{1, 22, 43, \dots\}$. These integers correspond to July 1, July 22, and August 12.
    $x = 2 \pmod{21}$ has solution set
    $\{2, 23, 44, \dots\}$. These integers correspond to July 2, July 23, and August 13.
    $x = 8 \pmod{21}$ has solution set
    $\{8, 29, 50, \dots\}$. These integers correspond to July 8, July 29, and August 19.

    For Carrie, $x = 23 \pmod{30}$ has solution set
    $\{23, 53, \dots\}$. These integers correspond to July 23 and August 22.
    $x = 29 \pmod{30}$ has solution set
    $\{29, 59, \dots\}$. These integers correspond to July 29 and August 28.
    $x = 30 \pmod{30}$ has solution set
    $\{30, 60, \dots\}$. These integers correspond to July 30 and August 29.

    The only days that they will both be in Chicago are July 23 and July 29.

    New Orleans
    For Eva, $x = 5 \pmod{21}$ has solution set
    $\{5, 26, 47, \dots\}$. These integers correspond to July 5, July 26, and August 15.
    $x = 12 \pmod{21}$ has solution set
    $\{12, 33, 54, \dots\}$. These integers correspond to July 12, August 6, and August 23.

    For Carrie, $x = 5 \pmod{30}$ has solution set
    $\{5, 35, \dots\}$. These integers correspond to July 5 and August 4.
    $x = 6 \pmod{30}$ has solution set
    $\{6, 36, \dots\}$. These integers correspond to July6 and August 5.
    $x = 17 \pmod{30}$ has solution set
    $\{17, 47, \dots\}$. These integers correspond to July 17 and August 16.

    The only days that they will both be in New Orleans are July 5 and August 16.

San Francisco

For Eva, $x = 6 \pmod{21}$ has solution set
$\{6, 27, 47, \dots\}$. These integers correspond to July 6,
July 27, and August 16.
$x = 18 \pmod{21}$ has solution set
$\{18, 39, 60, \dots\}$. These integers correspond to July 18,
August 8, and August 29.

For Carrie, $x = 8 \pmod{30}$ has solution set
$\{8, 38, \dots\}$. These integers correspond to July 8 and
August 7.
$x = 15 \pmod{30}$ has solution set
$\{15, 45, \dots\}$. These integers correspond to July 15 and
August 14.
$x = 20 \pmod{30}$ has solution set
$\{20, 50, \dots\}$. These integers correspond to July 20 and
August 19. Finally, $x \equiv 25 \pmod{30}$ has the solution
set $\{25, 55, \dots\}$, which means July 25 and August 24.

The only day that they will both be in San Francisco is
August 9.

29. (a)

+	0	1	2	3	4
0	0	1	2	3	4
1	1	2	3	4	0
2	2	3	4	0	1
3	3	4	0	1	2
4	4	0	1	2	3

(b) All four properties are satisfied.

(c) Since 0 is the identity element, two elements will be
inverses if their sum is 0. From the table we see that 0 is
its own inverse. The numbers 1 and 4 are inverses of
each other; 2 and 3 are inverses of each other.

31. (a)

×	0	1
0	0	0
1	0	1

(b) All properties are satisfied.
(c) The number 1 is its own inverse.

33. (a)

×	0	1	2	3
0	0	0	0	0
1	0	1	2	3
2	0	2	0	2
3	0	3	2	1

(b) There is no inverse property. The number 1 is the
identity element. 2 has an inverse.

(c) The number 1 is its own inverse and 3 is its own
inverse.

35.

+	0	1	2	3	4
0	0	1	2	3	4
1	1	2	3	4	0
2	2	3	4	0	1
3	3	4	0	1	2
4	4	0	1	2	3

−	0	1	2	3	4
0	0	4	3	2	1
1	1	0	4	3	2
2	2	1	0	4	3
3	3	2	2	0	4
4	4	3	2	1	0

Read the subtraction table to see that $1 - 3 \equiv 3$.
Then, given the information in the text, for the formula,

$$a - b = d,$$

$a = 1, b = 3$, and $d = 3$.
Read the addition table to see that $3 + 3 \equiv 1$.
Given the information in the text, for the formula,

$$b + d = a,$$

the values for a, b and d are the same.
Therefore, $1 - 3 \equiv 3 \pmod{5}$.

37. Read the subtraction table in Exercise 35 to see that
$3 - 4 \equiv 4$. Then, given the information in the text, for
the formula,

$$a - b = d,$$

$a = 3, b = 4$, and $d = 4$.
Read the addition table to see that $4 + 4 \equiv 3$.
Given the information in the text, for the formula,

$$b + d = a,$$

the values for a, b and d are the same.
Therefore, $3 - 4 \equiv 4 \pmod{5}$.

39. Given the information in the text including the formulas
$a \div b = q$ if and only if $b \times q = a$, in this exercise
$a = 3$ and $b = 1$. Use the multiplication table in
Exercise 39 to find $b \times ? = a$; that is, $1 \times ? = 3$. Read
the table to see that the answer is 3.

41. No. See Exercise 38 to understand that $1 \div 3 = 2$. See
Exercise 39 to understand that $3 \div 1 = 3$. Therefore,

$$1 \div 3 \neq 3 \div 1.$$

43. $1400 + 500 = 1900$.

45. $0750 + 1630 = 2380$. This number is equivalent to
$2400 \div 2380 = 1$, with a remainder of 20. Therefore,
the time is 0020.

47. The first statement simply shows the sum of the two
numbers. The second statement is a representation of
12-hour clock arithmetic. The third statement is a
representation of 24-hour clock arithmetic.

49. i^{47}

$$47 \equiv 3 \pmod{4}$$
$$i^{47} = i^3 = -i.$$

51. i^{137}

$$137 \equiv 1 \ (\text{mod} \ 4)$$
$$i^{137} = i^1 = i.$$

53. Substitute 1865 for y in the formula given in the text.

$$a = 1865 + [1864/4] - [1864/100] + [1864/400]$$
$$= 1865 + [466] - [18.64] + [4.66]$$
$$= 1865 + 466 - 18 + 4$$
$$= 2317$$

Since

$$2317 = (7 \times 331) + 0$$
$$2317 \equiv 0 \ (\text{mod} \ 7)$$

and 0 corresponds to Sunday, January 1, 1865 was a Sunday.

55. Substitute 1999 for y in the formula given in the text.

$$a = 1999 + [1998/4] - [1998/100] + [1998/400]$$
$$= 1999 + [499.5] - [19.98] + [4.995]$$
$$= 1999 + 499 - 19 + 4$$
$$= 2483$$

Since

$$2483 = (7 \times 354) + 5$$
$$2483 \equiv 5 \ (\text{mod} \ 7)$$

and 5 corresponds to Friday, January 1, 1999 was a Friday.

57. 1995
Substitute 1995 for y in the formula given in the text for Exercises 52–55.

$$a = 1995 + [1994/4] - [1994/100] + [1994/400]$$
$$= 1995 + [498.5] - [19.94] + [4.985]$$
$$= 1995 + 498 - 19 + 4$$
$$= 2478$$

Since

$$2478 = (7 \times 354) + 0$$
$$2478 \equiv 0 \ (\text{mod} \ 7)$$

and 0 corresponds to Sunday, the first day of the year is Sunday, and 1995 is not a leap year. Therefore, the table in the textbook shows that Friday the thirteenth occurs in January and October.

59. 2200
Substitute 2200 for y in the formula given in the text for Exercises 52–55.

$$a = 2200 + [2199/4] - [2199/100] + [2199/400]$$
$$= 2200 + [549.75] - [21.99] + [5.4975]$$
$$= 2200 + 549 - 21 + 5$$
$$= 2733$$

Since

$$2733 = (7 \times 390) + 3$$
$$2733 \equiv 3 \ (\text{mod} \ 7)$$

and 0 corresponds to Wednesday, the first day of the year is Wednesday, and 2200 is not a leap year. Therefore, the table in the textbook shows that Friday the thirteenth occurs in June.

61. 0–399–13615–4

$$(10 \times 0) + (9 \times 3) + (8 \times 9) + (7 \times 9) + (6 \times 1)$$
$$+ (5 \times 3) + (4 \times 6) + (3 \times 1) + (2 \times 5) = 220.$$

Because 220 is a multiple of 11, the check digit is 0; 4 is incorrect.

63. 0–8027–1344–

$$(10 \times 0) + (9 \times 8) + (8 \times 0) + (7 \times 2) + (6 \times 7)$$
$$+ (5 \times 1) + (4 \times 3) + (3 \times 4) + (2 \times 4) = 165.$$

Because 165 is a multiple of 11, the check digit is 0.

65. 0–394–58640–

$$(10 \times 0) + (9 \times 3) + (8 \times 9) + (7 \times 4) + (6 \times 5)$$
$$+ (5 \times 8) + (4 \times 6) + (3 \times 4) + (2 \times 0) = 233.$$

Because $233 + 9 = 242$, a multiple of 11, the check digit is 9.

5.5 EXERCISES

1. Find the sum of the sixteenth and seventeenth terms to obtain the eighteenth

$$987 + 1597 = 2584.$$

3. Find the difference between F_{25} and F_{23}

$$75025 - 28657 = 46,368.$$

5. $\dfrac{1 + \sqrt{5}}{2}$

7. The pattern on the left side of the equations consists of the beginning consecutive terms of the Fibonacci sequence. The pattern on the right side of the equations consists of consecutive individual odd terms of the Fibonacci sequence less one. On both sides add the next consecutive term of the Fibonacci sequence to produce the next equation. The next term on the left side is 144, and the next term on the right side is 233. Therefore, the next equation is

$$1 + 1 + 2 + 3 + 5 + 8 = 21 - 1$$

which checks.

9. The pattern on the left side of the equations consists of the beginning consecutive odd terms of the Fibonacci sequence. The pattern on the right side of the equations consists of consecutive individual even terms of the Fibonacci sequence. On both sides add the next consecutive term of the Fibonacci sequence to produce the next equation. The next term on the left side is 89, and the next term on the right side is 144. Therefore, the next equation is

$$1 + 2 + 5 + 13 + 34 + 89 = 144$$

which checks.

11. The pattern on the left side of the equations consists of the two terms of the Fibonacci sequence in reverse order. The pattern on the right side of the equations consists of consecutive individual even terms of the Fibonacci sequence. It is the difference of the squares of the terms on the left side that equals the term on the right side. Notice also that on the left side, the differences alternate between the squares of consecutive odd terms and consecutive even term. On the left side, then, write the difference of squares of the next consecutive odd terms, in reverse order to produce the next equation. The next terms on the left side are 13^2 and 5^2, and the next term on the right side is 144. Therefore, the next equation is

$$13^2 - 5^2 = 144$$

which checks.

13. The left side of each equation is built with alternating terms of the Fibonacci sequence with alternating signs. The right side of each equation consists of Fibonacci numbers squared with alternating signs. The next equation should be $1 - 2 + 5 - 13 + 34 - 89 = -8^2$. The left side of this equation has a sum of -64. The right side of the equations also equals -64.

15. (a) $37 = 34 + 3$. Another possibility is $37 = 21 + 8 + 5 + 3$. Can you find more?

   (b) $40 = 34 + 3 + 2 + 1$.

   (c) $52 = 21 + 13 + 8 + 5 + 3 + 2$.

17. (a) $m = 10$ and $n = 4$. The greatest common factor of 10 and 4 is 2. Also, $F_{10} = 55$ and $F_4 = 3$; $F_2 = 1$, which is the greatest common factor of 55 and 3.

   (b) $m = 12$ and $n = 6$. The greatest common factor of 12 and 6 is 6. Also $F_{12} = 144$ and $F_6 = 8$; The greatest common factor of 144 and 8 is 8.

   (c) $m = 14$ and $n = 6$. The greatest common factor of 14 and 6 is 2. Also, $F_{14} = 377$ and $F_6 = 8$; $F_2 = 1$, which is the greatest common factor of 377 and 8.

19. (a) Square 13. Multiply the terms of the sequence two positions away from 13 (i.e., 5 and 34). Subtract the smaller result from the larger, and record your answer. $13^2 = 1699$; $5 \cdot 34 = 170$; $170 - 169 = 1$.

   (b) Square 13. Multiply the terms of the sequence three positions away from 13. Once again subtract the smaller result from the larger, and record your answer. $3^2 = 169$; $3 \cdot 55 = 165$; $169 - 165 = 4$.

   (c) Repeat the process, moving four terms away from 13. $13^2 = 169$; $2 \cdot 89 = 178$; $178 - 169 = 9$.

   (d) Make a conjecture about what will happen when you repeat the process, moving five terms away. Verify your answer. The results in parts a, b, and d are 1, 4, and 9. These are the squares of the Fibonacci numbers. Because the next Fibonacci number is 5, the result should be $5^2 = 25$. Verification: $13^2 = 169$; $1 \cdot 144 = 144$. $169 - 144 = 25$.

21. This term is found by adding the ninth and tenth terms: $76 + 123 = 199$.

23. $1, 3, 4, 7, 11, 18, \ldots$

   Add the first and third terms: $1 + 4 = 5$.
   Add the first, third, and fifth terms: $1 + 4 + 11 = 16$.
   Add the first, third, fifth, and seventh terms:
   $1 + 4 + 11 + 29 = 45$.
   Also, $1 + 4 + 11 + 29 + 76 = 121$.

   Each of the sums is 2 less than a Lucas number.

25. (a)
$$1 \cdot 1 = 1$$
$$1 \cdot 3 = 3$$
$$2 \cdot 4 = 8$$
$$3 \cdot 7 = 21$$
$$5 \cdot 11 = 55$$

   Notice that reading downward, the first numbers in each equation are the first five members of the Fibonacci sequence. The products, reading downward, are the 2nd, 4th, 6th, 8th, and 10th terms of the Fibonacci sequence. The next equation should be $8 \cdot 18 = 144$.

(b)
$$1 + 2 = 3$$
$$1 + 3 = 4$$
$$2 + 5 = 7$$
$$3 + 8 = 11$$
$$5 + 13 = 18$$

Reading downward, the first terms on the left side of the equation are the first five Fibonacci numbers. The second terms on the left side are also ascending Fibonacci numbers, beginning with 2. The next equation should be $8 + 21 = 29$.

(c)
$$1 + 1 = 2 \cdot 1$$
$$1 + 3 = 2 \cdot 2$$
$$2 + 4 = 2 \cdot 3$$
$$3 + 7 = 2 \cdot 5$$
$$5 + 11 = 2 \cdot 8$$

Read downward to see the first five Fibonacci numbers on the extreme left side of the equation. Read downward on the extreme right side to see the second through sixth Fibonacci numbers. The next equation should be $8 + 18 = 2 \cdot 13$.

27. $1, 1, 2, 3$

Multiply the first and fourth: $1 \cdot 3 = 3$.
Double the product of the second and third:
$2(1 \cdot 2) = 4$.
Add the squares of the second and third: $1^2 + 2^2 = 5$.
We have obtained the triple 3, 4, 5. We can substitute these numbers into the Pythagorean theorem to verify that this is a Pythagorean triple:

$$3^2 + 4^2 = 5^2$$
$$9 + 16 = 25.$$

29. $2, 3, 5, 8$

Multiply the first and fourth: $2 \cdot 8 = 16$.
Double the product of the second and third:
$2(3 \cdot 5) = 30$.
Add the squares of the second and third: $3^2 + 5^2 = 34$.
We have obtained the triple 16, 30, 34. We can substitute these numbers into the Pythagorean theorem to verify that this is a Pythagorean triple:

$$16^2 + 30^2 = 34^2$$
$$256 + 900 = 1156.$$

31. The sums of the terms on the diagonals are the Fibonacci numbers: $1, 1, 2, 3, 5, 8, 13, \ldots$

33. $\dfrac{1 + \sqrt{5}}{2} \approx 1.618033989; \dfrac{1 - \sqrt{5}}{2} \approx -0.618033989$

The digits in the decimal positions are the same.

35. $n = 14$

$$\frac{\left(\frac{1+\sqrt{5}}{2}\right)^{14} - \left(\frac{1-\sqrt{5}}{2}\right)^{14}}{\sqrt{5}}$$

$$\approx \frac{(1.618033988)^{14} - (-0.618033988)^{14}}{\sqrt{5}}$$

$$\approx \frac{842.9988137 - 0.001186241}{\sqrt{5}}$$

$$\approx 377$$

37. $n = 22$

Use a calculator to find

$$\frac{\left(\frac{1+\sqrt{5}}{2}\right)^{22} - \left(\frac{1-\sqrt{5}}{2}\right)^{22}}{\sqrt{5}} \approx 17,711.$$

The 22nd Fibonacci number is 17,711.

Term	All	Even	Odd
F_1	1		1
F_2	1	1	
F_3	2		2
F_4	3	3	
F_5	5		5
F_6	8	8	
F_7	13		13
F_8	21	21	
F_9	34		34
F_{10}	55	55	
F_{11}	89		89
F_{12}	144	144	

Reference table for Exercises 39–42.

39. Observe the pattern in Exercise 9:

$$1 = 1$$
$$1 + 2 = 3$$
$$1 + 2 + 5 = 8$$
$$1 + 2 + 5 + 13 = 21$$
$$1 + 2 + 5 + 13 + 34 = 55$$

Using the Reference Table above, notice that the odd terms of the Fibonacci sequence are added on the left. Also notice that the numbers on the right are the even terms. Letting $n = 1$, the subscript "$2n$" represents an even number and "$2n + 1$" represents an odd number.

Therefore, the formula is:

$$F_1 + F_3 + \ldots + F_{2n-1} = F_{2n}.$$

41. Observe the pattern in Exercise 11:

$$2^2 - 1^2 = 3$$
$$3^2 - 1^2 = 8$$
$$5^2 - 2^2 = 21$$
$$8^2 = 3^2 = 55.$$

The numbers on the left side of each equation are squares of the Fibonacci numbers, beginning with $F_3^2 - F_{3-2}^2 = F_3^2 - F_1^2$. The next difference of squares is $F_4^2 - F_{4-2}^2 = F_4^2 - F_2^2$, etc. Refer to the Reference Table to see that this pattern continues. The totals on the right are all Fibonacci numbers beginning with F_4, then F_6, F_8, and F_{10}. When $n = 3$, these numbers are represented by $2n - 2$. Note that $2 \cdot 3 = 2 = 4$, $2 \cdot 4 - 2 = 6$, etc. The formula then, when $n = 3$, is:

$$F_n^2 - F_{n-2}^2 = F_{2n-2}.$$

EXTENSION: MAGIC SQUARES

1. 180° in a clockwise direction
    Imagine a straight line (180°) from the top left corner to the bottom right corner of the magic square in Figure 5.5. That moves the 8 from its original position to the bottom right corner, and the other numbers follow.

2	7	6
9	5	1
4	3	8

3. 90° in a clockwise direction
    Imagine the top left corner of the box that contains the number 17 as an origin. Rotate the entire square 90° clockwise. Then the 17 will be in the top right box and the 11 will be in the top left. All the other numbers follow.

11	10	4	23	17
18	12	6	5	24
25	19	13	7	1
2	21	20	14	8
9	3	22	16	15

5. 90° in a counterclockwise direction

    Use the upper right corner of the box containing 15 as the pivot point. Rotate 90° in a counterclockwise direction. Then the 9 will be in the upper right box, and 15 will be in the upper left.

15	16	22	3	9
8	14	20	21	2
1	7	13	19	25
24	5	6	12	18
17	23	4	10	11

7. Figure 5, multiply by 3

24	9	12
3	15	27
18	21	6

    $$\begin{aligned} \text{MS} &= \frac{n(n^2 + 1)}{2} \cdot 3 \\ &= \frac{3(3^2 + 1)}{2} \cdot 3 \\ &= \frac{3(10)}{2} \cdot 3 \\ &= 45 \end{aligned}$$

9. Figure 7, divide by 2

$\frac{17}{2}$	12	$\frac{1}{2}$	4	$\frac{15}{2}$
$\frac{23}{2}$	$\frac{5}{2}$	$\frac{7}{2}$	7	8
2	3	$\frac{13}{2}$	10	11
5	6	$\frac{19}{2}$	$\frac{21}{2}$	$\frac{3}{2}$
$\frac{11}{2}$	9	$\frac{25}{2}$	1	$\frac{9}{2}$

    $$\begin{aligned} \text{MS} &= \frac{n(n^2 + 1)}{2} \div 2 \\ &= \frac{5(5^2 + 1)}{2} \div 2 \\ &= \frac{5(26)}{2} \div 2 \\ &= \frac{65}{2} \end{aligned}$$

11. Using the third row the magic sum is

    $$281 + 467 + 59 = 807.$$

    Then

    $$807 - (71 + 257) = 479.$$

    The missing entry is 479.

13. Using the first column the magic sum is

$$389 + 71 + 347 = 807.$$

Then

$$807 - (191 + 149) = 467.$$

The missing entry is 467.

15. Using the first column the magic sum is

$$401 + 17 + 389 = 807.$$

Then

$$807 - (257 + 281) = 269.$$

The missing entry is 269.

17. Using the second column the magic sum is

$$68 + 72 + 76 = 216.$$

(a) $216 - (75 + 68) = 73$

(b) $216 - (75 + 71) = 70$

(c) Use the answer from (b) to find

$$216 - (72 + 70) = 74.$$

(d) $216 - (71 + 76) = 69$

19. Using the second column to obtain the magic sum

$$20 + 14 + 21 + 8 + 2 = 65.$$

(a) $65 - (3 + 20 + 24 + 11) = 7$

(b) $65 - (14 + 1 + 18 + 10) = 22$

(c) $65 - (9 + 21 + 13 + 17) = 5$

(d) $65 - (16 + 8 + 25 + 12) = 4$

(e) Use the first column and (b):

$$65 - (3 + 9 + 16 + 22) = 15.$$

(f) Use the third column and (a):

$$65 - (1 + 13 + 25 + 7) = 19.$$

(g) Use the fourth column and (c):

$$65 - (24 + 18 + 12 + 5) = 6.$$

(h) Use the fifth column and (d):

$$65 - (11 + 10 + 17 + 4) = 23.$$

21. Use the "staircase method" to construct a magic square of order 7, containing the entries 1, 2, 3, ...

	31	40	49	2	11	20	
30	39	48	1	10	19	28	30
38	47	7	9	18	27	29	38
46	6	8	17	26	35	37	46
5	14	16	25	34	36	45	5
13	15	24	33	42	44	4	13
21	23	32	41	43	3	12	21
22	31	40	49	2	11	20	

23. The sum of the entries in the four corners is
$16 + 3 + 4 + 1 = 34.$

25. The entries in the diagonals are

$$16 + 10 + 7 + 1 + 13 + 11 + 6 + 4 = 68.$$

The entries not in the diagonals are

$$3 + 2 + 5 + 8 + 9 + 12 + 15 + 14 = 68.$$

27. Sum of cubes of diagonal entries:

$16^3 + 10^3 + 7^3 + 1^3 + 13^3 + 11^3 + 6^3 + 4^3 =$
$4096 + 1000 + 343 + 1 + 2197 + 1331 + 216 + 64 =$
$9248.$

Sum of cubes of entries not in the diagonals:

$3^3 + 2^3 + 5^3 + 8^3 + 9^3 + 12^3 + 15^3 + 14^3 =$
$27 + 8 + 125 + 572 + 729 + 1728 + 3375 + 2749 =$
$9248.$

29. $16^2 + 3^2 + 2^2 + 13^2 + 9^2 + 6^2 + 7^2 + 12^2 =$
$256 + 9 + 4 + 169 + 81 + 36 + 49 + 144 = 748;$

$5^2 + 10^2 + 11^2 + 8^2 + 4^2 + 15^2 + 14^2 + 1^2 =$
$25 + 100 + 121 + 64 + 16 + 196 + 1 = 748.$

31.

→	2	3	→
5	→	→	8
9	→	→	12
→	14	15	→

16	2	3	13
5	11	10	8
9	7	6	12
4	14	15	1

The second and third columns are reversed.

33. $a = 16, b = 2, c = -6$

Replace a, b, and c with these numbers to find the entries in the magic square.

$$a + b = 16 + 2 = 18$$
$$a - b - c = 16 - 2 - (-6) = 20$$
$$a + c = 16 + (-6) = 10$$
$$a - b + c = 16 - 2 + (-6) = 8$$
$$a = 16$$
$$a + b - c = 16 + 2 - (-6) = 24$$
$$a - c = 16 - (-6) = 22$$
$$a + b + c = 16 + 2 + (-6) = 12$$
$$a - b = 16 - 2 = 14$$

The magic square is then as follows.

18	20	10
8	16	24
22	12	14

35.

39	48	57	10	19	28	37
47	56	16	18	27	36	38
55	15	17	26	35	44	46
14	23	25	34	43	45	54
22	24	33	42	51	53	13
30	32	41	50	52	12	21
31	40	49	58	11	20	29

$$\begin{aligned} MS &= \frac{7(2 \cdot 10 + 7^2 - 1)}{2} \\ &= \frac{7(20 + 49 - 1)}{2} \\ &= \frac{7(68)}{2} \\ &= 238 \end{aligned}$$

37. There are many ways to find the magic sum. One of them is by adding the top row:

$$52 + 61 + 4 + 13 + 20 + 29 + 36 + 45 = 260.$$

39. $52 + 45 + 16 + 17 + 54 + 43 + 10 + 23 = 260.$

41. Start by placing 1 in the fourth row, second column. Move up two, right one and place 2 in the second row, third column. If we now move up two, right one again, we will go outside the square. Drop down 5 cells (one complete column) to the bottom of the fourth column to place 3. Move up two, right one, to place 4 in the third row, fifth column. Moving up two, right one, we again go outside the square; move 5 cells to the left to place 5 in the first row, first column. Moving up two, right one, takes us two rows outside the square. Counting downward 5 cells, find that the fourth row, second

column is blocked with a 1 already there. The number 6 is then placed just below the entry 5. Continue in this manner until all 24 numbers have been placed. Notice that in trying to enter the number 21, it is blocked by 16 which is already in the cell. Because 20 is already in the bottom cell of the last row, dropping a cell just "below" this one, moves it to the top row, third column. The completed magic square is shown here.

	14	22	10	18	
	20	3	16	24	
5	13	21	9	17	5
6	19	2	15	23	11
12	25	8	16	4	12
18	1	14	22	10	
24	7	20	3	11	

CHAPTER 5 TEST

1. False; 2 and 3 differ by one and both are prime numbers.

2. True

3. True, because 9 is divisible by 3

4. True

5. True; 1 is a factor of all numbers. The smallest multiple of a given number is found by multiplying that number by 1.

6. 331,153,470

(a) It is divisible by 2 because it is even.

(b) It is divisible by 3 because the sum of the digits is 27, a number divisible by 3.

(c) It is not divisible by 4 because the last two digits taken as the number 70, is not divisible by 4.

(d) It is divisible by 5 because it ends in 0.

(e) It is divisible by 6 because it is divisible by both 2 and 3.

(f) It is not divisible by 8 because it is not divisible by both 2 and 4; it is only divisible by 2.

(g) It is divisible by 9 because the sum of the digits, 27, is divisible by 9.

(h) It is divisible by 10 because it ends in 0.

(i) It is not divisible by 12 because it is not divisible by both 3 and 4; it is only divisible by 3.

7. (a) The number 93 is composite because it has factors other than itself and 1. $3 \cdot 31 = 93$.

(b) The number 1 is neither prime nor composite.

(c) The number 59 is prime because its only factors are 59 and 1.

8.

$$\begin{array}{r} 2 \,\lfloor\,1440 \\ 2 \,\lfloor\,720 \\ 2 \,\lfloor\,360 \\ 2 \,\lfloor\,180 \\ 2 \,\lfloor\,90 \\ 3 \,\lfloor\,45 \\ 3 \,\lfloor\,15 \\ 5 \end{array}$$

The prime factorization of 1440 is $2^5 \cdot 3^2 \cdot 5$.

9. Writing exercise

10. (a) The only proper factor of 17 is 1, so 17 is deficient.

(b) The proper factors of 6 are 1, 2, and 3. Because $1 + 2 + 3 = 6$, the number is perfect.

(c) The proper factors of 24 are 1, 2, 3, 4, 6, 8, and 12. Their sum, $1 + 2 + 3 + 4 + 6 + 8 + 12 = 36$, is greater than 24, so the number is abundant.

11. Statement c is false. Goldbach's Conjecture for the number 8 is verified by the equation $8 = 3 + 5$. The number 1 is not prime.

12. A pair of twin primes between 40 and 50 is 41 and 43.

13.
$$270 = 2 \cdot 3^3 \cdot 5$$
$$450 = 2 \cdot 3^2 \cdot 5^2$$

Then $2 \cdot 3^2 \cdot 5 = 90$, the greatest common factor.

14.
$$24 = 2^3 \cdot 3$$
$$36 = 2^2 \cdot 3^2$$
$$60 = 2^2 \cdot 3 \cdot 5$$

Then $2^3 \cdot 3^2 \cdot 5 = 360$, the least common multiple.

15. This exercise is similar to Exercise 27 in section 5.4. If Sherrie is off every 6 days and Della is off every 4 days, this corresponds to modulo 6 and modulo 4. Also, the days of the week, beginning with Sunday, correspond to modulo 7.

For Sherrie, start with $x = 3 \pmod 6$. This is the set of integers $\{3, 9, 15, 21, \ldots\}$. For Della, $x = 3 \pmod 4$. This is the set of integers $\{3, 7, 11, 15, 19, \ldots\}$.

The next common number for Sherrie and Della is 15, which corresponds to Monday.

16. $17,711 + 28,657 = 46,368$

17. Notice that the numbers that are being subtracted are four Fibonacci numbers, in order, and each "set" of them deletes the first and adds another as the equations progress. The final values on the right side of the equation are also Fibonacci numbers. Finally, the first number in each equation is also a Fibonacci number, with the first several of them omitted. The next equation should be

$$89 - (8 + 13 + 21 + 34) = 13.$$

18. (b) Sometimes. See Section 5.1.

19. $(12 + 16)(\bmod 5)$

First add 12 and 16 to get 28. Then divide 28 by 5. The remainder is 3, so $(12 + 16) \pmod 5 \equiv 3$.

20. $(4x - 7) \equiv 2 \pmod 3$

The value on the left must be congruent to 2, mod 3. Replace x with integer values beginning with zero, to see if it makes the equation true.

$x = 0$: Is it true that $4 \cdot 0 - 7 \equiv 2 \pmod 3$? No
$x = 1$: Is it true that $4 \cdot 1 - 7 \equiv 2 \pmod 3$? No
$x = 2$: Is it true that $4 \cdot 2 - 7 \equiv 2 \pmod 3$? No
$x = 3$: Is it true that $4 \cdot 3 - 7 \equiv 2 \pmod 3$? Yes

Then the set of positive solutions for x is $\{3, 6, 9, 12, \ldots\}$, all the multiples of 3.

21. (a) The sequence is obtained by adding two successive terms into order to obtain the next:
$$1, 5, 6, 11, 17, 28, 45, 73.$$

(b) First let us choose 11.

$$11^2 = 121$$
$$6 \cdot 17 = 102$$
$$121 - 102 = 19$$

Now let us choose 45.

$$45^2 = 2025$$
$$28 \cdot 73 = 2044$$
$$2044 - 2025 = 19$$

It appears that this process yields 19 each time.

22. Choice (a) is the exact value of the golden ratio. Options (c) and (d) are approximations.

23. Writing exercise

6   THE REAL NUMBERS AND THEIR REPRESENTATIONS

6.1 Real Numbers, Order, and Absolute Value
6.2 Operations, Properties, and Applications of Real Numbers
6.3 Rational Numbers and Decimal Representation
6.4 Irrational Numbers and Decimal Representation
6.5 Applications of Decimals and Percents
Extension: Complex Number
Chapter 6 Test

Chapter Goals

This chapter includes the study of the subsets of the real numbers: natural numbers, whole numbers, integers, rational numbers and irrational numbers.   After completing the chapter, the student should be able to

- Identify the various types of numbers in the set of real numbers.
- Identify the properties of real numbers.
- Understand and apply the fundamental order of operations.
- Add, subtract, multiply, and divide rational numbers.
- Convert a rational number to decimal form and vice versa.
- Add, subtract, multiply and divide rational numbers in decimal form.
- Round decimal numbers to the nearest tenth and the nearest hundredth.
- Convert decimal numbers to percent form and vice versa.
- Work problems involving percent.
- Simplify irrational numbers.
- Simplify complex numbers.

Chapter Summary

The student should gain an understanding of the structure of the set of real numbers and their properties. The structure of the subsets of the real numbers can be seen by examining

Figure 4 in the text. A few examples of each type of number are shown there.

This diagram shows that the set of real numbers can be described in terms of the following subsets: rational numbers, integers, whole numbers, natural numbers, and irrational numbers. The set of real numbers is partitioned into two sets, rational numbers and irrational numbers.   There are no elements in common between the two sets.

The set of rational numbers has structure as shown in Figure 4. The set of natural numbers, also called the counting numbers, is a proper subset of the whole numbers. The set of whole numbers contains zero in addition to the natural numbers. The set of whole numbers is a proper subset of the integers. The set of integers contains negative natural numbers in addition to whole numbers. The set of integers is a proper subset of the rational numbers.   The set of rational numbers contains fractions in addition to integers.   And finally, the set of rational numbers is a subset of the real numbers.

It is often necessary to carefully think about the interrelationships between numbers and what the names of numbers really mean. The need to distinguish between the different kinds of real numbers can be seen in this example.

EXAMPLE A   Rational numbers

Decide whether the following statement is true or false: Every rational number is an integer.

Solution
False. A rational number is a number that can be written as the ratio of two integers. Some rational numbers that are not integers are 2/3 and -1.5. The second number -1.5 can also be written as $-3/2$.

The properties of real numbers are explained in section 6.2. They are summarized here.

The properties of real numbers a, b, and c can be shown as follows:

PROPERTY	ADDITION
Closure	The sum of $a + b$ is a real number.
Commutative	$a + b = b + a$
Associative	$(a + b) + c = a + (b + c)$
Identity	$a + 0 = a$ and $0 + a = a$
Inverse	If $a \neq 0$, there exists a number $-a$ such that $a + (-a) = 0$ and $(-a) + a = 0$.

PROPERTY	MULTIPLICATION
Closure	The product ab is a real number.
Commutative	$ab = ba$
Associative	$(ab)c = a(bc)$
Identity	$a \cdot 1 = a$ and $1 \cdot a = a$
Inverse	If $a \neq 0$, there exists a number $\frac{1}{a}$ such that $a \cdot \frac{1}{a} = 1$ and $\frac{1}{a} \cdot a = 1$.

The Distributive Property combines the two operations of multiplication and addition. It can be show algebraically as:

$$a(b + c) = ab + ac \text{ or } (b + c)a = ba + ca.$$

The order of operations is important when more than one operation is indicated. If parentheses or fraction bars are present, simplify within parentheses, innermost first, and above and below fraction bars separately, in the following order.

1. Apply all exponents.
2. Perform multiplication or division in the order in which they occur, working from left to right.
3. Perform additions or subtractions in the order in which they occur, working from left to right.

If no parentheses or fraction bars are present, start with Step 1.

EXAMPLE B   Order of operations

$$\frac{-9 \cdot (-2) - (-4) \cdot (-2)}{-2(3) - (-2) \cdot (2)} = \frac{18 - 8}{-6 - (-4)}$$
$$= \frac{10}{-6 + 4}$$
$$= \frac{10}{-2}$$
$$= -5$$

The fraction bar acts as a grouping symbol. Perform multiplications within the numerator and denominator separately. Then perform subtractions. Simplify the fraction.

As stated earlier, rational numbers are numbers that can be written as the ratio of one integer to another. Fractions such as 1/2 and 4/3 are just two examples of rational numbers. Below are examples of multiplying, dividing, and adding rational numbers.

EXAMPLE C   Product of rational numbers

$$\frac{1}{10} \cdot \frac{6}{3} = \frac{1 \cdot (2 \cdot 3)}{(5 \cdot 2) \cdot 3}$$
$$= \frac{1 \cdot (2 \cdot 3)}{5 \cdot (2 \cdot 3)}$$
$$= \frac{1}{5} \cdot 1 \cdot 1$$
$$= \frac{1}{5}$$

The 6 and the 10 are factored in order to simplify the fraction readily. Because $\frac{2}{2} = 1$ and $\frac{3}{3} = 1$, the final answer is $\frac{1}{5}$.

EXAMPLE D   Quotient of rational numbers

$$\frac{3}{8} \div \frac{5}{4} = \frac{3}{8} \cdot \frac{4}{5}$$
$$= \frac{12}{40}$$
$$= \frac{4 \cdot 3}{4 \cdot 10}$$
$$= \frac{3}{10}$$

To find the quotient of rational numbers the divisor is inverted and the two fractions are then multiplied. The numerator 12 and the denominator 40 are factored in order to simplify the fraction.

To find the sum or difference a common denominator must be found. See Chapter 5, Section 3 for an explanation of least common denominators or least common multiples.

EXAMPLE E   Sum of rational numbers

$$\frac{5}{9} + \frac{4}{15} = \frac{5}{9} \cdot \frac{5}{5} + \frac{4}{15} \cdot \frac{3}{3}$$
$$= \frac{25}{45} + \frac{12}{45}$$
$$= \frac{37}{45}$$

The LCM is 45; each fraction must be "built" to obtain 45 in the denominator. Then the numerators are added: $25 + 12 = 37$.

Rational numbers can also be expressed as decimals by dividing the numerator of the fraction by the denominator.

EXAMPLE F Convert a fraction to a decimal.

To change 3/4 to a decimal numeral, divide 3 by 4.

$$
\begin{array}{r}
.75 \\
4\overline{)3.00} \\
\underline{2\ 8} \\
20 \\
\underline{20} \\
\end{array}
$$

$$\frac{3}{4} = .75$$

If the digits in the answer continue to repeat, it is called a repeating decimal instead of a terminating decimal.

EXAMPLE G Repeating decimal

The fraction $\frac{3}{11}$ becomes a repeating decimal when 3 is divided by 11.

$$
\begin{array}{r}
.2727\ldots \\
11\overline{)3.0000} \\
\underline{2\ 2} \\
80 \\
\underline{77} \\
30 \\
\underline{22} \\
80 \\
\underline{77} \\
\end{array}
$$

There are various methods of converting from the decimal form of a number to its percent form. These techniques can be studied in more depth in the text. Here are some examples.

EXAMPLE H Decimal to percent

$$.42 = \frac{42}{100} = 42\%.$$

Think of .42 as forty-two hundredths or "forty-two per hundred." Remember that percent means "per hundred."

EXAMPLE I Percent to decimal

$$8\% \text{ means } \frac{8}{100} = .08.$$

Percent problems can be done in a variety of ways. One technique is shown here.

EXAMPLE J Percent problem

28% of 596 means $(.28)(596) = 166.88$
In mathematics the word "of" translates as "times."

Many students find the ratio method of solving percent problems a more convenient method to use. The percent is expressed as a ratio or fraction in which the numerator is the value of the percent and the denominator is 100. This fraction is set equal to the ratio of the other two numbers.

EXAMPLE K Percent problem

What number is 118% of 36?

Let $n = $ the number

$$
\begin{aligned}
\frac{118}{100} &= \frac{n}{36} \\
100 \cdot n &= 118 \cdot 36 \\
100n &= 4248 \\
\frac{100n}{100} &= \frac{4248}{100} \\
n &= 42.48
\end{aligned}
$$

Numbers that are not rational are called irrational. They cannot be written as the ratio of one integer to another. Irrational numbers are often written as radicals that must be simplified.

EXAMPLE L Simplifying irrational numbers

To simplify $\sqrt{12}$ factor 12 into $4 \cdot 3$. Because 4 is a perfect square the radical can be written as $\sqrt{4 \cdot 3}$ or $\sqrt{4} \cdot \sqrt{3}$. This simplifies to $2\sqrt{3}$.

6.1 EXERCISES

1. An integer between 3.5 and 4.5 is 4.

3. A whole number that is not positive and is less than 1 is 0.

5. An irrational number that is between $\sqrt{11}$ and $\sqrt{13}$ is $\sqrt{12}$. There are many others.

7. It is true that every natural number is positive. The natural numbers consist of $\{1, 2, 3, 4, \ldots\}$.

9. True. The set of integers is included in the set of rational numbers.

11. (a) $3, 7$

    (b) $0, 3, 7$

    (c) $-9, 0, 3, 7$

    (d) $-9, -1\frac{1}{4}, -\frac{3}{5}, 0, 3, 5.9, 7$

    (e) $-\sqrt{7}, \sqrt{5}$

    (f) All are real numbers.

13. Writing exercise.

15. The change is 93,000 housing starts.

17. $-30°$.

19. $-31,532$.

21. -8.

23. (a) Pacific Ocean, Indian Ocean, Caribbean Sea, South China Sea, Gulf of California

    (b) Point Success, Ranier, Matlalcueyetl, Steele, McKinley

    (c) This statement is true because the absolute value of each number is its nonnegative value.

    (d) This statement is false. The absolute value of the depth of the Gulf of California is 2375; the absolute value of the depth of the Caribbean Sea is 8448.

25.

    -6 -4 -2 0 2 4

27.

    $-3\frac{4}{5}$ $-1\frac{5}{8}$ $\frac{1}{4}$ $2\frac{1}{2}$

    -4 -2 0 2 4

29. (a) $|-7| = 7$, which is choice A.

    (b) $-(-7) = 7$, which is choice A.

(c) $-|-7| = -7$, which is choice B.

(d) $-|-(-7)| = -|7| = -7$, which is choice B.

31. -2

    (a) Additive inverse is 2.

    (b) Absolute value is 2.

33. 6

    (a) Additive inverse is -6.

    (b) Absolute value is 6.

35. $7 - 4 = 3$

    (a) Additive inverse is -3.

    (b) Absolute value is 3.

37. $7 - 7 = 0$

    (a) Additive inverse is 0.

    (b) Absolute value is 0.

39. If $a - b > 0$, then the absolute value of $a - b$ in terms of a and b is $a - b$ because this expression produces a nonnegative number.

41. -12.

43. -8.

45. The smaller number is 3 because $|-4| = 4$.

47. $|-3|$ is the smaller number.

49. $-|-6| = -6$, the smaller number.

51. $|5 - 3| = |2| = 2$; $|6 - 2| = |4| = 4$. The first is the smaller number.

53. $6 > -(-2)$ is a true statement because $6 > 2$.

55. $-4 \leq -(-5)$ is true because $-4 \leq 5$.

57. $|-6| \leq |-9|$ is true because $6 < 9$.

59. $-|8| > |-9|$ is false because $-8 < 9$.

61. $-|-5| \geq -|-9|$ is true because $-5 \geq -9$.

63. $|6 - 5| \geq |6 - 2|$ is false because $1 \leq 4$.

65. The greatest drop is for the commodity softwood plywood during the year 1995 to 1996—a drop of 14.1.

67. True. $|11.3| < |-14.1|$.

69. Writing exercise

71. Three positive real numbers but not integers between -6 and 6 are 5.1, 0.25, and $4\frac{1}{2}$.

73. Three real numbers but not whole numbers between -6 and 6 are -5.1, 0.25, and $4\frac{1}{2}$.

75. Three real numbers but not rational numbers between -6 and 6 are $-\sqrt{6}$, $\sqrt{2}$, and $\sqrt{5}$.

6.2 EXERCISES

1. The sum of two negative numbers will always be a <u>negative</u> number.

3. To simplify the expression $8 + [-2 + (-3 + 5)]$, I should begin by adding $\underline{-3}$ and $\underline{5}$, according to the rule for order of operations.

5. Writing exercise

7. $-12 + (-8) = -20$

9. $12 + (-16) = -4$

11. $-12 - (-1) = -12 + 1 = -11$

13. $-5 + 11 + 3 = 6 + 3 = 9$

15. $12 - (-3) - (-5) = 12 + 3 + 5 = 20$

17. $-9 - (-11) - (4 - 6) = -9 + 11 - (-2)$
$$= -9 + 11 + 2$$
$$= 2 + 2$$
$$= 4$$

19. $(-12)(-2) = 24$

21. $9(-12)(-4)(-1)3 = -1296$

23. $\dfrac{-18}{-3} = 6$

25. $\dfrac{36}{-6} = -6$

27. $\dfrac{0}{12} = 0$

29. $-6 + [5 - (3 + 2)] = -6 + [5 - 5]$
$$= -6 + 0$$
$$= -6$$

31. $-8(-2) - [(4^2) + (7 - 3)] = 16 - [16 + (4)]$
$$= 16 - [20]$$
$$= -4$$

33. $-4 - 3(-2) + 5^2 = -4 + 6 + 25$
$$= 2 + 25$$
$$= 27$$

35. $(-8 - 5)(-2 - 1) = (-13)(-3)$
$$= 39$$

37. $\dfrac{(-6 + 3) \cdot (-4)}{-5 - 1} = \dfrac{(-3) \cdot (-4)}{-6}$
$$= \dfrac{12}{-6}$$
$$= -2$$

39. $\dfrac{2(-5) + (-3)(-2^2)}{-6 + 5 + 1} = \dfrac{-10 + (-3)(-4)}{-6 + 6}$
$$= \dfrac{-10 + (-3)(-4)}{0}$$

Division by zero is undefined.

41. $-\dfrac{1}{4}[3(-5) + 7(-5) + 1(-2)] = -\dfrac{1}{4}[-15 - 35 - 2]$
$$= -\dfrac{1}{4}[-52]$$
$$= 13$$

43. Division by zero is undefined, so A, B, and C are all undefined.

45. Commutative property of addition

47. Associative property of addition

49. Inverse property of addition

51. Identity property of multiplication

53. Identity property of addition

55. Distributive property

57. Closure property of addition

59. (a) $6 - 8 = -2$ and $8 - 6 = 2$

(b) By the results of part (a), we may conclude that subtraction is not a(n) <u>commutative</u> operation.

(c) When $a = b$, it is a true statement. For example, let $a = b = 5$. Then $5 - 5 = 5 - 5$ or 0.

61. (a) The inverse of cleaning up your room would be messing up your room.

(b) The inverse of earning money would be spending money.

(c) The inverse of increasing the volume on your portable CD player would be decreasing the volume.

63. Jack recognized the identity property for addition.

65. Use the given hint: Let $a = 2$, $b = 3$, $c = 4$. Now test $a + (b \cdot c) = (a + b) \cdot (a + c)$.

$$a + (b \cdot c) = 2 + (3 \cdot 4) = 2 + 12$$
$$= 14$$

but,

$$(a + b) \cdot (a + c) = (2 + 3) \cdot (2 + 4) = 5 \cdot 6$$
$$= 30$$

The two expressions are not equivalent. The distributive property for multiplication with respect to addition does not hold.

67. Writing exercise

69. $-3^4 = -81$ The notation indicates the opposite of 3^4.

71. $(-3)^4 = (-3)(-3)(-3)(-3) = 81$

73. $-(-3)^4 = -81$

75. $-[-(-3)]^4 = -[3]^4 = -81$

77. (a) The change in outlay from 1988 to 1989 was

$$303.6 - 290.4 = \$13.2 \text{ billion.}$$

(b) The change in outlay from 1986 to 1987 was

$$282.0 - 273.4 = \$8.6 \text{ billion.}$$

(c) The change in outlay from 1990 to 1991 was

$$273.3 - 299.3 = -\$26.0 \text{ billion.}$$

(d) The change in outlay from 1993 to 1994 was

$$279.8 - 291.1 = -\$11.3 \text{ billion.}$$

79. (a) The change in the number of stolen bases from 1991 to 1992: $48 - 58 = -10$.

(b) The change in the number of stolen bases from 1992 to 1993: $53 - 48 = 5$.

(c) The change in the number of stolen bases from 1989 to 1990: $65 - 77 = -12$.

(d) The change in the number of stolen bases from 1993 to 1994: $22 - 53 = -31$.

81. Shalita's new balance is $54 - 89 = -\$35.00$.

83. $-4 + 49 = 45°F$

85. $-27 - 14 = -41°F$

87. $14,494 - (-282) = 14,494 + 282 = 14,776$ feet

89. $660 - 2(45) - 205 = 660 - 90 - 205 = 365$ pounds

91. $6 - 12 + 43 = -6 + 43 = 37$ yards

93. $-19 + 28 - 5 + 13$ Sometimes it is easier to use the commutative property of addition to change the order of the terms so that the negative numbers can be combined separately from the positive numbers as follows.

$$\begin{aligned} -19 + 28 - 5 + 13 &= -19 - 5 + 28 + 13 \\ &= -24 + 41 \\ &= 17 \end{aligned}$$

95.

$$\begin{aligned} -870 + 35.90 &+ 150 - 82.50 - 2(10) + 500 - 37.23 \\ &= -870 - 82.50 - 20 - 37.23 + 35.90 + 150 + 500 \\ &= -1009.73 + 685.90 \\ &= -323.83 \end{aligned}$$

This means that she still owes $323.83.

6.3 EXERCISES

1. $\dfrac{4}{8} = \dfrac{1}{2} = .5 = .5\overline{0}$ which are choices A, C, and D.

The fraction $\frac{4}{8}$ can be simplified to its equivalent fraction $\frac{1}{2}$. It can also be changed to decimal form by dividing 4 by 8. Remember that the overline on zero indicates that this digit repeats indefinitely.

3. $\dfrac{5}{9} = .\overline{5}$, which is choice C. When 5 is divided by 9, the digit 5 repeats.

5. $\dfrac{16}{48} = \dfrac{16 \cdot 1}{16 \cdot 3} = \dfrac{1}{3}$

7. $-\dfrac{15}{35} = -\dfrac{5 \cdot 3}{5 \cdot 7} = -\dfrac{3}{7}$

9. $\dfrac{3}{8} = \dfrac{3 \cdot 2}{8 \cdot 2} = \dfrac{6}{16}$
   $\dfrac{3}{8} = \dfrac{3 \cdot 3}{8 \cdot 3} = \dfrac{9}{24}$
   $\dfrac{3}{8} = \dfrac{3 \cdot 4}{8 \cdot 4} = \dfrac{12}{32}$

11. $-\dfrac{5}{7} = -\dfrac{5 \cdot 2}{7 \cdot 2} = -\dfrac{10}{14}$
    $-\dfrac{5}{7} = -\dfrac{5 \cdot 3}{7 \cdot 3} = -\dfrac{15}{21}$
    $-\dfrac{5}{7} = -\dfrac{5 \cdot 4}{7 \cdot 4} = -\dfrac{20}{28}$

13. (a) $\dfrac{2}{6} = \dfrac{1}{3}$
    (b) $\dfrac{2}{8} = \dfrac{1}{4}$
    (c) $\dfrac{4}{10} = \dfrac{2}{5}$
    (d) $\dfrac{3}{9} = \dfrac{1}{3}$

15. There are two dots in the intersection of the triangle and the rectangle. This represents 2 out of the 24 dots. As a fraction this is expressed as 2/24 or 1/12.

17. (a) Christine O'Brien had 12 hits out of 36 at-bats. The fraction 12/36 simplifies to 1/3.

(b) Brenda Bravener had 5 hits out of 11 at-bats. The

fraction 5/11 is a little less than 1/2.

(c) Brenda Bravener had 1 home run out of 11 at-bats. The fraction 1/11 is just less than 1/10.

(d) Bill Poole made 9 hits out of 40 times at bat. The fraction 9/40 is just less than 10/40, which equals 1/4.

(e) Otis Taylor made 8 hits out of 16 times at bat; Carol Britz made 10 hits out of 20 times at bat. The fractions 8/16 and 10/20 both reduce to 1/2.

19.
$$\frac{3}{8} + \frac{1}{8} = \frac{3+1}{8}$$
$$= \frac{4}{8}$$
$$= \frac{4 \cdot 1}{4 \cdot 2}$$
$$= \frac{1}{2}$$

21. $\dfrac{5}{16} \cdot \dfrac{3}{3} + \dfrac{7}{12} \cdot \dfrac{4}{4} = \dfrac{15}{48} + \dfrac{28}{48} = \dfrac{43}{48}$

23. $\dfrac{2}{3} \cdot \dfrac{8}{8} - \dfrac{7}{8} \cdot \dfrac{3}{3} = \dfrac{16}{24} - \dfrac{21}{24} = -\dfrac{5}{24}$

25. $\dfrac{5}{8} \cdot \dfrac{7}{7} - \dfrac{3}{14} \cdot \dfrac{4}{4} = \dfrac{35}{56} - \dfrac{12}{56} = \dfrac{23}{56}$

27. $\dfrac{3}{4} \cdot \dfrac{9}{5} = \dfrac{27}{20}$

29. $-\dfrac{2}{3} \cdot -\dfrac{5}{8} = \dfrac{10}{24} = \dfrac{5 \cdot 2}{12 \cdot 2} = \dfrac{5}{12}$

31.
$$\frac{5}{12} \div \frac{15}{4} = \frac{5}{12} \cdot \frac{4}{15}$$
$$= \frac{20}{180}$$
$$= \frac{20 \cdot 1}{20 \cdot 9}$$
$$= \frac{1}{9}$$

33.
$$-\frac{9}{16} \div -\frac{3}{8} = -\frac{9}{16} \cdot -\frac{8}{3}$$
$$= \frac{72}{48}$$
$$= \frac{3 \cdot 24}{2 \cdot 24}$$
$$= \frac{3}{2}$$

35.
$$\left(\frac{1}{3} \div \frac{1}{2}\right) + \frac{5}{6} = \left(\frac{1}{3} \cdot \frac{2}{1}\right) + \frac{5}{6}$$
$$= \frac{2}{3} + \frac{5}{6}$$
$$= \frac{2}{3} \cdot \frac{2}{2} + \frac{5}{6}$$
$$= \frac{4}{6} + \frac{5}{6}$$
$$= \frac{9}{6}$$

The fraction $\frac{9}{6}$ can be simplified: $\frac{3 \cdot 3}{3 \cdot 2} = \frac{3}{2}$.

37. (a) $6 \cdot \dfrac{3}{4} = \dfrac{6}{1} \cdot \dfrac{3}{4} = \dfrac{18}{4}$ or $4\dfrac{1}{2}$ cups

(b) $\dfrac{1}{2}$ of $\left(\dfrac{3}{4} + 1\right) = \dfrac{1}{2} \cdot \left(\dfrac{3}{4} + \dfrac{4}{4}\right) = \dfrac{1}{2} \cdot \dfrac{7}{4} = \dfrac{7}{8}$ cup

39. $4 + \dfrac{1}{3} = \dfrac{4}{1} + \dfrac{1}{3} = \dfrac{12}{3} + \dfrac{1}{3} = \dfrac{13}{3}$

41. $2\dfrac{9}{10} = 2 + \dfrac{9}{10} = \dfrac{2}{1} + \dfrac{9}{10} = \dfrac{20}{10} + \dfrac{9}{10} = \dfrac{29}{10}$

43. $27 \div 4 = 6\dfrac{3}{4}$

45.
$$3\frac{1}{4} + 2\frac{7}{8} = \frac{13}{4} + \frac{23}{8}$$
$$= \frac{26}{8} + \frac{23}{8}$$
$$= \frac{49}{8}$$
$$= 6\frac{1}{8}$$

47.
$$-4\frac{7}{8} \cdot 3\frac{2}{3} = -\frac{39}{8} \cdot \frac{11}{3}$$
$$= -\frac{429}{24}$$
$$= -17\frac{21}{24}$$
$$= -17\frac{7}{8}$$

49. Find the numbers in the first two columns of the first row. These are the high and low prices for AirProduct (APD). Find the difference:
$$59\frac{5}{8} - 42\frac{1}{2} = 59\frac{5}{8} - 42\frac{4}{8} = 17\frac{1}{8}$$

51. Using Method 1

$$\frac{\frac{1}{2} + \frac{1}{4}}{\frac{1}{2} - \frac{1}{4}} = \frac{3}{4} \div \frac{1}{4}$$

$$= \frac{3}{4} \cdot \frac{4}{1}$$

$$= \frac{12}{4}$$

$$= 3$$

53. Using Method 1

$$\frac{\frac{5}{8} - \frac{1}{4}}{\frac{1}{8} + \frac{3}{4}} = \frac{3}{8} \div \frac{7}{8}$$

$$= \frac{3}{8} \cdot \frac{8}{7}$$

$$= \frac{24}{56}$$

$$= \frac{3}{7}$$

55. Using Method 2

$$\frac{\frac{7}{11} + \frac{3}{10}}{\frac{1}{11} - \frac{9}{10}} \cdot \frac{110}{110} = \frac{70 + 33}{10 - 99}$$

$$= \frac{103}{-89}$$

$$= -\frac{103}{89}$$

57. $\dfrac{28}{13} = 2 + \dfrac{2}{13}$

$$= 2 + \frac{1}{\frac{13}{2}}$$

$$= 2 + \frac{1}{6 + \frac{1}{2}}$$

59. $\dfrac{52}{11} = 4 + \dfrac{8}{11}$

$$= 4 + \frac{1}{\frac{11}{8}}$$

$$= 4 + \frac{1}{1 + \frac{3}{8}}$$

$$= 4 + \frac{1}{1 + \frac{1}{\frac{8}{3}}}$$

$$= 4 + \frac{1}{1 + \frac{1}{2 + \frac{2}{3}}}$$

$$= 4 + \frac{1}{1 + \frac{1}{2 + \frac{1}{\frac{3}{2}}}}$$

$$= 4 + \frac{1}{1 + \frac{1}{2 + \frac{1}{1 + \frac{1}{2}}}}$$

61. $2 + \dfrac{1}{1 + \frac{1}{3 + \frac{1}{2}}} = 2 + \dfrac{1}{1 + \frac{1}{\frac{6}{2} + \frac{1}{2}}}$

$$= 2 + \frac{1}{1 + \frac{1}{\frac{7}{2}}}$$

$$= 2 + \frac{1}{1 + \frac{2}{7}}$$

$$= 2 + \frac{1}{\frac{7}{7} + \frac{2}{7}}$$

$$= 2 + \frac{1}{\frac{9}{7}}$$

$$= 2 + \frac{7}{9}$$

$$= \frac{18}{9} + \frac{7}{9}$$

$$= \frac{25}{9}$$

63. $\dfrac{\frac{1}{2} + \frac{3}{4}}{2} = \dfrac{\frac{2}{4} + \frac{3}{4}}{2}$

$$= \frac{\frac{5}{4}}{2}$$

$$= \frac{5}{4} \div \frac{2}{1}$$

$$= \frac{5}{4} \cdot \frac{1}{2}$$

$$= \frac{5}{8}$$

65. $\dfrac{\frac{3}{5} + \frac{2}{3}}{2} = \dfrac{\frac{9}{15} + \frac{10}{15}}{2}$

$$= \frac{\frac{19}{15}}{2}$$

$$= \frac{19}{15} \div \frac{2}{1}$$

$$= \frac{19}{15} \cdot \frac{1}{2}$$

$$= \frac{19}{30}$$

67. $\dfrac{-\frac{2}{3} + \left(-\frac{5}{6}\right)}{2} = \dfrac{-\frac{4}{6} + \left(-\frac{5}{6}\right)}{2}$

$$= \frac{-\frac{9}{6}}{2}$$

$$= -\frac{9}{6} \div \frac{2}{1}$$

$$= -\frac{9}{6} \cdot \frac{1}{2}$$

$$= -\frac{9}{12}$$

$$= -\frac{3}{4}$$

69. $\dfrac{15.30 + 15.08 + 12.37 + 14.23 + 12.43 + 7.69 + 12.33 + 11.39}{8}$

$= \dfrac{100.82}{8}$

$\approx \$12.60$

71. $\dfrac{5 + 9}{6 + 13} = \dfrac{14}{19}$

73. $\dfrac{4 + 9}{13 + 16} = \dfrac{13}{29}$

75. $\dfrac{2 + 3}{1 + 1} = \dfrac{5}{2}$

77. Using the consecutive integers 6 and 7

$$\dfrac{6 + 7}{1 + 1} = \dfrac{13}{2} \text{ or } 6\dfrac{1}{2}$$

The number will be halfway between the integers.

79. $\dfrac{3}{4} = .75$

81. $\dfrac{3}{16} = .1875$

83. $\dfrac{3}{11} = .\overline{27}$

85. $\dfrac{2}{7} = .\overline{285714}$

87. $.4 = \dfrac{4}{10} = \dfrac{2 \cdot 2}{2 \cdot 5} = \dfrac{2}{5}$

89. $.85 = \dfrac{85}{100} = \dfrac{5 \cdot 17}{5 \cdot 20} = \dfrac{17}{20}$

91. $.934 = \dfrac{934}{1000} = \dfrac{2 \cdot 467}{2 \cdot 500} = \dfrac{467}{500}$

93. $\qquad\qquad \dfrac{8}{15}: 15 = 3 \cdot 5$

Because 3 is one of the prime factors of the denominator, the fraction will yield a repeating decimal.

95. $\dfrac{13}{125}: 125 = 5^3$

Because 5 is the only prime number that is a factor of the denominator, the fraction will yield a terminating decimal.

97. $\dfrac{22}{55} = \dfrac{2 \cdot 11}{5 \cdot 11} = \dfrac{2}{5}$

Because 5 is the only prime number that is a factor of the denominator, the fraction will yield a terminating decimal.

99. (a) The decimal representation for 1/3 is .333....

(b) The decimal representation for 2/3 is .666....

(c) .333... + .666... = .999....

(d) Writing exercise

6.4 EXERCISES

1. This number is rational because it can be written as the ratio of one integer to another.

3. This number is irrational because it cannot be written as the ratio of one integer to another; only an approximation of the number can be written in this form.

5. $.37 = \dfrac{37}{100}$, a rational number

7. $.\overline{41}$, a rational number. Use Example 9 from Section 6.3 to show that it is equivalent to the rational number $\frac{41}{99}$.

9. This number is irrational; it is nonterminating and nonrepeating.

11. This number is rational. It can be written as the ratio of one integer to another

$$3\dfrac{14159}{100000} \text{ or } \dfrac{314159}{100000}$$

13. The number symbolized by π is irrational. Its value is a nonterminating, nonrepeating decimal. The values given in Exercises 11 and 12 are approximations of the value of π.

15. (a) $\qquad .272772777277772...$
$\qquad + \quad .616116111611116...$
$\qquad\qquad \overline{.888888888888888...}$

(b) Based on the result of part (a), we can conclude that the sum of two <u>irrational</u> numbers may be a(n) <u>rational</u> number.

17. $\sqrt{39} \approx 6.244997998$

19. $\sqrt{15.1} \approx 3.885871846$

21. $\sqrt{884} \approx 29.73213749$

23. First find $9 \div 8 = 1.125$ on your calculator. Then take the square root of the quotient.

$$\sqrt{1.125} \approx 1.060660172$$

25. $\sqrt{50} = \sqrt{25 \cdot 2} = \sqrt{25} \cdot \sqrt{2} = 5\sqrt{2}$

Using a calculator,

$$\sqrt{50} \approx 7.071067812$$

and

$$5\sqrt{2} \approx 7.071067812.$$

27. $\sqrt{75} = \sqrt{25 \cdot 3} = \sqrt{25} \cdot \sqrt{3} = 5\sqrt{3}$

Using a calculator,

$$\sqrt{75} \approx 8.660254038$$

and

$$5\sqrt{3} \approx 8.660254038.$$

29. $\sqrt{288} = \sqrt{144 \cdot 2} = \sqrt{144} \cdot \sqrt{2} = 12\sqrt{2}$

Using a calculator,

$$\sqrt{288} \approx 16.97056275$$

and

$$12\sqrt{2} \approx 16.97056275.$$

31. $\dfrac{5}{\sqrt{6}} = \dfrac{5}{\sqrt{6}} \cdot \dfrac{\sqrt{6}}{\sqrt{6}} = \dfrac{5\sqrt{6}}{6}$

Using a calculator,

$$\frac{5}{\sqrt{6}} \approx 2.041241452$$

and

$$\frac{5\sqrt{6}}{6} \approx 2.041241452.$$

33. $\sqrt{\dfrac{7}{4}} = \dfrac{\sqrt{7}}{\sqrt{4}} = \dfrac{\sqrt{7}}{2}$

Using a calculator,

$$\sqrt{\frac{7}{4}} \approx 1.322875656$$

and

$$\frac{\sqrt{7}}{2} \approx 1.322875656.$$

35. $\sqrt{\dfrac{7}{3}} = \dfrac{\sqrt{7}}{\sqrt{3}} \cdot \dfrac{\sqrt{3}}{\sqrt{3}} = \dfrac{\sqrt{21}}{3}$

Using a calculator,

$$\sqrt{\frac{7}{3}} \approx 1.527525232$$

and

$$\frac{\sqrt{21}}{3} \approx 1.527525232.$$

37. $\sqrt{17} + 2\sqrt{17} = (1 + 2)\sqrt{17} = 3\sqrt{17}$

39. $5\sqrt{7} - \sqrt{7} = (5 - 1)\sqrt{7} = 4\sqrt{7}$

41. $\begin{aligned} 3\sqrt{18} + \sqrt{2} &= 3\sqrt{9 \cdot 2} + \sqrt{2} \\ &= 3\sqrt{9} \cdot \sqrt{2} + \sqrt{2} \\ &= 3 \cdot 3\sqrt{2} + \sqrt{2} \\ &= 9\sqrt{2} + \sqrt{2} \\ &= (9 + 1)\sqrt{2} \\ &= 10\sqrt{2} \end{aligned}$

43. $\begin{aligned} -\sqrt{12} + \sqrt{75} &= -\sqrt{4 \cdot 3} + \sqrt{25 \cdot 3} \\ &= -\sqrt{4} \cdot \sqrt{3} + \sqrt{25} \cdot \sqrt{3} \\ &= -2\sqrt{3} + 5\sqrt{3} \\ &= (-2 + 5)\sqrt{3} \\ &= 3\sqrt{3} \end{aligned}$

45. (a)

May		I	have	a	large	container	of	coffee
3	.	1	4	1	5	9	2	6

Verify (b) and (c) in a similar manner by mapping the number of letters in each word of the sentence to the digits of the value of π.

47. $355 \div 113 = 3.1415929$, which agrees with the first seven digits in the decimal for π.

49. Using 3.14 for π and the given formula

$$P = 2\pi\sqrt{\frac{L}{32}}$$

$$P \approx 2(3.14)\sqrt{\frac{5.1}{32}}$$

$$P \approx 6.28\sqrt{.159375}$$

$$P \approx 2.5 \text{ seconds}$$

51. Add 6 and 28 to get the total height.
$$H = 6 + 28 = 34$$

Substitute the given values into the formula and follow the order of operations.
$$D = \sqrt{2H}$$
$$D = \sqrt{2 \cdot 34}$$
$$D = \sqrt{68}$$
$$D \approx 8.2 \text{ miles}$$

53. The semiperimeter, s, of the Bermuda triangle is $\frac{1}{2}(850 + 925 + 1300)$ or 1537.5 miles.
$$\sqrt{1537.5(1537.5 - 850)(1537.5 - 925)(1537.5 - 1300)}$$
$$= \sqrt{1537.5(687.5)(612.5)(237.5)}$$
$$\approx 392,000 \text{ square miles}$$

55. Substitute the given values into the formula and follow the order of operations with $L = 4$, $W = 3$, and $H = 2$.
$$D = \sqrt{L^2 + W^2 + H^2}$$
$$D = \sqrt{4^2 + 3^2 + 2^2}$$
$$D = \sqrt{16 + 9 + 4}$$
$$D = \sqrt{29}$$
$$D \approx 5.4 \text{ feet}$$

57. (a) Substitute the given values into the formula and follow the order of operations.
$$s = 30\sqrt{\frac{a}{p}}$$
$$s = 30\sqrt{\frac{862}{156}}$$
$$s \approx 30\sqrt{5.525641026}$$
$$s \approx 70.5 \text{ miles per hour}$$

   (b) Substitute the given values into the formula and follow the order of operations.
$$s = 30\sqrt{\frac{a}{p}}$$
$$s = 30\sqrt{\frac{382}{96}}$$
$$s \approx 30\sqrt{5.525641026}$$
$$s \approx 59.8 \text{ miles per hour}$$

   (c) Substitute the given values into the formula and follow the order of operations.

$$s = 30\sqrt{\frac{a}{p}}$$
$$s = 30\sqrt{\frac{84}{26}}$$
$$s \approx 30\sqrt{3.230769231}$$
$$s \approx 53.9 \text{ miles per hour}$$

59. (a) $\sqrt[3]{125} = 5$

   (b) $\sqrt[3]{1000} = 10$

   (c) $\sqrt[3]{13,824} = 24$

61. (a) Use a calculator to verify that:
$$2^{\frac{1}{2}} = \sqrt{2} \approx 1.414213562.$$

   (b) $7^{\frac{1}{2}} = \sqrt{7} \approx 2.645751311$

   (c) $13.2^{\frac{1}{2}} = \sqrt{13.2} \approx 3.633180425$

   (d) $25^{\frac{1}{2}} = \sqrt{25} \approx 5$

63. $(1.1)^{10} \approx 2.59374246$

   $(1.01)^{100} \approx 2.704813829$

   $(1.001)^{1000} \approx 2.716923932$

   $(1.0001)^{10,000} \approx 2.718145927$

   $(1.00001)^{100,000} \approx 2.718268273$

   The computed values seem to be approaching the value of e.

6.5 EXERCISES

1. True. $3.00(12) = 36$

3. False. When 759.367 is rounded to the nearest hundredth, the result is 759.37.

5. True. $50\% = .5 = \frac{1}{2}$ and multiplying by one half yields the same result as dividing by 2.

7. True. $.70(50) = 35$

9. False. $.99 \, \cent = \frac{99}{100}$ cent, meaning that it is less than the value of one penny.

11. $8.53 + 2.785 = 11.315$

13. $8.74 - 12.955 = -4.215$

15. $25.7 \times .032 = .8224$

17. $1019.825 \div 21.47 = 47.5$

19. $\dfrac{118.5}{1.45 + 2.3} = \dfrac{118.5}{3.75} = 31.6$

21. The 10.2% change in 1988 showed the most drastic increase.

23. The indicated price is .33¢, which means $\dfrac{33}{100}$ of a penny. You should be able to purchase three stamps for one cent.

25. (a) $\$1742.18 + \$9271.94 + \$28.37 = \$11,042.49$

(b) $\$7195.14 + \$511.09 + \$1291.03 = \$8997.26$

(c) $(\$1856.12 + \$11,042.49) - \$8997.26 = \$3901.35$

27. (a) $\text{BAC} = \dfrac{[11.52]}{190} - (.03) \approx .031$

(b) $\text{BAC} = \dfrac{[10.8]}{135} - (.045) = .035$

29. Substitute given values and follow order of operations.

$$\text{Horsepower} = \frac{195 \times 302 \times 4000}{792,000}$$
$$\text{Horsepower} = \frac{235,560,000}{792,000}$$
$$\text{Horsepower} \approx 297$$

31. (a) 78.4

(b) 78.41

33. (a) .1

(b) .08

35. (a) 12.7

(b) 12.69

37. $.42 = \dfrac{42}{100} = 42\%$

39. $.365 = \dfrac{365}{1000} \div \dfrac{10}{10} = \dfrac{36.5}{100} = 36.5\%$

41. $.008 = \dfrac{8}{1000} \div \dfrac{10}{10} = \dfrac{.8}{100} = .8\%$

43. $2.1 = 2\dfrac{1}{10} \cdot \dfrac{10}{10} = 2\dfrac{10}{100} = \dfrac{210}{100} = 210\%$

45. $\dfrac{1}{5} = 1 \div 5 = .2$, which is 20%

47. $\dfrac{1}{100} = 1 \div 100 = .01$, which is 1%

49. $\dfrac{3}{8} = 3 \div 8 = .375$, which is 37.5%

51. $\dfrac{3}{2} = 3 \div 2 = 1.5$, which is 150%

53. Writing exercise

55. (a) 5% means 5 in every 100.

(b) 25% means 6 in every 24.

(c) 200% means 8 for every 4.

(d) .5% means .5 in every 100.

(e) 600 % means 12 for every 2.

57. No. If the item is discounted 20%, its new price is $\$60 - .2 \times \$60 = \$48$. Then, if the new price is marked up 20%, the price becomes $\$48 + .2 \times \$48 = \$57.60$.

59. $10.5 \div 16 = .656$.

61. Using Method 1: $(.26)(480) = 124.8$.

63. Using Method 1: $(.105)(28) = 2.94$.

65. Using Method 2

$$\frac{x}{100} = \frac{45}{30}$$
$$30x = 4500$$
$$x = \frac{4500}{30}$$
$$x = 150\%$$

67. Using Method 2

$$\frac{25}{100} = \frac{150}{x}$$
$$25x = 15000$$
$$x = \frac{15000}{25}$$
$$x = 600$$

69. Using Method 1

$$(x)(28) = .392$$
$$x = \frac{.392}{28}$$
$$x = .014$$
$$x = 1.4\%$$

71. Choice A. The difference between $5 and $4 is $1; $1 compared to the original $4 is 1/4 or 25%.

73. Choice C. Rounding, the population of Alabama is approximately 4,000,000. About 25% of this number is $1/4 \times 4,000,000 = 1,000,000$.

75. The amount of increase from 1991 to 1994 was
$23.4 - .597 = 22.803$. The percent increase is found by
comparing this difference to the original amount in 1991
and expressing that fraction as a percent.

$$\frac{22.803}{.597} \approx 38.20$$

Expressed as a percent, this is 3820%.

77. (a) $\dfrac{20.48}{4151.29} \approx .005$ This is .5%.

    (b) $\dfrac{71.36}{11,489.36} \approx .006$ This is .6%.

79. $0.164(8450) = \$1385.80$

81. $\dfrac{175}{1500} = .11\dfrac{2}{3}$

$$= 11\frac{2}{3}\%$$

83. $29.57
1. Rounded to the nearest dollar, the amount of the bill
is $30.
2. 10% of $30 is $3.
3. 1/2 of $3 is $1.50. $3 + 1.50 = \$4.50$.

85. $5.15
1. Rounded to the nearest dollar, the amount of the bill
is $5.
2. 10% of $5 is $.50.
3. 1/2 of $.50 is $.25. $.50 + .25 = \$.75$.

87. (a) In algebra, we learn that multiplying powers of the
same number is accomplished by adding the exponents.
Thus, $10^3 \cdot 10^2 = 10^{\underline{3+2}} = 10^{\underline{5}}$.

    (b) The __5__ places to the right of the decimal point in
the product are the result of division by $10^{\underline{5}}$.

89. Writing exercise

EXTENSION: COMPLEX NUMBERS

1. $\sqrt{-144} = i\sqrt{144} = 12i$

3. $-\sqrt{-225} = -i\sqrt{225} = -15i$

5. $\sqrt{-3} = i\sqrt{3}$

7. $\sqrt{-75} = i\sqrt{25 \cdot 3} = i\sqrt{25} \cdot \sqrt{3} = 5i\sqrt{3}$

9. $\sqrt{-5} \cdot \sqrt{-5} = i\sqrt{5} \cdot i\sqrt{5}$
$$= i^2\sqrt{5 \cdot 5}$$
$$= i^2\sqrt{25}$$
$$= 5i^2$$
$$= 5(-1)$$
$$= -5$$

11. $\sqrt{-9} \cdot \sqrt{-36} = i\sqrt{9} \cdot i\sqrt{36}$
$$= 3i \cdot 6i$$
$$= 18i^2$$
$$= 18(-1)$$
$$= -18$$

13. $\sqrt{-16} \cdot \sqrt{-100} = i\sqrt{16} \cdot i\sqrt{100}$
$$= i^2 \cdot 4 \cdot 10$$
$$= 40i^2$$
$$= 40(-1)$$
$$= -40$$

15. $\dfrac{\sqrt{-200}}{\sqrt{-100}} = \dfrac{i\sqrt{200}}{i\sqrt{100}}$
$$= \sqrt{\frac{200}{100}}$$
$$= \sqrt{2}.$$

17. $\dfrac{\sqrt{-54}}{\sqrt{6}} = \dfrac{i\sqrt{54}}{\sqrt{6}}$
$$= i\sqrt{\frac{54}{6}}$$
$$= i\sqrt{9}$$
$$= 3i$$

19. $\dfrac{\sqrt{-288}}{\sqrt{-8}} = \dfrac{i\sqrt{288}}{i\sqrt{8}}$
$$= \sqrt{\frac{288}{8}}$$
$$= \sqrt{36}$$
$$= 6$$

21. Writing exercise

23. $i^8 = i^4 \cdot i^4 = 1 \cdot 1 = 1$

25. i^{42}
$42 \div 4 = 10$, with a remainder of 2. This means that
$i^{42} = i^2$, which is -1.

27. i^{47}
$47 \div 4 = 11$ with a remainder of 3. This means that
$i^{47} = i^3$, which is $-i$.

29. i^{101}
$101 \div 4 = 25$ with a remainder of 1. This means that
$i^{101} = i$.

31. Writing exercise

CHAPTER 6 TEST

1. $\left\{ -4, -\sqrt{5}, -3/2, -.5, 0, \sqrt{3}, 4.1, 12 \right\}$

    (a) The only natural number is 12.

    (b) Whole numbers are 0 and 12.

    (c) Integers are -4, 0, and 12.

    (d) Rational numbers are -4, $-3/2$, $-.5$, 0, 4.1, and 12.

    (e) Irrational numbers are $-\sqrt{5}$ and $\sqrt{3}$.

    (f) All the numbers in the set are real numbers.

2. (a) C

    (b) B

    (c) D

    (d) A

3. (a) False. The absolute value of a number is always nonnegative; the absolute value of zero is zero.

    (b) True. Both sides of the equation are equal to 7.

    (c) True.

    (d) False. Zero, a real number, is neither positive nor negative.

4. $$6^2 - 4(9 - 1) = 36 - 4(8)$$
$$= 36 - 32$$
$$= 4$$

5. $$(-3)(-2) - [5 + (8 - 10)] = 6 - [5 + (-2)]$$
$$= 6 - [3]$$
$$= 3$$

6. $$\frac{(-8 + 3) - (5 + 10)}{7 - 9} = \frac{-5 - 15}{-2}$$
$$= \frac{-20}{-2}$$
$$= 10$$

7. (a) $-6439 - 5039 = -11,478$

    (b) $2284 - 20,060 = -17,776$

8. $\underset{\text{After}}{(225 + 3852)} - \underset{\text{Before}}{(-1299 + 80)} = 5296$ feet.

9. $97,069 - 88,140 = 8929.$
$86,133 - 97,069 = -10,936$
$71,558 - 86,133 = -14,575$
$71,128 - 71,558 = -430$
$71,931 - 71,128 = 803.$

10. (a) E, Associative property

    (b) A, Distributive property

    (c) B, Identity property

    (d) D, Commutative property

    (e) F, Inverse property

    (f) C, Closure property

11. (a) Whitney made 4 out of 6, which is more than 1/2.

    (b) Moura made 13 out of 40; because 13 out of 39 would be 1/3, this is just less than 1/3. Dawkins made 2 out of 7; because 2 out of 6 would be 1/3, this is just less than 1/3.

    (c) Whitney made 4 out of 6, which is the same ratio as 2 out of 3.

    (d) Pritchard made 4 out of 10; Miller made 8 out of 20.

$$\frac{4}{10} = \frac{8}{20} = \frac{2}{5}$$

12. $$\frac{3}{16} + \frac{1}{2} = \frac{3}{16} + \frac{1}{2} \cdot \frac{8}{8}$$
$$= \frac{3}{16} + \frac{8}{16}$$
$$= \frac{11}{16}$$

13. $$\frac{9}{20} - \frac{3}{32} = \frac{9}{20} \cdot \frac{8}{8} - \frac{3}{32} \cdot \frac{5}{5}$$
$$= \frac{72}{160} - \frac{15}{160}$$
$$= \frac{57}{160}$$

14. $$\frac{3}{8} \cdot \left(-\frac{16}{15} \right) = -\frac{48}{120}$$
$$= -\frac{2 \cdot 24}{5 \cdot 24}$$
$$= -\frac{2}{5}$$

15. $\dfrac{7}{9} \div \dfrac{14}{27} = \dfrac{7}{9} \cdot \dfrac{27}{14}$

$\phantom{\dfrac{7}{9} \div \dfrac{14}{27}} = \dfrac{7 \cdot 3 \cdot 9}{7 \cdot 2 \cdot 9}$

$\phantom{\dfrac{7}{9} \div \dfrac{14}{27}} = \dfrac{3}{2}$

16. $\dfrac{9}{20} = .45$

17. $\dfrac{5}{12} = .41\overline{6}$

18. $.72 = \dfrac{72}{100} = \dfrac{4 \cdot 18}{4 \cdot 25} = \dfrac{18}{25}$

19. $.\overline{58}$

Let $x = .\overline{58}$

$$100x = 58.585858\ldots$$
$$\underline{-x = .585858\ldots}$$
$$99x = 58$$
$$x = \dfrac{58}{99}$$

20. (a) $\sqrt{10}$ is irrational because it cannot be written as the ratio of two integers.

(b) $\sqrt{16}$ is rational. Its value is 4, which can be written as 4/1.

(c) $.01 = 1/100$, a rational number.

(d) $.\overline{01}$ can be converted to 1/99, a rational number.

(e) $.0101101110\ldots$ is an irrational number.

21. (a) $\sqrt{150} \approx 12.24744871$

(b) $\sqrt{150} = \sqrt{25} \cdot \sqrt{6} = 5\sqrt{6}$

22. (a) $\dfrac{13}{\sqrt{7}} \approx 4.913538149$

(b) $\dfrac{13}{\sqrt{7}} \cdot \dfrac{\sqrt{7}}{\sqrt{7}} = \dfrac{13\sqrt{7}}{\sqrt{49}}$

$\phantom{\dfrac{13}{\sqrt{7}} \cdot \dfrac{\sqrt{7}}{\sqrt{7}}} = \dfrac{13\sqrt{7}}{7}$

23. (a) $2\sqrt{32} - 5\sqrt{128} \approx -45.254834$

(b) $2\sqrt{32} - 5\sqrt{128} = 2\sqrt{16 \cdot 2} - 5\sqrt{64 \cdot 2}$

$\phantom{2\sqrt{32} - 5\sqrt{128}} = 2 \cdot 4\sqrt{2} - 5 \cdot 8\sqrt{2}$

$\phantom{2\sqrt{32} - 5\sqrt{128}} = 8\sqrt{2} - 40\sqrt{2}$

$\phantom{2\sqrt{32} - 5\sqrt{128}} = -32\sqrt{2}$

24. Writing exercise

25. $4.6 + 9.21 = 13.81$

26. $12 - 3.725 - 8.59 = -.315$

27. $86(.45) = 38.7$

28. $236.439 \div (-9.73) = -24.3$

29. (a) 9.04

(b) 9.045

30. $.185(90) = 16.65$

31. $\dfrac{145}{100} = \dfrac{x}{70}$

$100x = 10150$

$x = \dfrac{10150}{100}$

$x = 101.5$

32. (a) 4 out of 15; $4 \div 15 = .26\frac{2}{3} = 26\frac{2}{3}\%.$

(b) 10 out of 15; $10 \div 15 = .66\frac{2}{3} = 66\frac{2}{3}\%.$

33. Choice B. 300,000 out of 900,000 is the same as 3 out of 9, which is 1/3 or 33 1/3%.

34. (a) 3% of 50,000: $.03(50,000) = 1500.$

(b) 21% of 110,000: $.21(110,000) = 23,100.$

(c) 35% of 35,000: $.35(35,000) = 12,250.$

(d) 12% of 65,000: $.12(65,000) = 7800.$

35. $\dfrac{1611 - 1532}{1611} = \dfrac{79}{1611}$

$\phantom{\dfrac{1611 - 1532}{1611}} \approx .049$, which is 4.9%.

7 THE BASIC CONCEPTS OF ALGEBRA

7.1 **Linear Equations**
7.2 **Applications of Linear Equations**
7.3 **Ratio, Proportion, and Variation**
7.4 **Linear Inequalities**
7.5 **Properties of Exponents and Scientific Notation**
7.6 **Polynomials and Factoring**
7.7 **Quadratic Equations and Applications**
CHAPTER 7 TEST

Chapter Goals

After completion of this chapter the student should be able to

- Solve linear equations and inequalities and apply the techniques to solve word problems.
- Translate verbal phrases into mathematical expressions.
- Solve problems involving ratio, proportion, and variation.
- Evaluate exponential expressions.
- Add, subtract, and multiply polynomials.
- Factor polynomials.
- Multiply and divide numbers that are expressed in scientific notation.
- Solve quadratic equations by the methods of: factoring, the square root property, and the quadratic formula.
- Use graphs, set-builder notation, and interval notation to specify an inequality.

Chapter Summary

A linear equation is an equation that can be written in the form $ax + b = c$, in which a, b, and c are real numbers and $a \neq 0$. Linear equations are solved by a series of steps in which an operation is performed to both sides of the equation to keep it "balanced." In the following example note that the same value is added to both sides of the equation in order to isolate the term containing the variable. Then in order to obtain $1z$ or z, both sides are divided by 6.

EXAMPLE A   Solving a linear equation

$$6z + 1 = 43$$
$$\underline{-1 = -1}$$
$$6z = 42$$
$$\frac{6z}{6} = \frac{42}{6}$$
$$z = 7$$

In more complicated equations the distributive property is used first to combine like terms on either side of the equal sign.

EXAMPLE B   Solving a linear equation

$$5(2m - 1) = 4(2m + 1) + 7$$
$$5(2m) + 5(-1) = 4(2m) + 4(1) + 7$$
$$10m + (-5) = 8m + 11$$
$$\underline{-8m \qquad = -8m}$$
$$2m - 5 = 11$$
$$\underline{+5 = +5}$$
$$2m = 16$$
$$\frac{2m}{2} = \frac{16}{2}$$
$$m = 8$$

Summarizing the steps to solve a linear equation:

1. Clear fractions, usually by multiplying both sides by the LCM.
2. Simplify each side of the equation separately; i. e., use the distributive property as in Example B, and combine like terms.
3. Isolate the terms containing the variable on one side of the equation.

4. Apply appropriate operations to obtain the variable with a coefficient of 1.
5. Check.

When solving problems it is necessary to translate or restate the problem using mathematical expressions. Then the techniques of solving an equation can be applied. Often a picture and/or formula is used.

EXAMPLE C   Solving a word problem

Antonio has a board 44 inches long. He wishes to cut it into two pieces so that one piece will be six inches longer than the other. How long should the shorter piece be?

$$x \qquad\qquad x + 6$$

1. Use the diagram to make algebraic statements.
2. Replace words with algebraic terms.
3. Simplify.
4. Subtract 6 from both sides.
5. Divide both sides by 2.

1.	Shorter piece + Longer piece = 44 inches
2.	$x + (x + 6) = 44$
3.	$2x + 6 = 44$
4.	$2x = 38$
5.	$x = 19$

Reread the example to check: If the shorter piece is 19 inches, the longer piece must be $19 + 6 = 25$ inches.

Is 44 the sum of 19 and 25? Yes.

A ratio is a quotient of two quantities. If one ratio is set equal to another, it is called a proportion. To determine whether a proportion is true, the cross-product method can be used:

$\frac{a}{b} = \frac{c}{d}$ is true if $a \cdot d = b \cdot c$.

EXAMPLE D   Proportion

Is this a true statement?

$$\frac{4}{7} = \frac{12}{21}$$

Use cross multiplication to show that

$$4 \cdot 21 = 7 \cdot 12 = 84.$$

Therefore, it is a true statement.

This technique can be used to solve for the unknown in a proportion.

EXAMPLE E   Solving a proportion

State problem.	$\dfrac{10}{5} = \dfrac{z}{20}$
Find the cross product.	$5 \cdot z = 10 \cdot 20$
Simplify.	$5x = 200$
Divide both sides by 5.	$z = 40$

Variation problems can be categorized as direct, inverse, and combined. To solve a variation problem first find a constant of variation. This is often symbolized by "k" (because mathematicians cannot spell!). Then the equation can be solved to answer the question. Here are two examples.

EXAMPLE F   Direct variation

If p varies directly as y, and $p = 4$ when $y = 3$, find p when y is 7.

Solution

Translate "p varies directly as y."	$p = ky$
Substitute values given for p and y.	$4 = k \cdot 3$
Solve for k.	$\dfrac{4}{3} = k$
Replace k with $\dfrac{4}{3}$; replace y with 7.	$p = \dfrac{4}{3} \cdot 7$
Perform the operation.	$p = \dfrac{28}{3}$

EXAMPLE G   Inverse variation

If t varies inversely as s, and $t = 3$ when $s = 5$, find t when $s = 3$.

Solution

Translate "t varies inversely as s."	$t = \dfrac{k}{s}$
Substitute values given for t and s.	$3 = \dfrac{k}{5}$
Solve for k by multiplying both sides by 5.	$15 = k$
Replace k with 15; replace s with 3.	$t = \dfrac{15}{3}$
Perform the operation.	$t = 5$

Inequalities can be specified by set-builder notation, graphs, or interval notation. Here is a simple statement that describes an inequality: "x is less than or equal to negative four." Now here are the three methods of representing this statement mathematically.

EXAMPLE H   Set-builder notation

Set-builder notation for "x is less than or equal to negative four" is $\{x \mid x \leq -4\}$. This mathematical statement is read, "All x such that x is less than or equal to negative four." The symbol " $\mid$ " is translated as "such that" when used in set-builder notation.

EXAMPLE I   Solve and graph an inequality

$$5t + 2 \geq 52$$
$$5t + 2 - 2 \geq 52 - 2$$
$$5t \geq 50$$
$$\frac{5t}{5} \geq \frac{50}{5}$$
$$t \geq 10$$

Interval notation: $[10, \infty)$

Graph:

The mathematical statement "$t \geq 10$" can be shown on a number line. This is a graph of the algebraic inequality. A close bracket is drawn to show that ten is included in the set of numbers. If the statement were $t > 10$, an open parenthesis, "(", would be used.

EXAMPLE J   Interval notation

The inequality $x \leq -4$ is stated in interval notation as $(-\infty, -4]$. This mathematical statement indicates that the interval on the number line that is described is the interval bounded by negative four on the right. If the inequality were $x < -4$, interval notation would be $(-\infty, -4)$. The parenthesis indicates an open interval; the bracket indicates a closed interval.

Inequalities are solved using techniques that are similar to those for equations with one important difference. When multiplying or dividing both sides by a negative quantity, the sense of the inequality changes: $<$ becomes $>$, $\leq$ becomes $\geq$, etc.

EXAMPLE K   Solving an inequality

1. State inequality.
2. Use the distributive property to remove parentheses.
3. Add $6y$ to both sides.
4. Add 10 to both sides.
5. Divide both sides by 26.
6. Restate the answer with the variable on the left side.

1. $-2(3y - 8) \geq 5(4y - 2)$
2. $-6y + 16 \geq 20y - 10$
3. $16 \geq 26y - 10$
4. $26 \geq 26y$
5. $1 \geq y$
6. $y \leq 1$

The rules of operating with exponents are explained in section 7.5, along with applying these rules to work with numbers that are expressed in scientific notation. These properties are summarized in the following table.

For all integers m and n and all real numbers a and b:

Product Rule $a^m \cdot a^n = a^{m+n}$

Quotient Rule $\dfrac{a^m}{a^n} = a^{m-n} (a \neq 0)$

Zero Exponent $a^0 = 1, (a \neq 0)$

Negative Exponent $a^{-n} = \dfrac{1}{a^n} (a \neq 0)$

Power Rules $(a^m)^n = a^{mn}$
$(ab)^m = a^m b^m$
$\left(\dfrac{a}{b}\right)^m = \dfrac{a^m}{b^m}, (b \neq 0)$

EXAMPLE L   Using exponent rules

Use exponent rules to simplify

$$k^{-5} k^{-3} k^4.$$

Solution

$$k^{-5+(-3)+4} = k^{-4}$$

Change the final answer to a positive exponent by using the appropriate rule:

$$\frac{1}{k^4}$$

First the product rule is applied; the bases are the same so the exponents can be added. Then the negative exponent rule is applied to simplify the expression.

A polynomial is a term or many terms that have only nonnegative integer exponents attached to the variables. For example, $2n^3$ is a term; it is also called a monomial because it is a polynomial of only one term. The polynomial $6y^2 - 3y + 1$ is a polynomial of three terms. Polynomials can be added and subtracted by combining like terms. Here is an example from the exercises.

EXAMPLE M   Adding polynomials

$$\left(4m^3 - 3m^2 + 5\right) + \left(-3m^3 - m^2 + 5\right) =$$
$$4m^3 + \left(-3m^3\right) - 3m^2 + \left(-m^2\right) + 5 + 5 =$$
$$m^3 - 4m^2 + 10$$

Solution
Like terms are added by combining their numerical coefficients. $4m^3$ and $-3m^3$ are like terms because they have m^3 in common; both the variable and its exponent must be the same. The numerical coefficients 4 and -3 are added to obtain 1. When the coefficient of a variable is 1, the 1 is not written; i. e., "$1m$" is simply "m." Similarly, if the coefficient of a variable is -1, only the negative sign is written; i. e., "$-1m$" is written "$-m$." The remaining like terms are added in a similar way.

Finding the product of polynomials is also described in section 7.6. The patterns for special products can be found throughout this section or summarized at the end of the chapter. The techniques of factoring are also explained, including greatest common factor, factoring by grouping, factoring a trinomial, and some special cases of factoring. Here is one of the exercises on factoring.

EXAMPLE N   Factoring a polynomial

Factor the trinomial $36a^2 + 60a + 25$.

Solution
This trinomial is one of the special cases. Notice that $36a^2$ can be expressed as $(6a)^2$ and that 25 can be written as $(5)^2$. Also notice that $2 \cdot 6a \cdot 5 = 60a$, which is the middle term. That is, when the product of $6a$ and 5 is doubled, the middle term is obtained. This kind of polynomial is called a perfect square trinomial and is factored as follows: $(6a + 5)(6a + 5)$ or simply $(6a + 5)^2$.

An equation that can be written in the form $ax^2 + bx + c = 0$, where $a \neq 0$, is a quadratic equation. When the equation factors readily, then one method of solving a quadratic equation is by the zero-factor property. Each factor is set equal to 0 and each equation is then solved for the variable.

EXAMPLE O   Solving an equation using the zero-factor property

Set each factor equal to zero and solve each equation.

$$(m + 6)(m + 4) = 0$$

$$m + 6 = 0 \quad \text{or} \quad m + 4 = 0$$
$$m = -6 \qquad\qquad m = -4$$

The solution set is $\{-6, -4\}$.

In the square-root property method, the square root of both sides of the equation must be found.

EXAMPLE P   Solving an equation by the square-root property method

$$x^2 = 64 \qquad \text{implies that } x = \pm\sqrt{64}.$$
$$\sqrt{x^2} = \pm\sqrt{64} \quad \text{Take the square root of both sides.}$$
$$x = \pm 8. \qquad \text{Simplify.}$$

The solution set is $\{8, -8\}$ or simply $\{\pm 8\}$.

Quadratic equations can also be solved by the quadratic formula

$$x = \frac{-b \pm \sqrt{b^2 - 4ac}}{2a}$$

The formula is derived from the standard form of a quadratic equation $ax^2 + bx + c = 0$. When the quadratic formula is used, the equation must first be written in standard form in order to identify a, b, and c.

EXAMPLE Q   Solving an equation by the quadratic formula

For the equation $m^2 + 2m - 5 = 0$, $a = 1$, $b = 2$, and $c = -5$. Substitute these values into the quadratic formula.

$$m = \frac{-(2) \pm \sqrt{(2)^2 - 4 \cdot 1 \cdot -5}}{2 \cdot 1}$$
$$= \frac{-2 \pm \sqrt{4 + 20}}{2}$$
$$= \frac{-2 \pm \sqrt{24}}{2}$$
$$= \frac{-2 \pm \sqrt{4 \cdot 6}}{2}$$
$$= \frac{-2 \pm 2\sqrt{6}}{2}$$
$$= \frac{2\left(-1 \pm 1\sqrt{6}\right)}{2}$$
$$= -1 \pm \sqrt{6}$$

7.1 EXERCISES

1. Equations A and C are linear in x because the variable x is of degree one.

3.
$$3(x + 4) = 5x$$
$$3(6 + 4) = 5 \cdot 6$$
$$3(10) = 30$$
$$30 = 30$$

Six is a solution because the left side of the equation equals the right side when 6 is substituted for x.

5. If two equations are equivalent, they have the same solution set.

7.
$$.06(10 - x)(100) = (100).06(10 - x)$$
$$= 6(10 - x)$$
$$= 6 \cdot 10 - 6 \cdot x$$
$$= 60 - 6x$$

The left side of the equation is equivalent to letter B.

9.
$$7k + 8 = 1$$
$$7k + 8 - 8 = 1 - 8$$
$$7k = -7$$
$$\frac{7k}{7} = \frac{-7}{7}$$
$$k = -1$$

11.
$$8 - 8x = -16$$
$$-8 + 8 - 8x = -8 + (-16)$$
$$-8x = -24$$
$$\frac{-8x}{-8} = \frac{-24}{-8}$$
$$x = 3$$

13.
$$7y - 5y + 15 = y + 8$$
$$2y + 15 = y + 8$$
$$2y + 15 - 15 = y + 8 - 15$$
$$2y = y - 7$$
$$2y - y = y - y - 7$$
$$y = -7$$

15.
$$12w + 15w - 9 + 5 = -3w + 5 - 9$$
$$27w - 4 = -3w - 4$$
$$27w + 3w - 4 = -3w + 3w - 4$$
$$30w - 4 = -4$$
$$30w - 4 + 4 = -4 + 4$$
$$30w = 0$$
$$\frac{30w}{30} = \frac{0}{30}$$
$$w = 0$$

17.
$$2(x + 3) = -4(x + 1)$$
$$2 \cdot x + 2 \cdot 3 = -4 \cdot x - 4 \cdot 1$$
$$2x + 6 = -4x - 4$$
$$2x + 6 - 6 = -4x - 4 - 6$$
$$2x = -4x - 10$$
$$2x + 4x = -4x + 4x - 10$$
$$6x = -10$$
$$\frac{6x}{6} = \frac{-10}{6}$$
$$x = \frac{-2 \cdot 5}{2 \cdot 3}$$
$$x = \frac{-5}{3}$$

19.
$$3(2w + 1) - 2(w - 2) = 5$$
$$3 \cdot 2w + 3 \cdot 1 - 2 \cdot w + 2 \cdot 2 = 5$$
$$6w + 3 - 2w + 4 = 5$$
$$6w - 2w + 3 + 4 = 5$$
$$4w + 7 = 5$$
$$4w + 7 - 7 = 5 - 7$$
$$4w = -2$$
$$\frac{4w}{4} = \frac{-2}{4}$$
$$w = -\frac{2 \cdot 1}{2 \cdot 2}$$
$$w = -\frac{1}{2}$$

21.
$$2x + 3(x - 4) = 2(x - 3)$$
$$2x + 3 \cdot x - 3 \cdot 4 = 2 \cdot x - 2 \cdot 3$$
$$2x + 3x - 12 = 2x - 6$$
$$5x - 12 = 2x - 6$$
$$5x - 12 + 12 = 2x - 6 + 12$$
$$5x = 2x + 6$$
$$5x - 2x = 2x - 2x + 6$$
$$3x = 6$$
$$\frac{3x}{3} = \frac{6}{3}$$
$$x = 2$$

23.
$$6p - 4(3 - 2p) = 5(p - 4) - 10$$
$$6p - 4 \cdot 3 + 4 \cdot 2p = 5 \cdot p - 5 \cdot 4 - 10$$
$$6p - 12 + 8p = 5p - 20 - 10$$
$$6p + 8p - 12 = 5p - 30$$
$$14p - 12 = 5p - 30$$
$$14p - 12 + 12 = 5p - 30 + 12$$
$$14p = 5p - 18$$
$$14p - 5p = 5p - 5p - 18$$
$$9p = -18$$
$$\frac{9p}{9} = \frac{-18}{9}$$
$$x = -2$$

25.
$$-[2z - (5z + 2)] = 2 + (2z + 7)$$
$$-[2z - 5z - 2] = 2 + 2z + 7$$
$$-[-3z - 2] = 2 + 7 + 2z$$
$$-[-3z - 2] = 9 + 2z$$
$$+3z + 2 = 9 + 2z$$
$$+3z + 2 - 2 = 9 - 2 + 2z$$
$$3z = 7 + 2z$$
$$3z - 2z = 7 + 2z - 2z$$
$$z = 7$$

27.
$$-3m + 6 - 5(m - 1) = 4m - (2m - 4) - 9m + 5$$
$$-3m + 6 - 5 \cdot m + 5 \cdot 1 = 4m - 2m + 4 - 9m + 5$$
$$-3m + 6 - 5m + 5 = 4m - 2m - 9m + 4 + 5$$
$$-3m - 5m + 6 + 5 = 4m - 2m - 9m + 4 + 5$$
$$-8m + 11 = -7m + 9$$
$$-8m + 11 - 11 = -7m + 9 - 11$$
$$-8m = -7m - 2$$
$$-8m + 7m = -7m + 7m - 2$$
$$-m = -2$$
$$\frac{-m}{-1} = \frac{-2}{-1}$$
$$m = 2$$

29.
$$-[3y - (2y + 5)] = -4 - [3(2y - 4) - 3y]$$
$$-[3y - 2y - 5] = -4 - [3 \cdot 2y - 3 \cdot 4 - 3y]$$
$$-[y - 5] = -4 - [6y - 12 - 3y]$$
$$-y + 5 = -4 - [6y - 3y - 12]$$
$$-y + 5 = -4 - [3y - 12]$$
$$-y + 5 = -4 - 3y + 12$$
$$-y + 5 = -4 + 12 - 3y$$
$$-y + 5 = 8 - 3y$$
$$-y + 5 - 5 = 8 - 5 - 3y$$
$$-y = 3 - 3y$$
$$-y + 3y = 3 - 3y + 3y$$
$$2y = 3$$
$$\frac{2y}{2} = \frac{3}{2}$$
$$y = \frac{3}{2}$$

31.
$$-(9 - 3a) - (4 + 2a) - 3 = -(2 - 5a) + (-a) + 1$$
$$-9 + 3a - 4 - 2a - 3 = -2 + 5a - a + 1$$
$$3a - 2a - 9 - 4 - 3 = -2 + 1 + 5a - a$$
$$a - 16 = -1 + 4a$$
$$a - 16 + 16 = -1 + 16 + 4a$$
$$a = 15 + 4a$$
$$a - 4a = 15 + 4a - 4a$$
$$-3a = 15$$
$$\frac{-3y}{-3} = \frac{15}{-3}$$
$$y = -5$$

33.
$$2(-3 + m) - (3m - 4) = -(-4 + m) - 4m + 6$$
$$2 \cdot -3 + 2 \cdot m - 3m + 4 = 4 - m - 4m + 6$$
$$-6 + 2m - 3m + 4 = 4 - m - 4m + 6$$
$$+4 - 6 + 2m - 3m = +6 + 4 - m - 4m$$
$$-2 - m = 10 - 5m$$
$$-2 + 2 - m = 10 + 2 - 5m$$
$$-m = 12 - 5m$$
$$-m + 5m = 12 - 5m + 5m$$
$$4m = 12$$
$$\frac{4m}{4} = \frac{12}{4}$$
$$m = 3$$

35. Writing exercise

37.
$$12 \cdot \left(\frac{8y}{3} - \frac{2y}{4}\right) = (-13) \cdot 12$$
$$\frac{12}{1} \cdot \frac{8y}{3} - \frac{12}{1} \cdot \frac{2y}{4} = -156$$
$$4 \cdot 8y - 3 \cdot 2y = -156$$
$$32y - 6y = -156$$
$$26y = -156$$
$$\frac{26y}{26} = \frac{-156}{26}$$
$$y = -6$$

39.
$$21 \cdot \left(\frac{2r - 3}{7} + \frac{3}{7}\right) = -\frac{r}{3} \cdot 21$$
$$\frac{21}{1} \cdot \frac{2r - 3}{7} + \frac{21}{1} \cdot \frac{3}{7} = -\frac{r}{3} \cdot \frac{21}{1}$$
$$3 \cdot (2r - 3) + 3 \cdot 3 = -7 \cdot 7$$
$$6r - 9 + 9 = -7r$$
$$6r = -7r$$
$$6r + 7r = -7r + 7r$$
$$13r = 0$$
$$\frac{13r}{13} = \frac{0}{13}$$
$$r = 0$$

41.
$$10 \cdot \frac{2x + 5}{5} = \left(\frac{3x + 1}{2} + \frac{-x + 7}{2}\right) \cdot 10$$
$$\frac{10}{1} \cdot \frac{2x + 5}{5} = \frac{10}{1} \cdot \frac{3x + 1}{2} + \frac{10}{1} \cdot \frac{-x + 7}{2}$$
$$2 \cdot (2x + 5) = 5 \cdot (3x + 1) + 5 \cdot (-x + 7)$$
$$4x + 10 = 15x + 5 - 5x + 35$$
$$4x + 10 = 10x + 40$$
$$4x + 10 - 10 = 10x + 40 - 10$$
$$4x = 10x + 30$$
$$4x - 10x = 10x - 10x + 30$$
$$-6x = 30$$
$$\frac{-6x}{-6} = \frac{30}{-6}$$
$$x = -5$$

43.
$$100 \cdot [.09k + .13(k + 300)] = 61 \cdot 100$$
$$100 \cdot .09k + 100 \cdot .13 \cdot (k + 300) = 6100$$
$$9k + 13 \cdot (k + 300) = 6100$$
$$9k + 13k + 3900 = 6100$$
$$22k + 3900 = 6100$$
$$22k + 3900 - 3900 = 6100 - 3900$$
$$22k = 2200$$
$$\frac{22k}{22} = \frac{2200}{22}$$
$$k = 100.$$

45.
$$100 \cdot [.20(14,000) + .14t] = .18(14,000 + t) \cdot 100$$
$$100 \cdot .20(14,000) + 100 \cdot .14t = 100 \cdot .18(14,000 + t)$$
$$20(14,000) + 14t = 18(14,000 + t)$$
$$280,000 + 14t = 252,000 + 18t$$
$$280,000 + 14t - 18t = 252,000 + 18t - 18t$$
$$280,000 - 4t = 252,000$$
$$280,000 - 280,000 - 4t = 252,000 - 280,000$$
$$-4t = -28,000$$
$$\frac{-4t}{-4} = \frac{-28,000}{-4}$$
$$t = 7000$$

47.
$$100 \cdot [.08x + .12(260 - x)] = .48x \cdot 100$$
$$100 \cdot .08x + 100 \cdot .12(260 - x) = 48x$$
$$8x + 12(260 - x) = 48x$$
$$8x + 3120 - 12x = 48x$$
$$3120 - 4x = 48x$$
$$3120 - 4x + 4x = 48x + 4x$$
$$3120 = 52x$$
$$\frac{3120}{52} = \frac{52x}{52}$$
$$60 = x$$

49. Letter A is the only conditional equation.

B. This is an identity:

$$x = 3x - 2x$$
$$x = x.$$

C. This is a contradiction:

$$2(x + 2) = 2x + 2$$
$$2x + 4 = 2x + 2$$

D. This is an identity:

$$5x - 3 = 4x + x - 5 + 2$$
$$5x - 3 = 5x - 3.$$

51.
$$-2p + 5p - 9 = 3(p - 4) - 5$$
$$3p - 9 = 3p - 12 - 5$$
$$3p - 9 = 3p - 17$$
$$3p - 9 + 9 = 3p - 17 + 9$$
$$3p = 3p - 8$$
$$3p - 3p = 3p - 3p - 8$$
$$0 = -8$$

This is a contradiction; the solution is the empty set, which can be symbolized by ϕ.

53.
$$6x + 2(x - 2) = 9x + 4$$
$$6x + 2x - 4 = 9x + 4$$
$$8x - 4 = 9x + 4$$
$$8x - 4 + 4 = 9x + 4 + 4$$
$$8x = 9x + 8$$
$$8x - 9x = 9x - 9x + 8$$
$$-x = 8$$
$$\frac{-x}{-1} = \frac{8}{-1}$$
$$x = -8.$$

This is a conditional equation.

55.
$$-11m + 4(m - 3) + 6m = 4m - 12$$
$$-11m + 4m - 12 + 6m = 4m - 12$$
$$-11m + 4m + 6m - 12 = 4m - 12$$
$$-11m + 4m + 6m - 12 = 4m - 12$$
$$-m - 12 = 4m - 12$$
$$-m - 12 + 12 = 4m - 12 + 12$$
$$-m = 4m$$
$$-m - 4m = 4m - 4m$$
$$-5m = 0$$
$$\frac{-5m}{-5} = \frac{0}{-5}$$
$$m = 0.$$

This is a conditional equation.

57.
$$7[2 - (3 + 4r)] - 2r = -9 + 2(1 - 15r)$$
$$7[2 - 3 - 4r] - 2r = -9 + 2 - 30r$$
$$7[-1 - 4r] - 2r = -7 - 30r$$
$$-7 - 28r - 2r = -7 - 30r$$
$$-7 - 30r = -7 - 30r$$

This equation is an identity. The solution set is {all real numbers}.

59. The following algebraic manipulations show that letter A, B, and C are all equivalent.

A. $h = 2\left(\dfrac{A}{b}\right)$

$h = \dfrac{2}{1} \cdot \dfrac{A}{b}$

$h = \dfrac{2A}{b}$

B. $h = 2A\left(\dfrac{1}{b}\right)$

$h = \dfrac{2A}{1} \cdot \dfrac{1}{b}$

$h = \dfrac{2A}{b}$

C. $h = \dfrac{A}{\frac{1}{2}b}$

$h = \dfrac{A}{1} \div \dfrac{1}{2}b$

$h = \dfrac{A}{1} \div \dfrac{b}{2}$

$h = \dfrac{A}{1} \cdot \dfrac{2}{b}$

$h = \dfrac{2A}{b}$

D. Here is the simplification of this equation:

$h = \dfrac{\frac{1}{2}A}{b}$

$h = \dfrac{1}{2}A \div b$

$h = \dfrac{A}{2} \div \dfrac{b}{1}$

$h = \dfrac{A}{2} \cdot \dfrac{1}{b}$

$h = \dfrac{A}{2b}.$

This equation is not equivalent to the given equation.

61. $d = rt$

$\dfrac{d}{r} = \dfrac{rt}{r}$

$\dfrac{d}{r} = t$

63. $A = bh$

$\dfrac{A}{h} = \dfrac{bh}{h}$

$\dfrac{A}{h} = b$

65.
$P = a + b + c$

$P - b = a + b - b + c$

$P - b - c = a + c - c$

$P - b - c = a$

67. $A = \dfrac{1}{2}bh$

$2 \cdot A = 2 \cdot \dfrac{1}{2}bh$

$2A = bh$

$\dfrac{2A}{b} = \dfrac{bh}{b}$

$\dfrac{2A}{b} = h$

69.
$S = 2\pi rh + 2\pi r^2$

$S - 2\pi r^2 = 2\pi rh + 2\pi r^2 - 2\pi r^2$

$S - 2\pi r^2 = 2\pi rh$

$\dfrac{S - 2\pi r^2}{2\pi r} = \dfrac{2\pi rh}{2\pi r}$

$\dfrac{S - 2\pi r^2}{2\pi r} = h$

The last equation can be simplified further:

$h = \dfrac{S - 2\pi r^2}{2\pi r}$

$= \dfrac{S}{2\pi r} - \dfrac{2\pi r^2}{2\pi r}$

$= \dfrac{S}{2\pi r} - r.$

71. $C = \dfrac{5}{9}(F - 32)$

$\dfrac{9}{5} \cdot C = \dfrac{9}{5} \cdot \dfrac{5}{9}(F - 32)$

$\dfrac{9}{5}C = (F - 32)$

$\dfrac{9}{5}C + 32 = F - 32 + 32$

$\dfrac{9}{5}C + 32 = F$

73.
$A = 2HW + 2LW + 2LH$

$A - 2LW = 2HW + 2LW - 2LW + 2LH$

$A - 2LW = 2HW + 2LH$

$A - 2LW = H(2W + 2L)$

$\dfrac{A - 2LW}{2W + 2L} = \dfrac{H(2W + 2L)}{2W + 2L}$

$\dfrac{A - 2LW}{2W + 2L} = H$

75. (a) $y = .55x - 42.5$
$$y = .55(95) - 42.5$$
$$= 52.25 - 42.5$$
$$= 9.75 \text{ million tickets}$$

(b)
$$y = .55x - 42.5$$
$$7.9 = .55x - 42.5$$
$$7.9 + 42.5 = .55x - 42.5 + 42.5$$
$$50.4 = .55x$$
$$\frac{50.4}{.55} = \frac{.55x}{.55}$$
$$91.6 \approx x$$

This corresponds to 1991–1992.

77. To estimate the number of youths attending in 1993–1994, substitute 93 for x in the model:

$$y = .1x - 8.5$$
$$y = .1(93) - 8.5$$
$$= 9.3 - 8.5$$
$$= 0.8 \text{ million youths}$$

To find in which season the ticket sales amounted to .75 million, substitute .75 for y and solve for x:

$$y = .1x - 8.5$$
$$.75 = .1x - 8.5$$
$$.75 + 8.5 = .1x - 8.5 + 8.5$$
$$9.25 = .1x$$
$$\frac{9.25}{.1} = \frac{.1x}{.1}$$
$$92.5 = x$$

This corresponds to 1992–1993.

7.2 EXERCISES

1. Expression

3. Equation: $\frac{2}{3} \cdot x = 12$

5. Expression

7. Let $x =$ the number of five-dollar bills
$x - 25 =$ the number of ten-dollar bills

Number of bills	Denomination	Dollar Value
x	5	$5x$
$x - 25$	10	$10(x - 25)$

$$5x + 10(x - 25) = 200$$
$$5x + 10x - 250 = 200$$
$$15x - 250 = 200$$
$$15x - 250 + 250 = 200 + 250$$
$$15x = 450$$
$$\frac{15x}{15} = \frac{450}{15}$$
$$x = 30$$

The number of five-dollar bills is 30; the number of ten-dollar bills is $30 - 25 = 5$. The answer has not changed.

9. $x - 18$

11. $(x - 9)(x + 6)$

13. $\frac{12}{x}, x \neq 0$

15. Writing exercise

17. $\frac{x}{6} + 2x = x - 8$

19. $12 - \frac{2}{3}x = 10$

21. *Step 2*: $x - 70 = $ the number of shoppers at small chain/independent bookstores.

Step 3: $x + x - 70 = 442$.

Step 4:

$$2x - 70 = 442$$
$$2x - 70 + 70 = 442 + 70$$
$$2x = 512$$
$$\frac{2x}{2} = \frac{512}{2}$$
$$x = 256.$$

Step 5: There were 256 large chain bookstore shoppers and $256 - 70$ or 186 small chain/independent shoppers.

Step 6: The number of large chain shoppers was 70 more than the number of small chain/independent shoppers, and the total number of these two bookstore types was 442.

23. Let x = the sales for Ingram Industries in billions of dollars.
Let $x - 9.3$ = the sales for TLC Beatrice, International, in billions of dollars.

$$x + x - 9.3 = 13.7$$
$$2x - 9.3 = 13.7$$
$$2x - 9.3 + 9.3 = 13.7 + 9.3$$
$$2x = 23$$
$$\frac{2x}{2} = \frac{23}{2}$$
$$x = 11.5$$

The sales for Ingram was $11.5 billion; the sales for TLC Beatrice, International was $11.5 - 9.3 = \$2.2$ billion.

25. Let x = the number of base hits for Hornsby.
Let $x - 57$ = the number of base hits for Ruth.

$$x + x - 57 = 5803$$
$$2x - 57 = 5803$$
$$2x - 57 + 57 = 5803 + 57$$
$$2x = 5860$$
$$\frac{2x}{2} = \frac{5860}{2}$$
$$x = 2930$$

The number of hits for Hornsby was 2930; the number of hits for Ruth was $2930 - 57 = 2873$ hits.

27. Let x = the revenue from video sales in billions of dollars.
Let $2x + .27$ = the revenue from video rentals in billions of dollars.

$$x + 2x + .27 = 9.81$$
$$3x + .27 - .27 = 9.81 - .27$$
$$3x = 9.54$$
$$\frac{3x}{3} = \frac{9.54}{3}$$
$$x = 3.18$$

The total revenue for video sales was $3.18 billion; the total revenue for rentals was $2(3.18) + .27 = \$6.63$ billion.

29. The values seem to correspond to the numbers along the vertical axis of the graph.

31. 14% of 250

$$.14(250) = 35 \text{ ml}$$

33. 3.5% of 10,000

$$.035(10,000) = \$350$$

35. $283(.05) = \$14.15$

37. Let x = the number of liters of 20% solution.

Strength	L of solution	L of alcohol
12%	12	.12(12)
20%	x	.20(x)
14%	$12 + x$	.14$(12 + x)$

Create an equation by adding the first two algebraic expressions in the last column to total the third:

$$.12(12) + .20(x) = .14(12 + x)$$
$$100[.12(12) + .20(x)] = 100[.14(12 + x)]$$
$$12(12) + 20x = 14(12 + x)$$
$$144 - 144 + 20x = 168 - 144 + 14x$$
$$20x - 14x = 24 + 14x - 14x$$
$$6x = 24$$
$$\frac{6x}{6} = \frac{24}{6}$$
$$x = 4 \text{ liters of 20\% solution.}$$

39. Let x = the number of liters of pure alcohol.

Strength	L of solution	L of alcohol
70%	50	.70(50)
100%	x	1.00(x)
78%	$50 + x$	.78$(50 + x)$

Create an equation by adding the first two algebraic expressions in the last column to total the third:

$$.70(50) + 1.00(x) = .78(50 + x)$$
$$100[.70(50) + 1.00(x)] = 100[.78(50 + x)]$$
$$70(50) + 100(x) = 78(50 + x)$$
$$3500 + 100x = 3900 + 78x$$
$$3500 + 100x - 78x = 3900 + 78x - 78x$$
$$3500 - 3500 + 22x = 3900 - 3500$$
$$22x = 400$$
$$\frac{22x}{22} = \frac{400}{22}$$
$$x = 18\frac{2}{11} \text{ liters of pure alcohol.}$$

41. Let x = number of liters of 20% solution to be replaced with 100% solution. The amount of 20% solution that is replaced must be subtracted while the amount of 100% solution must be added.

Strength	L of solution	L of chemical
20%	20	.20(20)
20%	x	$-.20(x)$
100%	x	1.00(x)
40%	20	.40(20)

Create an equation by adding the first three algebraic expressions in the last column to total the last. To clear the decimal in this equation, multiply by 10 to move the decimal point only one place to the right:

$$.20(20) - .20(x) + 1.00(x) = .40(20)$$
$$10[.20(20) - .20(x) + 1.00(x)] = 10[.40(20)]$$
$$2(20) - 2x + 10x = 4(20)$$
$$40 - 2x + 10x = 80$$
$$40 + 8x = 80$$
$$40 - 40 + 8x = 80 - 40$$
$$8x = 40$$
$$\frac{8x}{8} = \frac{40}{8}$$
$$x = 5 \text{ liters.}$$

43. Let $x =$ the amount invested at 3%.

% as decimal	Amount Invested	Interest in one year
.03	x	$.03x$
.04	$12,000 - x$	$.04(12,000 - x)$
Totals	$12,000$	440

Create an equation by adding the first two algebraic expressions in the last column to total the third:

$$.03x + .04(12,000 - x) = 440$$
$$100[.03x + .04(12,000 - x)] = 100(440)$$
$$3x + 4(12,000 - x) = 44,000$$
$$3x + 48,000 - 4x = 44,000$$
$$48,000 - x = 44,000$$
$$48,000 - 48,000 - x = 44,000 - 48,000$$
$$-x = -4000$$
$$\frac{-x}{-1} = \frac{-4000}{-1}$$
$$x = 4000.$$

The amount invested at 3% is $4000; the amount invested at 4% is: $12,000 - 4000 = \$8000$.

45.

% as decimal	Amount Invested	Interest in one year
.045	x	$.045x$
.03	$2x - 1000$	$.03(2x - 1000)$
Total		1020

Create an equation by adding the first two algebraic expressions in the last column to total the third. To clear the decimals, both sides of the equation must be multiplied by 1000 to move the decimal point 3 places to the right:

$$.045x + .03(2x - 1000) = 1020$$
$$1000[.045x + .03(2x - 1000)] = 1000(1020)$$
$$45x + 30(2x - 1000) = 1,020,000$$
$$45x + 60x - 30.000 = 1,020,000$$
$$105x - 30,000 = 1,020,000$$
$$105x - 30,000 + 30,000 = 1,020,000 + 30,000$$
$$105x = 1,050,000$$
$$\frac{105x}{105} = \frac{1,050,000}{105}$$
$$x = 10,000.$$

The amount invested at 4.5% is $10,000; the amount invested at 3% is: $2 \cdot 10,000 - 1000 = \$19,000$.

47.

% as decimal	Amount Invested	Interest in one year
.05	$29,000$	$.05(29,000)$
.02	x	$.02x$
.03	$29.000 + x$	$.03(29.000 + x)$

Create an equation by adding the first two algebraic expressions in the last column to total the third:

$$.05(29,000) + .02x = .03(29,000 + x)$$
$$100[.05(29,000) + .02x] = 100[.03(29,000 + x)]$$
$$5(29,000) + 2x = 3(29,000 + x)$$
$$145,000 + 2x = 87,000 + 3x$$
$$145,000 + 2x - 2x = 87,000 + 3x - 2x$$
$$145,000 - 87,000 = 87,000 - 87,000 + x$$
$$58,000 = x.$$

He should invest $58,000 at 2% in order to have an average return of 3%.

49.

Number of coins	Denomination	Value
x	$.01$	$.01x$
x	$.10$	$.10x$
$44 - 2x$	$.25$	$.25(44 - 2x)$

Because there are 44 coins altogether, the number of quarters is represented by subtracting the total number of pennies and dimes from 44. Now create an equation by adding the algebraic expressions in the last column; the total value of all the coins is $4.37.

$$.01x + .10x + .25(44 - 2x) = 4.37$$
$$100[.01x + .10x + .25(44 - 2x)] = 100(4.37)$$
$$x + 10x + 25(44 - 2x) = 437$$
$$x + 10x + 1100 - 50x = 437$$
$$x + 10x - 50x + 1100 = 437$$
$$-39x + 1100 - 1100 = 437 - 1100$$
$$-39x = -663$$
$$\frac{-39x}{-39} = \frac{-663}{-39}$$
$$x = 17$$

Sam has 17 pennies, 17 dimes, and $44 - 2 \cdot 17 = 10$ quarters.

51.

Number of tickets	Value/ticket	Total Value
x	3	$3x$
$410 - x$	7	$7(410 - x)$

Now create an equation by adding the algebraic expressions in the last column; the total value of all the tickets is $1650.

$$3x + 7(410 - x) = 1650$$
$$3x + 2870 - 7x = 1650$$
$$2870 - 2870 - 4x = 1650 - 2870$$
$$-4x = -1220$$
$$\frac{-4x}{-4} = \frac{-1220}{-4}$$
$$x = 305$$

There were 305 students who attended and $410 - 305 = 105$ non students who attended.

53.

Number of tickets	Value/ticket	Total Value
x	35	$35x$
$105 - x$	30	$30(105 - x)$

Now create an equation by adding the algebraic expressions in the last column; the total value of all the tickets is $3420.

$$35x + 30(105 - x) = 3420$$
$$35x + 3150 - 30x = 3420$$
$$5x + 3150 = 3420$$
$$5x + 3150 - 3150 = 3420 - 3150$$
$$5x = 270$$
$$\frac{5x}{5} = \frac{270}{5}$$
$$x = 54$$

There were 54 seats sold in Row 1 and $105 - 54 = 51$ seats sold in Row 2.

55.

Number of coins	Denomination	Value
x	10	$10x$
$80 - x$	20	$20(80 - x)$

Now create an equation by adding the algebraic expressions in the last column; the total value of all the coins is $1060.

$$10x + 20(80 - x) = 1060$$
$$10x + 1600 - 20x = 1060$$
$$-10x + 1600 = 1060$$
$$-10x + 1600 - 1600 = 1060 - 1600$$
$$-10x = -540$$
$$\frac{-10x}{-10} = \frac{-540}{-10}$$
$$x = 54$$

Peg has 54 $10 coins and $80 - 54 = 26$ $20 coins.

57. Solve the formula $rt = d$ for t.

$$rt = d$$
$$\frac{rt}{r} = \frac{d}{r}$$
$$t = \frac{d}{r}$$

Now substitute the given values for d and r.

$$t = \frac{500}{134.479} \approx 3.718 \text{ hours}$$

59. Substitute the given values for d and r.

$$t = \frac{225}{148.725} \approx 1.715 \text{ hours}$$

61. Solve the formula $rt = d$ for r.

$$rt = d$$
$$\frac{rt}{t} = \frac{d}{t}$$
$$r = \frac{d}{t}$$

Now substitute the given values for d and t.

$$r = \frac{400}{53.23} \approx 7.51 \text{ meters per second}$$

63. Substitute the given values for d and t.

$$r = \frac{100}{9.92} \approx 10.08 \text{ meters per second}$$

65. Use the formula $rt = d$.

$$53 \cdot 10 = 530 \text{ miles}$$

67. Writing exercise

69.

	Rate	Time	Distance
First train	85	t	$85t$
Second train	95	t	$95t$

Because the trains are traveling in opposite directions, the sum of their distances will equal the total distance apart of 315 kilometers. Use this information to create an equation.

$$85t + 95t = 315$$
$$180t = 315$$
$$\frac{180t}{180} = \frac{315}{180}$$
$$t = 1.75 \text{ hours}$$

71.

	Rate	Time	Distance
Nancy	35	t	$35t$
Mark	40	$t - \frac{1}{4}$	$40\left(t - \frac{1}{4}\right)$

Because Nancy and Mark are traveling in opposite directions, the sum of their distances will equal the total distance apart of 140 miles. Use this information to create an equation.

$$35t + 40\left(t - \frac{1}{4}\right) = 140$$
$$35t + 40t - 10 = 140$$
$$75t - 10 = 140$$
$$75t - 10 + 10 = 140 + 10$$
$$75t = 150$$
$$\frac{75t}{75} = \frac{150}{75}$$
$$t = 2 \text{ hours}$$

The question asks at what time will they be 140 miles apart. The value of t tells us that in two hours they will be 140 miles apart. Because Nancy left the house at 8:00, the time would be 10:00 A.M.

73.

	Rate	Time	Distance
Car	r	$\frac{1}{2}$	$\frac{1}{2}r$
Bus	$r - 12$	$\frac{3}{4}$	$\frac{3}{4}(r - 12)$

The distance to and from work is the same, so set the distances equal to each other.

$$\frac{1}{2}r = \frac{3}{4}(r - 12)$$
$$\frac{4}{1} \cdot \frac{1}{2}r = \frac{4}{1} \cdot \frac{3}{4}(r - 12)$$
$$2r = 3(r - 12)$$
$$2r = 3r - 36$$
$$2r - 3r = 3r - 3r - 36$$
$$-r = -36$$
$$\frac{-r}{-1} = \frac{-36}{-1}$$
$$r = 36 \text{ miles per hour}$$

Tri's rate of speed when driving is 36 mph. The distance to work is $\frac{1}{2} \cdot 36 = 18$ miles.

75.

	Rate	Time	Distance
First part	10	t	$10t$
Second part	15	t	$15t$

The sum of the distances is 100.

$$10t + 15t = 100$$
$$25t = 100$$
$$\frac{25t}{25} = \frac{100}{25}$$
$$t = 4 \text{ hours.}$$

The time for each part of the trip is 4 hours, so the total time for the trip is 8 hours.

7.3 EXERCISES

1. $\dfrac{25}{40} = \dfrac{5 \cdot 5}{5 \cdot 8} = \dfrac{5}{8}$

3. $\dfrac{18}{72} = \dfrac{18 \cdot 1}{18 \cdot 4} = \dfrac{1}{4}$

5. The units of measure must be the same in order to factor the numerator and denominator to simplify the fraction. This simplification can be done by dimensional analysis:
$$\frac{144 \text{ inches}}{6 \text{ feet}} \cdot \frac{1 \text{ foot}}{12 \text{ inches}} = \frac{12 \cdot 12}{12 \cdot 6} = \frac{2}{1}.$$

7. $\dfrac{5 \text{ days}}{40 \text{ hours}} \cdot \dfrac{24 \text{ hours}}{1 \text{ day}} = \dfrac{5 \cdot 8 \cdot 3}{40} = \dfrac{3}{1}$

9. (a) $.4 = \dfrac{4}{10} = \dfrac{2}{5}$

   (b) 4 to 10 means $\dfrac{4}{10}$, which is equivalent to $\dfrac{2}{5}$

   (c) 20 to 50 means $\dfrac{20}{50}$, which simplifies to $\dfrac{2}{5}$

   (d) 5 to 2 means $\dfrac{5}{2}$, which is not the same ratio as 2 to 5.

11. Writing exercise

In exercises 13 through 18, check to see if the cross products are equal to determine if the proportions are true or false.

13. $5 \cdot 56 = 280$; $35 \cdot 8 = 280$. The proportion is true.

15. $120 \cdot 10 = 1200$; $82 \cdot 7 = 574$. The proportion is false.

17. $\dfrac{1}{2} \cdot 10 = 5$; $5 \cdot 1 = 5$ The proportion is true.

19. $\dfrac{k}{4} = \dfrac{175}{20}$
$$20k = 4 \cdot 175$$
$$20k = 700$$
$$\frac{20k}{20} = \frac{700}{20}$$
$$k = 35$$

21.
$$\frac{x}{6} = \frac{18}{4}$$
$$4x = 6 \cdot 18$$
$$4x = 108$$
$$\frac{4x}{4} = \frac{108}{4}$$
$$x = 27$$

23.
$$\frac{3y - 2}{5} = \frac{6y - 5}{11}$$
$$5(6y - 5) = 11(3y - 2)$$
$$30y - 25 = 33y - 22$$
$$30y - 25 + 25 = 33y - 22 + 25$$
$$30y - 33y = 33y - 33y + 3$$
$$-3y = 3$$
$$\frac{-3y}{-3} = \frac{3}{-3}$$
$$y = -1$$

25. Let x = number of fluid ounces of oil.

$$\frac{2.5 \text{ fl oz.}}{1 \text{ gallon}} = \frac{x}{2.75 \text{ gallons}}$$
$$1 \cdot x = (2.5)(2.75)$$
$$x = 6.875 \text{ ounces}$$

27. Let x = amount of U.S. money exchanged.

$$\frac{1 \text{ pound}}{\$1.6762} = \frac{400 \text{ pounds}}{x}$$
$$1 \cdot x = (1.6762)(400)$$
$$x = \$670.48$$

29. Let c = cost to fill tank.

$$\frac{6 \text{ gal}}{\$3.72} = \frac{15 \text{ gal}}{c}$$
$$6 \cdot c = (3.72)(15)$$
$$6c = 55.8$$
$$\frac{6c}{6} = \frac{55.8}{6}$$
$$x = \$9.30$$

31. Let x = number of feet.

$$\frac{600 \text{ miles}}{2.4 \text{ feet}} = \frac{1000 \text{ miles}}{x}$$
$$600 \cdot x = (2.4)(1000)$$
$$600x = 2400$$
$$\frac{600x}{600} = \frac{2400}{600}$$
$$x = 4 \text{ feet}$$

33. Let x = number of fish in lake.

$$\frac{250 \text{ tagged}}{x} = \frac{7 \text{ tagged}}{350 \text{ fish}}$$
$$7 \cdot x = (250)(350)$$
$$7x = 87,500$$
$$\frac{7x}{7} = \frac{87,500}{7}$$
$$x = 12,500 \text{ fish}$$

35. (a)
$$\frac{26}{100} = \frac{x}{350}$$
$$100 \cdot x = 26 \cdot 350$$
$$100x = 9100$$
$$\frac{100x}{100} = \frac{9100}{100}$$
$$x = \$91 \text{ million}$$

(b)
$$\frac{32}{100} = \frac{x}{350}$$
$$100 \cdot x = 32 \cdot 350$$
$$100x = 11,200$$
$$\frac{100x}{100} = \frac{11,200}{100}$$
$$x = \$112 \text{ million}$$

The amount provided by sponsorship was \$112 million. Next \$112 divided equally among 10 sponsors is \$11.2 million per sponsor.

(c)
$$\frac{34}{100} = \frac{x}{350}$$
$$100 \cdot x = 34 \cdot 350$$
$$100x = 11,900$$
$$\frac{100x}{100} = \frac{11,900}{100}$$
$$x = \$119 \text{ million}$$

37.
$$\frac{\$3.09}{20 \text{ bags}} = \$0.1545 \text{ per bag}$$

$$\frac{\$4.59}{30 \text{ bags}} = \$0.153 \text{ per bag}$$

The 30-count size is the better buy.

39.
$$\frac{\$2.99}{15 \text{ oz.}} = \$0.199 \text{ per ounce}$$

$$\frac{\$4.49}{25 \text{ oz.}} = \$0.180 \text{ per ounce}$$

$$\frac{\$5.49}{31 \text{ oz.}} = \$0.177 \text{ per ounce}$$

The 31-ounce size is the best buy.

41. $\dfrac{\$0.89}{14\text{ oz.}} = \$0.064$ per ounce

$\dfrac{\$1.19}{32\text{ oz.}} = \$0.037$ per ounce

$\dfrac{\$2.95}{64\text{ oz.}} = \$0.046$ per ounce

The 32-ounce size is the best buy.

43. $\dfrac{12}{x} = \dfrac{9}{3}$

$9 \cdot x = 12 \cdot 3$

$9x = 36$

$\dfrac{9x}{9} = \dfrac{36}{9}$

$x = 4$

45. $\dfrac{3}{x} = \dfrac{6}{2}$

$6 \cdot x = 3 \cdot 2$

$6x = 6$

$\dfrac{6x}{6} = \dfrac{6}{6}$

$x = 1$

47. (a)

(b) Let $x =$ height of the chair.

$$\dfrac{18}{4} = \dfrac{x}{12}$$

$$4 \cdot x = 18 \cdot 12$$

$$4x = 216$$

$$\dfrac{4x}{4} = \dfrac{216}{4}$$

$$x = 54\text{ feet}$$

49. $\dfrac{225}{130.7} = \dfrac{x}{140.3}$

$130.7x = 225(140.3)$

$130.7x = 31567.5$

$\dfrac{130.7x}{130.7} = \dfrac{31567.5}{130.7}$

$x \approx \$242$

51. $\dfrac{225}{130.7} = \dfrac{x}{148.2}$

$130.7x = 225(148.2)$

$130.7x = 33345$

$\dfrac{130.7x}{130.7} = \dfrac{33345}{130.7}$

$x \approx \$255$

53. $\dfrac{3000}{90.5} = \dfrac{x}{146.3}$

$90.5x = 3000(146.3)$

$90.5x = 438900$

$\dfrac{90.5x}{90.5} = \dfrac{438900}{90.5}$

$x \approx \$4850$

55. If

$$x = ky$$

$$27 = k \cdot 6$$

$$\dfrac{27}{6} = k,$$

then

$$x = \dfrac{27}{6} \cdot \dfrac{2}{1} = 9.$$

57. If

$$m = kp^2$$

$$20 = k \cdot 2^2$$

$$20 = 4k$$

$$\dfrac{20}{4} = \dfrac{4k}{4}$$

$$5 = k,$$

then

$$m = 5 \cdot 5^2 = 5 \cdot 25 = 125.$$

59. If

$$p = \dfrac{k}{q^2}$$

$$4 = \dfrac{k}{\left(\frac{1}{2}\right)^2}$$

$$4 = \dfrac{k}{\frac{1}{4}}$$

$$4 = k \div \dfrac{1}{4}$$

$$4 = k \cdot \dfrac{4}{1}$$

$$4 = 4k$$

$$1 = k,$$

then

$$p = \frac{1}{\left(\frac{3}{2}\right)^2}$$

$$p = 1 \div \frac{9}{4}$$

$$p = 1 \cdot \frac{4}{9}$$

$$p = \frac{4}{9}.$$

61. Assume that the constant of variation, k, is positive.
(a) If y varies directly as x, then as y increases, x <u>increases.</u>

(b) If y varies inversely as x, then as y increases, x <u>decreases.</u>

63. Let $i =$ the amount of interest and $r =$ rate of interest.

$$i = kr$$
$$48 = k(.05)$$
$$\frac{48}{.05} = k$$
$$960 = k$$

Then

$$i = 960(.042) = \$40.32$$

65. Let $r =$ speed or rate and $t =$ time.

$$r = \frac{k}{t}$$
$$30 = \frac{k}{.5}$$
$$30(.5) = \frac{k}{.5} \cdot \frac{.5}{1}$$
$$15 = k$$

Then

$$r = \frac{15}{.75} = 20 \text{ miles per hour.}$$

67. Let $m =$ the weight of an object on the moon and $e =$ the weight of the object on the earth.

$$m = ke$$
$$59 = k \cdot 352$$
$$\frac{59}{352} = \frac{k \cdot 352}{352}$$
$$\frac{59}{352} = k$$

Then

$$m = \frac{59}{352} \cdot 1800 \approx 302 \text{ pounds.}$$

69. Let $p =$ pressure and $d =$ depth.

$$p = kd$$
$$50 = k \cdot 10$$
$$\frac{50}{10} = \frac{k \cdot 10}{10}$$
$$5 = k$$

Then

$$p = 5 \cdot 20 = 100 \text{ pounds per square inch.}$$

71. Let $p =$ pressure and $v =$ volume.

$$p = \frac{k}{v}$$
$$10 = \frac{k}{3}$$
$$10 \cdot 3 = \frac{k}{3} \cdot 3$$
$$30 = k$$

Then

$$p = \frac{30}{1.5} = 20 \text{ pounds per square foot}$$

73. Let $d =$ distance and $t =$ time.

$$d = kt^2$$
$$400 = k \cdot 5^2$$
$$400 = k \cdot 25$$
$$\frac{400}{25} = \frac{k \cdot 25}{25}$$
$$16 = k$$

Then

$$d = 16 \cdot 3^2 = 16 \cdot 9 = 144 \text{ feet.}$$

7.4 EXERCISES

1. $\{x \mid x \leq 3\}$ matches with letter D, $(-\infty, 3]$. The close bracket is used in interval notation because 3 is included in the set.

3. $\{x \mid x < 3\}$ matches with letter B, the number line showing all values less than 3.

5. $\{x \mid -3 \leq x \leq 3\}$ matches with letter F, $[-3, 3]$. The brackets indicate that the endpoints, -3 and 3, are included.

7. Writing exercise

9.
$$4x + 1 \geq 21$$
$$4x + 1 - 1 \geq 21 - 1$$
$$4x \geq 20$$
$$\frac{4x}{4} \geq \frac{20}{4}$$
$$x \geq 5$$

Interval notation: $[5, \infty)$
Graph:

11.
$$\frac{3k - 1}{4} > 5$$
$$\frac{4}{1} \cdot \frac{3k - 1}{4} > 5 \cdot 4$$
$$3k - 1 > 20$$
$$3k - 1 + 1 > 20 + 1$$
$$3k > 21$$
$$\frac{3k}{3} > \frac{21}{3}$$
$$k > 7$$

Interval notation: $(7, \infty)$
Graph:

13. In this exercise remember to reverse the inequality sign when multiplying or dividing by a negative number.
$$-4x < 16$$
$$\frac{-4x}{-4} > \frac{16}{-4}$$
$$x > -4$$

Interval notation: $(-4, \infty)$
Graph:

15. In this exercise remember to reverse the inequality sign when multiplying or dividing by a negative number.
$$-\frac{3}{4}r \geq 30$$
$$-\frac{4}{3} \cdot -\frac{3}{4}r \leq \frac{30}{1} \cdot -\frac{4}{3}$$
$$r \leq -\frac{30 \cdot 4}{3}$$
$$r \leq -40$$

Interval notation: $(-\infty, -40]$
Graph:

17.
$$-1.3m \geq -5.2$$
$$\frac{-1.3m}{-1.3} \leq \frac{-5.2}{-1.3}$$
$$y \leq 4$$

Interval notation: $(-\infty, 4]$
Graph:

19.
$$\frac{2k - 5}{-4} > 5$$
$$\frac{-4}{1} \cdot \frac{2k - 5}{4} < 5 \cdot -4$$
$$2k - 5 < -20$$
$$2k - 5 + 5 < -20 + 5$$
$$2k < -15$$
$$\frac{2k}{2} < \frac{-15}{2}$$
$$k < \frac{-15}{2}$$

Interval notation: $(-\infty, \frac{-15}{2})$
Graph:

21.
$$y + 4(2y - 1) \geq y$$
$$y + 8y - 4 \geq y$$
$$9y - 4 \geq y$$
$$9y - 4 + 4 \geq y + 4$$
$$9y \geq y + 4$$
$$9y - y \geq y - y + 4$$
$$8y \geq 4$$
$$\frac{8y}{8} \geq \frac{4}{8}$$
$$y \geq \frac{4 \cdot 1}{4 \cdot 2}$$
$$y \geq \frac{1}{2}$$

Interval notation: $[\frac{1}{2}, \infty)$
Graph:

23.
$$-(4 + r) + 2 - 3r < -14$$
$$-4 - r + 2 - 3r < -14$$
$$-2 - 4r < -14$$
$$-2 + 2 - 4r < -14 + 2$$
$$-4r < -12$$
$$\frac{-4m}{-4} > \frac{-12}{-4}$$
$$m > 3$$

Interval notation: $(3, \infty)$
Graph:

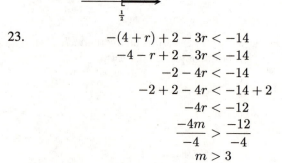

25.
$$-3(z - 6) > 2z - 2$$
$$-3z + 18 > 2z - 2$$
$$-3z - 2z + 18 > 2z - 2z - 2$$
$$-5z + 18 > -2$$
$$-5z + 18 - 18 > -2 - 18$$
$$-5z > -20$$
$$\frac{-5z}{-5} < \frac{-20}{-5}$$
$$z < 4$$

Interval notation: $(-\infty, 4)$
Graph:

27. Clear the fractions by multiplying both sides of the equation by the LCM, 6.

$$\frac{2}{3}(3k - 1) \geq \frac{3}{2}(2k - 3)$$
$$\frac{6}{1} \cdot \frac{2}{3}(3k - 1) \geq \frac{6}{1} \cdot \frac{3}{2}(2k - 3)$$
$$2 \cdot 2(3k - 1) \geq 3 \cdot 3(2k - 3)$$
$$4(3k - 1) \geq 9(2k - 3)$$
$$12k - 4 \geq 18k - 27$$
$$12k - 4 + 4 \geq 18k - 27 + 4$$
$$12k \geq 18k - 23$$
$$12k - 18k \geq 18k - 18k - 23$$
$$-6k \geq -23$$
$$\frac{-6k}{-6} \leq \frac{-23}{-6}$$
$$k \leq \frac{23}{6}$$

Interval notation: $(-\infty, \frac{23}{6}]$
Graph:

29. Clear the fractions by multiplying both sides of the equation by the LCM, 8.

$$-\frac{1}{4}(p + 6) + \frac{3}{2}(2p - 5) < 10$$
$$\frac{8}{1} \cdot -\frac{1}{4}(p + 6) + \frac{8}{1} \cdot \frac{3}{2}(2p - 5) < 10 \cdot 8$$
$$2 \cdot -1(p + 6) + 4 \cdot 3(2p - 5) < 80$$
$$-2(p + 6) + 12(2p - 5) < 80$$
$$-2p - 12 + 24p - 60 < 80$$
$$22p - 72 < 80$$
$$22p - 72 + 72 < 80 + 72$$
$$22p < 152$$
$$\frac{22p}{22} < \frac{152}{22}$$
$$p < \frac{76 \cdot 2}{11 \cdot 2}$$
$$p < \frac{76}{11}$$

Interval notation: $(-\infty, \frac{76}{11})$
Graph:

31.
$$3(2x - 4) - 4x < 2x + 3$$
$$6x - 12 - 4x < 2x + 3$$
$$2x - 12 < 2x + 3$$
$$2x - 2x - 12 < 2x - 2x + 3$$
$$-12 < 3$$

Because it is always true that negative 12 is less than 3, x can be replaced by any real number and the inequality will be true. This is called an identity.

Interval notation: $(-\infty, +\infty)$
Graph:

33.
$$8\left(\frac{1}{2}x + 3\right) < 8\left(\frac{1}{2}x - 1\right)$$
$$4x + 24 < 4x - 8$$
$$4x - 4x + 24 < 4x - 4x - 8$$
$$24 < -8$$

Because it is not true that 24 is less than negative 8, there is no real number that can replace x to obtain a true statement. This is called a contradiction and there is no solution. This can be symbolized by the empty set: $\emptyset$.

35. Writing exercise

37. In a three-part inequality, remember to work toward isolating the variable by applying the same operation to all three parts of the inequality.

$$-4 < x - 5 < 6$$
$$-4 + 5 < x - 5 + 5 < 6 + 5$$
$$1 < x < 11$$

Interval notation: $(1, 11)$
Graph:

39.
$$-9 \leq k + 5 \leq 15$$
$$-9 - 5 \leq k + 5 - 5 \leq 15 - 5$$
$$-14 \leq k \leq 10$$

Interval notation: $[-14, 10]$
Graph:

41.

$$-6 \le 2z + 4 \le 16$$
$$-6 - 4 \le 2z + 4 - 4 \le 16 - 4$$
$$-10 \le 2z \le 12$$
$$\frac{-10}{2} \le \frac{2z}{2} \le \frac{12}{2}$$
$$-5 \le z \le 6$$

Interval notation: $[-5, 6]$
Graph:

43.

$$-19 \le 3x - 5 \le 1$$
$$-19 + 5 \le 3x - 5 + 5 \le 1 + 5$$
$$-14 \le 3x \le 6$$
$$\frac{-14}{3} \le \frac{3x}{3} \le \frac{6}{3}$$
$$\frac{-14}{3} \le x \le 2$$

Interval notation: $\left[\frac{-14}{3}, 2\right]$
Graph:

45.

$$-1 \le \frac{2x - 5}{6} \le 5$$
$$6 \cdot -1 \le \frac{6}{1} \cdot \frac{2x - 5}{6} \le 6 \cdot 5$$
$$-6 \le 2x - 5 \le 30$$
$$-6 + 5 \le 2x - 5 + 5 \le 30 + 5$$
$$-1 \le 2x \le 35$$
$$\frac{-1}{2} \le \frac{2x}{2} \le \frac{35}{2}$$
$$\frac{-1}{2} \le x \le \frac{35}{2}$$

Interval notation: $\left[\frac{-1}{2}, \frac{35}{2}\right]$
Graph:

47. Remember to reverse the inequality symbols in this exercise when dividing by -9.

$$4 \le 5 - 9x < 8$$
$$4 - 5 \le 5 - 5 - 9x < 8 - 5$$
$$-1 \le -9x < 3$$
$$\frac{-1}{-9} \ge \frac{-9x}{-9} > \frac{3}{-9}$$
$$\frac{1}{9} \ge x > -\frac{1}{3}$$

To make this more meaningful, restate the inequality as

$$-\frac{1}{3} < x \le \frac{1}{9}$$

Interval notation: $\left(-\frac{1}{3}, \frac{1}{9}\right]$
Graph:

49. The following months show percentages greater than 7.7%: April, 12.9%; May, 22.1%; June, 20.7%, and July, 11.1%.

51. First find what percent 1500 is of 17,252.
Let $r =$ the percent in decimal form

$$r \cdot 17,252 = 1500$$
$$\frac{r \cdot 17,252}{17,252} = \frac{1500}{17,252}$$
$$r = .087$$

This decimal value, .087 is equivalent to 8.7%. Look for the months for which the percent is less than 8.7%. These months are: January, February, March, August, September, October, November, and December.

53. Read the vertical axis of the graph to find the heights of the bars that are at or below the 8 million mark. They are the four shortest bars that represent the following years: 1990–1991, 1991–1992, 1992–1993, 1993–1994.

55. Let x represent the number of tickets in millions that need to be sold during the 1993–1994 season. Solve the inequality:

$$\frac{7.3 + 7.5 + 7.9 + x}{4} \ge 7.75$$
$$\frac{4}{1} \cdot \frac{7.3 + 7.5 + 7.9 + x}{4} \ge 7.75 \cdot 4$$
$$7.3 + 7.5 + 7.9 + x \ge 31$$
$$22.7 + x \ge 31$$
$$22.7 - 22.7 + x \ge 31 - 22.7$$
$$x \ge 8.3$$

This means that at least 8.3 million tickets must be sold.

57. Let $d =$ number of additional 1/5 miles. Dantrell must pay $1.50 plus $.25 for each additional 1/5 mile, and the amount cannot exceed $3.75. Solve the inequality.

$$1.50 + .25d \le 3.75$$
$$1.50 - 1.50 + .25d \le 3.75 - 1.50$$
$$.25d \le 2.25$$
$$\frac{.25d}{.25} \le \frac{2.25}{.25}$$
$$d \le 9$$

Then Dantrell can travel $9 \cdot \frac{1}{5} = \frac{9}{5}$ or $1\frac{4}{5}$ miles in addition to the first $\frac{1}{5}$ mile for a total of 2 miles.

59. Let x = score needed on the third test. Solve the inequality:

$$\frac{90 + 82 + x}{3} \geq 84$$
$$\frac{3}{1} \cdot \frac{90 + 82 + x}{3} \geq 84 \cdot 3$$
$$90 + 82 + x \geq 252$$
$$172 + x \geq 252$$
$$172 - 172 + x \geq 252 - 172$$
$$x \geq 80$$

Margaret must score at least a grade of 80 on the third test.

61. Let m = number of miles and the cost to rent each vehicle is:
Ford $= 35 + .14m$
Chevy $= 34 + .16m$
To determine the number of miles at which the cost to rent a Chevy will exceed the cost for a Ford, solve the inequality:

$$\text{Cost for Chevy} > \text{Cost for Ford}$$
$$34 + .16m > 35 + .14m$$
$$34 - 34 + .16m > 35 - 34 + .14m$$
$$.16m > 1 + .14m$$
$$.16m - .14m > 1 + .14m - .14m$$
$$.02m > 1$$
$$\frac{.02m}{.02} > \frac{1}{.02}$$
$$m > 50 \text{ miles.}$$

The price to rent the Chevrolet would exceed the price to rent the Ford after 50 miles.

63. To find the smallest number of units x that must be sold for the business to show a profit, set up an inequality in which the revenue is greater than the cost, that is, $R > C$.

$$24x > 20x + 100$$
$$24x - 20x > 20x - 20x + 100$$
$$4x > 100$$
$$\frac{4x}{4} > \frac{100}{4}$$
$$x > 25$$

Because x must exceed 25, the smallest whole number of videotapes is 26.

65. Because $a > b$, the expression $b - a$ is a negative quantity. When multiplying an inequality by a negative quantity, the inequality symbol must be reversed.

7.5 EXERCISES

1. $\left(\frac{5}{3}\right)^2 = \frac{25}{9}$, which is choice A.

3. $\left(-\frac{3}{5}\right)^{-2} = \frac{1}{\left(-\frac{3}{5}\right)^2}$
$$= \frac{1}{\frac{9}{25}}$$
$$= 1 \div \frac{9}{25}$$
$$= 1 \cdot \frac{25}{9}$$
$$= \frac{25}{9}, \text{ which is choice A}$$

5. $-\left(-\frac{3}{5}\right)^2 = -\left(-\frac{3}{5} \cdot -\frac{3}{5}\right)$
$$= -\left(\frac{9}{25}\right)$$
$$= -\frac{9}{25}, \text{ which is choice D}$$

7. $5^4 = 5 \cdot 5 \cdot 5 \cdot 5 = 625$

9. $(-2)^5 = -2 \cdot -2 \cdot -2 \cdot -2 \cdot -2 = -32$

11. $-2^3 = -(2 \cdot 2 \cdot 2) = -8$

13. $-(-3)^4 = -(-3 \cdot -3 \cdot -3 \cdot -3) = -81$

15. $7^{-2} = \frac{1}{7^2} = \frac{1}{49}$

17. $-7^{-2} = -\frac{1}{7^2} = -\frac{1}{49}$

19. $\frac{2}{(-4)^{-3}} = 2 \div (-4)^{-3}$
$$= 2 \div \frac{1}{(-4)^3}$$
$$= 2 \cdot (-4)^3$$
$$= 2 \cdot -64$$
$$= -128$$

21. $\frac{5^{-1}}{4^{-2}} = \frac{\frac{1}{5}}{\frac{1}{4^2}}$
$$= \frac{1}{5} \div \frac{1}{4^2}$$
$$= \frac{1}{5} \cdot \frac{4^2}{1}$$
$$= \frac{16}{5}$$

23. $\left(\dfrac{1}{5}\right)^{-3} = \dfrac{1^{-3}}{5^{-3}}$

$= \dfrac{5^3}{1^3}$

$= 125$

25. $\left(\dfrac{4}{5}\right)^{-2} = \dfrac{4^{-2}}{5^{-2}}$

$= \dfrac{5^2}{4^2}$

$= \dfrac{25}{16}$

27. $4^{-1} + 5^{-1} = \dfrac{1}{4} + \dfrac{1}{5}$

$= \dfrac{5}{5} \cdot \dfrac{1}{4} + \dfrac{1}{5} \cdot \dfrac{4}{4}$

$= \dfrac{5}{20} + \dfrac{4}{20}$

$= \dfrac{9}{20}$

29. $12^0 = 1$

31. $(-4)^0 = 1$

33. $3^0 - 4^0 = 1 - 1 = 0$

35. In order to raise a fraction to a negative power, we may change the fraction to its <u>reciprocal</u> and change the exponent to <u>additive</u> <u>inverse</u> of the original exponent.

37. (a) Simplify each side of the equation to see if the statement is true:

$$-\dfrac{3}{4} = \left(\dfrac{3}{4}\right)^{-1}$$

$$= \dfrac{3^{-1}}{4^{-1}}$$

$$= \dfrac{4}{3}.$$

This equation is not correct.

(b) $$\dfrac{3^{-1}}{4^{-1}} = \left(\dfrac{4}{3}\right)^{-1}$$

$$\dfrac{4}{3} = \dfrac{4^{-1}}{3^{-1}}$$

$$= \dfrac{3}{4}$$

This equation is not correct.

(c) $$\dfrac{3^{-1}}{4} = \dfrac{3}{4^{-1}}$$

$$\dfrac{1}{3 \cdot 4} = 3 \cdot 4$$

$$\dfrac{1}{12} = 12$$

This equation is not correct.

(d) $$\dfrac{3^{-1}}{4^{-1}} = \left(\dfrac{3}{4}\right)^{-1}$$

$$= \dfrac{3^{-1}}{4^{-1}}$$

This equation is correct.

39. $x^{12} \cdot x^4 = x^{12+4} = x^{16}$

41. $\dfrac{5^{17}}{5^{16}} = 5^{17-16} = 5$

43. $\dfrac{3^{-5}}{3^{-2}} = 3^{-5-(-2)} = 3^{-5+2} = 3^{-3} = \dfrac{1}{3^3} = \dfrac{1}{27}$

45. $\dfrac{9^{-1}}{9} = 9^{-1-1} = 9^{-2} = \dfrac{1}{9^2} = \dfrac{1}{81}$

47. $t^5 t^{-12} = t^{5+(-12)} = t^{-7} = \dfrac{1}{t^7}$

49. $(3x)^2 = 3^2 \cdot x^2 = 9x^2$

51. $a^{-3} a^2 a^{-4} = a^{-3+2+(-4)} = a^{-3+2-4} = a^{-5} = \dfrac{1}{a^5}$

53. $\dfrac{x^7}{x^{-4}} = x^{7-(-4)} = x^{7+4} = x^{11}$

55. $\dfrac{r^3 r^{-4}}{r^{-2} r^{-5}} = \dfrac{r^{3+(-4)}}{r^{-2+(-5)}} = \dfrac{r^{-1}}{r^{-7}} = r^{-1-(-7)} = r^{-1+7} = r^6$

57. $7k^2(-2k)(4k^{-5}) = (7 \cdot -2 \cdot 4)k^{2+1+(-5)}$

$$= -56k^{-2}$$

$$= \dfrac{-56}{k^2}$$

59. $(z^3)^{-2} z^2 = z^{3 \cdot -2} z^2 = z^{-6} z^2 = z^{-6+2} = z^{-4} = \dfrac{1}{z^4}$

61. $-3r^{-1}(r^{-3})^2 = -3r^{-1} r^{-3 \cdot 2}$

$$= -3r^{-1} r^{-6}$$

$$= -3r^{-1+(-6)}$$

$$= -3r^{-7}$$

$$= \dfrac{-3}{r^7}$$

63. $(3a^{-2})^3 (a^3)^{-4} = 3^3 a^{-6} \cdot a^{-12}$

$$= 27a^{-6+(-12)}$$

$$= 27a^{-18}$$

$$= \dfrac{27}{a^{18}}$$

65. $(x^{-5} y^2)^{-1} = x^{-5 \cdot -1} y^{2 \cdot -1}$

$$= x^5 y^{-2}$$

$$= \dfrac{x^5}{y^2}$$

67. The reciprocal of x is $\frac{1}{x}$.

(a) $x^{-1} = \frac{1}{x}$, by definition.

(b) $\frac{1}{x}$ is the reciprocal.

(c) $\left(\frac{1}{x^{-1}}\right)^{-1} = \frac{1^{-1}}{x^1} = \frac{1}{x}$

(d) $-x$ is the opposite of x. It is not the reciprocal.

69. $230 = 2.3 \times 10^2$

71. $.02 = 2 \times 10^{-2}$

73. $6.5 \times 10^3 = 6500$

75. $1.52 \times 10^{-2} = .0152$

77.
$$\frac{.002 \times 3900}{.000013} = \frac{(2 \times 10^{-3}) \times (3.9 \times 10^3)}{1.3 \times 10^{-5}}$$
$$= \frac{(2 \times 3.9) \times (10^{-3} \times 10^3)}{1.3 \times 10^{-5}}$$
$$= \frac{(7.8) \times (10^0)}{1.3 \times 10^{-5}}$$
$$= \frac{7.8}{1.3} \times \frac{10^0}{10^{-5}}$$
$$= 6 \times 10^{0-(-5)}$$
$$= 6 \times 10^{0+5}$$
$$= 6 \times 10^5$$

79.
$$\frac{.0004 \times 56,000}{.000112} = \frac{(4 \times 10^{-4}) \times (5.6 \times 10^4)}{1.12 \times 10^{-4}}$$
$$= \frac{(4 \times 5.6) \times (10^{-4} \times 10^4)}{1.12 \times 10^{-4}}$$
$$= \frac{22.4 \times 10^0}{1.12 \times 10^{-4}}$$
$$= \frac{22.4}{1.12} \times \frac{10^0}{10^{-4}}$$
$$= 20 \times 10^{0-(-4)}$$
$$= 20 \times 10^4$$
$$= 2.0 \times 10^5$$

Note that 20×10^4 is not in scientific notation. As 20 is reduced by a power of ten, the exponent on 10 must increase by a power of ten.

81.
$$\frac{840,000 \times .03}{.00021 \times 600} = \frac{(8.4 \times 10^5) \times (3 \times 10^{-2})}{(2.1 \times 10^{-4}) \times (6 \times 10^2)}$$
$$= \frac{(8.4 \times 3) \times (10^5 \times 10^{-2})}{(2.1 \times 6) \times (10^{-4} \times 10^2)}$$
$$= \frac{25.2 \times 10^3}{12.6 \times 10^{-2}}$$
$$= 2 \times 10^{3-(-2)}$$
$$= 2 \times 10^{3+2}$$
$$= 2 \times 10^5$$

83. $\$7,326,000,000 = \$7.326 \times 10^9$

85. $\$30,262,000,000 = \$3.0262 \times 10^{10}$

87. 10 billion $= 10,000,000,000 = 1 \times 10^{10}$

89. $2 \times 10^9 = 2,000,000,000$.

91.
$$\frac{3.9 \times 10^8}{15,000} = \frac{3.9 \times 10^8}{1.5 \times 10^4} = \frac{3.9}{1.5} \times 10^{8-4} = 2.6 \times 10^4$$

93.
$$\frac{1.8 \times 10^7}{19 \times 10^{12}} = \frac{1.8}{19} \times 10^{7-12} \approx .09474 \times 10^{-5}$$

Written in scientific notation, this is 9.474×10^{-7}

95.
$$\frac{9 \times 10^{12}}{3 \times 10^{10}} = \frac{9}{3} \times 10^{12-10} = 3 \times 10^2, \text{ which is 300 sec.}$$

97.
$$\frac{1.86 \times 10^5 \text{mi}}{1 \text{ s}} \cdot \frac{60s}{1\text{min}} \cdot \frac{60 \text{ min}}{1 \text{ hr}} \cdot \frac{24 \text{ hr}}{1 \text{ day}} \cdot \frac{365 \text{ days}}{1 \text{ year}}.$$

This simplifies to $58,665,960 \times 10^5$ miles; in scientific notation this is approximately 5.87×10^{12}.

99. (a) Use the formula $d = rt$. The total distance that the spacecraft must travel is the difference between 6.7×10^7 and 3.6×10^7. The rate of the spacecraft is 1.55×10^3. Insert this information into the formula and solve for t.

$$6.7 \times 10^7 - 3.6 \times 10^7 = (1.55 \times 10^3)t$$
$$3.1 \times 10^7 = (1.55 \times 10^3)t$$
$$\frac{3.1 \times 10^7}{1.55 \times 10^3} = \frac{(1.55 \times 10^3)t}{1.55 \times 10^3}$$
$$\frac{3.1}{1.55} \times \frac{10^7}{10^3} = t$$
$$2 \times 10^4 = t$$

The time would be $20,000$ hours.

101. $123,000 = 1.23 \times 10^5$

103. $424,000 = 4.24 \times 10^5$

105. $440,000 = 4.4 \times 10^5$

7.6 EXERCISES

1. $(3x^2 - 4x + 5) + (-2x^2 + 3x - 2) =$
$\quad 3x^2 - 2x^2 - 4x + 3x + 5 - 2 = x^2 - x + 3$

3. Remember that the negative in front of the parenthesis affects all the terms within the grouping symbol; i. e., all the signs change.

$(12y^2 - 8y + 6) - (3y^2 - 4y + 2) =$
$\quad 12y^2 - 3y^2 - 8y + 4y + 6 - 2 = 9y^2 - 4y + 4$

5.

$$\left(6m^4 - 3m^2 + m\right) - \left(2m^3 + 5m^2 + 4m\right) + \left(m^2 - m\right) =$$
$$6m^4 - 3m^2 + m - 2m^3 - 5m^2 - 4m + m^2 - m =$$
$$6m^4 - 2m^3 - 3m^2 - 5m^2 + m^2 + m - 4m - m =$$
$$6m^4 - 2m^3 - 7m^2 - 4m$$

7. $5\left(2x^2 - 3x + 7\right) - 2\left(6x^2 - x + 12\right) =$
$$10x^2 - 15x + 35 - 12x^2 + 2x - 24 =$$
$$10x^2 - 12x^2 - 15x + 2x + 35 - 24 =$$
$$-2x^2 - 13x + 11$$

9. Use the FOIL method:

$$(x+3)(x-8) = x \cdot x + x \cdot (-8) + 3 \cdot x + 3 \cdot (-8)$$
$$= x^2 - 8x + 3x - 24$$
$$= x^2 - 5x - 24.$$

11. $(4r - 1)(7r + 2) = 4r \cdot 7r + 4r \cdot 2 + (-1) \cdot 7r + (-1) \cdot 2$
$$= 28r^2 + 8r - 7r - 2$$
$$= 28r^2 + r - 2$$

13. Use the distributive property. Also remember to add exponents when multiplying variables that have the same base.

$$4x^2\left(3x^3 + 2x^2 - 5x + 1\right) =$$
$$4x^2 \cdot 3x^3 + 4x^2 \cdot 2x^2 + 4x^2 \cdot (-5x) + 4x^2 \cdot 1 =$$
$$12x^5 + 8x^4 - 20x^3 + 4x^2$$

15. The FOIL method can always be used to multiply two binomials.

$$(2m + 3)(2m - 3) = 2m \cdot 2m + 2m \cdot (-3) + 3 \cdot 2m + 3 \cdot (-3)$$
$$= 4m^2 - 6m + 6m - 9$$
$$= 4m^2 - 9$$

However, it is helpful to recognize that it is the product of the sum and difference of two terms.

17. It is important to remember that the binomial $4m + 2n$ is the base that is being squared.

$$(4m + 2n)^2 = (4m + 2n)(4m + 2n)$$
$$= 4m \cdot 4m + 4m \cdot 2n + 4m \cdot 2n + 2n \cdot 2n$$
$$= 16m^2 + 16mn + 4n^2$$

It is also helpful to recognize the pattern of a binomial squared.

19. It is important to remember that the binomial $5r + 3t^2$ is the base that is being squared:

$$\left(5r + 3t^2\right)^2 = \left(5r + 3t^2\right)\left(5r + 3t^2\right)$$
$$= 5r \cdot 5r + 5r \cdot 3t^2 + 3t^2 \cdot 5r + 3t^2 \cdot 3t^2$$
$$= 25r^2 + 15rt^2 + 15rt^2 + 9t^4$$
$$= 25r^2 + 30rt^2 + 9t^4$$

It is also helpful to recognize the pattern of a binomial squared.

21. Vertical multiplication is often less confusing when multiplying two polynomials of more than two terms. Multiply from the right as in number multiplication; i.e., start with -1 times -4. Line up like terms in columns.

$$
\begin{array}{r}
-z^2 + 3z - 4 \\
2z - 1 \\
\hline
z^2 - 3z + 4 \\
-2z^3 + 6z^2 - 8z \\
\hline
-2z^3 + 7z^2 - 11z + 4
\end{array}
$$

23. Using vertical multiplication, these polynomials have been multiplied beginning at the right: First multiply $-3k$ from the second polynomial times each term of the first polynomial; then multiply $+2n$ times each term of the first polynomial; finally multiply $+m$ times each term of the first polynomial. Place like terms in columns as you multiply to simplify being added later.

$$
\begin{array}{r}
m - n + k \\
m + 2n - 3k \\
\hline
-3km + 3kn \qquad - 3k^2 \\
+2mn \qquad + 2kn \quad - 2n^2 \\
- mn + km \qquad + m^2 \\
\hline
+ mn - 2km + 5kn + m^2 - 2n^2 - 3k^2
\end{array}
$$

25. Vertical multiplication might be the preferred method as in the previous exercise.

$$
\begin{array}{r}
a - b + 2c \\
a - b + 2c \\
\hline
+2ac - 2bc + 4c^2 \\
- ab + b^2 \qquad - 2bc \\
+ a^2 - ab \qquad + 2ac \\
\hline
a^2 - 2ab + b^2 + 4ac - 4bc + 4c^2
\end{array}
$$

27. The answer is (A). Letter (B) is a trinomial of degree 6; letter (C) is a trinomial of degree 6, but it is not in descending powers; letter (D) is not a trinomial.

29. Writing exercise

31.
$$8m^4 + 6m^3 - 12m^2 =$$
$$2m^2 \cdot 4m^2 + 2m^2 \cdot 3m - 2m^2 \cdot 6 =$$
$$2m^2\left(4m^2 + 3m - 6\right)$$

33.
$$4k^2m^3 + 8k^4m^3 - 12k^2m^4 =$$
$$4k^2m^3 \cdot 1 + 4k^2m^3 \cdot 2k^2 - 4k^2m^3 \cdot 3m =$$
$$4k^2m^3\left(1 + 2k^2 - 3m\right)$$

35. In this exercise the greatest common factor is $2(a + b)$.

$$2(a + b) + 4m(a + b) =$$
$$2 \cdot (a + b) + 2m \cdot 2 \cdot (a + b) =$$
$$2(a + b)(1 + 2m)$$

37. In this exercise the greatest common factor is $m - 1$.

$$2(m - 1) - 3(m - 1)^2 + 2(m - 1)^3 =$$
$$(m - 1) \cdot 2 - (m - 1) \cdot 3(m - 1)^1 + (m - 1) \cdot 2(m - 1)^2 =$$
$$(m - 1)\left[2 - 3(m - 1) + 2(m - 1)^2\right] =$$
$$(m - 1)[2 - 3m + 3 + 2(m - 1)(m - 1)] =$$
$$(m - 1)[2 - 3m + 3 + 2(m^2 - 2m + 1)] =$$
$$(m - 1)[2 - 3m + 3 + 2m^2 - 4m + 2] =$$
$$(m - 1)[2m^2 - 7m + 7]$$

39.
$$6st + 9t - 10s - 15 =$$
$$3t \cdot (2s + 3) - 5 \cdot (2s + 3) =$$
$$(2s + 3)(3t - 5)$$

Remember that the binomial factors can be written either as shown above or as $(3t - 5)(2s + 3)$.

41.
$$rt^3 + rs^2 - pt^3 - ps^2 =$$
$$r \cdot (t^3 + s^2) - p \cdot (t^3 + s^2) =$$
$$(t^3 + s^2)(r - p)$$

43.
$$16a^2 + 10ab - 24ab - 15b^2 =$$
$$2a \cdot (8a + 5b) - 3b \cdot (8a + 5b) =$$
$$(8a + 5b)(2a - 3b)$$

45.
$$20z^2 - 8zx - 45zx + 18x^2 =$$
$$4z \cdot (5z - 2x) - 9x \cdot (5z - 2x) =$$
$$(5z - 2x)(4z - 9x)$$

47. Recall that the mental process is to think of two numbers whose product is -15 and whose sum is -2. Use these numbers, -5 and $+3$ to rename the middle term of the trinomial. After creating the 4-term polynomial in the second line, factor by grouping.

$$x^2 - 2x - 15 = x^2 - 5x + 3x - 15$$
$$= x(x - 5) + 3(x - 5)$$
$$= (x - 5)(x + 3)$$

49. $y^2 + 2y - 35 = y^2 + 7y - 5y - 35$
$$= y(y + 7) - 5(y + 7)$$
$$= (y + 7)(y - 5)$$

51. First, factor out the greatest common factor 6; then proceed to factor the trinomial.

$$6a^2 - 48a - 120 = 6(a^2 - 8a - 20)$$
$$= 6[a^2 - 10a + 2a - 20]$$
$$= 6[a(a - 10) + 2(a - 10)]$$
$$= 6(a - 10)(a + 2)$$

53. $3m^3 + 12m^2 + 9m = 3m[m^2 + 4m + 3]$
$$= 3m[m^2 + 3m + m + 3]$$
$$= 3m[m(m + 3) + 1(m + 3)]$$
$$= 3m(m + 3)(m + 1)$$

55. When the leading coefficient is not 1, remember that the product to consider is found by multiplying the leading coefficient by the last term. In this exercise multiply 6 times -6, so that a product of -36 is needed along with a sum of $+5$.

$$6k^2 + 5kp - 6p^2 = 6k^2 + 9kp - 4kp - 6p^2$$
$$= 3k(2k + 3p) - 2p(2k + 3p)$$
$$= (2k + 3p)(3k - 2p)$$

57. $5a^2 - 7ab - 6b^2 = 5a^2 - 10ab + 3ab - 6b^2$
$$= 5a(a - 2b) + 3b(a - 2b)$$
$$= (a - 2b)(5a + 3b)$$

59. $9x^2 - 6x^3 + x^4 = x^2[9 - 6x + x^2]$
$$= x^2[9 - 3x - 3x + x^2]$$
$$= x^2[3(3 - x) - x(3 - x)]$$
$$= x^2(3 - x)(3 - x).$$

The final answer can also be written as $x^2(3 - x)^2$.

61. $24a^4 + 10a^3b - 4a^2b^2 = 2a^2[12a^2 + 5ab - 2b^2]$
$$= 2a^2[12a^2 - 3ab + 8ab - 2b^2]$$
$$= 2a^2[3a(4a - b) + 2b(4a - b)]$$
$$= 2a^2(4a - b)(3a + 2b)$$

63. Writing exercise

65. $9m^2 - 12m + 4 = (3m)^2 - 2(3m)(2) + (2)^2$
$$= (3m - 2)^2$$

67. $32a^2 - 48ab + 18b^2 = 2[16a^2 - 24ab + 9b^2]$
$$= 2[(4a)^2 - 2(4a)(3b) + (3b)^2]$$
$$= 2(4a - 3b)^2$$

69. $4x^2y^2 + 28xy + 49 = (2xy)^2 + 2(2xy)(7) + (7)^2$
$$= (2xy + 7)^2$$

71. $x^2 - 36 = (x)^2 - (6)^2 = (x + 6)(x - 6)$

73. $y^2 - w^2 = (y)^2 - (w)^2 = (y + w)(y - w)$

75. $9a^2 - 16 = (3a)^2 - (4)^2 = (3a + 4)(3a - 4)$

77. $25s^4 - 9t^2 = (5s^2)^2 - (3t)^2 = (5s^2 + 3t)(5s^2 - 3t)$

79. This exercise requires factoring twice.

$$p^4 - 625 = (p^2)^2 - (25)^2$$
$$= (p^2 + 25)(p^2 - 25)$$
$$= (p^2 + 25)[(p)^2 - (5)^2]$$
$$= (p^2 + 25)(p + 5)(p - 5)$$

81. $8 - a^3 = (2)^3 - (a)^3$

$= (2 - a)(2^2 + 2 \cdot a + a^2)$

$= (2 - a)(4 + 2a + a^2)$

83. $125x^3 - 27 = (5x)^3 - (3)^3$

$= (5x - 3)\left[(5x)^2 + 5x \cdot 3 + 3^2\right]$

$= (5x - 3)(25x^2 + 15x + 9)$

85. $27y^9 + 125z^6 = (3y^3)^3 + (5z^2)^3$

$= (3y^3 + 5z^2)\left((3y^3)^2 - 3y^3 \cdot 5z^2 + (5z^2)^2\right)$

$= (3y^3 + 5z^2)(9y^6 - 15y^3z^2 + 25z^4)$

87. Factor by grouping:

$x^2 + xy - 5x - 5y =$
$x \cdot (x + y) - 5 \cdot (x + y) =$
$(x + y)(x - 5)$

89. Factor out the greatest common factor, $(m - 2n)$.

$p^4(m - 2n) + q(m - 2n) = (m - 2n)(p^4 + q)$

91. This is a perfect square trinomial that can be factored by following the pattern.

$4z^2 + 28z + 49 =$
$(2z)^2 + 2(2z)(7) + (7)^2 =$
$(2z + 7)^2$

93. This is a sum of cubes.

$1000x^3 + 343y^3 = (10x)^3 + (7y)^3$

$= (10x + 7y)((10x)^2 - 10x \cdot 7y + (7y)^2)$

$= (10x + 7y)(100x^2 - 70xy + 49y^2)$

95. This is the difference of cubes.

$125m^6 - 216 = (5m^2)^3 - (6)^3$

$= (5m^2 - 6)\left((5m^2)^2 + 5m^2 \cdot 6 + (6)^2\right)$

$= (5m^2 - 6)(25m^4 + 30m^2 + 36)$

97. Factor this trinomial by multiplying the leading coefficient, 12, by -35 to obtain a product of -420. Then, find two numbers whose product is -420 and whose sum is $+16$.

$12m^2 + 16mn - 35n^2 =$
$12m^2 - 14mn + 30mn - 35n^2 =$
$2m \cdot (6m - 7n) + 5n \cdot (6m - 7n) =$
$(6m - 7n)(2m + 5n)$

99. Replace x with 4 and y with 2 to show that the left side of the equation does not equal the right side.

$x^2 + y^2 \neq (x + y)(x + y)$
$4^2 + 2^2 \neq (4 + 2)(4 + 2)$
$16 + 4 \neq (6)(6)$
$20 \neq 36$

7.7 EXERCISES

1. For the quadratic equation $4x^2 + 5x - 9 = 0$, the values of a, b, and c are respectively $\underline{4}$, $\underline{5}$. and $\underline{-9}$.

3. When using the quadratic formula, when $b^2 - 4ac$ is positive, the equation has $\underline{\text{two}}$ real solution(s). Consider the quadratic formula,

$$x = \frac{-b \pm \sqrt{b^2 - 4ac}}{2a}$$

If the quantity under the radical is positive, there will be two real solutions:

$$x_1 = \frac{-b + \sqrt{\text{positive number}}}{2a}$$

$$x_2 = \frac{-b - \sqrt{\text{positive number}}}{2a}$$

5. Set each factor equal to zero and solve each equation.

$$(x + 3)(x - 9) = 0$$

$x + 3 = 0$ or $x - 9 = 0$
$x = -3$ $x = 9$

The solution set is $\{-3, 9\}$.

7. Set each factor equal to zero and solve each equation.

$$(2t - 7)(5t + 1) = 0$$

$2t - 7 = 0$ or $5t + 1 = 0$
$2t = 7$ $5t = -1$
$t = \frac{7}{2}$ $t = -\frac{1}{5}$

The solution set is $\left\{\frac{7}{2}, -\frac{1}{5}\right\}$.

9. Factor the trinomial to obtain the two factors; then set each one equal to zero.

$$x^2 - x - 12 = 0$$
$$(x - 4)(x + 3) = 0$$

$x - 4 = 0$ or $x + 3 = 0$
$x = 4$ $x = -3$

The solution set is $\{4, -3\}$.

11. Factor the trinomial to obtain the two factors; then set each one equal to zero.

$$y^2 + 9y + 14 = 0$$
$$(y + 2)(y + 7) = 0$$

$$y + 2 = 0 \quad \text{or} \quad y + 7 = 0$$
$$y = -2 \qquad\qquad y = -7$$

The solution set is $\{-2, -7\}$.

13. Add -1 to both sides of the equation and then factor the left side.

$$12x^2 + 4x = 1$$
$$12x^2 + 4x - 1 = 1$$
$$(2x + 1)(6x - 1) = 0$$

$$2x + 1 = 0 \quad \text{or} \quad 6x - 1 = 0$$
$$2x = -1 \qquad\qquad 6x = 1$$
$$x = -\tfrac{1}{2} \qquad\qquad x = \tfrac{1}{6}$$

The solution set is $\left\{-\tfrac{1}{2}, \tfrac{1}{6}\right\}$.

15. FOIL the left side of the equation and combine like terms. Then add 16 to both sides to set the equation equal to zero.

$$(x + 4)(x - 6) = -16$$
$$x^2 - 6x + 4x - 24 = -16$$
$$x^2 - 2x - 24 = -16$$
$$x^2 - 2x - 8 = 0$$
$$(x - 4)(x + 2) = 0$$

$$x - 4 = 0 \quad \text{or} \quad x + 2 = 0$$
$$x = 4 \qquad\qquad x = -2$$

The solution set is $\{4, -2\}$.

17. $x^2 = 64$

$$\sqrt{x^2} = \pm\sqrt{64}$$
$$x = \pm 8$$

19. $x^2 = 24$

$$\sqrt{x^2} = \pm\sqrt{24}$$
$$x = \pm\sqrt{4 \cdot 6}$$
$$x = \pm 2\sqrt{6}$$

21. $r^2 = -5$. The solution set is $\emptyset$. There is no real number that will produce a negative number when it is squared.

23. $$(x - 4)^2 = 3$$
$$\sqrt{(x - 4)^2} = \pm\sqrt{3}$$
$$x - 4 = \pm\sqrt{3}$$
$$x - 4 + 4 = 4 \pm \sqrt{3}$$
$$x = 4 \pm \sqrt{3}$$

25. $$(2x - 5)^2 = 13$$
$$\sqrt{(2x - 5)^2} = \pm\sqrt{13}$$
$$2x - 5 = \pm\sqrt{13}$$
$$2x - 5 + 5 = 5 \pm \sqrt{13}$$
$$2x = 5 \pm \sqrt{13}.$$
$$x = \frac{5 \pm \sqrt{13}}{2}$$

27. For the equation $4x^2 - 8x + 1 = 0$, $a = 4$, $b = -8$, and $c = 1$. Substitute these values into the quadratic formula.

$$x = \frac{-b \pm \sqrt{b^2 - 4ac}}{2a}$$

$$x = \frac{-(-8) \pm \sqrt{(-8)^2 - 4 \cdot 4 \cdot 1}}{2 \cdot 4}$$
$$= \frac{8 \pm \sqrt{64 - 16}}{8}$$
$$= \frac{8 \pm \sqrt{48}}{8}$$
$$= \frac{8 \pm \sqrt{16 \cdot 3}}{8}$$
$$= \frac{8 \pm 4\sqrt{3}}{8}$$
$$= \frac{4\left(2 \pm 1\sqrt{3}\right)}{8}$$
$$= \frac{2 \pm \sqrt{3}}{2}$$

29. First, write the equation in standard form which is
$2y^2 - 2y - 1 = 0$. Then $a = 2, b = -2, c = -1$.
Substitute these values into the quadratic formula.

$$x = \frac{-b \pm \sqrt{b^2 - 4ac}}{2a}$$

$$y = \frac{-(-2) \pm \sqrt{(-2)^2 - 4 \cdot 2 \cdot -1}}{2 \cdot 2}$$

$$= \frac{2 \pm \sqrt{4 + 8}}{4}$$

$$= \frac{2 \pm \sqrt{12}}{4}$$

$$= \frac{2 \pm \sqrt{4 \cdot 3}}{4}$$

$$= \frac{2 \pm 2\sqrt{3}}{4}$$

$$= \frac{2\left(1 \pm \sqrt{3}\right)}{4}$$

$$= \frac{1 \pm \sqrt{3}}{2}$$

31. First, write the equation in standard form which is
$q^2 - q - 1 = 0$. Then, $a = 1, b = -1, c = -1$.
Substitute these values into the quadratic formula.

$$x = \frac{-b \pm \sqrt{b^2 - 4ac}}{2a}$$

$$q = \frac{-(-1) \pm \sqrt{(-1)^2 - 4 \cdot 1 \cdot -1}}{2 \cdot 1}$$

$$= \frac{1 \pm \sqrt{1 + 4}}{2}$$

$$= \frac{1 \pm \sqrt{5}}{2}$$

33. Write the equation in standard form by expanding the left
side of the equation to obtain $4k^2 + 4k = 1$. Then
subtract 1 from both sides to obtain $4k^2 + 4k - 1 = 0$.
Now $a = 4, b = 4, c = -1$. Substitute these values into
the quadratic formula.

$$x = \frac{-b \pm \sqrt{b^2 - 4ac}}{2a}$$

$$k = \frac{-(4) \pm \sqrt{(4)^2 - 4 \cdot 4 \cdot -1}}{2 \cdot 4}$$

$$= \frac{-4 \pm \sqrt{16 + 16}}{8}$$

$$= \frac{-4 \pm \sqrt{32}}{8}$$

$$= \frac{-4 \pm \sqrt{16 \cdot 2}}{8}$$

$$= \frac{-4 \pm 4\sqrt{2}}{8}$$

$$= \frac{4\left(-1 \pm 1\sqrt{2}\right)}{8}$$

$$= \frac{-1 \pm \sqrt{2}}{2}$$

35. FOIL the left side of the equation to obtain
$g^2 - g - 6 = 1$. Then subtract 1 from both sides to
obtain $g^2 - g - 7 = 0$. Now $a = 1, b = -1, c = -7$.
Substitute these values into the quadratic formula.

$$x = \frac{-b \pm \sqrt{b^2 - 4ac}}{2a}$$

$$g = \frac{-(-1) \pm \sqrt{(-1)^2 - 4 \cdot 1 \cdot -7}}{2 \cdot 1}$$

$$= \frac{1 \pm \sqrt{1 + 28}}{2}$$

$$= \frac{1 \pm \sqrt{29}}{2}$$

37. Write the equation in standard form by adding 14 to both
sides of the equation: $m^2 - 6m + 14 = 0$. Now $a = 1$,
$b = -6, c = 14$. Substitute these values into the
quadratic formula.

$$x = \frac{-b \pm \sqrt{b^2 - 4ac}}{2a}$$

$$m = \frac{-(-6) \pm \sqrt{(-6)^2 - 4 \cdot 1 \cdot 14}}{2 \cdot 1}$$

$$= \frac{6 \pm \sqrt{36 - 56}}{2}$$

$$= \frac{6 \pm \sqrt{-20}}{2}.$$

A negative under the radical creates an imaginary
number. The final solutions to the equation are not real
numbers. This can be symbolized by $\emptyset$.

39. Writing exercise

41. Writing exercise

43. $a = 1, b = 6, c = 9$

$$b^2 - 4ac = 6^2 - 4 \cdot 1 \cdot 9$$
$$= 36 - 36$$
$$= 0$$

(c) The equation has one rational solution because the quantity under the radical is zero. The solution will be

$$x = \frac{-b}{2a}$$

45. $a = 6, b = 7, c = -3$

$$b^2 - 4ac = 7^2 - 4 \cdot 6 \cdot -3$$
$$= 49 + 72$$
$$= 121$$

(a) The equation has two different rational solutions.
$\sqrt{121} = 11$.

47. $a = 9, b = -30, c = 15$
$$b^2 - 4ac = (-30)^2 - 4 \cdot 9 \cdot 15$$
$$= 900 - 540$$
$$= 360$$

(b) The equation has two different irrational solutions because 360 is not a perfect square.

49. Replace s with 200:

$$200 = -16t^2 + 45t + 400$$
$$200 - 200 = -16t^2 + 45t + 400 - 200$$
$$0 = -16t^2 + 45t - 200$$

Then $a = -16, b = 45, c = 200$

$$t = \frac{-45 \pm \sqrt{(45)^2 - 4 \cdot -16 \cdot 200}}{2 \cdot -16}$$
$$t = \frac{-45 \pm \sqrt{2025 + 12,800}}{-32}$$
$$t = \frac{-45 \pm \sqrt{14,825}}{-32}$$
$$t \approx \frac{-45 \pm 121.76}{-32}$$

Then $t \approx \frac{-45+121.76}{-32}$ or $t \approx \frac{-45-121.76}{-32}$. The first equation produces a negative value of t which is not meaningful. The second equation produces $t \approx 5.2$ seconds.

51. At time zero, the object is at ground level. Therefore, by letting $t = 0$, the value of s can be found. This is the height of the building.

53. (a) Replace s with 128. Obtain standard form for the equation by adding $16t^2$ and subtracting $144t$ to and from both sides of the equation.

$$s = 144t - 16t^2$$
$$128 = 144t - 16t^2$$
$$16t^2 - 144t + 128 = 0$$

Now solve for t by factoring.

$$16t^2 - 144t + 128 = 0$$
$$16(t^2 - 9t + 8) = 0$$
$$16(t - 1)(t - 8) = 0$$

Set each factor containing t equal to zero.

$$t - 1 = 0 \quad \text{or} \quad t - 8 = 0$$
$$t = 1 \qquad\qquad t = 8$$

The object will be 128 feet above the ground at 1 second and at 8 seconds. As it travels upward it will reach this height at 1 second; as it falls back to earth, it will be at the same height at 8 seconds.

(b) The object will strike the ground when $s = 0$.

$$0 = 144t - 16t^2$$
$$0 = 16t(9 - t)$$

Set each factor equal to zero.

$$16t = 0 \quad \text{or} \quad 9 - t = 0$$
$$t = 0 \qquad\qquad 9 = t$$

The object is on the ground at time zero when it is first projected, and it falls back to the ground 9 seconds later.

55. Use the Pythagorean Theorem with the legs represented by the algebraic expressions $2m$ and $2m + 3$. The longest side or hypotenuse has the value $5m$.

$$(2m)^2 + (2m + 3)^2 = (5m)^2$$
$$4m^2 + (4m^2 + 12m + 9) = 25m^2$$
$$8m^2 + 12m + 9 = 25m^2$$
$$8m^2 - 25m^2 + 12m + 9 = 0$$
$$-17m^2 + 12m + 9 = 0$$

Now $a = -17, b = 12, c = 9$

$$m = \frac{-(12) \pm \sqrt{(12)^2 - 4 \cdot -17 \cdot 9}}{2 \cdot -17}$$
$$= \frac{-12 \pm \sqrt{144 + 612}}{-34}$$
$$= \frac{-12 \pm \sqrt{756}}{-34}$$
$$\approx \frac{-12 \pm 27.495}{-34}$$

Now $m \approx \dfrac{-12 + 27.495}{-34}$ yields a negative number, which is not meaningful; however,

$$m \approx \dfrac{-12 - 27.495}{-34} \approx 1.16.$$

Then, $2m = 2(1.16)$ or approximately 2.3; $2m + 3 = 2.3 + 3$ or 5.3; and $5m = 5(1.16)$ or 5.8.

57. Let $100 =$ length of the shorter leg
$400 =$ length of the longer leg (height of the Mart)
$c =$ length of the hypotenuse.

Use the Pythagorean Theorem $a^2 + b^2 = c^2$.

$$(100)^2 + (400)^2 = c^2$$
$$10,000 + 160,000 = c^2$$
$$170,000 = c^2$$
$$\sqrt{170,000} = \sqrt{c^2}$$
$$412.3 \approx c$$

The length of the wire is about 412.3 feet.

59. Examine the figure in the text to see that the two legs of the right triangle can be represented by x and $x + 70$. If the two ships are 170 miles apart, this value is the length of the hypotenuse. Again use the Pythagorean Theorem with $a = x$, $b = x + 70$, and $c = 170$.

$$x^2 + (x + 70)^2 = (170)^2$$
$$x^2 + (x^2 + 140x + 4900) = 28900$$
$$2x^2 + 140x + 4900 = 28900$$
$$2x^2 + 140x - 24000 = 0$$

Now use the quadratic formula with $a = 2$, $b = 140$, $c = -24,000$.

$$x = \dfrac{-(140) \pm \sqrt{(140)^2 - 4 \cdot 2 \cdot -24,000}}{2 \cdot 2}$$
$$= \dfrac{-140 \pm \sqrt{19,600 + 192,000}}{4}$$
$$= \dfrac{-140 \pm \sqrt{211,600}}{4}$$
$$\approx \dfrac{-140 \pm 460}{4}$$

Use $\dfrac{-140+460}{4} \approx 80$ miles. Then the ship traveling due east had traveled 80 miles, and the ship traveling south had traveled $80 + 70$ or 150 miles.

61. Let $a =$ length of the shorter leg
$2a + 2 =$ length of the longer leg
$2a + 2 + 1 =$ length of the hypotenuse
Use the Pythagorean Theorem $a^2 + b^2 = c^2$.

$$a^2 + (2a + 2)^2 = (2a + 2 + 1)^2$$
$$a^2 + (4a^2 + 8a + 4) = (2a + 3)^2$$
$$5a^2 + 8a + 4 = 4a^2 + 12a + 9$$
$$a^2 - 4a - 5 = 0$$

Now use the quadratic formula with $a = 1$, $b = -4$, and $c = -5$.

$$a = \dfrac{-(-4) \pm \sqrt{(4)^2 - 4 \cdot 1 \cdot -5}}{2 \cdot 1}$$
$$= \dfrac{4 \pm \sqrt{16 + 20}}{2}$$
$$= \dfrac{4 \pm \sqrt{36}}{2}$$
$$= \dfrac{4 \pm 6}{2}$$

Use $\dfrac{4+6}{2} = 5$ to obtain a positive value for a. Then the shorter leg is 5 cm, the longer leg is $2 \cdot 5 + 2 = 12$ cm, and the hypotenuse is $12 + 1 = 13$ cm.

63. Let $w =$ width of the rectangle. Then $2w - 1 =$ length of rectangle. Use the Pythagorean Theorem with the width and length as the legs and 2.5 as the hypotenuse of the right triangle.

$$w^2 + (2w - 1)^2 = (2.5)^2$$
$$w^2 + (4w^2 - 4w + 1) = 6.25$$
$$5w^2 - 4w + 1 = 6.25$$
$$5w^2 - 4w - 5.25 = 0$$

Now use the quadratic formula with $a = 5$, $b = -4$, and $c = -5.25$.

$$a = \dfrac{-(-4) \pm \sqrt{(-4)^2 - 4 \cdot 5 \cdot -5.25}}{2 \cdot 5}$$
$$= \dfrac{4 \pm \sqrt{16 + 105}}{10}$$
$$= \dfrac{4 \pm \sqrt{121}}{10}$$
$$= \dfrac{4 \pm 11}{10}$$

Use $\dfrac{4+11}{10} = 1.5$ to obtain a positive value for w. The second value $\dfrac{4-11}{10}$ is negative, which is not meaningful. The width is 1.5 cm and the length is $2(1.5) - 1 = 2$ cm.

65. The area of the floor is $15 \cdot 20$ or 300 square feet, and the area of the rug is given as 234 square feet. The remaining area of the strip around the rug will be $300 - 234 = 66$ square feet. Draw a sketch of the rug surrounded by the flooring. Let w equal the width of the border; that is, the distance from the rug to the wall.

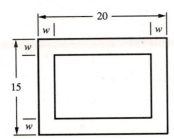

Four rectangles can be formed: The two rectangles on the left and right sides of the figure each have an area of $15 \cdot w$. The rectangles on the top and bottom of the figure each have an area of $(20 - 2w) \cdot w$. Because the total area of these rectangles is 66, an equation can be written:

$$2(15w) + 2[(20 - 2w) \cdot w] = 66$$
$$30w + 2[20w - 2w^2] = 66$$
$$30w + 40w - 4w^2 = 66$$
$$70w - 4w^2 = 66$$
$$-4w^2 + 70w - 66 = 66 - 66$$
$$-4w^2 + 70w - 66 = 0$$

Before using the quadratic formula, both sides of the equation can be divided by 2 or -2 in order to make the coefficients smaller.

$$\frac{-4w^2}{-2} + \frac{70w}{-2} - \frac{66}{-2} = \frac{0}{-2}$$
$$2w^2 - 35w + 33 = 0$$

Now use the quadratic formula with $a = 2$, $b = -35$, $c = 33$.

$$w = \frac{-(-35) \pm \sqrt{(-35)^2 - 4 \cdot 2 \cdot 33}}{2 \cdot 2}$$
$$= \frac{35 \pm \sqrt{1225 - 264}}{4}$$
$$= \frac{35 \pm \sqrt{961}}{4}$$
$$= \frac{35 \pm 31}{4}$$

The first value of w, $\frac{35+31}{4}$ yields $\frac{66}{4}$ or $16\frac{1}{2}$ feet. The second value of w, $\frac{35-31}{4}$ yields $\frac{4}{4}$ or 1 foot. Only the second value is meaningful in the context of the problem.

That is, the width of the border is 1 foot rather than $16\frac{1}{2}$ feet.

67. It is helpful to make a sketch of the proposed garden and grass strip surrounding it. In the figure, the inner rectangle that represents the garden has dimensions $20 - 2x$ meters by $30 - 2x$ meters, where x represents the width of strip of grass surrounding the garden. The outer rectangle, which represents the backyard has dimensions 20 by 30. Also the area of the garden added to the area of the border will equal the total area of the backyard, or the area of the surrounding rectangle. This can be written as an equation.

$$(30 - 2x)(20 - 2x) + 184 = 30 \cdot 20$$
$$600 - 100x + 4x^2 + 184 = 600$$
$$784 - 100x + 4x^2 = 600$$
$$784 - 600 - 100x + 4x^2 = 600 - 600$$
$$184 - 100x + 4x^2 = 0$$

To make computations with the quadratic formula a little easier, divide both sides of the equation by 4 to give

$$46 - 25x + x^2 = 0$$

Now use the quadratic formula with $a = 1$; $b = -25$; $c = 46$.

$$x = \frac{-(-25) \pm \sqrt{(-25)^2 - 4 \cdot 1 \cdot 46}}{2 \cdot 1}$$
$$= \frac{25 \pm \sqrt{625 - 184}}{2}$$
$$= \frac{25 \pm \sqrt{441}}{2}$$
$$= \frac{25 \pm 21}{2}$$

The first fraction yields 23 and the second fraction yields 2. However one dimension of the garden, $20 - 2x$, would be a negative number if 23 is used to replace x. When 2 is used to replace x, the dimensions are $20 - 2 \cdot 2 = 16$ meters by $30 - 2 \cdot 2 = 26$ meters, which are reasonable answers.

69. Replace the variables in the formula with the given number values and solve for r. The computations will be easier if both sides of the equation are divided by 2000 before expanding $(1 + r)^2$.

$$A = P(1 + r)^2$$
$$2142.25 = 2000(1 + r)^2$$
$$1.071125 = (1 + r)^2$$
$$1.071125 = 1 + 2r + r^2$$
$$0 = r^2 + 2r - .071125$$

Now use the quadratic formula with $a = 1$, $b = 2$, $c = -.071125$.

$$r = \frac{-(2) \pm \sqrt{(2)^2 - 4 \cdot 1 \cdot -.071125}}{2 \cdot 1}$$

$$= \frac{-2 \pm \sqrt{4 + .2845}}{2}$$

$$= \frac{-2 \pm \sqrt{4.2845}}{2}$$

$$\approx \frac{-2 \pm 2.07}{2}$$

The solution $\dfrac{-2 + 2.07}{2}$ yields approximately .035.

71. Set these two expressions equal to each other and solve for p. Begin by multiplying both sides of the equation by p to clear the fractions.

$$\frac{3200}{p} = 3p - 200$$

$$3200 = 3p^2 - 200p$$

$$0 = 3p^2 - 200p - 3200$$

Now use the quadratic formula with $a = 3$, $b = -200$, $c = -3200$.

$$p = \frac{-(-200) \pm \sqrt{(-200)^2 - 4 \cdot 3 \cdot -3200}}{2 \cdot 3}$$

$$= \frac{200 \pm \sqrt{40,000 + 38,400}}{6}$$

$$= \frac{200 \pm \sqrt{78,400}}{6}$$

$$= \frac{200 \pm 280}{6}$$

Use $p = \dfrac{200 + 280}{6} = 80$ cents.

73. Substitute the given values into the formula for the Froude number and solve for v. Clear the fractions by multiplying both sides of the equation by the lowest common denominator, $(9.8)(1.2)$. Then take the square root of both sides and use only the principal or positive root.

$$F = \frac{v^2}{gl}$$

$$2.57 = \frac{v^2}{(9.8)(1.2)}$$

$$30.2232 = v^2$$

$$5.5 \approx v$$

The value of v is approximately 5.5 meters per second.

75. To find the length of side AC, use proportions to write the following equation:

$$\frac{AC}{DF} = \frac{BC}{EF}$$

$$\frac{3x - 19}{x - 3} = \frac{x - 4}{4}$$

$$4(3x - 19) = (x - 3)(x - 4)$$

$$12x - 76 = x^2 - 7x + 12$$

$$12x - 12x - 76 = x^2 - 7x - 12x + 12$$

$$-76 + 76 = x^2 - 19x + 12 + 76$$

$$0 = x^2 - 19x + 88$$

Now use the quadratic formula with $a = 1$, $b = -19$, $c = 88$.

$$x = \frac{-(-19) \pm \sqrt{(-19)^2 - 4 \cdot 1 \cdot 88}}{2 \cdot 1}$$

$$= \frac{19 \pm \sqrt{361 - 352}}{2}$$

$$= \frac{19 \pm \sqrt{9}}{2}$$

$$\approx \frac{19 \pm 3}{2}$$

Using $x = \dfrac{19 + 3}{2} = 11$ or $x = \dfrac{19 - 3}{2} = 8$, there are two possible values for side AC:

$$3 \cdot 11 - 19 = 14 \quad \text{or} \quad 3 \cdot 8 - 19 = 5$$

The possible values for AC are 14 or 5.

CHAPTER 7 TEST

1.
$$5x - 3 + 2x = 3(x - 2) + 11$$
$$7x - 3 = 3x - 6 + 11$$
$$7x - 3 = 3x + 5$$
$$7x - 3 + 3 = 3x + 5 + 3$$
$$7x = 3x + 8$$
$$7x - 3x = 3x - 3x + 8$$
$$4x = 8$$
$$\frac{4x}{4} = \frac{8}{4}$$
$$x = 2$$

2.
$$\frac{2p-1}{3} + \frac{p+1}{4} = \frac{43}{12}$$

$$\frac{12}{1} \cdot \frac{2p-1}{3} + \frac{12}{1} \cdot \frac{p+1}{4} = \frac{12}{1} \cdot \frac{43}{12}$$

$$4(2p-1) + 3(p+1) = 43$$

$$8p - 4 + 3p + 3 = 43$$

$$11p - 1 = 43$$

$$11p - 1 + 1 = 43 + 1$$

$$11p = 44$$

$$\frac{11p}{11} = \frac{44}{11}$$

$$p = 4$$

3.
$$3x - (2-x) + 4x = 7x - 2 - (-x)$$

$$3x - 2 + x + 4x = 7x - 2 + x$$

$$8x - 2 = 8x - 2$$

This is an identity; the solution set is {all real numbers}.

4.
$$S = vt - 16t^2$$

$$S + 16t^2 = vt - 16t^2 + 16t^2$$

$$S + 16t^2 = vt$$

$$\frac{S + 16t^2}{t} = \frac{vt}{t}$$

$$\frac{S + 16t^2}{t} = v$$

5. Let k = the area of Kauai; $k + 177$ = the area of Maui; and $(k + 177) + 3293$ = the area of Hawaii

$$k + k + 177 + (k + 177) + 3293 = 5300$$

$$3k + 3647 = 5300$$

$$3k + 3647 - 3647 = 5300 - 3647$$

$$3k = 1653$$

$$\frac{3k}{3} = \frac{1653}{3}$$

$$k = 551$$

Then the area of Kauai is 551 square miles; the area of Maui is $551 + 177 = 728$ square miles; and the area of Hawaii is $728 + 3293 = 4021$ square miles.

6.

Strength	L of solution	L of alcohol
20%	x	$.20(x)$
50%	10	$.50(10)$
40%	$x + 10$	$.40(x+10)$

Create an equation by adding the first two algebraic expressions in the last column to total the third:

$$.2(x) + .5(10) = .4(x + 10)$$

$$10[.2(x) + .5(10)] = 10[.4(x + 10)]$$

$$2x + 50 = 4(x + 10)$$

$$2x + 50 = 4x + 40$$

$$2x + 50 - 50 = 4x + 40 - 50$$

$$2x = 4x - 10$$

$$2x - 4x = 4x - 4x - 10$$

$$-2x = -10$$

$$\frac{-2x}{-2} = \frac{-10}{-2}$$

$$x = 5 \text{ liters of 20\% solution.}$$

7.

	Rate	Time	Distance
Passenger train	60	t	$60t$
Freight train	75	t	$75t$

Because the trains are traveling in opposite directions, the sum of their distances will equal the total distance apart of 297 miles. Use this information to create an equation:

$$60t + 75t = 297$$

$$135t = 297$$

$$\frac{135t}{135} = \frac{297}{135}$$

$$t = 2.2 \text{ hours.}$$

8.
$$\frac{\$2.19}{8 \text{ slices}} \approx \$0.274 \text{ per slice}$$

$$\frac{\$3.30}{12 \text{ bags}} \approx \$0.275 \text{ per slice}$$

The 8-count size is the better buy.

9. Let x = the actual distance between Seattle and Cincinnati.

$$\frac{z}{46} = \frac{1050}{21}$$

$$21z = 46 \cdot 1050$$

$$21z = 48,300$$

$$\frac{21z}{21} = \frac{48,300}{21}$$

$$z = 2300 \text{ miles}$$

10.
$$I = \frac{k}{r}$$
$$80 = \frac{k}{30}$$
$$30 \cdot 80 = k$$
$$2400 = k$$

Then

$$I = \frac{2400}{12}$$
$$I = 200 \text{ amps}$$

11.
$$-4x + 2(x - 3) \geq 4x - (3 + 5x) - 7$$
$$-4x + 2x - 6 \geq 4x - 3 - 5x - 7$$
$$-2x - 6 \geq -x - 10$$
$$-2x - 6 + 6 \geq -x - 10 + 6$$
$$-2x \geq -x - 4$$
$$-2x + x \geq -x + x - 4$$
$$-x \geq -4$$
$$\frac{-x}{-1} \leq \frac{-4}{-1}$$
$$x \leq 4$$

Interval notation: $(-\infty, 4]$
Graph:

12.
$$-10 < 3k - 4 \leq 14$$
$$-10 + 4 < 3k - 4 + 4 \leq 14 + 4$$
$$-6 < 3k \leq 18$$
$$\frac{-6}{3} < \frac{3k}{3} \leq \frac{18}{3}$$
$$-2 < k \leq 6$$

Interval notation: $(-2, 6]$
Graph:

13. (a)
$$-3x < 9$$
$$\frac{-3x}{-3} > \frac{9}{-3}$$
$$x > -3$$

This is not equivalent.

(b)
$$-3x > -9$$
$$\frac{-3x}{-3} < \frac{-9}{-3}$$
$$x < 3$$

This is not equivalent.

(c)
$$-3x > 9$$
$$\frac{-3x}{-3} < \frac{9}{-3}$$
$$x < -3$$

This is equivalent.

(d)
$$-3x < -9$$
$$\frac{-3x}{-3} > \frac{-9}{-3}$$
$$x > 3$$

This is not equivalent.

14.   Let $x =$ the possible scores on the fourth test.

$$\frac{83 + 76 + 79 + x}{4} \geq 80$$
$$\frac{4}{1} \cdot \frac{238 + x}{4} \geq 80 \cdot 4$$
$$238 + x \geq 320$$
$$x \geq 82$$

He must score an 82 or better.

15. $$\left(\frac{4}{3}\right)^2 = \frac{4}{3} \cdot \frac{4}{3} = \frac{16}{9}$$

16. $$-(-2)^6 = -(-2 \cdot -2 \cdot -2 \cdot -2 \cdot -2 \cdot -2)$$
$$= -(64)$$
$$= -64$$

17. $$\left(\frac{3}{4}\right)^{-3} = \frac{3^{-3}}{4^{-3}}$$
$$= \frac{\frac{1}{3^3}}{\frac{1}{4^3}}$$
$$= \frac{\frac{1}{27}}{\frac{1}{64}}$$
$$= \frac{1}{27} \div \frac{1}{64}$$
$$= \frac{1}{27} \cdot \frac{64}{1}$$
$$= \frac{64}{27}$$

18. $$-5^0 + (-5)^0 = -1 + 1 = 0$$

19. $$9(4p^3)(6p^{-7}) = 9 \cdot 4 \cdot 6 \cdot p^{3+(-7)}$$
$$= 216p^{-4}$$
$$= \frac{216}{1} \cdot \frac{1}{p^4}$$
$$= \frac{216}{p^4}$$

20. $$\frac{m^{-2}(m^3)^{-3}}{m^{-4}m^7} = \frac{m^{-2} \cdot m^{-9}}{m^{-4+7}}$$
$$= \frac{m^{-11}}{m^3}$$
$$= m^{-11-3}$$
$$= m^{-14}$$
$$= \frac{1}{m^{14}}$$

21.
$$\frac{(2,500,000)(.00003)}{(.05)(5,000,000)} = \frac{(2.5 \times 10^6)(3 \times 10^{-5})}{(5 \times 10^{-2})(5 \times 10^6)}$$
$$= \frac{(2.5 \times 3) \times (10^{6-5})}{(5 \times 5) \times (10^{-2+6})}$$
$$= \frac{7.5 \times 10^1}{25 \times 10^4}$$
$$= \frac{7.5}{25} \times \frac{10^1}{10^4}$$
$$= .3 \times 10^{1-4}$$
$$= .3 \times 10^{-3}$$
$$= 3 \times 10^{-4}$$

Remember that scientific notation has one nonzero digit to the left of the decimal point. As .3 is made larger by a power of ten to become 3, 10^{-3} must become smaller by a power of ten.

22. Solve the formula $D = rt$ for t by dividing both sides of the equation by r.
$$\frac{D}{r} = \frac{rt}{r} = t.$$

Then replace the given distance and rate and simplify.
$$\frac{4.58 \times 10^9}{3.00 \times 10^5} = \frac{4.58}{3.00} \times \frac{10^9}{10^5}$$
$$\approx 1.53 \times 10^{9-5}$$
$$= 1.53 \times 10^4$$
$$= 15,300 \text{ seconds}$$

23. *For 1995*

106.3 billion is 106,300,000,000. In scientific notation this is 1.063×10^{11}.

For 1996

112.4 billion is 112,400,000,000. In scientific notation this is 1.124×10^{11}.

For 1997

125.0 billion is 125,000,000,000. In scientific notation this is 1.25×10^{11}.

24.
$$\left(3k^3 - 5k^2 + 8k - 2\right) - \left(3k^3 - 9k^2 + 2k - 12\right) =$$
$$3k^3 - 5k^2 + 8k - 2 - 3k^3 + 9k^2 - 2k + 12 =$$
$$\left(3k^3 - 3k^3\right) + \left(-5k^2 + 9k^2\right) + (8k - 2k) + (-2 + 12) =$$
$$4k^2 + 6k + 10$$

25.
$$(5x + 2)(3x - 4) = 5x \cdot 3x + 5x \cdot -4 + 2 \cdot 3x + 2 \cdot -4$$
$$= 15x^2 - 20x + 6x - 8$$
$$= 15x^2 - 14x - 8$$

26.
$$\left(4x^2 - 3\right)\left(4x^2 + 3\right) = \left(4x^2\right)^2 - (3)^2$$
$$= 16x^4 - 9$$

If the pattern is recognized, this multiplication of binomials can be done quickly. Otherwise, use the FOIL method as in Exercise 25.

27. Using vertical multiplication:

$$
\begin{array}{r}
3x^2 + 8x - 9 \\
x + 4 \\
\hline
12x^2 + 32x - 36 \\
3x^3 + 8x^2 - 9x \\
\hline
3x^3 + 20x^2 + 23x - 36
\end{array}
$$

28. One of many possibilities is
$$2y^5 + 8y^4 + y^3 - 7y^2 + 2y + 1.$$

29. Find two quantities whose product is $2 \cdot 3q^2 = 6q^2$ and whose sum is $-5q$, the coefficient of the middle term. The two quantities are $-2q$ and $-3q$. Use these two algebraic expressions as coefficients for p in place of the middle term; then factor the four-term polynomial by grouping.

$$2p^2 - 5pq + 3q^2 = 2p^2 - 2pq - 3pq + 3q^2$$
$$= 2p(p - q) - 3q(p - q)$$
$$= (p - q)(2p - 3q)$$

Remember that the two factors can also be expressed as
$$(2p - 3q)(p - q).$$

30. This is the difference of squares.
$$(10x)^2 - (7y)^2 = (10x + 7y)(10x - 7y)$$

31. This is the difference of cubes.
$$(3y)^3 - (5x)^3 = (3y - 5x)\left[(3y)^2 + 3y \cdot 5x + (5x)^2\right]$$
$$= (3y - 5x)\left(9y^2 + 15xy + 25x^2\right)$$

32. Factor by grouping.
$$4x + 4y - mx - my = 4(x + y) - m(x + y)$$
$$= (x + y)(4 - m)$$

33. In this equation $a = 6, b = 7, c = -3$. Substitute these values into the quadratic formula.

$$x = \frac{-b \pm \sqrt{b^2 - 4ac}}{2a}$$

$$x = \frac{-(7) \pm \sqrt{(7)^2 - 4 \cdot 6 \cdot -3}}{2 \cdot 6}$$

$$= \frac{-7 \pm \sqrt{49 + 72}}{12}$$

$$= \frac{-7 \pm \sqrt{121}}{12}$$

$$= \frac{-7 \pm 11}{12}$$

The two solutions are

$$\frac{-7 + 11}{12} = \frac{4}{12} = \frac{1}{3}$$

or

$$\frac{-7 - 11}{12} = \frac{-18}{12} = \frac{-3}{2}$$

This equation could also be solved by factoring.

34. First write the equation in standard form.

$$x^2 - x - 7 = 0$$

In this equation $a = 1, b = -1, c = -7$. Substitute these values into the quadratic formula.

$$x = \frac{-b \pm \sqrt{b^2 - 4ac}}{2a}$$

$$x = \frac{-(-1) \pm \sqrt{(-1)^2 - 4 \cdot 1 \cdot -7}}{2 \cdot 1}$$

$$= \frac{1 \pm \sqrt{1 + 28}}{2}$$

$$= \frac{1 \pm \sqrt{29}}{2}$$

The two solutions are

$$\frac{1 + \sqrt{29}}{2}$$

or

$$\frac{1 - \sqrt{29}}{2}.$$

35. Replace s with 25 and solve for t.

$$25 = 16t^2 + 15t$$
$$25 - 25 = 16t^2 + 15t - 25$$
$$0 = 16t^2 + 15t - 25$$

In this equation $a = 16, b = 15, c = -25$. Substitute these values into the quadratic formula.

$$x = \frac{-b \pm \sqrt{b^2 - 4ac}}{2a}$$

$$x = \frac{-(15) \pm \sqrt{(15)^2 - 4 \cdot 16 \cdot -25}}{2 \cdot 16}$$

$$= \frac{-15 \pm \sqrt{225 + 1600}}{32}$$

$$= \frac{-15 \pm \sqrt{1825}}{32}$$

$$\approx \frac{-15 \pm 42.7}{32}$$

The two potential solutions are

$$\frac{-15 + 42.7}{32}$$

or

$$\frac{-15 - 42.7}{32}.$$

The first fraction simplifies to about .87 seconds. The second fraction produces a negative number, which is not meaningful.

8 | GRAPHS, FUNCTIONS, AND SYSTEMS OF EQUATIONS AND INEQUALITIES

Chapter Goals

The general goal of this chapter is to provide an overview for the sketching of graphs of linear and quadratic functions and some of the associated applications. Upon completion of this chapter the reader should be able to

- Sketch the graphs of linear equations.
- Develop an understanding of the concept of functions.
- Sketch the graphs of parabolas (quadratic functions).
- Find and interpret the slope of a line.
- Write the equations of lines passing through a given point with a given slope.
- Solve a system of linear equations in two unknowns.
- Graph the solution set of linear inequalities.
- Apply the graphing of linear inequalities to the solution of linear programming problems.

Chapter Summary

We have all heard the expression that "a picture is worth a 1000 words." This is especially true in the areas of mathematics and statistics. Graphs are "pictures" of relationships, equations, and correspondences between related variables. They convey information and characteristics and model the behavior of mathematical expressions—all at a glance. Graphs also help us to solve problems in a direct fashion, giving us approximate and sometimes exact solutions to a wide variety of problems ranging from the simple to the complex.

To graph an equation means to make a drawing of its solutions. Such a drawing is called the graph of the equation.

LINEAR EQUATIONS IN TWO VARIABLES

Linear equations in two unknowns, as the name suggests, will yield a straight line when graphed. Since two points determine a straight line, we are required to find at least two ordered pairs that satisfy the equation of interest. A third point is useful as a check, since it also must lie on the same line.

Two convenient points which are often used to graph equations are the x-intercept and y-intercept. The x-intercept is the x-value (if any) of the point where the line crosses (or, intercepts) the x-axis. The ordered pair which locates this point will be of the form $(a, 0)$ which implies that if we let $y = 0$, the corresponding x value, $x = a$, is the x-intercept. Similarly, the y-intercept is the y-value (if any) of the point where the line crosses the y-axis. The ordered pair which locates this point will be of the form $(0, b)$ which implies that if we let $x = 0$, the corresponding y value, $y = b$, is the y-intercept. In summary:

To compute the x-intercept, let $y = 0$ and solve for x.
To compute the y-intercept, let $x = 0$ and solve for y.

Any equation of the form (or which can be written in the form)

$$ax + by = c \quad \textbf{\textit{standard form}}$$

will have a straight line as its graph (if a and b are not both 0 and x and y can be any real number). Such equations are called linear equations in two variables (x and y).

Once we know that any equation of the form $ax + by = c$ produces a straight-line graph, along with the fact that two points determine a straight line, graphing linear equations becomes a simple process. Although any two points whose coordinates satisfy the equation can be used, the intercepts are generally the easiest to compute or find. As indicated above, the computation of a third ordered pair is useful as a check.

The following example shows several situations that may be encountered.

EXAMPLE A

Graph the following equations.

(a) $4x + 3y = 12$ **Using intercepts with check point**

To find the x-intercept, let $y = 0$. Then,

$$4x + 3(0) = 12$$
$$4x = 12$$
$$x = 3.$$

Thus the point $(3, 0)$ is on the line.

To find the y-intercept, let $x = 0$. Then,

$$4(0) + 3y = 12$$
$$3y = 12$$
$$y = 4.$$

Thus, the point $(0, 4)$ is on the line. A third point should be used as a check. We substitute any arbitrary value for x and solve for y. If we let $x = -2$, then

$$4(0) + 3y = 12$$
$$4(-2) + 3y = 12$$
$$-8 + 3y = 12$$
$$3y = 4 + 8.$$
$$y = \frac{20}{3} \text{ or } 6\frac{2}{3}$$

We see that the point $\left(-2, 6\frac{2}{3}\right)$ is on the graph, so the graph is probably correct.

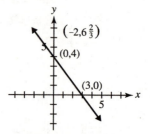

(b) $y = -3$ **Equation with missing variable**

Any ordered pair $(x, -3)$ is a solution (where x represents any real number). We can think of the equation as $0 \cdot x + y = -3$. No matter what number we choose for x, we find that $y = -3$.

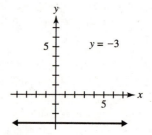

(c) $x = 2$ **Equation with missing variable**

Any ordered pair $(2, y)$ is a solution. That is, y can be any real number. So the line is parallel to the y-axis with x-intercept at $(2, 0)$.

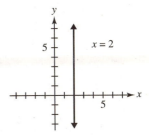

It is worth remembering that the graph of $y = k$ is a horizontal line. The graph of $x = k$ is a vertical line (where k represents some constant).

In summary:

To Graph Linear Equations

1. Is the equation of the type $x = k$ or $y = k$? If so, the graph will be a line parallel to an axis.

2. If the line is not of the type $x = k$ or $y = k$ (the equation contains both an x-term and a y-term), find any two points. The intercept(s) are usually the easiest to locate.

3. In either case, use a third point as a check.

THE SLOPE AND ALTERNATE FORMS FOR EQUATIONS OF LINES

The slope of a line is one way to indicate the steepness or inclination of the line. If we know two points on a line, we can find the slope of the line from the following definition.

The slope, m, of the line through the different points (x_1, y_1) and (x_2, y_2) is given by

$$m = \frac{\text{change in } y}{\text{change in } x} = \frac{y_2 - y_1}{x_2 - x_1}, \quad x_2 \neq x_1.$$

The following summary from the text shows several useful forms for equations for straight lines:

Forms of equations for lines

$ax + by = c$ *Standard form*

$y - y_1 = m(x - x_1)$ *Point-slope form*
  slope is m; passes through (x_1, y_1)

$y = mx + b$ *Slope-intercept form*
  slope is m; y-intercept is b

$x = k$ *Vertical line*
  through $(k, 0)$: no [finite] slope (infinite slope)

$y = k$ *Horizontal line*
  through $(0, k)$: slope is zero

The **point-slope form** of the line is especially useful if we know (or can easily find) a point on the line and the slope. With this information we can easily write the equation of the line. See Examples in the text and/or the 8.3 EXERCISES of the *Study Guide and Solutions Manual*.

The **slope-intercept form** of the line is especially useful if we are given an equation of the line and want to quickly identify the slope of the corresponding graph. For example if we are given an equation such as $2x + 5y = 6$, we can find the slope by solving for y and reading the coefficient of the x-term which represents the slope. That is

$$2x + 5y = 35$$
$$5y = -2x + 35$$
$$y = -\frac{2}{5}x + 7$$

Thus, by inspection, the slope of the line is $-\frac{2}{5}$.

FUNCTIONS

The concept of **correspondence** and the very special correspondence referred to as a **function** plays an important role in mathematics and represents a common thread in many higher level mathematics topics. One typical way to represent a correspondence is by an equation. For example, the equation $y = 2x + 4$ represents a correspondence between the set of values for x and the set of corresponding values for y. If we let $x = -1$, then the corresponding y value is $y = 2(-1) + 4 = -2 + 4 = 2$. Since we are getting "one and only one value" of y [$y = 2$] for the given value [$x = -1$]— and this is the case each time we replace x by some value—we call the correspondence a **function**. The **domain** element -1 corresponds to the **range** element 2. The following are (two) commonly used definitions of function:

Function   A function is a rule of correspondence between two sets X and Y that assign to each element x of set X one, and only one, element y of set Y. The set X is called the **domain** of the function, and the set of corresponding values in Y is called the **range** of the function. Letters such as f, g, or h are used to name functions. The only difference is that they are used to distinguish one function from another. Given a function f, for each value of x in the function's domain, there is exactly one value of y found in the range. Since there is only one such y, it can be represented by the single symbol $f(x)$. Range elements are often referred to as the values of the function and the notation $f(x)$, read "the value of f at x," or "f at x," or "f of x," is used. Observe that we thus have two ways to suggest a range element y and also $f(x)$. In other words $y = f(x)$.

We have used equations like the example above to generate ordered pairs (x, y) for the purpose of graphing. That is, the ordered pair $(-1, 2)$ represents a solution to the equation $y = 2x + 4$ and a point of the graph of the same equation. Because we can think of correspondences as ordered pairs, we are led to the second (alternative) definition of function.

Function (alternative definition)   A function is a set of ordered pairs (x, y) such that no first component is repeated (i.e., there are no two ordered pairs with the same first element). The set of all first components is the domain of the function, while the set of all second components is the **range** of the function.

If we are given (or can sketch) the graph of an equation or inequality, then it is an easy matter to determine if the correspondence is a function or not. The idea is summarized in the statement of the vertical line test.

Vertical line test   A graph is the graph of the function if no vertical line intersects the graph in more than one point.

EXAMPLE B   Functional notation, generation of ordered pairs, range elements

Given the function $y = f(x) = \frac{-3}{x-1}$, use a domain consisting of the elements $\left\{ \frac{7}{8}, \frac{1}{2}, 0, -2 \right\}$ to generate a set of ordered pairs which satisfy the equation and correspond to the graph. What are the range elements?

Solution
Replace successive values of x from the domain to find the range element and corresponding ordered pairs.

Range element	Ordered pair
$y = f\left(\dfrac{7}{8} \right)$	
$= \dfrac{-3}{\frac{7}{8} - 1}$	
$= \dfrac{-3}{\frac{7}{8} - \frac{8}{8}}$	
$= \dfrac{-3}{-\frac{1}{8}}$	
$= -3\left(-\dfrac{8}{1} \right)$	
$= 24$	$\left(\dfrac{7}{8}, 24 \right)$

Range element	Ordered pair
$y = f\left(\dfrac{1}{2} \right)$	
$= \dfrac{-3}{\frac{1}{2} - 1}$	
$= \dfrac{-3}{\frac{1}{2} - \frac{2}{2}}$	
$= \dfrac{-3}{-\frac{1}{2}}$	
$= -3\left(-\dfrac{2}{1} \right)$	
$= 6$	$\left(\dfrac{1}{2}, 6 \right)$

Range element	Ordered pair
$y = f(0)$	
$= \dfrac{-3}{0 - 1}$	
$= \dfrac{-3}{-1}$	
$= 3$	$(0, 3)$

Range element	Ordered pair
$y = f(-2)$	
$= \dfrac{-3}{-2 - 1}$	
$= \dfrac{-3}{-3}$	
$= 1$	$(-2, 1)$

GRAPHING PARABOLAS

Not all graphs are straight lines. Equations where x or y (or both) is raised to the second power–and no greater power— have graphs called conic sections. The parabola is one example of a conic section. Since parabolas model many real world situations, they are very useful curves to study. Parabolas which are concave up or down (open upward or downward) represent **quadratic functions**. Note that if a parabola is concave left or right, then it would not be a function, by the vertical line test. Equations where the power (exponent) on any y term is 1 (or is understood to be 1) and the highest power on any x term is 2 will yield a parabola which is concave up or down (that is, they represent a quadratic function).

The following represents two forms of quadratic functions which are often encountered. The graph of either form will be a parabola which is concave up (if $a > 0$) or concave down (if $a < 0$). The second (standard) form has the advantage that we can recognize by inspection of the vertex or turning point of the parabola. The vertex has coordinates given by (h, k),

---

Forms of equations for parabolas

$y = f(x) = ax^2 + bx + c$ *general form*

$y = f(x) = a(x - h)^2 + k$ *standard form*
where a, b, and c are real numbers and a is not zero. If $a > 0$, then the parabola opens upward, and if $a < 0$, then the parabola opens downward.

The ordered pair (h, k) represents the coordinates of parabola's **vertex**.

Observe that if the coefficient "b" of the x term (general form) is zero (i.e., there is no x term), then the y-intercept of the graph will be the vertex of the parabola. This makes the vertex easy to find, in this case, since all we need to do is replace x by 0 in the equation in order to find the y-intercept.

The following examples show us how to sketch the graph of parabolas given an equation in either form.

EXAMPLE C   Graph of parabola; general form

Sketch the graph of $y = f(x) = 2x^2 - 2$.

Solution
Observe that the coefficient "a" of the x^2 terms is 2. Since it is positive, the graph opens upward. Also since there is no "x-term" ($b = 0$), the turning point (or, vertex) will lie on the y-axis. To find the y-intercept, let

$x = 0$. We get $y = 2 \cdot 0^2 - 2 = -2$. Thus the vertex has coordinates $(0, -2)$. Generate other ordered pairs by letting x take on several values. Completing a table of ordered pairs, as shown in the textbook examples, is helpful. Other ordered pairs which satisfy the equation include $(-1, 0)$, $(1, 0)$, $(-2, 6)$, $(2, 6)$, and so on.

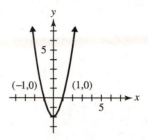

EXAMPLE D   Graph of parabola; standard form

Sketch the graph of $y = (x + 1)^2 - 2$.

Solution
Since the squared term is positive (coefficient $a = 1$), the graph opens upward. By inspection $h = -1$ [note: $(x + 1)$ is same as $[(x - (-1))]$ and $k = -2$. Thus the vertex is located at $(-1, -2)$. To find the y-intercept, let $x = 0$; then $y = -1$. Thus, the curve crosses the y-axis at $(0, -1)$. A symmetrical point (on other branch) would include $(-2, -1)$. You may want to plot several other points (including the vertex) to arrive at the indicated parabola.

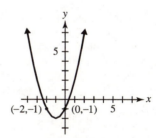

Alternative method (1) to find the vertex
Rather than identifying the values for h and k, one may find the vertex by the following analysis. Since we know the graph opens upward and thus the vertex represents the lowest point, we could identify the vertex alternatively by letting $x = -1$ which will make y as small as possible in the equation (i.e., $y = -2$), giving us the same ordered pair $(-1, -2)$ for the vertex. Note: Since $(x - 1)^2$ is squared, it must always be non-negative, so the smallest possible value it may take on is 0. In other words if $x = -1$, then $y = -2$.

Alternative method (2) to find the vertex
Another way to identify the coordinates of the vertex, or turning point, is to use the formula $x = \frac{-b}{2a}$, which gives the x-value of the vertex. Substitution of this value in the function will give the y-value. This works best if the quadratic function is stated in the general form: $y = f(x) = ax^2 + bx + c$ in order to inspect the values a, b, and c.

Graphing other functions   There are many types of functions used in mathematics. We have seen here several examples of **algebraic functions**. Examples of non-algebraic (**transcendental**) functions include **trigonometric, exponential, and logarithmic functions**. See Examples in the text and/or the EXERCISES (8.6) in the *Study Guide and Solutions Manual* for examples and suggestions on graphing exponential and logarithmic functions.

SOLVING SYSTEMS OF EQUATIONS

Many applications give rise to systems of equations. The simultaneous solution of any system is a set of values for the unknowns that satisfy each equation in the system at the same time. The following is an example of an application that can be solved by creating, then solving, a related "2 by 2" (2 equations in 2 unknowns) system of equations.

EXAMPLE E   Solving a system of equations - Application

On her new jet ski, Sami can travel upstream at top speed a distance of 36 miles in two hours. Returning, she finds that the trip downstream, still at top speed, takes only 1.5 hours. Find the speed of her jet ski and the speed of the current.

Solution
Let x = the speed of the boat in still water and
  y = the speed of the current.

Using the relationship $d = rt$, the distance traveled by boat upstream into the current is given by $36 = (x - y)(2)$ and the same distance, returning, with the current, is $(x + y)(1.5)$. The resulting two equations are

$$36 = (x - y)(2)$$
$$36 = (x + y)(1.5)$$

Write the equations in standard form.

$$2x - 2y = 36$$
$$1.5x + 1.5y = 36$$

Observe that the second equation may be simplified as follows before solving the system of equations.

$$1.5x + 1.5y = 36$$
$$15x + 15y = 360$$
$$x + y = 24$$

The resulting system is

$$2x - 2y = 36$$
$$x + y = 24.$$

Solve by multiplying the second equation by 2 and adding to the first.

$$2x - 2y = 36$$
$$\underline{2x + 2y = 48}$$
$$4x \quad\quad = 84$$
$$x = 21$$

Substitute this value of x into the equation $x + y = 24$ and solve for y.

$$(21) + y = 24$$
$$x = 3$$

Thus, the speed of the jet ski is 21 miles per hour in still water and the speed of the current is 3 miles per hour.

GRAPHING LINEAR SYSTEMS OF EQUATIONS

Graphing linear equations is sometimes a help in the solution of a system of equations, as in the following example.

EXAMPLE F   Use a graphing approach to solve the system:
$$x + y = -5$$
$$-2x + y = 1.$$

Solution
Graph the line $x + y = -5$ through its intercepts $(0, -5)$ and $(-5, 0)$ and the line $-2x + y = 1$ through its intercepts $(0, 1)$ and $\left(-\frac{1}{2}, 0\right)$. The lines appear to intersect at $(-2, -3)$.

Check this ordered pair in the system.

$$-2 + (-3) = -5$$
$$-5 = -5 \quad \text{True}$$
$$-2(-2) + -3 = 1$$
$$1 = 1 \quad \text{True}$$

The solution set is $\{(-2, -3)\}$.

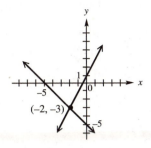

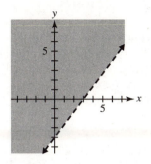

GRAPHING LINEAR INEQUALITIES

Graphing linear inequalities in two variables is almost as easy as graphing linear equations. The following discussion will lead us to a simple, step-by-step process. Consider the following equation and related inequalities.

$$x + y = 2 \qquad x + y > 2 \qquad x + y < 2$$

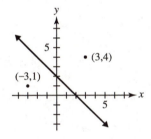

The straight line is the graph of $x + y = 2$. The line divides the plane into two **half-planes**, one *above* the line and one *below* the line. For each point in the half-plane above the line, the ordered pair (x, y) associated with the point satisfies the inequality $x + y > 2$. For example, the ordered pair $(3, 4)$ produces the true statement $3 + 4 > 2$. Likewise, for each point in the half-plane below the line, the ordered pair (x, y) satisfies the inequality $x + y < 2$. For example, $(-3, 1)$ produces the true statement $-3 + 1 < 2$. Observe that if the line itself is to be included in the answer, the relation would have to be " $\leq$ " or " $\geq$ " instead of " $<$." A dashed line is used if the relation does not include equality ($<$ or $>$), since all points in the solution will lie off of the line. A solid line is used if the relation includes equality ($\leq$ or $\geq$).

The following example illustrates the use of these ideas.

EXAMPLE G   Linear inequality

Graph $4x - 3y < 12$

Solution
Step 1. Graph the line $4x - 3y = 12$. The intercepts are $(3, 0)$ and $(0, -4)$. Use a dashed line for the graph since the relation is " $<$."

Step 2. Pick a point that does not belong to the line. Substitute to determine whether this point is a solution. The origin $(0, 0)$, when not on the line, is usually an easy one to use: $4 \cdot 0 - 3 \cdot 0 < 12$ is true, so the origin is a solution. This means that we must shade the left half-plane. Had the substitution given us a false inequality we would have shaded the other half-plane.

In summary:

> ### To Graph Linear Inequalities
>
> 1. Graph the corresponding equality. Use a solid line if equality is included in the original statement and a dashed line if equality is not included.
>
> 2. Choose a test point not on the line and substitute its coordinates into the inequality.
>   (The origin is a convenient point if it is not on the line.)
>
> 3. The graph of the original inequality is
>   (a) the half-plane containing the test point if the inequality is satisfied by that point, or
>   (b) the half-plane not containing the test point if the inequality is not satisfied by the point.

A **system of linear inequalities** consists of two or more linear inequalities. The solution of a system is the set of all ordered pairs that makes all inequalities of the system true at the same time. To find the solution, graph all the inequalities of the system on the same coordinate axes and locate the common points—the areas covered by all shadings simultaneously (i.e., the intersection)—of all the graphs. See examples in the text and/or exercises in the EXERCISES section of the *Study Guide and Solutions Manual*.

LINEAR PROGRAMMING

An important application of systems of linear inequalities is that of **linear programming**. Linear programming is of special interest in business and the social sciences where it is used to find such values as minimum cost, maximum profit, and the maximum amount of earning that can take place under given conditions.

The basic objective of a linear programming problem is to maximize (or minimize) some value or entity, subject to certain restrictions. The quantity for which an optimal solution is desired is represented by a mathematical equation (sometimes called the **objective function**). The restrictions on the variables associated with a problem are represented by linear inequalities and are called the **constraints** of the problem. We graph these constraints on the same set of axes to produce a region in the plane called the **feasible region**.

According to linear programming theory, if there is a unique solution that maximizes (or minimizes) an objective function, then the solution must occur at one of the corners of the feasible region. Thus, each corner is considered a possible solution to the problem; and one corner, the one that will produce a maximum (or minimum) value, is called the **optimal solution**. The following summarizes the necessary steps in the solution of a linear programming problem.

To solve a linear programming problem

1 Determine from the problem the objective function and the constraints and translate them into mathematical statements. (Remember that the objective function is an equation and the constraints are inequalities.)

2. Graph each inequality on the same set of axes in order to determine the feasible region. (Remember that this is the region formed by the intersection of the graphs of the inequalities.)

3. Find the corner points of the region by the simultaneous solution of corresponding equations and substitute these values into the objective function to determine the maximum (or minimum) value this function will take on.

EXAMPLE H   Linear programming problem

A wholesaler of party goods wishes to display her products at a convention of social secretaries in such a way that she gets the maximum number of inquiries about her whistles and hats. Her booth at the convention has 12 square meters of floor space to be used for display purposes. A display unit for hats requires 2 square meters; and for whistles, 4 square meters. Experience tells the wholesaler that she should never have more than a total of 5 units of whistles and hats on display at one time. If she receives three inquiries for each unit of hats and two inquiries for each unit of whistles on display, how many of each should she display in order to get the maximum number of inquiries?

Solution
Let $x =$ number of display units of hats and $y =$ number of display units of whistles. Then, the following constraints apply:

$$12 \leq 2x + 4y$$
$$x + y \leq 5 \quad \text{and by implication}$$
$$x \geq 0 \quad \text{(number of display}$$
$$x \geq 0 \quad \text{units must be positive)}$$

where $3x + 2y$ represents the number of inquiries which is to be maximized. (The objective function is "number of inquires $3x + 2y$.")

Sketch the graph representing the intersection of the constraints.

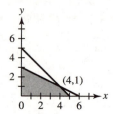

To find the upper right corner point solve:

$$2x + 4y = 12$$
$$x + y = 5.$$

Multiply the second equation by -2 and add to the first equation.

$$2x + 4y = 12$$
$$\underline{-2x - 2y = -10}$$
$$2y = 2$$
$$y = 1$$

Replace this value of x into the equation $x + y = 5$ and solve for x.

$$x + (1) = 5.$$
$$x = 4$$

The resulting ordered pair is $(4, 1)$. The other corner points are: $(0, 3), (0, 0),$ and $(5, 0)$.

The following table, indicating the corner points and the value of the objective function, is an easy way to organize the information (note that some of the corner points can be found by inspection).

Point	Number of inquired $= 3x + 2y$	
$(0, 3)$	$3(0) + 2(3)$	$= 6$
$(4, 1)$	$3(4) + 2(1)$	$= 14$
$(5, 0)$	$3(5) + 2(0)$	$= 15$ ← Maximum
$(0, 0)$	$3(0) + 2(0)$	$= 0$

Thus, the maximum number of inquiries is 15, which will result if 5 units of hats and no whistles are displayed.

8.1 EXERCISES

1. (a) The numbers increased between 1991–1992 and 1993–1994.

(b) The decrease was greatest between 1992 and 1993.

(c) The number was about 6000 in 1993.

3. For any value of x, the point $(x, 0)$ lies on the $\underline{x}$-axis.

5. The circle $x^2 + y^2 = 9$ has the point $\underline{(0,0)}$ as its center.

7. (a) The point $(1, 6)$ is located in quadrant I since the x-value and the y-value are both positive.

(b) The point $(-4, -2)$ is located in quadrant III since the x-value and the y-value are both negative.

(c) The point $(-3, 6)$ is located in quadrant II since the x-value is negative and the y-value is positive.

(d) The point $(7, -5)$ is located in quadrant IV since the x-value is positive and the y-value is negative.

(e) The point $(-3, 0)$ is located between quadrants II and III (i.e., on the negative x-axis) since the x-value is negative and the y-value is 0.

9. (a) If $xy > 0$, then x and y must have same signs. Thus, the point must lie in Quadrant I or III.

(b) If $xy < 0$, then x and y must have opposite signs. Thus, the point must lie in Quadrant II or IV. opposite signs.

(c) If $\frac{x}{y} < 0$, then x and y must have opposite signs. Thus, the point must lie in Quadrant II or IV.

(d) If $\frac{x}{y} > 0$, then x and y must have same signs. Thus, the point must lie in Quadrant I or III.

11–20. Note: graph shows even and odd answers.

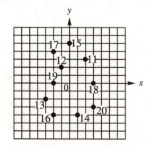

21. (a)
$$d = \sqrt{(x_2 - x_1)^2 + (y_2 - y_1)^2}$$
$$= \sqrt{(-2 - 3)^2 + (1-4)^2}$$
$$= \sqrt{(-5)^2 + (-3)^2}$$
$$= \sqrt{25 + 9}$$
$$= \sqrt{34}$$

(b) Using $\left(\dfrac{x_1 + x_2}{2}, \dfrac{y_1 + y_2}{2}\right)$, we have
$$\left(\frac{3 + (-2)}{2}, \frac{4 + 1}{2}\right) = \left(\frac{1}{2}, \frac{5}{2}\right).$$

23. (a)
$$d = \sqrt{(x_2 - x_1)^2 + (y_2 - y_1)^2}$$
$$= \sqrt{[3 - (-2)]^2 + (-2 - 4)^2}$$
$$= \sqrt{5^2 + (-6)^2}$$
$$= \sqrt{25 + 36}$$
$$= \sqrt{61}$$

(b) Using $\left(\dfrac{x_1 + x_2}{2}, \dfrac{y_1 + y_2}{2}\right)$, we have
$$\left(\frac{(-2) + 3}{2}, \frac{4 + (-2)}{2}\right) = \left(\frac{1}{2}, 1\right).$$

25. (a)
$$d = \sqrt{(x_2 - x_1)^2 + (y_2 - y_1)^2}$$
$$= \sqrt{[2 - (-3)]^2 + (-4 - 7)^2}$$
$$= \sqrt{(5)^2 + (-11)^2}$$
$$= \sqrt{25 + 121}$$
$$= \sqrt{146}$$

(b) Using $\left(\dfrac{x_1 + x_2}{2}, \dfrac{y_1 + y_2}{2}\right)$,
$$\left(\frac{-3 + 2}{2}, \frac{7 + (-4)}{2}\right) = \left(-\frac{1}{2}, \frac{3}{2}\right).$$

27. $(x - 3)^2 + (y - 2)^2 = 25$

The equation indicates a graph of a circle with a radius of 5 and centered at $(3, 2)$. Therefore, choose graph B.

29. $(x + 3)^2 + (y - 2)^2 = 25$
$[x - (-3)]^2 + (y - 2)^2 = 25$

The equation indicates a graph of a circle with a radius of 5 and centered at $(-3, 2)$. Thus, choose graph D.

31. Use the equation of a circle, where $h = 0$, $k = 0$, and $r = 6$.

$$(x - h)^2 + (y - k)^2 = r^2$$
$$(x - 0)^2 + (y - 0)^2 = 6^2$$
$$x^2 + y^2 = 36$$

33. Use the equation of a circle, where $h = -1$, $k = 3$, and $r = 4$.

$$(x - h)^2 + (y - k)^2 = r^2$$
$$[x - (-1)]^2 + (y - 3)^2 = 4^2$$
$$(x + 1)^2 + (y - 3)^2 = 16$$

35. Use the equation of a circle, where $h = 0$, $k = 4$, and $r = \sqrt{3}$.

$$(x - h)^2 + (y - k)^2 = r^2$$
$$(x - 0)^2 + (y - 4)^2 = \left(\sqrt{3}\right)^2$$
$$x^2 + (y - 4)^2 = 3$$

37. The equation, $x^2 + y^2 = r^2$, $r > 0$ or equivalently, $(x - 0)^2 + (y - 0)^2 = r^2$, $r > 0$, implies that the center is located at $(0, 0)$ and the radius is r.

39. To find the center and radius, complete the square on x and y.

$$x^2 + y^2 + 4x + 6y + 9 = 0$$
$$\left(x^2 + 4x + \quad\right) + \left(y^2 + 6y + \quad\right) = -9$$
$$\left(x^2 + 4x + 4\right) + \left(y^2 + 6y + 9\right) = -9 + 4 + 9$$
$$(x + 2)^2 + (y + 3)^2 = 4 \text{ or}$$
$$[x - (-2)]^2 + [y - (-3)]^2 = 4$$

Thus, by inspection, the center is located at $(-2, -3)$ and the radius is given by $r = \sqrt{4} = 2$.

Remember that the added constants, 4 and 9, come from squaring $\frac{1}{2}$ of the coefficients of each first degree term $\left(\text{i.e., } \left[\frac{1}{2}(4)\right]^2 = 4 \text{ and } \left[\frac{1}{2}(6)\right]^2 = 9\right)$.

41. To find the center and radius, complete the square on x and y.

$$x^2 + y^2 + 10x - 14y - 7 = 0$$
$$\left(x^2 + 10x + \quad\right) + \left(y^2 - 14y + \quad\right) = 7$$
$$\left(x^2 + 10x + 25\right) + \left(y^2 - 14y + 49\right) = 7 + 25 + 49$$
$$(x + 5)^2 + (y - 7)^2 = 81 \text{ or}$$
$$[x - (-5)]^2 + (y - 7)^2 = 9^2$$

Thus, by inspection, the center is located at $(-5, 7)$ and the radius is given by $r = 9$.

43. To find the center and radius, complete the square on x and y.

$$3x^2 + 3y^2 - 12x - 24y + 12 = 0$$
$$x^2 + y^2 - 4x - 8y + 4 = 0 \qquad \textit{Divide by 3}$$
$$\left(x^2 - 4x + \quad\right) + \left(y^2 - 8y + \quad\right) = -4$$
$$\left(x^2 - 4x + 4\right) + \left(y^2 - 8y + 16\right) = -4 + 4 + 16$$
$$(x - 2)^2 + (y - 4)^2 = 16$$

Thus, by inspection, the center is located at $(2, 4)$ and the radius is given by $r = \sqrt{16} = 4$.

45. The equation, $x^2 + y^2 = 36$, is equivalent to

$$(x - 0)^2 + (y - 0)^2 = 6^2.$$

Thus, the center of the graph is located at $(0, 0)$ and has a radius $r = 6$. The graph is as follows.

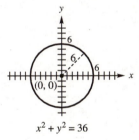

$$x^2 + y^2 = 36$$

47. The equation, $(x - 2)^2 + y^2 = 36$, is equivalent to

$$(x - 2)^2 + (y - 0)^2 = 36.$$

Thus, the center of the graph is located at $(2, 0)$ and has a radius $r = 6$. The graph is as follows.

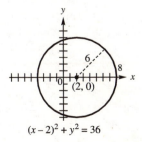

$$(x - 2)^2 + y^2 = 36$$

49. The equation, $(x + 2)^2 + (y - 5)^2 = 16$, is equivalent to

$$[x - (-2)]^2 + (y - 5)^2 = 4^2.$$

Thus, the center of the graph is located at $(-2, 5)$ and has a radius $r = 4$. The graph is as follows.

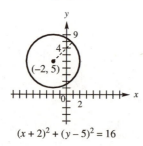

$(x + 2)^2 + (y - 5)^2 = 16$

51. The equation, $(x + 3)^2 + (y + 2)^2 = 36$, is equivalent to

$$[x - (-3)]^2 + [y - (-2)]^2 = 6^2$$

Thus, the center of the graph is located at $(-3, -2)$ and has a radius $r = 6$. The graph is as follows.

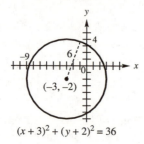

$(x + 3)^2 + (y + 2)^2 = 36$

53. Writing exercise

55. Writing exercise

57. Show algebraically that if three receiving stations at $(1, 4)$, $(-6, 0)$, and $(5, -2)$ record distances to an earthquake epicenter of 4 units, 5 units, and 10 units, respectively, the epicenter would lie at $(-3, 4)$.

Create equations of circles with centers at $(1, 4)$, $(-6, 0)$, and $(5, -2)$ and radii of 4 units, 5 units, and

$$(x - 1)^2 + (y - 4)^2 = 4^2 = 16$$
$$(x + 6)^2 + y^2 = 5^2 = 25$$
$$(x - 5)^2 + (y + 2)^2 = 10^2 = 100$$

If we replace x by -3 and y by 4 in each of these equations we see that $(-3, 4)$ is a solution in all three equations. That is, $(-3, 4)$ is the epicenter.

59. Writing exercise

61. If the endpoints of a line segment have coordinates (x_1, y_1) and (x_2, y_2), then

(a) $d_1 = \sqrt{\left(\dfrac{x_1 + x_2}{2} - x_1\right)^2 + \left(\dfrac{y_1 + y_2}{2} - y_1\right)^2}$

$\quad = \sqrt{\left(\dfrac{x_1 + x_2 - 2x_1}{2}\right)^2 + \left(\dfrac{y_1 + y_2 - 2y_1}{2}\right)^2}$

$\quad = \sqrt{\left(\dfrac{x_2 - x_1}{2}\right)^2 + \left(\dfrac{y_2 - y_1}{2}\right)^2}$

$d_2 = \sqrt{\left(\dfrac{x_1 + x_2}{2} - x_2\right)^2 + \left(\dfrac{y_1 + y_2}{2} - y_2\right)^2}$

$\quad = \sqrt{\left(\dfrac{x_1 + x_2 - 2x_2}{2}\right)^2 + \left(\dfrac{y_1 + y_2 - 2y_2}{2}\right)^2}$

$\quad = \sqrt{\left(\dfrac{x_1 - x_2}{2}\right)^2 + \left(\dfrac{y_1 - y_2}{2}\right)^2}$

$\quad = \sqrt{\left(\dfrac{x_2 - x_1}{2}\right)^2 + \left(\dfrac{y_2 - y_1}{2}\right)^2}.$

Thus, $d_1 = d_2$.

(b) From part (a)

$d_1 + d_2 = 2d_1$

$\quad = 2\sqrt{\left(\dfrac{x_2 - x_1}{2}\right)^2 + \left(\dfrac{y_2 - y_1}{2}\right)^2}$

$\quad = 2\sqrt{\dfrac{1}{4}\left[(x_2 - x_1^2) + (y_2 - y_1)^2\right]}$

$\quad = \sqrt{(x_2 - x_1)^2 + (y_2 - y_1)^2}$

$\quad = $ distance between (x_1, y_1) and (x_2, y_2).

(c) We can conclude that the point $\left(\frac{x_1 + x_2}{2}, \frac{y_1 + y_2}{2}\right)$ is the midpoint of the segment whose endpoints are (x_1, y_1) and (x_2, y_2).

63. Only option (B) can be written in the form $x^2 + y^2 = r^2$ where $r^2 > 0$, and hence, is the only equation that will represent a circle.

8.2 EXERCISES

1. The given equation is
$$2x + y = 5.$$

For $(0, \ )$:
$$2 \cdot 0 + y = 5$$
$$y = 5$$

or the ordered pair $(0, 5)$.

For $(\ , 0)$:
$$2x + 0 = 5$$
$$x = \frac{5}{2}$$

or the ordered pair $\left(\frac{5}{2}, 0\right)$.

For $(1, \ )$:
$$2 \cdot 1 + y = 5$$
$$y = 3$$

or the ordered pair $(1, 3)$.

For $(\ , 1)$:
$$2x + 1 = 5$$
$$2x = 4$$
$$x = 2$$

or the ordered pair $(2, 1)$.

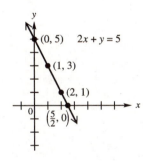

3. The given equation is
$$x - y = 4.$$

For $(0, \ )$:
$$0 - y = 4$$
$$-y = 4$$
$$y = -4$$

or the ordered pair $(0, -4)$.

For $(\ , 0)$:
$$x - 0 = 4$$
$$x = 4$$

or the ordered pair $(4, 0)$.

For $(2, \ )$:
$$2 - y = 4$$
$$-y = 2$$
$$y = -2$$

or the ordered pair $(2, -2)$.

For $(\ , -1)$:
$$x - (-1) = 4$$
$$x + 1 = 4$$
$$x = 3$$

or the ordered pair $(3, -1)$.

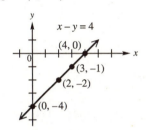

5. The given equation is
$$4x + 5y = 20.$$

For $(0, \ )$:
$$4 \cdot 0 + 5y = 20$$
$$5y = 20$$
$$y = 4$$

or the ordered pair $(0, 4)$.

For $(\ , 0)$:
$$4x + 5 \cdot 0 = 20$$
$$4x = 20$$
$$x = 5$$

or the ordered pair $(5, 0)$.

For $(3, \ )$:
$$4 \cdot 3 + 5y = 20$$
$$12 + 5y = 20$$
$$5y = 8$$
$$y = \frac{8}{5}$$

or the ordered pair $\left(3, \frac{8}{5}\right)$.

For (, 2):

$$4x + 5 \cdot 2 = 20$$
$$4x + 10 = 20$$
$$4x = 10$$
$$x = \frac{5}{2}$$

or the ordered pair $\left(\frac{5}{2}, 2\right)$.

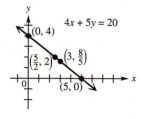

7.   The given equation is

$$3x + 2y = 8.$$

From the partially completed table in text we want to complete the evaluation of the ordered pairs $(0, \ )$, $(\ , 0)$, $(2, \ )$, $(\ , -2)$.

For $(0, \ )$:

$$3 \cdot 0 + 2y = 8$$
$$2y = 8$$
$$y = 4$$

or the ordered pair $(0, 4)$.

For $(\ , 0)$:

$$3x + 0 \cdot y = 8$$
$$3x = 8$$
$$x = \frac{8}{3}$$

or the ordered pair $\left(\frac{8}{3}, 0\right)$.

For $(2, \ )$:

$$3 \cdot 2 + 2y = 8$$
$$6 + 2y = 8$$
$$2y = 2$$
$$y = 1$$

or the ordered pair $(2, 1)$.

For $(\ , -2)$:

$$3x + 2 \cdot (-2) = 8$$
$$3x - 4 = 8$$
$$3x = 12$$
$$x = 4$$

or the ordered pair $(4, -2)$.

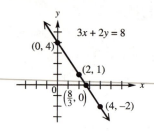

9.   Writing exercise

11.   Option A is correct since y is constant at $y = 3$.

13.   The given equation is

$$3x + 2y = 12.$$

To find the x-intercept, let $y = 0$.

$$3x + 2 \cdot 0 = 12$$
$$3x = 12$$
$$x = 4$$

The x-intercept is $(4, 0)$.

To find the y-intercept, let $x = 0$.

$$3 \cdot 0 + 2y = 12$$
$$2y = 12$$
$$y = 6$$

The y-intercept is $(0, 6)$.

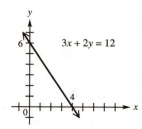

15.   The given equation is

$$5x + 6y = 10.$$

To find the x-intercept, let $y = 0$.

$$5x + 6 \cdot 0 = 10$$
$$5x = 10$$
$$x = 2$$

The x-intercept is $(2, 0)$.

To find the y-intercept, let $x = 0$.

$$5 \cdot 0 + 6y = 10$$
$$6y = 10$$
$$y = \frac{5}{3}$$

The y-intercept is $\left(0, \frac{5}{3}\right)$.

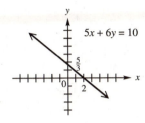

17. The given equation is

$$2x - y = 5.$$

To find the x-intercept, let $y = 0$.

$$2x - 0 = 5$$
$$x = \frac{5}{2}$$

The x-intercept is $\left(\frac{5}{2}, 0\right)$.

To find the y-intercept, let $x = 0$.

$$2 \cdot 0 - y = 5$$
$$-y = 5$$
$$y = -5$$

The y-intercept is $(0, -5)$.

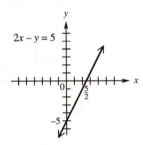

19. The given equation is

$$x - 3y = 2.$$

To find the x-intercept, let $y = 0$.

$$x - 3 \cdot 0 = 2$$
$$x = 2$$

The x-intercept is $(2, 0)$.

To find the y-intercept, let $x = 0$.

$$0 - 3y = 2$$
$$y = -\frac{2}{3}$$

The x-intercept is $\left(0, -\frac{2}{3}\right)$.

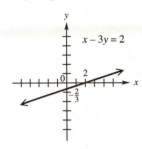

21. The given equation is

$$y + x = 0.$$

To find the y-intercept, let $x = 0$.

$$y + 0 = 0$$
$$y = 0$$

The y-intercept is $(0, 0)$. Observe that this is also the x-intercept. Thus, the graph runs through the origin.

To find a second point let x (or y) take on a value and solve for the other variable, i.e. let $x = 2$.

$$y + 2 = 0$$
$$y = -2$$

Thus, a second point would have coordinates $(2, -2)$.

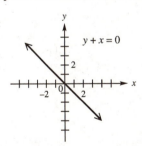

23. The given equation is

$$3x = y.$$

To find the y-intercept, let $x = 0$.

$$3 \cdot 0 = y$$
$$y = 0$$

The y-intercept is $(0, 0)$. Observe that this is also the x-intercept. Thus, the graph runs through the origin.

To find a second point let x (or y) take on a value and solve for the other variable, i.e. let $x = 1$, then

$$3 \cdot 1 = y$$
$$y = 3$$

Thus, a second point would have coordinates $(1, 3)$.

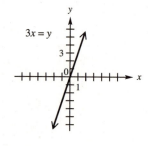

25. The given equation is

$$x = 2 .$$

The equation is represented by a vertical line where x is 2, for any value of y. Thus, when $y = 0$, x remains the value 2. The x-intercept is $(2, 0)$. There is no y-intercept.

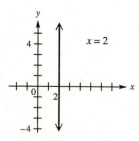

27. The given equation is

$$y = 4.$$

The equation is represented by a horizontal line where y is 4, for any value of x. Thus, when $x = 0$, y remains the value 4. The y-intercept is $(0, 4)$. There is no x-intercept.

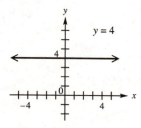

29. The graph of $y + 2 = 0$ is a horizontal line crossing the y-axis at $y = -2$. This fits option C.

31. The graph of $x + 3 = 0$ is a vertical line which crosses the x-axis at $x = -3$. This fits option A.

33. The graph of $y - 2 = 0$ is a horizontal line crossing the y-axis at $y = 2$. This fits option D.

35. The graph of $x - 3 = 0$ is a vertical line which crosses the x-axis at $x = 3$. This fits option B.

37. The diagram of the roof indicates a rise (change in y) of 6 feet and a run (change in x) of 20 feet. Thus, the slope (pitch) is given by

$$m = \frac{\text{rise}}{\text{run}} = \frac{6}{20} = \frac{3}{10}.$$

39. (a) The coordinates of the given points are $(-1, -4)$ and $(3, 2)$. Let $(x_1, y_1) = (-1, -4)$ and $(x_2, y_2) = (3, 2)$. Then,

$$m = \frac{y_2 - y_1}{x_2 - x_1}$$
$$= \frac{2 - (-4)}{3 - (-1)}$$
$$= \frac{3}{2}.$$

Observe that either point may be chosen as (x_1, y_1) or (x_2, y_2).

(b) The coordinates of the given points are $(-3, 5)$ and $(1, -2)$. Let $(x_1, y_1) = (-1, -4)$ and $(x_2, y_2) = (3, 2)$. Then,

$$m = \frac{y_2 - y_1}{x_2 - x_1}$$
$$= \frac{-2 - 5}{1 - (-3)}$$
$$= -\frac{7}{4}.$$

41. Let $(x_1, y_1) = (-2, -3)$ and $(x_2, y_2) = (-1, 5)$. Then,

$$m = \frac{y_2 - y_1}{x_2 - x_1}$$
$$= \frac{5 - (-3)}{-1 - (-2)}$$
$$= 8.$$

43. Let $(x_1, y_1) = (8, 1)$ and $(x_2, y_2) = (2, 6)$.
    Then,

$$m = \frac{y_2 - y_1}{x_2 - x_1}$$
$$= \frac{6 - 1}{2 - 8}$$
$$= -\frac{5}{6}.$$

45. Let $(x_1, y_1) = (2, 4)$ and $(x_2, y_2) = (-4, 4)$.
    Then,

$$m = \frac{y_2 - y_1}{x_2 - x_1}$$
$$= \frac{4 - 4}{(-4) - 2}$$
$$= \frac{0}{-6}$$
$$= 0.$$

47. Refer to graph (Figure A, in the text).
    (a) Let $(x_1, y_1) = (1990, 11{,}338)$ and
    $(x_2, y_2) = (2005, 14{,}818)$.
    Then,

$$m = \frac{y_2 - y_1}{x_2 - x_1}$$
$$= \frac{14{,}818 - 11{,}338}{2005 - 1990}$$
$$= \frac{3480}{15}$$
$$= 232 \ (\textit{thousand})$$
$$= 232{,}000$$

(b) The slope of the line in Figure A is <u>positive</u>. This means that during the period represented, enrollment <u>increased</u>.

(c) Since the slope, or *rate of change*, is

$$\frac{232{,}000 \text{ students}}{1 \text{ yr}}$$

the increase in students per year is $232{,}000$.

(d) Refer to graph (Figure B, in the text). Let $(x_1, y_1) = (1990, 12)$ and $(x_2, y_2) = (1996, 10)$.
Then,

$$m = \frac{y_2 - y_1}{x_2 - x_1}$$
$$= \frac{10 - 22}{1996 - 1990}$$
$$= \frac{-12}{6}$$
$$= -2.$$

(e) The slope of the line in Figure B is <u>negative</u>. This shows us that during the period represented, the number of students per computer <u>decreased</u>.

(f) Since the slope, or *rate of change*, is

$$\frac{-2 \text{ students per computer}}{1 \text{ yr}}$$

the decrease in students per computer *per year* is 2 students per computer.

49. Locate the point $(-3, 2)$. The slope is

$$m = \frac{1}{2} = \frac{\text{change in } y}{\text{change in } x}.$$

From $(-3, 2)$ move 1 units up and 2 units to the right. This brings you to another point, $(-1, 3)$. Draw the line through $(-3, 2)$ and $(-1, 3)$.

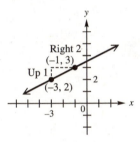

51. Locate the point $(-2, -1)$. The slope is

$$m = \frac{-5}{4} = \frac{\text{change in } y}{\text{change in } x}.$$

From $(-2, -1)$ move 5 units down and 4 units to the right. This brings you to another point, $(2, -6)$. Draw the line through $(-2, -1)$ and $(2, -6)$.

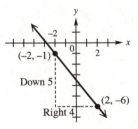

53. Locate the point $(-1, -4)$. The slope is

$$m = -2 = \frac{-2}{1} = \frac{\text{change in } y}{\text{change in } x}.$$

From $(-1, -4)$ move 2 units down and 1 unit to the right. This brings you to another point, $(0, -6)$. Draw the line through $(-1, -4)$ and $(0, -6)$.

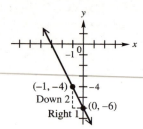

55. First locate the point $(2, -5)$. Use the definition of slope $(m = 0 = \frac{0}{n}$, for any non-zero integer, $n)$ to move up (or down) 0 units (change in y-values) and to the right (or left) n units (change in x-values) to locate another point on the graph. Draw a line through these two points.

Locate the point $(2, -5)$. The slope is

$$m = 0 = \frac{0}{n} = \frac{\text{change in } y}{\text{change in } x}, \text{ for any non-zero integer, } n.$$

From $(2, -5)$ move 0 units up and n units to the left or right to locate another point on the line. Draw the line through these two points.

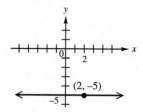

57. L_1 is through $(4, 6)$ and $(-8, 7)$, and L_2 is through $(7, 4)$ and $(-5, 5)$.

For L_1,

$$m = \frac{y_2 - y_1}{x_2 - x_1} = \frac{7 - 6}{-8 - 4} = \frac{-1}{12}.$$

For L_2,

$$m_2 = \frac{y_2 - y_1}{x_2 - x_1} = \frac{5 - 4}{-5 - 7} = \frac{-1}{12}.$$

Since the slopes are equal, the lines are parallel.

59. L_1 is through $(2, 0)$ and $(5, 4)$, and L_2 is through $(6, 1)$ and $(2, 4)$.

For L_1,

$$m = \frac{y_2 - y_1}{x_2 - x_1} = \frac{4 - 0}{5 - 2} = \frac{4}{3}.$$

For L_2,

$$m_2 = \frac{y_2 - y_1}{x_2 - x_1} = \frac{4 - 1}{2 - 6} = \frac{3}{-4}.$$

Since the slopes are negative reciprocals of each other, the lines are perpendicular.

61. L_1 is through $(0, 1)$ and $(2, -3)$, and L_2 is through $(10, 8)$ and $(5, 3)$.

For L_1,

$$m = \frac{y_2 - y_1}{x_2 - x_1} = \frac{-3 - 1}{2 - 0} = \frac{-4}{2} = -2.$$

For L_2,

$$m_2 = \frac{y_2 - y_1}{x_2 - x_1} = \frac{3 - 8}{5 - 10} = \frac{-5}{-5} = 1.$$

The slopes are not related and hence, the lines are neither parallel nor perpendicular.

63. The average rate of change in crude oil production for the following years:

1991 to 1992

$$m = \frac{y_2 - y_1}{x_2 - x_1} = \frac{15.15 - 15.64}{1992 - 1991} = \frac{-.49}{1} = -0.49.$$

1992 to 1993

$$m = \frac{y_2 - y_1}{x_2 - x_1} = \frac{14.66 - 15.15}{1993 - 1992} = \frac{-.49}{1} = -0.49.$$

1993 to 1994

$$m = \frac{y_2 - y_1}{x_2 - x_1} = \frac{14.17 - 14.66}{1994 - 1993} = \frac{-.49}{1} = -0.49.$$

1994 to 1995

$$m = \frac{y_2 - y_1}{x_2 - x_1} = \frac{13.68 - 14.17}{1994 - 1993} = \frac{-.49}{1} = -0.49.$$

Thus, the average rate of change is the same, no matter which two ordered pairs are selected to calculate it.

65. (a) The average rate of change in book sales for the following years:

1995–1996

$$m = \frac{y_2 - y_1}{x_2 - x_1} = \frac{20000 - 19000}{1996 - 1995} = \frac{1000}{1} = 1000,$$

or 1000 million per year.

1995 to 1999

$$m = \frac{y_2 - y_1}{x_2 - x_1} = \frac{23000 - 19000}{1999 - 1995} = \frac{4000}{4} = 1000,$$

or 1000 million per year.

1998 to 2000

$$m = \frac{y_2 - y_1}{x_2 - x_1} = \frac{24000 - 22000}{2000 - 1998} = \frac{2000}{2} = 1000,$$

or 1000 million per year.

Thus, the average rate of change is the same, no matter which two ordered pairs are selected to calculate it. This is true because the data points lie on a straight line.

(b)
1995–1996

$$r = \frac{1000}{19000} \approx 5.3\% \text{ increase per year.}$$

1995 to 1999

$$r = \frac{4000}{19000} \approx 21.05\% \text{ for the four years}$$

or, $21.05\%/4 \approx 5.3\%$ increase per year.

1998 to 2000

$$r = \frac{2000}{22000} \approx 9.09\% \text{ for the two years}$$

or, $9.09\%/2 \approx 4.5\%$ increase per year.

All are about 5%.

67. (a) The average annual rate of change in capacity of hydroelectric power for the successive years are as follows:

1991 to 1996

$$m = \frac{y_2 - y_1}{x_2 - x_1} = \frac{22.9 - 26.2}{1996 - 1991}$$
$$= \frac{-3.3}{5}$$
$$= -.66 \text{ million kilowatts.}$$

(b)
1993 to 1996

$$m = \frac{y_2 - y_1}{x_2 - x_1} = \frac{22.9 - 26.2}{1996 - 1993}$$
$$= \frac{-3.3}{3}$$
$$= -1.1 \text{ million kilowatts.}$$

(c) The graph is not a straight line, so the average rate of change varies for different pairs of years.

69. Since the maximum grade that an elephant can walk is 13%,

$$m = \frac{\text{vertical rise}}{\text{horizontal run}} = 13\%$$
$$= \frac{13}{100}.$$

Thus, $\dfrac{x}{150} = \dfrac{13}{100},$

$$x = 150 \times \frac{13}{100}.$$
$$x = 19.5 \text{ feet}$$

8.3 EXERCISES

1. The slope $= -2$, through the point $(4, 1)$ matches D, $y - 1 = -2(x - 4)$. Use point–slope form of line.

3. The line passing through the points $(0, 0)$ and $(4, 1)$ matches B, $y = \frac{1}{4}x$. Find the slope using the two points, then use the slope and either point in the point–slope form of line.

5. Using the two intercept points, the slope is given by

$$m = \frac{y_2 - y_1}{x_2 - x_1}$$
$$= \frac{0 - (-3)}{1 - 0}$$
$$= 3$$

and $b = -3$ (the ordinate of the y–intercept). Thus, the slope–intercept form of the line is

$$y = mx + b$$
$$= 3x + (-3), \text{ or}$$
$$y = 3x - 3.$$

7. Using the two intercept points, the slope is given by

$$m = \frac{y_2 - y_1}{x_2 - x_1}$$
$$= \frac{0 - 3}{3 - 0}$$
$$= \frac{-3}{3}$$
$$= -1,$$

and $b = 3$ (the ordinate of the y–intercept).
Thus, the slope–intercept form of the line is

$$y = mx + b$$
$$= (-1)x + 3, \text{ or }$$
$$y = -x + 3.$$

9. The equation $y = 2x + 3$ matches the graph A. Observe that the positive slope, 2, discounts options B, C, D, E, and G. Of the remaining options only A suggests the y–intercept as potentially the value 3.

11. The equation $y = -2x - 3$ matches the graph C. Observe that the negative slope, -2, discounts options A, B, E, F, and H. Of the remaining options only C suggests the y–intercept as potentially the value -3.

13. The equation $y = 2x$ matches the graph H. Observe that the y–intercept is the value 0. This means that the graph runs through the origin, $(0, 0)$. Only G and H satisfy this condition and G indicates a negative slope.

15. The equation $y = 3$ represents a horizontal line with a y–intercept at 3. These conditions match option B.

17. Use the point-slope form of the line to write the equation:

$$y - y_1 = m(x - x_1), \text{ or }$$
$$y - 4 = -\frac{3}{4}[x - (-2)].$$

To write the equation in slope–intercept form:

$$y - 4 = -\frac{3}{4}x + \left(-\frac{3}{4}\right) \cdot 2$$
$$y = -\frac{3}{4}x - \frac{3}{2} + 4$$
$$y = -\frac{3}{4}x - \frac{3}{2} + \frac{8}{2}$$
$$y = -\frac{3}{4}x + \frac{5}{2}.$$

19. Use the point-slope form of the line to write the equation:

$$y - y_1 = m(x - x_1), \text{ or }$$
$$y - 8 = -2(x - 5).$$

To write the equation in slope–intercept form:

$$y - 8 = -2x + 10$$
$$y = -2x + 10 + 8$$
$$y = -2x + 18.$$

21. Use the point-slope form of the line to write the equation:

$$y - y_1 = m(x - x_1), \text{ or }$$
$$y - 4 = \frac{1}{2}[x - (-5)].$$

To write the equation in slope–intercept form:

$$y - 4 = \frac{1}{2}x + \left(\frac{1}{2}\right) \cdot 5$$
$$y = \frac{1}{2}x + \frac{5}{2} + 4$$
$$y = \frac{1}{2}x + \frac{5}{2} + \frac{8}{2}$$
$$y = \frac{1}{2}x + \frac{13}{2}.$$

23. Use the point-slope form of the line to write the equation:

$$y - y_1 = m(x - x_1), \text{ or }$$
$$y - 0 = 4 \cdot (x - 3).$$

This simplifies to slope–intercept form:

$$y = 4x - 12.$$

25. Using the point-slope form of the line to write the equation,

$$y - y_1 = m(x - x_1), \text{ or }$$
$$y - 5 = 0 \cdot (x - 9).$$

This simplifies to slope–intercept form:

$$y - 5 = 0, \text{ or }$$
$$y = 5.$$

Alternatively, with $m = 0$ (horizontal line), the equation takes the form of $y = k$ and by inspection we can recognize $y = 5$ as the equation of the horizontal line.

27. An undefined slope indicates that the line is vertical and therefore, the equation is of the form $x = k$. Thus, by inspection, $x = 9$ is the equation of the vertical line.

29. A vertical line is of the form $x = k$. Thus, by inspection, $x = .5$ is the equation of the line.

31. A horizontal line is of the form $y = k$. Thus, by inspection, $y = 8$ is the equation of the line.

33. Using the two points, the slope is given by

$$m = \frac{y_2 - y_1}{x_2 - x_1}$$
$$= \frac{8 - 4}{5 - 3}$$
$$= \frac{4}{2}$$
$$= 2.$$

Use the point-slope form of the line (and either point) to write the equation:

$$y - y_1 = m(x - x_1), \text{ or}$$
$$y - 4 = 2(x - 3).$$

To write the equation in slope–intercept form:

$$y - 4 = 2x - 6$$
$$y = 2x - 6 + 4$$
$$y = 2x - 2.$$

35. Using the two points, the slope is given by

$$m = \frac{y_2 - y_1}{x_2 - x_1}$$
$$= \frac{5 - 1}{(-2) - 6}$$
$$= \frac{4}{-8}$$
$$= -\frac{1}{2}.$$

Use the point-slope form of the line (and either point) to write the equation:

$$y - y_1 = m(x - x_1), \text{ or}$$
$$y - 1 = -\frac{1}{2}(x - 6).$$

To write the equation in slope–intercept form:

$$y - 1 = -\frac{1}{2}x - \left(-\frac{1}{2}\right)6$$
$$y - 1 = -\frac{1}{2}x + 3$$
$$y = -\frac{1}{2}x + 3 + 1$$
$$y = -\frac{1}{2}x + 4.$$

37. Using the two points, the slope is given by

$$m = \frac{y_2 - y_1}{x_2 - x_1}$$
$$= \frac{(2/3) - (2/5)}{(4/3) - (-2/5)}$$
$$= \frac{(10/15) - (6/15)}{(20/15) + (6/15)}$$
$$= \frac{(4/15)}{(26/15)}$$
$$= \frac{4}{15} \cdot \frac{15}{26}$$
$$= \frac{2}{13}.$$

Use the point-slope form of the line (and either point) to write the equation:

$$y - y_1 = m(x - x_1), \text{ or}$$
$$y - \frac{2}{5} = \frac{2}{13}\left[x - \left(-\frac{2}{5}\right)\right].$$

To write the equation in slope–intercept form:

$$y - \frac{2}{5} = \frac{2}{13}\left(x + \frac{2}{5}\right)$$
$$y - \frac{2}{5} = \frac{2}{13}x + \left(\frac{2}{13}\right) \cdot \left(\frac{2}{5}\right)$$
$$y = \frac{2}{13}x + \frac{4}{65} + \frac{2}{5}$$
$$y = \frac{2}{13}x + \frac{4}{65} + \frac{26}{65}$$
$$y = \frac{2}{13}x + \frac{30}{65}$$
$$y = \frac{2}{13}x + \frac{6}{13}.$$

39. Using the two points, the slope is given by

$$m = \frac{y_2 - y_1}{x_2 - x_1}$$
$$= \frac{5 - 5}{1 - 2}$$
$$= \frac{0}{-1}$$
$$= 0.$$

Using the point-slope form of the line to write the equation,

$$y - y_1 = m(x - x_1), \text{ or}$$
$$y - 5 = 0 \cdot (x - 2).$$

This simplifies to slope–intercept form:

$$y - 5 = 0, \text{ or}$$
$$y = 5.$$

41. These points lie on a vertical line and the slope is undefined (since the denominator in the slope ratio, $x_2 - x_1 = 7 - 7$, is 0). Therefore, one can't use the point–slope form of the line. Rather, the equation of a vertical line is in the form $x = k$. Thus, $x = 7$.

43. Using the two points, the slope is given by

$$m = \frac{y_2 - y_1}{x_2 - x_1}$$
$$= \frac{-3 - (-3)}{-1 - 1}$$
$$= \frac{0}{-2}$$
$$= 0.$$

Instead of using the point-slope form of the line to write the equation, we offer an alternative solution. We know that with $m = 0$, the line is horizontal and is of the form $y = k$. Thus, $y = -3$ is the equation.

45. Using the slope–intercept form of the line:

$$y = mx + b, \text{ or}$$
$$y = 5x + 15.$$

47. Using the slope–intercept form of the line:

$$y = mx + b, \text{ or}$$
$$y = -\frac{2}{3}x + \frac{4}{5}.$$

49. Using the slope–intercept form of the line with $m = \frac{2}{5}$ and $b = 5$:

$$y = mx + b, \text{ or}$$
$$y = \frac{2}{5}x + 5.$$

51. Writing exercise

53. (a) To write $x + y = 12$ in slope–intercept form, solve for y:
$$x + y = 12$$
$$y = -x + 12.$$

    (b) By inspection, the slope, m, is given by $m = -1$ (the understood coefficient of the x–term).

    (c) By inspection, $b = 12$ (the constant), so the y–intercept is $(0, 12)$.

55. (a) To write $5x + 2y = 20$ in slope–intercept form, solve for y:
$$5x + 2y = 20$$
$$2y = -5x + 20$$
$$y = -\frac{5}{2}x + 10.$$

    (b) By inspection, the slope, m, is given by $m = -\frac{5}{2}$.

    (c) By inspection, $b = 10$, so the y–intercept is $(0, 10)$.

57. (a) To write $2x - 3y = 10$ in slope–intercept form, solve for y:
$$2x - 3y = 10$$
$$-3y = -2x + 10$$
$$y = \frac{2}{3}x - \frac{10}{3}.$$

    (b) By inspection, the slope, m, is given by $m = \frac{2}{3}$.

    (c) By inspection, $b = -\frac{10}{3}$, so the y–intercept is $\left(0, -\frac{10}{3}\right)$.

59. Write $3x - y = 8$ in slope–intercept form in order to identify the slope of the given line:
$$3x - y = 8$$
$$-y = -3x + 8$$
$$y = 3x - 8.$$

By inspection, $m = 3$. This is also the slope of the new line, since they are parallel. Using the point–slope form of the line,

$$y - y_1 = m(x - x_1), \text{ or}$$
$$y - 2 = 3(x - 7).$$

This simplifies to slope–intercept form

$$y - 2 = 3x - 21, \text{ or}$$
$$y = 3x - 19.$$

61. Write $-x + 2y = 10$ in slope–intercept form in order to identify the slope of the given line:

$$-x + 2y = 10$$
$$2y = x + 10$$
$$y = \frac{1}{2}x + 5.$$

By inspection, $m = \frac{1}{2}$. This is also the slope of the new line, since they are parallel. Using the point, $(-2, -2)$ and point–slope form of the line,

$$y - y_1 = m(x - x_1), \text{ or}$$
$$y - (-2) = \frac{1}{2}[x - (-2)]$$
$$y + 2 = \frac{1}{2}x + \frac{2}{2}$$
$$y = \frac{1}{2}x + 1 - 2$$
$$y = \frac{1}{2}x - 1.$$

63. Write $2x - y = 7$ in slope–intercept form in order to identify the slope of the given line:

$$2x - y = 7$$
$$-y = -2x + 7$$
$$y = 2x - 7.$$

By inspection, $m = 2$. Thus, the slope of any line perpendicular to the given line is $-\frac{1}{2}$.

Using the given point $(8, 5)$ and the point–slope form of the line,

$$y - y_1 = m(x - x_1), \text{ or}$$
$$y - 5 = -\frac{1}{2}(x - 8)$$
$$2y - 10 = -x + 8$$
$$2y = -x + 8 + 10$$
$$y = -\frac{1}{2}x + \frac{18}{2}$$
$$y = -\frac{1}{2}x + 9.$$

65. Since $x = 9$ is a vertical line, any line perpendicular to it will be horizontal and have slope $= 0$. To write the equation of a horizontal line through the point $(-2, 7)$ we can use the form $y = k$ with $k = 7$. Thus, the line is $y = 7$.

67. Choose any two data points to create slope. We will choose the points $(0, 229)$ and $(4, 279)$.
The slope is given by

$$m = \frac{y_2 - y_1}{x_2 - x_1}$$
$$= \frac{279 - 229}{4 - 0}$$
$$= \frac{50}{4}$$
$$= 12.5.$$

Use the vertical axis intercept $(0, 229)$ as the point $(0, b)$ and the slope–intercept form for the equation to arrive at

$$y = mx + b, \text{ or}$$
$$y = 12.5x + 229.$$

69. (a) The ordered pairs representing the other data values are $(5, 42)$, $(15, 61)$, and $(25, 76)$.

(b)

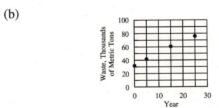

Yes, the points lie approximately in a line.

(c) Using the points $(0, 32)$ and $(25, 76)$, the slope is given by

$$m = \frac{y_2 - y_1}{x_2 - x_1}$$
$$= \frac{76 - 32}{25 - 0}$$
$$= \frac{44}{25}$$
$$= 1.76 .$$

Use the vertical axis intercept $(0, 32)$ as the point $(0, b)$ and the slope–intercept form for the equation to arrive at

$$y = mx + b, \text{ or}$$
$$y = 1.76x + 32.$$

(d) In the year 2005, $x = 10$. Therefore,

$$y = 1.7(10) + 32$$
$$= 17.6 + 32$$
$$= 49.6 .$$

Thus, there will be about 49.6 thousand $= 49.6(1000)$ or $49,600$ metric tons produced in the year 2005.

71. (a) When $C = 0°$, $F = \underline{32°}$; when $C = 100°$, $F = \underline{212°}$

(b) The two ordered pairs are $(0, 32)$ and $(100, 212)$.

(c) Using the points $(0, 32)$ and $(100, 212)$, the slope is given by

$$m = \frac{y_2 - y_1}{x_2 - x_1}$$
$$= \frac{212 - 32}{100 - 0}$$
$$= \frac{180}{100}$$
$$= \frac{9}{5} .$$

(d) Use the vertical axis intercept $(0, 32)$ as the point $(0, b)$ and the slope–intercept form for the equation to arrive at

$$F = mC + b, \text{ or}$$
$$F = \frac{9}{5}C + 32.$$

(e) To solve for C in terms of F:

$$F = \frac{9}{5}C + 32$$
$$F - 32 = \frac{9}{5}C$$
$$\frac{5}{9}(F - 32) = C, \text{ or}$$
$$C = \frac{5}{9}(F - 32).$$

(f) When the Celsius temperature is $50°$, the Fahrenheit temperature is $122°$.

8.4 EXERCISES

1.   Writing exercise

3.   The first element in an ordered pair is the independent variable.

5.   The relation, $\{(2, 5), (3, 7), (4, 9), (5, 11)\}$ is a function since, corresponding to each first component, there is a unique second component.
     The domain is $\{2, 3, 4, 5\}$.
     The range is $\{5, 7, 9, 11\}$.

7.   The input-output machine would not represent a function since there are two outputs (positive and negative square roots) for each input. The domain is $(0, \infty)$. The range is $(-\infty, 0) \cup (0, \infty)$.

9.   The table does represent a function, since there is only one price associated with each type of gas/oil. The domain is $\{$unleaded regular, unleaded premium, crude oil$\}$. The range is $\{1.22, 1.44, .21\}$.

11. The graph represents a function, since it passes the "vertical line test" (only one intersection). The domain is $(-\infty, \infty)$. The range is $(-\infty, \infty)$.

13. The graph does not represent a function, since it does not pass the "vertical line test." The domain is $[-4, 4]$. The range is $[-3, 3]$.

15. The equation, $y = x^2$, represent a function, since any value for x will yield exactly one value for y (for a number squared, there is only one answer). The domain is the set of reals numbers or $(-\infty, \infty)$.

17. To determine if the equation, $x = y^2$, is, or is not, a function, solve for y: $y = \pm\sqrt{x}$. Since any replacement for x will yield two values for y, the equation doesn't represent a function. In order to get "real" number values for y, x can be replaced only with values such that $x \geq 0$. These values represent the domain $[0, \infty)$.

19. Since ordered pairs such as $(2, 0)$, $(2, 1)$, and $(2, 3)$, etc., all satisfy the inequality, $x + y < 4$, it does not represent a function. Note that the graph of any linear inequality, such as this one, will be a half plane and hence, will not satisfy the "vertical line test." Any real number can be used for x. Therefore, the domain is $(-\infty, \infty)$.

21. Since any value for x in the domain of the relation, $y = \sqrt{x}$, will yield exactly one value for y, the principal square root of x, the equation represents a function. To keep y "real valued" (i.e., a real number), x must satisfy the inequality $x \geq 0$. Thus, the domain is given by $[0, \infty)$.

23. Solve the equation, $xy = 1$, for y: $y = \frac{1}{x}$. Since any value for x, in the domain, will yield exactly one value for y, the equation represents a function. Observe that in order to keep the fraction defined, $x \neq 0$, but all other (real) numbers will work. This implies that the domain is all real numbers except 0 or $(-\infty, 0) \cup (0, \infty)$.

25. The relation, $y = \sqrt{4x + 2}$, is a function, because, for any choice of x in the domain, there is exactly one corresponding value of y. The domain is the set of values that satisfies $4x + 2 \geq 0$. Solving the inequality we arrive at $x \geq \frac{-2}{4}$ or $x \geq -\frac{1}{2}$. Thus, the domain is $[-\frac{1}{2}, \infty)$.

27. The relation, $y = \frac{2}{x-9}$, is a function, because, for any choice of x in the domain, there is exactly one corresponding value of y. We may use any real value for x except those which make the denominator 0, i.e., $x \neq 9$. Thus, the domain is given by $(-\infty, 9) \cup (9, \infty)$.

29. (a) The values along the vertical axis, representing the dependent variable, are $[0, 3000]$.

(b) The water is increasing between 0 and 25 hours so the water is increasing for a total of $25 - 0 = 25$ hours. The water is decreasing between 50 and 75 hours so the water is decreasing for a total of $75 - 50 = 25$ hours.

(c) The graph shows that 2000 gallons of water are left in the pool after 90 hours.

(d) The value of $f(0) = 0$, represents the amount of water in the pool at 0 hours.

31. One example: The height of a child depends on her age, so height is a function of age.

33. If $f(x) = 3 + 2x$, then $f(1) = 3 + 2(1) = 5$.

35. If $g(x) = x^2 - 2$, then $g(2) = 2^2 - 2 = 2$.

37. If $g(x) = x^2 - 2$, then $g(-1) = (-1)^2 - 2 = -1$.

39. If $f(x) = 3 + 2x$, then $f(-8) = 3 + 2(-8) = -13$.

41. $f(x) = -2x + 5$

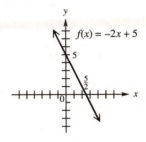

The domain and range is $(-\infty, \infty)$.

43. $h(x) = \frac{1}{2}x + 2$

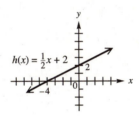

The domain and range is $(-\infty, \infty)$.

45. $G(x) = 2x$

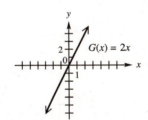

The domain and range is $(-\infty, \infty)$.

47. $f(x) = 5$

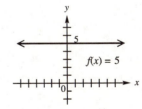

The domain is $(-\infty, \infty)$.
The range is $\{5\}$.

49. (a)
$$y + 2x^2 = 3$$
$$y = 3 - 2x^2$$
Thus, $f(x) = 3 - 2x^2$.

   (b) $f(3) = 3 - 2(3)^2 = -15$

51. (a)
$$4x - 3y = 8$$
$$-3y = 8 - 4x$$
$$y = \frac{8 - 4x}{-3}$$
Thus, $f(x) = \frac{8 - 4x}{-3}$.

   (b) $f(3) = \frac{8 - 4(3)}{-3} = \frac{4}{3}$

53. The equation $2x + 5y = 9$ has a straight <u>line</u> as its graph. One point that lies on the line is $(3, -2)$. If we solve the equation for y and use function notation, we have a linear function $f(x) = \underline{-2x + 4}$. For this function, $f(3) = -2(3) + 4 = \underline{-2}$, meaning that the point $\underline{(3, -2)}$ lies on the graph of the function.

55. (a)

x	$f(x)$
0	$0
1	$1.50
2	$3.00
3	$4.50

   (b) The linear function that gives a rule for the amount charged is $f(x) = \underline{1.50x}$.

   (c) The graph of this function for x, where x is an element of $\{0, 1, 2, 3\}$ is:

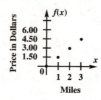

57. (a) Using $h(r) = 69.09 + 2.24r$, then
$$h(56) = 69.09 + 2.24(56) = 194.53 \text{ cm.}$$

   (b) Using $h(t) = 81.69 + 2.39t$, then
$$h(40) = 81.69 + 2.39(40) = 179.29 \text{ cm.}$$

   (c) Using $h(r) = 61.41 + 2.32r$, then
$$h(50) = 61.41 + 2.32(50) = 177.41 \text{ cm.}$$

   (d) Using $h(t) = 72.57 + 2.53t$, then
$$h(36) = 72.57 + 2.53(36) = 163.65 \text{ cm.}$$

59. Using $f(x) = -183x + 40034$, then

   (a) $f(1) = -183(1) + 40034 = 39,851.$

   (b) $f(3) = -183(3) + 40034 = 39,485.$

   (c) $f(5) = -183(5) + 40034 = 39,119.$

   (d) In 1992, there were 39,668 post offices in the U.S.

61. Let x represent the number of envelopes stuffed. Then,

   (a) $C(x) = .02x + 200.$

   (b) $R(x) = .04x.$

   (c) Set $C(x) = R(x)$ and solve for x.
$$.02x + 200 .04x = .04x$$
$$200 = .04x - .02x$$
$$200 = .02x$$
$$x = 10,000$$

   (d)

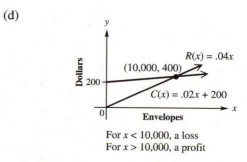

   For $x < 10,000$, a loss
   For $x > 10,000$, a profit

63. Let x represent the number of deliveries he makes. Then,

   (a) $C(x) = 3x + 2300.$

   (b) $R(x) = 5.50x.$

   (c) Set $C(x) = R(x)$ and solve for x.
$$3.00x + 2300 = 5.50x$$
$$2300 = 5.50x - 3.00x$$
$$2300 = 2.50x$$
$$x = 920$$

(d)

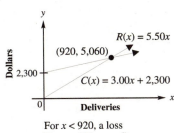

For $x < 920$, a loss
For $x > 920$, a profit

8.5 EXERCISES

1. The equation $g(x) = x^2 - 5$ matches F with a vertex $(0, -5)$ and opening up since the leading coefficient (1) is positive.

3. The equation $F(x) = (x - 1)^2$ matches C with a vertex $(1, 0)$ and opening up since the leading coefficient (1) is positive.

5. The equation $H(x) = (x - 1)^2 + 1$ matches E with a vertex $(1, 1)$ and opening up since the leading coefficient (1) is positive.

7. Writing exercise

9. Write the function, $f(x) = -3x^2$, in the form $f(x) = a(x - h)^2 + k$:
$$f(x) = -3(x - 0)^2 + 0.$$
Thus, the vertex, (h, k), is given by $(0, 0)$.

11. Write the function, $f(x) = x^2 + 4$, in the form $f(x) = a(x - h)^2 + k$:
$$f(x) = (x - 0)^2 + 4.$$
Thus, the vertex, (h, k), is given by $(0, 4)$.

13. Write the function, $f(x) = (x - 1)^2$, in the form $f(x) = a(x - h)^2 + k$:
$$f(x) = (x - 1)^2 + 0.$$
Thus, the vertex, (h, k), is given by $(1, 0)$.

15. Write the function, $f(x) = (x + 3)^2 - 4$, in the form $f(x) = a(x - h)^2 + k$:
$$f(x) = [x - (-3)]^2 + (-4).$$
Thus, the vertex, (h, k), is given by $(-3, -4)$.

17. Writing exercise

19. The graph $f(x) = -3x^2 + 1$ opens downward since the leading coefficient (-3) is negative. It is narrower since $|-3| > 1$.

21. The graph $f(x) = \frac{2}{3}x^2 - 4$ opens upward since the leading coefficient $\left(\frac{2}{3}\right)$ is positive. It is wider since $\left|\frac{2}{3}\right| < 1$.

23. (a) With $h > 0$, $k > 0$, both coordinates of the vertex are positive, $(+, +)$, which puts the vertex in quadrant I.

(b) With $h > 0$, $k < 0$, we have $(+, -)$ and the vertex is in quadrant IV.

(c) With $h < 0$, $k > 0$, we have $(-, +)$ and the vertex is in quadrant II.

(d) With $h < 0$, $k < 0$, we have $(-, -)$ and the vertex is in quadrant III.

25. Write the function, $f(x) = 3x^2$, in the form $f(x) = a(x - h)^2 + k$:
$$f(x) = 3(x - 0)^2 + 0.$$
Thus, the vertex, (h, k), is given by $(0, 0)$. Since $|a| = 3 > 1$, the graph has narrower branches than $f(x) = x^2$ and opens upward. To find two other points:
$$f(\pm 1) = 3(\pm 1)^2 = 3.$$
Thus, $(-1, 3)$ and $(1, 3)$ also lie on the graph.

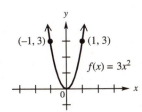

27. Write the function, $f(x) = -\frac{1}{4}x^2$, in the form $f(x) = a(x - h)^2 + k$:
$$f(x) = -\frac{1}{4}(x - 0)^2 + 0.$$
Thus, the vertex, (h, k), is given by $(0, 0)$. Since $|a| = \frac{1}{4} < 1$, the graph has wider branches than $f(x) = x^2$. It opens downward since, $a = -\frac{1}{4} < 0$. To find two other points:
$$f(\pm 4) = -\frac{1}{4}(\pm 4)^2 = -4.$$
Thus, $(-4, -4)$ and $(4, -4)$ also lie on the graph.

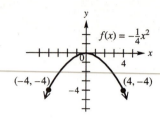

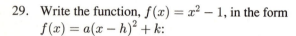

29. Write the function, $f(x) = x^2 - 1$, in the form $f(x) = a(x-h)^2 + k$:

$$f(x) = 1 \cdot (x - 0)^2 + (-1).$$

Thus, the vertex, (h, k), is given by $(0, -1)$. Since $|a| = 1$, the graph has the same branches as $f(x) = x^2$. It opens upward, since $a = 1 > 0$. To find two other points:

$$f(\pm 2) = 1 \cdot (\pm 2)^2 - 1 = 3.$$

Thus, $(-2, 3)$ and $(2, 3)$ also lie on the graph.

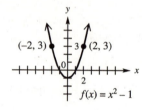

31. Write the function, $f(x) = -x^2 + 2$, in the form $f(x) = a(x-h)^2 + k$:

$$f(x) = -1 \cdot (x - 0)^2 + 2.$$

Thus, the vertex, (h, k), is given by $(0, 2)$. Since $|a| = 1$, the graph has the same branches as $f(x) = x^2$. It opens downward since, $a = -1 < 0$. To find two other points:

$$f(\pm 2) = -1 \cdot (\pm 2)^2 + 2 = -2.$$

Thus, $(-2, -2)$ and $(2, -2)$ also lie on the graph.

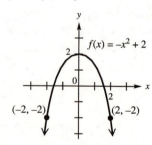

33. Write the function, $f(x) = 2x^2 - 2$, in the form $f(x) = a(x-h)^2 + k$:

$$f(x) = 2(x - 0)^2 + (-2).$$

Thus, the vertex, (h, k), is given by $(0, -2)$. Since $|a| = 2 > 1$, the graph has narrower branches than $f(x) = x^2$. It opens upward, since $a > 0$. To find two other points:

$$f(\pm 1) = 2(\pm 1)^2 - 2 = 0.$$

Thus, $(-1, 0)$ and $(1, 0)$ also lie on the graph.

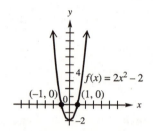

35. Write the function, $f(x) = (x - 4)^2$, in the form $f(x) = a(x-h)^2 + k$:

$$f(x) = 1 \cdot (x - 4)^2 + 0.$$

Thus, the vertex, (h, k), is given by $(4, 0)$. It opens upward, since $a > 0$. Find two other points, e.g. let $x = 3$ and $x = 5$:

$$f(3) = (3 - 4)^2 = 1.$$
$$f(5) = (5 - 4)^2 = 1.$$

Thus, $(3, 1)$ and $(5, 1)$ also lie on the graph.

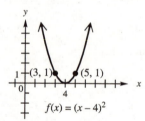

37. Write the function, $f(x) = 3(x + 1)^2$, in the form $f(x) = a(x-h)^2 + k$:

$$f(x) = 3[x - (-1)^2] + 0.$$

Thus, the vertex, (h, k), is given by $(-1, 0)$. It opens upward, since $a > 0$. Find two other points, e.g. let $x = -2$ and $x = 0$:

$$f(-2) = 3(-2 + 1)^2 = 3,$$
$$f(0) = 3(0 + 1)^2 = 3.$$

Thus, $(-2, 3)$ and $(0, 3)$ also lie on the graph.

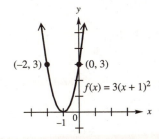

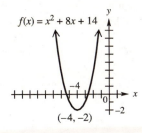

39. Write the function, $f(x) = (x + 1)^2 - 2$, in the form $f(x) = a(x - h)^2 + k$:

$$f(x) = 1 \cdot [x - (-1)]^2 - 2.$$

Thus, the vertex, (h, k), is given by $(-1, -2)$. It opens upward, since $a > 0$. Find two other points, e.g. let $x = -2$ and $x = 0$:

$$f(-2) = (-2 + 1)^2 - 2 = -1,$$
$$f(0) = (0 + 1)^2 - 2 = -1.$$

Thus, $(-2, -1)$ and $(0, -1)$ also lie on the graph.

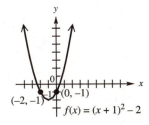

In exercises 41–46, we are finding the vertex by writing each function in the standard form of a quadratic function in order to identify directly the vertex, (h, k). Completing the square is the technique used. However, an alternate technique, using the formula $\left(-\frac{b}{2a}, f\left(-\frac{b}{2a}\right)\right)$ will yield the coordinates of each vertex.

41. Write the function, $f(x) = x^2 + 8x + 14$, in the form $f(x) = a(x - h)^2 + k$ by completing the square on x.

$$f(x) = x^2 + 8x + 14$$
$$= \left[x^2 + 8x + \left(\frac{8}{2}\right)^2\right] + 14 - \left(\frac{8}{2}\right)^2$$
$$= (x^2 + 8x + 16) + 14 - 16$$
$$= (x + 4)^2 - 2$$
$$f(x) = [x - (-4)]^2 - 2$$

Thus, the vertex, (h, k), is given by $(-4, -2)$. It opens upward, since $a > 0$.

43. Write the function, $f(x) = x^2 + 2x - 4$, in the form $f(x) = a(x - h)^2 + k$ by completing the square on x.

$$f(x) = x^2 + 2x - 4$$
$$= \left[x^2 + 2x + \left(\frac{2}{2}\right)^2\right] - 4 - \left(\frac{2}{2}\right)^2$$
$$= [x^2 + 2x + 1] - 4 - 1$$
$$= (x + 1)^2 - 5$$
$$f(x) = [x - (-1)]^2 - 5$$

Thus, the vertex, (h, k), is given by $(-1, -5)$. It opens upward, since $a > 0$.

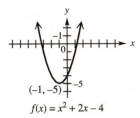

45. Write the function, $f(x) = -2x^2 + 4x + 5$, in the form $f(x) = a(x - h)^2 + k$ by completing the square on x.

$$f(x) = -2x^2 + 4x + 5$$
$$= -2(x^2 - 2x) + 5$$
$$= -2\left[x^2 - 2x + \left(\frac{2}{2}\right)^2\right] + 5 - (-2)\left(\frac{2}{2}\right)^2$$
$$= -2(x - 1)^2 + 5 + 2$$
$$f(x) = -2(x - 1)^2 + 7$$

Thus, the vertex, (h, k), is given by $(1, 7)$. It opens downward, since $a < 0$.

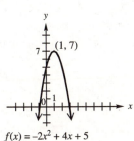

47. If we let x represent the width dimension, then $50 - x$ represents the length dimension (length plus width dimensions will equal half of the fencing needed for perimeter). Create a function representing the area.

$$A(x) = \text{length} \cdot \text{width}$$
$$= (50 - x) \cdot x$$
$$= -x^2 + 50x$$

Use the formula $\frac{-b}{2a}$ to create the x-value of the turning point. Thus,

$$x = \frac{-50}{2(-1)} = 25 \text{ meters}$$

49. Begin by writing the equation in standard form.

$$h = -16t^2 + 32t$$

Observe that the vertex $(t, h(t))$ of the parabola represents the time, t, when the object reaches its maximum and the maximum height, $h(t)$. Use the formula $\frac{-b}{2a}$ to find at what time the object reaches the maximum height. Thus,

$$t = \frac{-b}{2a} = \frac{-32}{2(-16)} = 1 \text{ sec.}$$

Then,

$$h(1) = -16(1)^2 + 32(1) = 16 \text{ feet}$$

is the maximum height. To find when the object hits the ground let $h = 0$. Then, solving for t, we have

$$32t - 16t^2 = 0$$
$$16t(2 - t) = 0$$
$$t = 0 \text{ and } t = 2.$$

Therefore, $t = 0$ seconds or $t = 2$ seconds. It is at ground level at $t = 0$ seconds (when it is thrown) and $t = 2$ seconds (when it hits the ground). Notice that it takes 1 second to reach maximum height and 1 more second to hit the ground.

51. The answer to both questions is given by the vertex (highest point) of the parabola suggested by the equation $s(t) = -4.9t^2 + 40t$. To find the vertex, use the formula, $\left(-\frac{b}{2a}, s\left(-\frac{b}{2a}\right)\right)$, or complete the square to reach standard form. Using the formula $\frac{-b}{2a}$ to find at what time the object reaches the maximum height,

$$\frac{-b}{2a} = \frac{-40}{2(-4.9)} \approx 4.1 \text{ seconds.}$$ Then,

$$s(4.1) = -4.9(4.1)^2 + 40(4.1)$$
$$\approx 81.6 \text{ meters}$$

is the maximum height.

53. For the function $f(x) = -20.57x^2 + 758.9x - 3140$:

(a) The coefficient, -20.57, of x^2 is negative because the parabola opens downward.

(b) To find the vertex, use the formula, $\left(-\frac{b}{2a}, f\left(-\frac{b}{2a}\right)\right)$.

$$-\frac{b}{2a} = -\frac{758.9}{2(-20.57)}$$
$$\approx 18.45.$$

$$f\left(-\frac{b}{2a}\right) = f(18.45)$$
$$= -20.57(18.45)^2 + 758.9(18.45) - 3140$$
$$\approx 3860.$$

Thus, the vertex is located at $(18.45, 3860)$.

(c) In 2018 social security assets will reach their maximum value of \$3860 billion.

55. (a) $R(x) = (\text{no. of seats})(\text{cost per seat})$
$$= (100 - x)(200 + 4x)$$
$$= 20,000 + 200x - 4x^2.$$

(b)

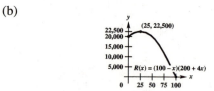

(c) The maximum revenue will occur at the vertex since the parabola opens downward $(a < 0)$. The formula, $-\frac{b}{2a}$, yields the number of unsold seats that will produce the maximum revenue while $f\left(-\frac{b}{2a}\right)$ yields the maximum revenue. Thus,

$$-\frac{b}{2a} = -\frac{200}{2(-4)}$$
$$= 25 \text{ seats.}$$

The maximum revenue is

$$f(25) = 20,000 + 200(25) - 4(25)^2$$
$$= \$22,500.$$

8.6 EXERCISES

1. For an exponential function $f(x) = a^x$, if $a > 1$, the graph <u>rises</u> from left to right. If $0 < a < 1$, the graph <u>falls</u> from left to right.

3. The graph of the exponential function $y = a^x$ <u>does not</u> have an x-intercept, since $a^0 = 1$ providing $a \neq 0..$

5. For a logarithmic function $g(x) = \log_a x$, if $a > 1$, the graph <u>rises</u> from left to right. If $0 < a < 1$, the graph <u>falls</u> from left to right.

7. The graph of the exponential function $g(x) = \log_a x$ <u>does not</u> have a y-intercept, since $\log_a 0$ does not exist.

9. $9^{\frac{3}{7}} \approx 2.56425419972$

11. $(.83)^{-1.2} \approx 1.25056505582$

13. $\left(\sqrt{6}\right)^{\sqrt{5}} \approx 7.41309466897$

15. $\left(\dfrac{1}{3}\right)^{9.8} \approx 2.10965628481 \times 10^{-5}$
$$= .0000210965628481$$

17. Generate several ordered pairs that satisfy the function $f(x) = 3^x$ (e.g., $\left(-1, \frac{1}{3}\right)$, $(0, 1)$, $(1, 3)$, etc.) and plot these values. Sketch a smooth curve through these points. Remember that the graph will rise from left to right since $b > 1$ and that the x-axis acts as an asymptote.

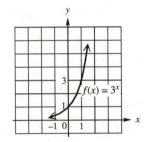

19. Generate several ordered pairs that satisfy the function $f(x) = \left(\frac{1}{4}\right)^x$ (e.g., $(-1, 4)$, $(0, 1)$, $\left(1, \frac{1}{4}\right)$, etc.) and plot these values. Sketch a smooth curve through these points. Remember that the graph will fall from left to right since $b < 1$ and that the x-axis acts as an asymptote.

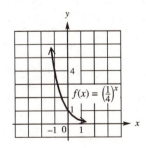

21. $e^3 \approx 20.0855369232$

23. e^{-4} or $1/e^4 \approx .018315638889$

25. $4^2 = 16$ is equivalent to $2 = \log_4 16$.

27. $\left(\dfrac{2}{3}\right)^{-3} = \dfrac{27}{8}$ is equivalent to $-3 = \log_{2/3}\left(\dfrac{27}{8}\right)$.

29. $5 = \log_2 32$ is equivalent to $2^5 = 32$.

31. $1 = \log_3 3$ is equivalent to $3^1 = 3$.

33. $\ln 4 \approx 1.38629436112$

35. $\ln .35 \approx -1.0498221245$

37. By inspecting the graph, the year 2000 corresponds
(a) to an approximate $.5°C$ increase on the exponential curve, and

(b) to an approximate $.35°C$ increase on the linear graph.

39. By inspecting the graph, the year 2020 corresponds
(a) to an approximate $\underline{1.6°C}$ increase on the exponential curve, and

(b) to an approximate $.5°C$ increase on the linear graph.

41. Since $g(x) = \log_3 x$ is the inverse of $f(x) = 3^x$ (Exercise 17), we can reflect the graph of $f(x)$ across the line $y = x$ to get the graph of $g(x)$. This may be accomplished by interchanging the roll of the x and y values that were generated to graph $f(x)$. Thus, some generated ordered pairs for $g(x)$ would include $\left(\frac{1}{3}, -1\right)$, $(1, 0)$, $(3, 1)$, etc.

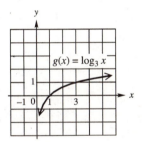

43. Since $g(x) = \log_{1/4} x$ is the inverse of $f(x) = \left(\frac{1}{4}\right)^x$ (Exercise 19), we can reflect the graph of $f(x)$ across the line $y = x$ to get the graph of $g(x)$. This may be accomplished by interchanging the roll of the x and y values that were generated to graph $f(x)$. Thus, some generated ordered pairs for $g(x)$ would include $\left(\frac{1}{4}, 1\right)$, $(1, 0)$, $(4, -1)$, etc. Observe that this will now give an asymptote along the y-axis.

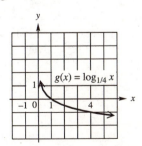

45. Using the compound interest formula $A = P\left(1 + \frac{r}{n}\right)^{nt}$ where $P = \$4,292$, $r = 6\% = .06$, $n = 1$ and $t = 10$ years, we have

$$A = 4292\left(1 + \frac{.06}{1}\right)^{1 \cdot 10}$$

$$= 4292(1.06)^{10}$$

$$\approx 4292 \cdot (1.79085)$$

$$\approx \$7686.32.$$

47. Use the compound interest formula $A = P\left(1 + \frac{r}{n}\right)^{nt}$ where $P = \$56,780$, $r = 5.3\% = .053$, $n = 4$ and $t = \frac{23}{4}$ years. Note that the total number of compoundings (nt) is 23.

$$A = 56780\left(1 + \frac{.053}{4}\right)^{23}$$

$$= 56780(1.01325)^{23}$$

$$\approx 56780 \cdot (1.3574255)$$

$$\approx \$76,855.95$$

49. Using the continuous compounding formula $A = Pe^{rt}$ where $P = \$25,000$ and $r = 5\% = .05$,

(a) For $n = 1$:
$$A = 25000e^{.05(1)}$$
$$\approx 25000 \cdot (1.05127)$$
$$\approx \$26,281.78.$$

(b) For $n = 5$:
$$A = 25000e^{.05(5)}$$
$$= 25000e^{.25}$$
$$\approx 25000 \cdot (1.284025)$$
$$\approx \$32,100.64.$$

(c) For $n = 10$:
$$A = 25000e^{.05(10)}$$
$$= 25000e^{.5}$$
$$\approx 25000 \cdot (1.64872)$$
$$\approx \$41,218.03.$$

51. Using the model
$$f(x) = 7147(1.0366)^x,$$
the approximate number of emissions in
(a) 1950, where $x = 0$, is given by

$$f(0) = 7147(1.0366)^0$$
$$= 7147(1)$$
$$= 7147 \text{ million short tons.}$$

(b) 1985, where $x = 35$, is given by

$$f(35) = 7147(1.0366)^{35}$$
$$\approx 7147(3.518780346)$$
$$\approx 25,149 \text{ million short tons.}$$

(c) 1990, where $x = 40$, is given by

$$f(40) = 7147(1.0366)^{40}$$
$$\approx 7147(4.21161031)$$
$$\approx 30,100 \text{ million short tons.}$$

The actual amount of emissions in 1990 was 25,010 million short tons, or less than that predicted by the model.

53. Using the model

$$P(x) = 70,967e^{.0526x},$$

the approximate total expenditure in
(a) 1987, where $x = 2$, is given by

$$P(2) = 70967e^{.0526(2)}$$
$$\approx 70967(1.110932775)$$
$$\approx 78,840 \text{ million dollars.}$$

(b) 1990, where $x = 5$, is given by

$$P(5) = 70967e^{.0526(5)}$$
$$\approx 70967(1.300826719)$$
$$\approx 92,316 \text{ million dollars.}$$

(c) 1993, where $x = 8$, is given by

$$P(8) = 70967e^{.0526(8)}$$
$$\approx 70967(1.523179612)$$
$$\approx 108,095 \text{ million dollars.}$$

(d) 1985, where $x = 0$, is given by

$$P(0) = 70967e^{.0526(0)}$$
$$\approx 70967(1)$$
$$\approx 70,967 \text{ million dollars.}$$

55. Use the model

$$B(x) = 8768e^{.072x},$$

to find the approximate consumer expenditure on all types of books in the US, in 1998. For the year 1998, $x = 18$, and

$$B(18) = 8768e^{.072(18)}$$
$$\approx 8768(3.65468796)$$
$$\approx 32,044 \text{ million dollars.}$$

57. Use the model

$$A(t) = 100(3.2)^{-.5t},$$

to approximate the amount of radioactive material present at any time t.

(a) The initial measurement amount, $t = 0$, is given by

$$A(0) = 100(3.2)^{-.5(0)}$$
$$= 100(1)$$
$$= 100 \text{ grams.}$$

(b) The amount 2 months later, $t = 2$, is given by

$$A(0) = 100(3.2)^{-.5(2)}$$
$$= 100(.3125)$$
$$= 31.25 \text{ grams.}$$

(c) The amount 10 months later, $t = 10$, is given by

$$A(0) = 100(3.2)^{-.5(10)}$$
$$\approx 100(.0029802322)$$
$$\approx .30 \text{ grams.}$$

The actual amount of emissions in 1990 was 25,010 million short tons, or less than that predicted by the model.

59. Using the model

$$f(x) = 11.34 + 317.01\log_2 x,$$

to the approximate number of hazardous waste sites in the U.S.:

(a) In 1984, where $x = 4$, the number is given by

$$f(4) = 11.34 + 317.01\log_2 4.$$

Note that you will want to

(1) Use the change of base formula to write the $\log_2$ either ln (base e) or log (base 10) in order to use your calculator: Choosing base 10:

$$\log_2 4 = \frac{\log_{10} 4}{\log_{10} 2}$$
$$= \frac{\log_{10} 4}{\log_{10} 2}$$
$$\approx \frac{.6020599913}{.3010299957}$$
$$= 2.$$

Or (2) simplify $\log_2 4$ by properties of logarithms as follows:

$$\log_2 4 = \log_2 (2^2)$$
$$= 2\log_2 (2)$$
$$= 2(1)$$
$$= 2.$$

Thus,

$$f(4) = 11.34 + 317.01(2)$$
$$= 11.34 + 317.01(2)$$
$$= 645 \text{ sites.}$$

(b) In 1988, where $x = 8$, the number is given by

$$f(8) = 11.34 + 317.01\log_2 8.$$

As above, we will want to choose a method to evaluate or simplify, $\log_2 8$. Simplifying by properties of logarithms results in the following:

$$\log_2 8 = \log_2 (2^3)$$
$$= 3\log_2 (2)$$
$$= 3(1)$$
$$= 3.$$

Thus,

$$f(8) = 11.34 + 317.01(3)$$
$$= 11.34 + 317.01(3)$$
$$= 962 \text{ sites.}$$

61. Using the exponential form of the Richter scale model,

$$x = 10^R x_0:$$

For the Lander's earthquake, the intensity is given by

$$x = 10^{7.3} x_0.$$

For the Northbridge earthquake, the intensity is given by

$$x = 10^{6.7} x_0.$$

Thus, the Lander's earthquake is

about

$$\frac{10^{7.3} x_0}{10^{6.7} x_0} = \frac{10^{7.3}}{10^{6.7}}$$
$$= 10^{7.3-6.7}$$
$$= 10^{.6}$$
$$\approx 4 \text{ times as powerful.}$$

8.7 EXERCISES

1. (a) The US economic growth matched that of Germany in the third quarter. The growth rate was about 1.5%.

    (b) The US economic growth matched that of Japan in the first quarter of 1992. The growth rate was about 2.2%.

    (c) The growth rate of Germany matched that of Japan two times. Germany had the larger growth rate between these two times.

3. $x + y = 6$
    $x - y = 4$

    To decide if $(5, 1)$ is a solution of the system, replace x by 5 and y by 1 in each equation to see if the results are true statements.

    $$5 + 1 = 6$$
    $$6 = 6 \quad True$$
    $$5 - 1 = 4$$
    $$4 = 4 \quad True$$

    Therefore, $(5, 1)$ is a solution to the above system.

5. $2x - y = 8$
    $3x + 2y = 20$

    To decide if $(5, 2)$ is a solution of the system, replace x by 5 and y by 2 in each equation to see if the results are true statements.

    $$2(5) - 2 = 8$$
    $$8 = 8 \quad True$$
    $$3(5) + 2(2) = 20$$
    $$19 = 20 \quad False$$

    Therefore, $(5, 2)$ is not a solution to the system.

7. $x + y = 4$
    $2x - y = 2$

    Graph the line $x + y = 4$ through its intercepts $(0, 4)$ and $(4, 0)$ and the line $2x - y = 2$ through its intercepts $(0, -2)$ and $(1, 0)$.

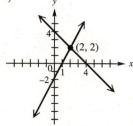

The lines appear to intersect at $(2, 2)$. Check this ordered pair in the system.

$$2 + 2 = 4$$
$$4 = 4 \quad True$$
$$2(2) - 2 = 2$$
$$2 = 2 \quad True$$

Thus, $\{(2, 2)\}$ is the solution.

9. $2x - 5y = 11$
    $3x + y = 8$

    Multiply the second equation by 5 then add to the first equation.

    $$2x - 5y = 11$$
    $$\underline{15x + 5y = 40}$$
    $$17x = 51$$
    $$x = 3$$

    Substitute this value for x into the first equation and solve for y.

    $$2(3) - 5y = 11$$
    $$-5y = 11 - 6$$
    $$-5y = 5$$
    $$y = -1$$

    Since the ordered pair $(3, -1)$ satisfies both equations, it checks. The solution set is $\{(3, -1)\}$.

11. $3x + 4y = -6$
     $5x + 3y = 1$

     Multiply the first equation by 3 and the second by -4 then add.

     $$9x + 12y = -18$$
     $$\underline{-20x - 12y = -4}$$
     $$-11x = -22$$
     $$x = 2$$

     Substitute this value for x into the first equation and solve for y.

     $$3(2) + 4y = -6$$
     $$4y = -6 - 6$$
     $$4y = -12$$
     $$y = -3$$

     Since the ordered pair $(2, -3)$ satisfies both equations, it checks. The solution set is $\{(2, -3)\}$.

13. $3x + 3y = 0$
    $4x + 2y = 3$

Simplify the first equation by dividing both sides by 3.

$$x + y = 0$$
$$4x + 2y = 3$$

Multiply the first equation by -2 then add. Solve the resulting equation for x.

$$-2x - 2y = 0$$
$$\underline{4x + 2y = 3}$$
$$2x = 3$$
$$x = \frac{3}{2}$$

Substitute this value for x into the equation $x + y = 0$ and solve for y.

$$\frac{3}{2} + y = 0$$
$$y = -\frac{3}{2}$$

Since the ordered pair $\left(\frac{3}{2}, -\frac{3}{2}\right)$ satisfies both equations, it checks. The solution set is $\left\{\left(\frac{3}{2}, -\frac{3}{2}\right)\right\}$.

15. $7x + 2y = 6$
    $-14x - 4y = -12$

To simplify the second equation, divide by 2. Then add both equations.

$$7x + 2y = 6$$
$$\underline{-7x - 2y = -6}$$
$$0 = 0 \quad \textit{True}$$

The equations are dependent and hence have an infinite number of solutions. Solve the first equation for x in term of y.

$$7x + 2y = 6$$
$$7x = 6 - 2y$$
$$x = \frac{6 - 2y}{7}$$

We will leave the answer as the ordered pair $\left(\frac{6-2y}{7}, y\right)$. The solution set is $\left\{\left(\frac{6-2y}{7}, y\right)\right\}$. Note that this allows one to create a specific solution as an ordered pair, for any real number replacement of y.

17. $\dfrac{x}{2} + \dfrac{y}{3} = -\dfrac{1}{3}$
    $\dfrac{x}{2} + 2y = -7$

Multiply the first equation by 6 and the second by -6 in order to eliminate fractions and to get opposite coefficients for the x-terms. Add the two resulting equations and solve for y.

$$3x + 2y = -2$$
$$\underline{-3x - 12y = 42}$$
$$-10y = 40$$
$$y = -4$$

Substitute this value for y in the equation $3x + 2y = -2$ and solve for x.

$$3x + 2(-4) = -2$$
$$3x - 8 = -2$$
$$3x = 6$$
$$x = 2$$

Since the ordered pair $(2, -4)$ satisfies both equations, it checks. The solution set is $\{(2, -4)\}$.

19. $5x - 5y = 3$
    $x - y = 12$

Multiply the second equation by -5 then add both equations.

$$5x - 5y = 3$$
$$\underline{-5x + 5y = 60}$$
$$0 = 63 \quad \textit{False}$$

The equations are inconsistent and thus, have no solutions. They are parallel lines with no points of intersection. The solution set is the empty set $\emptyset$.

21. $4x + y = 6$
    $y = 2x$

Substitute the value for y from the second equation into the first equation and solve for x.

$$4x + (2x) = 6$$
$$6x = 6$$
$$x = 1$$

Substitute this value of x into the equation $y = 2x$ and solve for y.

$$y = 2(1)$$
$$y = 2$$

Since the ordered pair $(1, 2)$ satisfies both equations, it checks. The solution set is $\{(1, 2)\}$.

23. $3x - 4y = -22$
     $-3x + y = 0$

Solve the second equation for y.

$$-3x + y = 0$$
$$y = 3x$$

Substitute this value for y into the first equation and solve for x.

$$3x - 4(3x) = -22$$
$$3x - 12x = -22$$
$$-9x = -22$$
$$x = \frac{22}{9}$$

Replace $x = \frac{22}{9}$ in the equation $y = 3x$ and evaluate for y.

$$y = 3\left(\frac{22}{9}\right)$$
$$y = \frac{22}{3}$$

Since the ordered pair $\left(\frac{22}{9}, \frac{22}{3}\right)$ satisfies both equations, it checks. The solution set is $\left\{\left(\frac{22}{9}, \frac{22}{3}\right)\right\}$.

25. $-x - 4y = -14$
     $2x = y + 1$

Solve the second equation for y.

$$2x = y + 1$$
$$2x - 1 = y$$

Substitute this value for y into the first equation and solve for x.

$$-x - 4(2x - 1) = -14$$
$$-x - 8x + 4 = -14$$
$$-9x = -18$$
$$x = 2$$

Substitute $x = 2$ into the equation $y = 2x - 1$ and evaluate y.

$$y = 2(2) - 1$$
$$y = 3$$

Since the ordered pair $(2, 3)$ satisfies both equations, it checks. The solution set is $\{(2, 3)\}$.

27. $5x - 4y = 9$
     $3 - 2y = -x$

Solve the second equation for x by multiplication of both sides by -1.

$$3 - 2y = -x$$
$$-3 + 2y = x$$

Substitute this value of x into the first equation and solve for y.

$$5(-3 + 2y) - 4y = 9$$
$$-15 + 10y - 4y = 9$$
$$6y = 24$$
$$y = 4$$

Substitute this value of y into the equation $x = -3 + 2y$ and evaluate for x.

$$x = -3 + 2(4)$$
$$x = 5$$

Since the ordered pair $(5, 4)$ satisfies both equations, it checks. The solution set is $\{(5, 4)\}$.

29. $x = 3y + 5$
     $x = \frac{3}{2}y$

Replace x in the first equation by $\frac{3}{2}y$, the value of x in the second equation. Multiply both sides by 2 in order to eliminate the fraction and solve for y.

$$\frac{3}{2}y = 3y + 5$$
$$3y = 6y + 10$$
$$-3y = 10$$
$$y = -\frac{10}{3}$$

Substitute this value for y into the second equation (or the first equation) and evaluate for x.

$$x = \frac{3}{2}y$$
$$x = \frac{3}{2}\left(-\frac{10}{3}\right)$$
$$x = -5$$

Since the ordered pair $\left(-5, -\frac{10}{3}\right)$ satisfies both equations, it checks. The solution set is $\left\{\left(-5, -\frac{10}{3}\right)\right\}$.

31. $\dfrac{1}{2}x + \dfrac{1}{3}y = 3$

$\qquad\qquad y = 3x$

Substitute the value for y from the second equation into the first equation and solve for x.

$$\frac{1}{2}x + \frac{1}{3}(3x) = 3$$
$$\frac{1}{2}x + x = 3$$
$$\frac{3}{2}x = 3$$
$$x = \left(\frac{2}{3}\right)3$$
$$x = 2$$

Substitute this value of x into the equation $y = 3x$ and solve for y.

$\qquad\qquad y = 3(2)$
$\qquad\qquad y = 6$

Since the ordered pair $(2, 6)$ satisfies both equations, it checks. The solution set is $\{(2, 6)\}$.

33. Writing exercise

35. $\quad 3x + 2y + z = 8$
$\quad 2x - 3y + 2z = -16$
$\quad\; x + 4y - z = 20$

Eliminate z from the first and third equations by adding.

$$\begin{array}{r} 3x + 2y + z = 8 \\ x + 4y - z = 20 \\ \hline 4x + 6y = 28 \end{array}$$

Eliminate z from the second and third equations by adding 2 times the third to the second.

$$\begin{array}{r} 2x - 3y + 2z = -16 \\ 2x + 8y - 2z = 40 \\ \hline 4x + 5y \quad\;\; = 24 \end{array}$$

We are left with two equations in x and y. Multiply the second by -1 and add to the first to eliminate the x-term.

$$\begin{array}{r} 4x + 6y = 28 \\ -4x - 5y = -24 \\ \hline y = 4 \end{array}$$

Substitute this value of y into the equation $4x + 5y = 24$ (Either equation in x and y may be used.) Solve for x.

$$4x + 5(4) = 24$$
$$4x = 4$$
$$x = 1$$

Replace x and y by these values in the equation $x + 4y - z = 20$ (Any one of the original 3 equations may be used.) Solve for z.

$$(1) + 4(4) - z = 20$$
$$17 - z = 20$$
$$-z = 3$$
$$z = -3$$

Since the ordered triple $(1, 4, -3)$ satisfies all three equations, it checks. The solution set is $\{(1, 4, -3)\}$.

37. $\quad 2x + 5y + 2z = 0$
$\quad 4x - 7y - 3z = 1$
$\quad 3x - 8y - 2z = -6$

Eliminate z from the first and third equations by adding.

$$\begin{array}{r} 2x + 5y + 2z = 0 \\ 3x - 8y - 2z = -6 \\ \hline 5x - 3y \quad\;\; = -6 \end{array}$$

Eliminate z from first and second equations by adding 3 times the first, to 2 times the second.

$$\begin{array}{r} 6x + 15y + 6z = 0 \\ 8x - 14y - 6z = 2 \\ \hline 14x + y \quad\;\; = 2 \end{array}$$

We are left with two equations in x and y. Multiply the second equation by 3 and add to the first.

$$\begin{array}{r} 5x - 3y = -6 \\ 42x + 3y = 6 \\ \hline 47x = 0 \\ x = 0 \end{array}$$

Substitute this value of x into the equation $14x + y = 2$ (Either equation in x and y may be used.) Solve for y.

$$14(0) + y = 2$$
$$y = 2$$

Replace x and y by these values in the equation $2x + 5y + 2z = 0$ (Any one of the original 3 equations may be used.) Solve for z.

$$2(0) + 5(2) + 2z = 0$$
$$10 + 2z = 0$$
$$2z = -10$$
$$z = -5$$

The solution set is $\{(0, 2, -5)\}$.

39.
$$x + y - z = -2$$
$$2x - y + z = -5$$
$$-x + 2y - 3z = -4$$

Add the first and second equations and solve for x.

$$x + y - z = -2$$
$$2x - y + z = -5$$
$$\overline{3x = -7}$$
$$x = -\frac{7}{3}$$

Add -3 times the first equation to the third.

$$-3x - 3y + 3z = 6$$
$$-x + 2y - 3z = -4$$
$$\overline{-4x - y = 2}$$

Substitute $x = -\frac{7}{3}$ into the equation $-4x - y = 2$.

$$-4\left(-\frac{7}{3}\right) - y = 2$$
$$\frac{28}{3} - y = 2$$
$$-y = \frac{6}{3} - \frac{28}{3}$$
$$-y = -\frac{22}{3}$$
$$y = \frac{22}{3}$$

Replace x and y by these values in the equation $x + y - z = -2$. Solve for z.

$$\left(-\frac{7}{3}\right) + \left(\frac{22}{3}\right) - z = -2$$
$$-z = -2 - \frac{15}{3}$$
$$-z = -\frac{6}{3} - \frac{15}{3}$$
$$-z = -\frac{21}{3}$$
$$z = 7$$

The solution set is $\left\{\left(-\frac{7}{3}, \frac{22}{3}, 7\right)\right\}$.

41.
$$2x - 3y + 2z = -1$$
$$x + 2y + z = 17$$
$$2y - z = 7$$

Add -2 times the second equation to the first equation.

$$2x - 3y + 2z = -1$$
$$-2x - 4y - 2z = -34$$
$$\overline{-7y = -35}$$
$$y = 5$$

Substitute $y = 5$ into the equation $2y - z = 7$.

$$2(5) - z = 7$$
$$10 - z = 7$$
$$-z = -3$$
$$z = 3$$

Substitute these values of y and z into the equation $x + 2y + z = 17$ and solve for x.

$$x + 2(5) + 3 = 17$$
$$x + 13 = 17$$
$$x = 4$$

The solution set is $\{(4, 5, 3)\}$.

43.
$$4x + 2y - 3z = 6$$
$$x - 4y + z = -4$$
$$-x + 2z = 2$$

Add the second equation and third equation.

$$x - 4y + z = -4$$
$$-x + 2z = 2$$
$$\overline{-4y + 3z = -2}$$

Add -4 times the second equation to the first equation.

$$4x + 2y - 3z = 6$$
$$-4x + 16y - 4z = 16$$
$$\overline{18y - 7z = 22}$$

We are left with two equations in y and z. Multiply the first equation by 7 and the second equation by 3.

$$-28y + 21z = -14$$
$$54y - 21z = 66$$
$$\overline{26y = 52}$$
$$y = 2$$

Replace this value of y in the equation $-4y + 3z = -2$.

$$-4(2) + 3z = -2$$
$$3z = -2 + 8$$
$$3z = 6$$
$$z = 2$$

Substitute this value for z, into $-x + 2z = 2$ and solve for x.

$$-x + 2(2) = 2$$
$$-x + 4 = 2$$
$$-x = -2$$
$$x = 2$$

The solution set is $\{(2, 2, 2)\}$.

45.
$$2x + y = 6$$
$$3y - 2z = -4$$
$$3x - 5z = -7$$

Multiply the first equation by -3 and add to the second equation in order to eliminate the y-term.

$$
\begin{array}{rcl}
-6x - 3y & = & -18 \\
3y - 2z & = & -4 \\
\hline
-6x \quad - 2z & = & -22 \text{ or} \\
-3x - z & = & -11
\end{array}
$$

Eliminate the x-term from this equation and the third original equation. Solve for z.

$$
\begin{array}{rcl}
3x - 5z & = & -7 \\
-3x - z & = & -11 \\
\hline
-6z & = & -18 \\
z & = & 3
\end{array}
$$

Substitute this value of z into the original equation $3x - 5z = -7$ to find x.

$$
\begin{aligned}
3x - 5(3) &= -7 \\
3x - 15 &= -7 \\
3x &= 8 \\
x &= \frac{8}{3}
\end{aligned}
$$

Substitute the values $z = 3$ into the equation $3y - 2z = -4$ and solve for y.

$$
\begin{aligned}
3y - 2z &= -4 \\
3y - 2(3) &= -4 \\
3y - 6 &= -4 \\
3y &= 2 \\
y &= \frac{2}{3}
\end{aligned}
$$

The solution set is $\left\{ \left(\frac{8}{3}, \frac{2}{3}, 3 \right) \right\}$.

47. Let $x =$ the number of basketball games won and
$y =$ the number of basketball games lost.

Write an equation to represent the total number of games played.

$$x + y = 82$$

Write a second equation to represent the relationship between games played and games lost.

$$x = y + 56$$

Substituting this value of x into the first equation, we arrive at a new equation in y only

$$(y + 56) + y = 82$$

Solve for y.

$$
\begin{aligned}
(y + 56) + y &= 82 \\
2y + 56 &= 82 \\
2y &= 26 \\
y &= 13
\end{aligned}
$$

Substitute this value for y into the equation $x = y + 56$ and solve for x.

$$
\begin{aligned}
x &= (13) + 56 \\
x &= 69
\end{aligned}
$$

Thus, there were 69 wins and 13 losses for the season.

49. Let $x =$ the number of hockey games won and
$y =$ the number of hockey games lost,
$z =$ the number of hockey games tied.

Write an equation to represent the total number of games played.

$$x + y + z = 82$$

Write a second equation to represent the total number of wins and loses.

$$x + y = 74$$

Write a third equation that relates their ties and their loses.

$$z = y - 18$$

Subtract the second equation from the first.

$$
\begin{array}{rcl}
x + y + z & = & 82 \\
x + y \quad & = & 74 \\
\hline
z & = & 8
\end{array}
$$

Substituting this value of z into the equation $z = y - 18$, we arrive at a new equation in y only.

$$
\begin{aligned}
(8) &= y - 18 \\
8 + 18 &= y \\
y &= 26
\end{aligned}
$$

Substitute this value of y into the equation $x + y = 74$ and solve for x.

$$
\begin{aligned}
x + (26) &= 74 \\
x &= 74 - 26 \\
y &= 48
\end{aligned}
$$

There were 48 wins, 26 losses and 8 ties for the season.

51. Let l = the length of the tennis court and
w = the width of the tennis court.

Use the relationship of the perimeter to the side lengths, $P = 2l + 2w$, to write the following equation.

$$228 = 2l + 2w \quad \text{or}$$
$$l + w = 114$$

Write a second equation to represent the relationship between the length and the width dimensions.

$$w = l - 42 \quad \text{or}$$
$$l - w = 42$$

Solve the resulting system of equations by adding to eliminate the w-term.

$$l + w = 114$$
$$\underline{l - w = 42}$$
$$2l = 156$$
$$l = 78$$

Substituting this value of l into the equation $w = l - 42$, we find the width dimension.

$$w = 78 - 42$$
$$= 36$$

The dimensions are: length = 78 feet and width = 36 feet.

53. Let l = the length of the rectangle and w = the width of the rectangle.

Write an equation to represent the relationship between the length and the width dimensions.

$$l = w + 7 \quad \text{or}$$
$$l - w = 7$$

Write a second equation to represent the relationship between the perimeter and the length and width dimensions and simplify.

$$p = 2(l - 3) + 2(w + 2)$$
$$32 = 2(l - 3) + 2(w + 2)$$
$$16 = (l - 3) + (w + 2)$$
$$16 = l + w - 1 \quad \text{or}$$
$$l + w = 17$$

Solve the resulting system of equations by adding to eliminate the w-term.

$$l - w = 7$$
$$\underline{l + w = 17}$$
$$2l = 24$$
$$l = 12$$

Substituting this value of l into the equation $l = w + 7$, we find the width dimension.

$$12 = w + 7$$
$$5 = w$$

The dimensions are: length = 12 feet and width = 5 feet.

55. Let x = the number of daily newspapers in Texas and
y = the number of daily newspapers in Florida.

Write an equation to represent the fact that Texas has 52 more daily newspapers than Florida.

$$x = y + 52$$

Write a second equation to represent the total number of daily newspapers.

$$x + y = 134$$

Substituting this value of x from the first equation into the second equation, we arrive at a new equation in y only.

$$(y + 52) + y = 134$$

Solve for y.

$$(y + 52) + y = 134$$
$$2y + 52 = 134$$
$$2y = 82$$
$$y = 41$$

Substitute this value for y into the equation $x = y + 52$ and solve for x.

$$x = (41) + 52$$
$$x = 93$$

Texas had 93 daily newspapers and Florida had 41 daily newspapers.

57. Let x = the FCI for the National Hockey League and y = the FCI for the National Basketball Association.

Write an equation to represent the total FCI prices.

$$x + y = 432.12$$

Write an equation to represent the relationship of the FCI prices for the two groups.

$$x = y + 16.36$$

Substituting this value of x from the second equation into the first equation, we arrive at a new equation in y only.

$$(y + 16.36) + y = 423.12$$

Solve for y.

$$(y + 16.36) + y = 423.12$$
$$2y + 16.36 = 423.12$$
$$2y = 406.76$$
$$y = 203.38$$

Substitute this value for y into the equation $x = y + 16.36$ and solve for x.

$$x = (203.38) + 16.36$$
$$x = 219.74$$

The FCI for the National Hockey League was $219.74. The FCI for the National Basketball Association was $203.38

59. Let $x = $ the cost of a CGA monitor and
    $y = $ the cost of a VGA monitor.

Write two equations to represent the two sets of purchase prices.

$$4x + 6y = 4600$$
$$6x + 4y = 4400$$

Simplify both equations by dividing out the greatest common factor, 2.

$$2x + 3y = 2300$$
$$3x + 2y = 2200$$

Multiply the first equation by 2 and the second equation by -3 and add to eliminate the y-term.

$$4x + 6y = 4600$$
$$\underline{-9x - 6y = -6600}$$
$$-5x = -2000$$
$$x = 400$$

Substitute this value of x into the equation $2x + 3y = 2300$ and solve for y.

$$2(400) + 3y = 2300$$
$$800 + 3y = 1500$$
$$3y = 1500$$
$$y = 500$$

The CGA monitors cost $400 for each monitor and the VGA monitors cost $500 for each monitor.

61. Let $x = $ the number of units of yarn and
    $y = $ the number of units of thread.

Write two equations representing the number of hours per day each machine runs.

$$\text{Machine A:} \quad 1x + 1y = 8$$
$$\text{Machine B:} \quad 2x + 1y = 14$$

Multiply the first equation by -1 and add to the second equation in order to eliminate the y-term.

$$-x - y = -8$$
$$\underline{2x + y = 14}$$
$$x = 6$$

Substitute this value of x into the equation $x + y = 8$ and solve for y.

$$(6) + y = 8$$
$$y = 2$$

Thus, 6 units of yarn and 2 units of thread will keep both machines running to capacity.

63. Let $x = $ the number of gallons of 25% alcohol needed and $y = $ the number of gallons of 35% alcohol needed.

Write an equation to represent the total gallons of solution.

$$x + y = 20$$

Write another equation to represent the total amount of alcohol.

$$.25x + .35y = .32(20), \quad \text{or equivalently}$$
$$25x + 35y = 32(20)$$
$$5x + 7y = 128$$

Solve the resulting system of equations.

$$x + y = 20$$
$$5x + 7y = 128$$

Multiply the first equation by -5 and add to the second equation in order to eliminate the x-term.

$$-5x - 5y = 100$$
$$\underline{5x + 7y = 128}$$
$$2y = 28$$
$$y = 14$$

Substitute this value of y into the equation $x + y = 20$ and solve for x.

$$x + (14) = 20$$
$$x = 6$$

It requires 6 gallons of 25% alcohol and 14 gallons of 35% alcohol to obtain 20 gallons of 32% alcohol.

65. Let $x =$ the number of liters of 100% (pure) acid and $y =$ the number of liters of 10% acid needed.

Write an equation to represent the total gallons of solution.

$$x + y = 27$$

Write another equation to represent the total amount of acid.

$$1.00x + .10y = .20(27)$$
$$10.0x + 1.0y = 2.0(27)$$
$$10x + 1y = 54$$

Solve the resulting system of equations.

$$x + y = 27$$
$$10x + y = 54$$

Multiply the first equation by -1 and add to the second equation in order to eliminate the y-term.

$$-x - y = -27$$
$$\underline{10x + y = 54}$$
$$9x = 27$$
$$x = 3$$

Substitute this value of y into the equation $x + y = 27$ and solve for y.

$$(3) + y = 27$$
$$y = 24$$

Three liters of pure acid and 24 liters of 10% acid should be used.

67. Let $x =$ the number of pounds of $3.60 pecan clusters and $y =$ the number of pounds of $7.20 chocolate truffles.

Write an equation to represent the total number of pounds of the mixture.

$$x + y = 80$$

Write (and simplify) another equation to represent the total value of the candy.

$$3.60x + 7.20y = 4.95(80)$$
$$360x + 720y = 495(80)$$
$$x + 2y = 110$$

Solve the resulting system of equations.

$$x + y = 80$$
$$x + 2y = 110$$

Multiply the first equation by -1 and add to the second equation in order to eliminate the x-term.

$$-x - y = -80$$
$$\underline{x + 2y = 110}$$
$$y = 30$$

Substitute this value of y into the equation $x + y = 80$ and solve for y.

$$x + (30) = 80$$
$$x = 50$$

Thus, 50 pounds of $3.60 pecan clusters and 30 pounds of $7.20 chocolate truffles are to be used.

69. Let $x =$ the number of general admission ($2.50) tickets sold and $y =$ the number of student admission ($2.00) tickets sold.

Write an equation to represent the total number of people who saw the performance.

$$x + y = 184$$

Write (and simplify) another equation to represent the total amount of money collected.

$$2.50x + 2.00y = 406$$
$$25x + 20y = 4060$$
$$5x + 4y = 812$$

Solve the resulting system of equations.

$$x + y = 184$$
$$5x + 4y = 812$$

Multiply the first equation by -4 and add to the second equation in order to eliminate the y-term.

$$-4x - 4y = -736$$
$$\underline{5x + 4y = 812}$$
$$x = 76$$

Substitute this value of x into the equation $x + y = 184$ and solve for y.

$$(76) + y = 184$$
$$y = 108$$

Thus, 76 with general admission tickets and 108 tickets with student identification were sold.

71. Let x = the number of dimes and
y = the number of quarters.

Write an equation to represent the total number of coins.

$$x + y = 94$$

Write (and simplify) another equation to represent the total value of money.

$$.10x + .25y = 19.30$$
$$10x + 25y = 1930$$
$$2x + 5y = 386$$

Solve the resulting system of equations.

$$x + y = 94$$
$$2x + 5y = 386$$

Multiply the first equation by -2 and add to the second equation in order to eliminate the x-term.

$$-2x - 2y = -188$$
$$\underline{2x + 5y = 386}$$
$$3y = 198$$
$$y = 66$$

Substitute this value of y into the equation $x + y = 94$ and solve for x.

$$x + (66) = 94$$
$$x = 28$$

She has 28 dimes and 66 quarters.

73. Let x = the amount of money invested at 2% and
y = the amount of money invested at 4% .

Write an equation to represent the total number of dollars invested.

$$x + y = 3000$$

Write another equation to represent the total return (interest earned) on the investments.

$$.02x + .04y = 100$$
$$2x + 4y = 10000$$
$$x + 2y = 5000$$

Solve the resulting system of equations.

$$x + y = 3000$$
$$x + 2y = 5000$$

Multiply the first equation by -1 and add to the second equation in order to eliminate the x-term.

$$-x - y = -3000$$
$$\underline{x + 2y = 5000}$$
$$y = 2000$$

Substitute this value of y into the equation $x + y = 3000$ and solve for x.

$$x + (2000) = 3000$$
$$x = 1000$$

There was $1000 deposited at 2% and $2000 deposited at 4%.

75. Let x = the speed of the freight train and
y = the speed of the express train.

Write an equation to show the relationship of the speed of the freight train compared to that of the express train.

$$x = y - 30$$

The distance traveled, $d = rt$, by the freight train during the 3 hours is $3x$ and that of the express train is $3y$. Since the total distance traveled by both trains is 390 kilometers, construct an equation showing total distance.

$$3x + 3y = 390$$
$$x + y = 130$$

Solve the resulting system of equations.

$$x = y - 30$$
$$x + y = 130$$

Substitute the value for x into the second equation $x + y = 130$ and solve for y

$$(y - 30) + y = 130$$
$$\underline{2y - 30 = 130}$$
$$2y = 160$$
$$y = 80$$

Substitute this value of y into the equation $x = y - 30$ and solve for x.

$$x = (80) - 30$$
$$x = 50$$

Thus, the speed of the freight train was 50 kilometers per hour and that of the express train was 80 kilometers per hour.

77. Let $x =$ the top speed of the snow speeder and
$y =$ the speed of the wind.

Remembering the relationship $d = rt$, the distance traveled by the snow speeder into the wind is given by $3600 = (x - y)(2)$ and the same distance, returning, with the wind, is $(x + y)(1.5)$. The resulting two equations are

$$3600 = (x - y)(2)$$
$$3600 = (x + y)(1.5).$$

Write the equations in standard form.

$$2x - 2y = 3600$$
$$1.5x + 1.5y = 3600$$

Observe that the second equation may be simplified as follows before solving the system of equations.

$$1.5x + 1.5y = 3600$$
$$15x + 15y = 36000$$
$$x + y = 2400$$

The resulting system is

$$2x - 2y = 3600$$
$$x + y = 2400.$$

Solve by multiplying the second equation by 2 and adding to the first.

$$2x - 2y = 3600$$
$$2x + 2y = 4800$$
$$\overline{4x \qquad = 8400}$$
$$x = 2100$$

Substitute this value of x into the equation $x + y = 2400$ and solve for y.

$$(2100) + y = 2400$$
$$x = 300$$

Thus, the top speed is 2100 miles per hour and the wind speed is 300 miles per hour.

79. Let $x =$ the number of $20 fish,
$y =$ the number of $40 fish, and
$z =$ the number of $65 fish

Write an equation to represent the total number of fish.

$$x + y + z = 29$$

Write a second equation to represent the relationship between the number of $20 fish and the $40 fish.

$$y = 2x - 1$$

Write this equation in standard form.

$$-2x + y = -1$$

Write a third equation to represent the total value of the fish.

$$20x + 40y + 65z = 1150$$

This equations simplifies to

$$4x + 8y + 13z = 230$$

The resulting system equations is as follows.

$$x + y + z = 29$$
$$-2x + y \qquad = -1$$
$$4x + 8y + 13z = 230$$

Multiply the first equation by -13 and add to the last in order to eliminate the z-term.

$$-13x - 13y - 13z = -377$$
$$4x + 8y + 13z = 230$$
$$\overline{-9x - 5y \qquad = -147}$$

The resulting two equations in x and y are as follows:

$$-2x + y = -1$$
$$-9x - 5y = -147$$

Solve by multiplying the first equation by 5 and adding to the second.

$$-10x + 5y = -5$$
$$-9x - 5y = -147$$
$$\overline{-19x = -152}$$
$$x = 8$$

Substitute this value of x into the equation $y = 2x - 1$.

$$y = 2(8) - 1$$
$$y = 15$$

Substitute these values for x and for y into the equation $x + y + z = 29$ and solve for z.

$$(8) + (15) + z = 29$$
$$23 + z = 29$$
$$z = 6$$

Thus, there are 8 fish at $20, 15 fish at $40, and 6 fish at $65.

81. Let $x =$ the length of the shortest side,
$y =$ the length of the middle side, and
$z =$ the length of the longest side.

Write an equation to represent the perimeter.

$$x + y + z = 70$$

Write a second equation to represent the relationship between the longest side and the others.

$$z = (x + y) - 4, \text{ or}$$
$$-x - y + z = -4$$

Write a third equation to to represent the relationship between the longest and shortest sides.

$$z = 2x + 9, \text{ or}$$
$$-2x + z = 9$$

The resulting system equations is as follows.

$$x + y + z = 70$$
$$-x - y + z = -4$$
$$-2x + z = 9$$

Add the first and second equation to eliminate the x and y-terms.

$$\begin{aligned} x + y + z &= 70 \\ -x - y + z &= -4 \\ \hline 2z &= 66 \\ z &= 33 \end{aligned}$$

Substitute this value of z into the equation $-2x + z = 9$ to find x.

$$-2x + (33) = 9$$
$$-2x = -24$$
$$x = 12$$

Subtitute the values for x and z into the equation $x + y + z = 70$ to find the value for y.

$$(12) + y + (33) = 70$$
$$y + 45 = 70$$
$$y = 25$$

The shortest side is 12 centimeters, the middle side is 25 centimeters, and the longest side is 33 centimeters.

83. Let $x =$ the number of cases sent to wholesaler A,
$y =$ the number of cases sent to wholesaler B, and
$z =$ the number of cases sent to wholesaler C.

Write an equation to represent the total cases sent.

$$x + y + z = 320$$

Write a second equation to represent the relationship between the number of wholesaler A and wholesale B's cases of trinkets.

$$x = 3y, \text{ or}$$
$$x - 3y = 0$$

Write a third equation to represent the relationship between the number of cases sent to wholesaler C and those to wholesalers A and B

$$z = (x + y) - 160, \text{ or}$$
$$-x - y + z = -160$$

The resulting system of equations is as follows:

$$x + y + z = 320$$
$$x - 3y \quad = 0$$
$$-x - y + z = -160.$$

Add the first and third equation to eliminate the x and y-terms.

$$\begin{aligned} x + y + z &= 320 \\ -x - y + z &= -160 \\ \hline 2z &= 160 \\ z &= 80 \end{aligned}$$

Substitute this value of z into the equation $x + y + z = 320$ and use with the other equation $x - 3y = 0$ in x and y.

$$x + y + (80) = 320$$
$$x + y = 240$$
$$x - 3y = 0$$

Multiply the equation $x - 3y = 0$ by -1 and add to the equation $x + y = 240$.

$$\begin{aligned} x + y &= 240 \\ -x + 3y &= 0 \\ \hline 4y &= 240 \\ y &= 60 \end{aligned}$$

Substitute the values for y and z into the equation $x + y + z = 320$ and solve for x.

$$x + (60) + (80) = 320$$
$$x + 240 = 320$$
$$x = 180$$

She must send 180 cases to wholesaler A, 60 cases to wholesaler B, and 80 cases to wholesaler C.

85. Let $x =$ the number of pounds of jelly beans,
     $y =$ the number of pounds of chocolate eggs, and
     $z =$ the number of pounds of marshmallow chicks.

Write an equation to represent the total pounds of candy mixture.

$$x + y + z = 15$$

Write and simplify a second equation to represent the total selling price.

$$.80x + 2.00y + 1.00z = (1.00)15$$
$$8x + 20y + 10z = 150$$
$$4x + 10y + 5z = 75$$

The two remaining relationships indicated are the following:

$$x = 2(y + z)$$
$$x = 5y.$$

Since both of these equations represent x set them equal to each other.

$$2(y + z) = 5y, \text{ or}$$
$$2y + 2z = 5y$$
$$-3y + 2z = 0$$

The resulting system of equations is as follows.

$$x + y + z = 15$$
$$4x + 10y + 5z = 75$$
$$-3y + 2z = 0$$

Multiply the first equation by -4 and add to the second to eliminate the x-terms.

$$-4x - 4y - 4z = -60$$
$$\underline{4x + 10y + 5z = 75}$$
$$6y + z = 15$$

Solve the resulting equations in y and z.

$$-3y + 2z = 0$$
$$6y + z = 15$$

Multiply the first equation by 2 and add to the second.

$$-6y + 4z = 0$$
$$\underline{6y + z = 15}$$
$$5z = 15$$
$$z = 3$$

Substitute into the equation $6y + z = 15$ and solve for y.

$$6y + (3) = 15$$
$$6y = 12$$
$$y = 2$$

Replace these values of y and z into the equation $x + y + z = 15$.

$$x + (2) + (3) = 15$$
$$x + 5 = 15$$
$$x = 10$$

She will use 10 pounds of jelly beans, 2 pounds of chocolate eggs, and 3 pounds of marshmallow chicks.

EXTENSION: USING MATRIX ROW OPERATIONS TO SOLVE SYSTEMS

1. $x + y = 5$
     $x - y = -1$

Write the augmented matrix.

$$\begin{bmatrix} 1 & 1 & | & 5 \\ 1 & -1 & | & -1 \end{bmatrix}$$

Multiply row 1 by -1 and add to row 2.

$$\begin{bmatrix} 1 & 1 & | & 5 \\ 0 & -2 & | & -6 \end{bmatrix}$$

Multiply row 2 by $-1/2$.

$$\begin{bmatrix} 1 & 1 & | & 5 \\ 0 & 1 & | & 3 \end{bmatrix}$$

Multiply row 2 by -1 and add to row 1.

$$\begin{bmatrix} 1 & 0 & | & 2 \\ 0 & 1 & | & 3 \end{bmatrix}$$

The resulting matrix represents the following system of equations.

$$1x + 0y = 2$$
$$0x + 1y = 3$$

That is,

$$x = 2$$
$$y = 3.$$

Thus, $\{(2, 3)\}$ is the solution set.

3.　$x + y = -3$
$2x - 5y = -6$

Form the augmented matrix.

$$\begin{bmatrix} 1 & 1 & | & -3 \\ 2 & -5 & | & -6 \end{bmatrix}$$

Multiply row 1 by -2 and add to row 2.

$$\begin{bmatrix} 1 & 1 & | & -3 \\ 0 & -7 & | & 0 \end{bmatrix}$$

Multiply row 2 by $-1/7$.

$$\begin{bmatrix} 1 & 1 & | & -3 \\ 0 & 1 & | & 0 \end{bmatrix}$$

Multiply row 2 by -1 and add to row 1.

$$\begin{bmatrix} 1 & 0 & | & -3 \\ 0 & 1 & | & 0 \end{bmatrix}$$

That is,

$$x = -3$$
$$y = 0.$$

Thus, $\{(-3, 0)\}$ is the solution set.

5.　$2x - 3y = 10$
$2x + 2y = 5$

Form the augmented matrix.

$$\begin{bmatrix} 2 & -3 & | & 10 \\ 2 & 2 & | & 5 \end{bmatrix}$$

Multiply row 1 by $1/2$.

$$\begin{bmatrix} 1 & -3/2 & | & 5 \\ 2 & 2 & | & 5 \end{bmatrix}$$

Multiply row 1 by -2 and add to row 2.

$$\begin{bmatrix} 1 & -3/2 & | & 5 \\ 0 & 5 & | & -5 \end{bmatrix}$$

Multiply row 2 by $1/5$.

$$\begin{bmatrix} 1 & -3/2 & | & 5 \\ 0 & 1 & | & -1 \end{bmatrix}$$

Multiply row 2 by 3/2 and add to row 1.

$$\begin{bmatrix} 1 & 0 & | & 7/2 \\ 0 & 1 & | & -1 \end{bmatrix}$$

That is,

$$x = 7/2$$
$$y = -1.$$

Thus, $\{(7/2, -1)\}$ is the solution set.

7.　$3x - 7y = 31$
$2x - 4y = 18$

Form the augmented matrix.

$$\begin{bmatrix} 3 & -7 & | & 31 \\ 2 & -4 & | & 18 \end{bmatrix}$$

Multiply row 1 by $1/3$.

$$\begin{bmatrix} 1 & -7/3 & | & 31/3 \\ 2 & -4 & | & 18 \end{bmatrix}$$

Multiply row 1 by -2 and add to row 2.

$$\begin{bmatrix} 1 & -7/3 & | & 31/3 \\ 0 & 2/3 & | & -8/3 \end{bmatrix}$$

Multiply row 2 by 3/2.

$$\begin{bmatrix} 1 & -7/3 & | & 31/3 \\ 0 & 1 & | & -4 \end{bmatrix}$$

Multiply row 2 by 7/3 and add to row 1.

$$\begin{bmatrix} 1 & 0 & | & 1 \\ 0 & 1 & | & -4 \end{bmatrix}$$

That is,

$$x = 1$$
$$y = -4.$$

Thus, $\{(1, -4)\}$ is the solution set.

9.　$x + y - z = 6$
$2x - y + z = -9$
$x - 2y + 3z = 1$

Form the augmented matrix.

$$\begin{bmatrix} 1 & 1 & -1 & | & 6 \\ 2 & -1 & 1 & | & -9 \\ 1 & -2 & 3 & | & 1 \end{bmatrix}$$

Multiply row 1 by -1 and add to row 3.
Multiply row 1 by -2 and add to row 2.

$$\begin{bmatrix} 1 & 1 & -1 & | & 6 \\ 0 & -3 & 3 & | & -21 \\ 0 & -3 & 4 & | & -5 \end{bmatrix}$$

Multiply row 2 by -1 and add to row 3.

$$\begin{bmatrix} 1 & 1 & -1 & | & 6 \\ 0 & -3 & 3 & | & -21 \\ 0 & 0 & 1 & | & 16 \end{bmatrix}$$

Multiply row 2 by $-1/3$.

$$\begin{bmatrix} 1 & 1 & -1 & | & 6 \\ 0 & 1 & -1 & | & 7 \\ 0 & 0 & 1 & | & 16 \end{bmatrix}$$

Multiply row 2 by -1 and add to row 1.

$$\begin{bmatrix} 1 & 0 & 0 & | & -1 \\ 0 & 1 & -1 & | & 7 \\ 0 & 0 & 1 & | & 16 \end{bmatrix}$$

Add row 3 to row 2.

$$\begin{bmatrix} 1 & 0 & 0 & | & -1 \\ 0 & 1 & 0 & | & 23 \\ 0 & 0 & 1 & | & 16 \end{bmatrix}$$

That is,

$$x = -1$$
$$y = 23$$
$$z = 16.$$

Thus, $\{(-1, 23, 16)\}$ is the solution set.

11. $2x - y + 3z = 0$
$x + 2y - z = 5$
$2y + z = 1$

Form the augmented matrix.

$$\begin{bmatrix} 2 & -1 & 3 & | & 0 \\ 1 & 2 & -1 & | & 5 \\ 0 & 2 & 1 & | & 1 \end{bmatrix}$$

Interchange row 1 and row 2.

$$\begin{bmatrix} 1 & 2 & -1 & | & 5 \\ 2 & -1 & 3 & | & 0 \\ 0 & 2 & 1 & | & 1 \end{bmatrix}$$

Multiply row 1 by -2 and add to row 2.

$$\begin{bmatrix} 1 & 2 & -1 & | & 5 \\ 0 & -5 & 5 & | & -10 \\ 0 & 2 & 1 & | & 1 \end{bmatrix}$$

Multiply row 2 by $-1/5$.

$$\begin{bmatrix} 1 & 2 & -1 & | & 5 \\ 0 & 1 & -1 & | & 2 \\ 0 & 2 & 1 & | & 1 \end{bmatrix}$$

Multiply row 2 by -2 and add to row 3.

$$\begin{bmatrix} 1 & 2 & -1 & | & 5 \\ 0 & 1 & -1 & | & 2 \\ 0 & 0 & 3 & | & -3 \end{bmatrix}$$

Multiply row 3 by $1/3$.

$$\begin{bmatrix} 1 & 2 & -1 & | & 5 \\ 0 & 1 & -1 & | & 2 \\ 0 & 0 & 1 & | & -1 \end{bmatrix}$$

Add row 3 to row 1 and to row 2.

$$\begin{bmatrix} 1 & 2 & 0 & | & 4 \\ 0 & 1 & 0 & | & 1 \\ 0 & 0 & 1 & | & -1 \end{bmatrix}$$

Add row -2 times row 2 to row 1.

$$\begin{bmatrix} 1 & 0 & 0 & | & 2 \\ 0 & 1 & 0 & | & 1 \\ 0 & 0 & 1 & | & -1 \end{bmatrix}$$

That is,

$$x = 2$$
$$y = 1$$
$$z = -1.$$

Thus, $\{(2, 1, -1)\}$ is the solution set.

13. $-x + y = -1$
$y - z = 6$
$x + z = -1$

Form the augmented matrix.

$$\begin{bmatrix} -1 & 1 & 0 & | & -1 \\ 0 & 1 & -1 & | & 6 \\ 1 & 0 & 1 & | & -1 \end{bmatrix}$$

Multiply row 1 by -1.

$$\begin{bmatrix} 1 & -1 & 0 & | & 1 \\ 0 & 1 & -1 & | & 6 \\ 1 & 0 & 1 & | & -1 \end{bmatrix}$$

Multiply row 1 by -1 and add to row 3.

$$\begin{bmatrix} 1 & -1 & 0 & | & 1 \\ 0 & 1 & -1 & | & 6 \\ 0 & 1 & 1 & | & -2 \end{bmatrix}$$

Add row 2 to row 1.

$$\begin{bmatrix} 1 & 0 & -1 & | & 7 \\ 0 & 1 & -1 & | & 6 \\ 0 & 1 & 1 & | & -2 \end{bmatrix}$$

Multiply row 2 by -1 and add to row 3.

$$\begin{bmatrix} 1 & 0 & -1 & | & 7 \\ 0 & 1 & -1 & | & 6 \\ 0 & 0 & 2 & | & -8 \end{bmatrix}$$

Multiply row 3 by 1/2.

$$\begin{bmatrix} 1 & 0 & -1 & | & 7 \\ 0 & 1 & -1 & | & 6 \\ 0 & 0 & 1 & | & -4 \end{bmatrix}$$

Add row 3 to row 1.

$$\begin{bmatrix} 1 & 0 & 0 & | & 3 \\ 0 & 1 & -1 & | & 6 \\ 0 & 0 & 1 & | & -4 \end{bmatrix}$$

Add row 3 to row 2.

$$\begin{bmatrix} 1 & 0 & 0 & | & 3 \\ 0 & 1 & 0 & | & 2 \\ 0 & 0 & 1 & | & -4 \end{bmatrix}$$

That is,

$$x = 3$$
$$y = 2$$
$$z = -4.$$

Thus, $\{(3, 2, -4)\}$ is the solution set.

15. $$2x - y + 4z = -1$$
$$-3x + 5y - z = 5$$
$$2x + 3y + 2z = 3$$

Form the augmented matrix.

$$\begin{bmatrix} 2 & -1 & 4 & | & -1 \\ -3 & 5 & -1 & | & 5 \\ 2 & 3 & 2 & | & 3 \end{bmatrix}$$

Add row 2 to row 1.

$$\begin{bmatrix} -1 & 4 & 3 & | & 4 \\ -3 & 5 & -1 & | & 5 \\ 2 & 3 & 2 & | & 3 \end{bmatrix}$$

Multiply row 1 by 2 and add to row 3.

$$\begin{bmatrix} -1 & 4 & 3 & | & 4 \\ -3 & 5 & -1 & | & 5 \\ 0 & 11 & 8 & | & 11 \end{bmatrix}$$

Multiply row 1 by -1.

$$\begin{bmatrix} 1 & -4 & -3 & | & -4 \\ -3 & 5 & -1 & | & 5 \\ 0 & 11 & 8 & | & 11 \end{bmatrix}$$

Multiply row 1 by 3 and add to row 2.

$$\begin{bmatrix} 1 & -4 & -3 & | & -4 \\ 0 & -7 & -10 & | & -7 \\ 0 & 11 & 8 & | & 11 \end{bmatrix}$$

Multiply row 2 by $-1/7$.

$$\begin{bmatrix} 1 & -4 & -3 & | & -4 \\ 0 & 1 & 10/7 & | & 1 \\ 0 & 11 & 8 & | & 11 \end{bmatrix}$$

Multiply row 2 by 4 and add to row 1.

$$\begin{bmatrix} 1 & 0 & 19/7 & | & 0 \\ 0 & 1 & 10/7 & | & 1 \\ 0 & 11 & 8 & | & 11 \end{bmatrix}$$

Multiply row 2 by -11 and add to row 3.

$$\begin{bmatrix} 1 & 0 & 19/7 & | & 0 \\ 0 & 1 & 10/7 & | & 1 \\ 0 & 0 & -54/7 & | & 0 \end{bmatrix}$$

Multiply row 3 by $-7/54$.

$$\begin{bmatrix} 1 & 0 & 19/7 & | & 0 \\ 0 & 1 & 10/7 & | & 1 \\ 0 & 0 & 1 & | & 0 \end{bmatrix}$$

Multiply row 3 by $-19/7$ and add to row 1.
Multiply row 3 by $-10/7$ and add to row 2.

$$\begin{bmatrix} 1 & 0 & 0 & | & 0 \\ 0 & 1 & 0 & | & 1 \\ 0 & 0 & 1 & | & 0 \end{bmatrix}$$

That is,

$$x = 0$$
$$y = 1$$
$$z = 0.$$

Thus, $\{(0, 1, 0)\}$ is the solution set.

8.8 EXERCISES

1. The answer is C since this represents the region where $x \geq 5$ and (at the same time) $y \leq -3$.

3. The answer is B since this represents the region where $x > 5$ and (at the same time) $y < -3$.

5. $x + y \leq 2$
   Graph the boundary line

$$x + y = 2.$$

Since the inequality is of the form " $\leq$," i.e., includes the equality, the graph is a solid line. Next, try a test point not on the line, such as $(0, 0)$, in the inequality.

$$x + y \leq 2$$
$$0 + 0 \leq 2 \quad True$$

Thus, shade the region containing the point $(0, 0)$, and, all the other points in the region below and including the line itself.

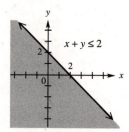

7. $4x - y \leq 5$
   Graph the boundary line

$$4x - y = 5.$$

Try a test point, such as $(0, 0)$, in the inequality.

$$4x - y \leq 5$$
$$0 - 0 \leq 5 \quad True$$

Thus, the half plane including $(0, 0)$ is to be shaded. Observe that the line itself is a part of the solution set.

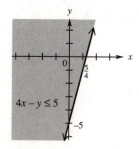

9. $x + 3y \geq -2$
   Graph the boundary line

$$x + 3y = -2$$

Try a test point, such as $(0, 0)$, in the inequality.

$$x + 3y \geq -2$$
$$0 + 3(0) \geq -2 \quad True$$

Thus, the half plane including $(0, 0)$ is to be shaded. The line itself is also a part of the solution set.

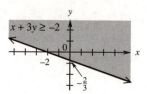

11. $x + 2y \leq -5$
    Graph the boundary line

$$x + 2y = -5.$$

Try a test point, such as $(0, 0)$, in the inequality.

$$x + 2y \leq -5$$
$$0 + 2(0) \leq -5$$
$$0 \leq -5 \quad False$$

Since the point does not work, the other half plane represents the solution and is to be shaded. The line itself is a part of the solution set.

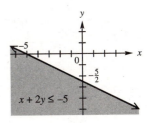

$$x + 2y \leq -5$$

13. $4x - 3y < 12$
Graph the boundary line

$$4x - 3y = 12.$$

Try a test point, such as $(0, 0)$, in the inequality.

$$4x - 3y < 12$$
$$4(0) - 3(0) < 12$$
$$0 < 12 \quad True$$

Thus, the half plane including $(0, 0)$ is to be shaded. The line itself is not a part of the solution set and therefore is indicated with a dashed line.

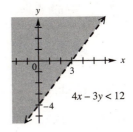

$$4x - 3y < 12$$

15. $y > -x$
Use the alternate equivalent form, $y + x > 0$. Graph the boundary line

$$y + x = 0.$$

Try a test point, such as $(1, 1)$, which does not lie on the line.

$$1 + 1 > 0$$
$$2 > 0 \quad True$$

Thus, the half plane to be shaded is that above the line and includes the point $(1, 1)$. The line must be dashed since "=" is not included in the strict inequality " $>$."

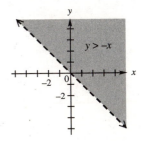

$$y > -x$$

17. $x + y \leq 1$
    $x \geq 0$

Using the boundary equations

$$x + y = 1 \quad and$$
$$x = 0, \ (y\text{-axis}).$$

sketch the graph of each individual inequality, shading the appropriate half planes. The intersection of these regions (area in common) represents the solution set as below.

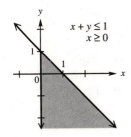

$$x + y \leq 1$$
$$x \geq 0$$

19. $2x - y \geq 1$
    $3x + 2y \geq 6$

Using the boundary equations

$$2x - y = 1 \quad and$$
$$3x + 2y = 6,$$

sketch the graph of each individual inequality, shading the appropriate half planes. The intersection of these regions represents the solution set as shown below.

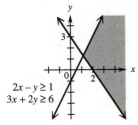

$$2x - y \geq 1$$
$$3x + 2y \geq 6$$

21. $-x - y < 5$
    $x - y \leq 3$

Use the boundary equations

$$-x - y = 5 \quad and$$
$$x - y = 3,$$

to sketch the graph of each individual inequality, shading the appropriate half planes. Note that the line itself is not included with the first inequality. The intersection of these regions represents the solution set as shown below.

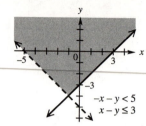

$$-x - y < 5$$
$$x - y \leq 3$$

23. $3x + 5y$.

Evaluate $3x + 5y$ at each vertex.

Point	Value = $3x + 5y$	
$(1, 1)$	$3(1) + 5(1) = 8$	←minimum
$(6, 3)$	$3(6) + 5(3) = 33$	
$(5, 10)$	$3(5) + 5(10) = 65$	←maximum
$(2, 7)$	$3(2) + 5(7) = 41$	

Thus, there is a maximum of 65 at $(5, 10)$ and a minimum of 8 at $(1, 1)$.

25. Find $x \geq 0$ and $y \geq 0$ such that

$$2x + 3y \leq 6$$
$$4x + y \leq 6$$

and $5x + 2y$ is maximized.

To find the vertex points, solve the system of boundary equations.

$$2x + 3y = 6$$
$$4x + y = 6$$

Multiply the second equation by -3 and add to the first equation.

$$2x + 3y = 6$$
$$\underline{-12x - 3y = -18}$$
$$-10x = -12$$
$$x = \frac{-12}{-10} = \frac{6}{5}$$

Substitute for this value for x in the second equation $4x + y = 6$.

$$4\left(\frac{6}{5}\right) + y = 6$$

$$y = 6 - \frac{24}{5}$$

$$y = \frac{30}{5} - \frac{24}{5}$$

$$y = \frac{6}{5}$$

Thus, a corner point is $\left(\frac{6}{5}, \frac{6}{5}\right)$.

Sketch the graph of the feasible region representing the intersection of all of the constraints (i.e., system of inequalities).

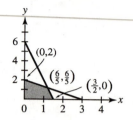

Evaluate $5x + 2y$ at each vertex.

Point	Value = $5x + 2y$	
$(0, 2)$	$5(0) + 2(2) = 4$	
$\left(\frac{6}{5}, \frac{6}{5}\right)$	$5\left(\frac{6}{5}\right) + 2\left(\frac{6}{5}\right) = \frac{42}{5}$	←maximum
$\left(\frac{3}{2}, 0\right)$	$5\left(\frac{3}{2}\right) + 2(0) = \frac{15}{2}$	

Thus, the maximum value occurs at the vertex point $\left(\frac{6}{5}, \frac{6}{5}\right)$ and has the value $\frac{42}{5}$.

27. Find $x \geq 2$ and $y \geq 5$ such that

$$3x - y \geq 12$$
$$x + y \leq 15$$

and $2x + y$ is minimized.

Sketch a graph of the solution to the system (feasible region).

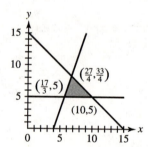

The vertices of the feasible region are the intersection points of each pair of boundary lines. The top vertex point is the intersection of the lines $3x - y = 12$ and $x + y = 15$.

Solve this system.

$$3x - y = 12$$
$$x + y = 15$$

Add the equations to eliminate y.

$$3x - y = 12$$
$$\underline{x + y = 15}$$
$$4x = 27$$
$$x = \frac{27}{4}$$

Substitute this value of x into the second equation $x + y = 15$ and solve for y.

$$\left(\frac{27}{4}\right) + y = 15$$
$$y = 15 - \frac{27}{4}$$
$$y = \frac{60}{4} - \frac{27}{4} = \frac{33}{4}$$

The resulting vertex point is $\left(\frac{27}{4}, \frac{33}{4}\right)$.

The bottom left vertex can be found by solving the system.

$$3x - y = 12$$
$$y = 5$$

Substitute this value of y into the first equation $3x - y = 12$ and solve for x.

$$3x - (5) = 12$$
$$3x = 17$$
$$x = \frac{17}{3}$$

This gives the vertex $\left(\frac{17}{3}, 5\right)$.

To find the remaining vertex solve the following system.
$$x + y = 15$$
$$y = 5$$

Substitute this value of y into the first equation $x + y = 15$ and solve for x.

$$x + (5) = 15$$
$$x = 10$$

This gives the vertex $(10, 5)$.

Evaluate $2x + y$ at each vertex.

Point	Value $= 2x + y$
$\left(\frac{27}{4}, \frac{33}{4}\right)$	$2\left(\frac{27}{4}\right) + \left(\frac{33}{4}\right) = \frac{87}{4} = 21.75$
$\left(\frac{17}{3}, 5\right)$	$2\left(\frac{17}{3}\right) + (5) = \frac{49}{3} \approx 16.33$ ←minimum
$(10, 5)$	$2(10) + (5) = 25$

Thus, the vertex point $\left(\frac{17}{3}, 5\right)$ gives the minimum value, $\frac{49}{3}$.

Note: in the following exercises $x \geq 0$ and $y \geq 0$ are understood constraints.

29. Let $x =$ the number of refrigerators shipped to warehouse A, and
$y =$ the number of refrigerators shipped to Warehouse B.

Since at least 100 refrigerators must be shipped,
$$x + y \geq 100.$$

Since warehouse A holds a maximum of 100 refrigerators and has 25 already, it has room for, at most, 75 more, so
$$0 \leq x \leq 75.$$

Similarly, warehouse B has room for, at most, 80 more refrigerators, so
$$0 \leq y \leq 80.$$

The cost function is given by $12x + 10y$.

The linear programming problem may be stated as follows:

Find $x \geq 0$ and $y \geq 0$ such that

$$x + y \geq 100$$
$$x \leq 75$$
$$y \leq 80$$

and $12x + 10y$ is minimized.

Graph the feasible region.

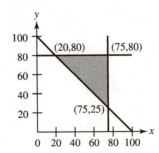

Evaluate the objective function at each vertex.

Point	Cost $= 12x + 10y$	
$(20, 80)$	$12(20) + 10(80) = 1040$	←minimum
$(75, 80)$	$12(75) + 10(80) = 1700$	
$(75, 25)$	$12(75) + 10(25) = 1150$	

The minimum value of the objective function is 1040 occurring at $(20, 80)$. Therefore, 20 refrigerators should be shipped to warehouse A and 80 to warehouse B, for a minimum cost of $1040.

31. Let $x =$ the number of red pills and
     $y =$ the number of blue pills.

Use a table to summarize the given information.

	Vitamin A	Vitamin B_1	Vitamin C
Red pills	8	1	2
Blue pills	2	1	7
Daily requirement	16	5	20

The linear programming problem may be stated as follows:

Find $x \geq 0$ and $y \geq 0$ such that

$$8x + 2y \geq 16 \quad \text{(Vitamin A)}$$
$$x + y \geq 5 \quad \text{(Vitamin } B_1)$$
$$2x + 7y \geq 20 \quad \text{(Vitamin C)}$$

and the Cost $= .1x + .2y$ is minimized.

Solve the corresponding equations to find the vertex points and graph the feasible region.

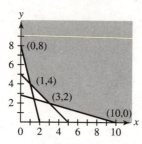

Evaluate the objective function at each vertex.

Point	Cost $= .1x + .2y$
$(0, 8)$	$.1(0) + .2(8) = 1.6$
$(1, 4)$	$.1(1) + .2(4) = .9$
$(3, 2)$	$.1(3) + .2(2) = .7$ ←minimum
$(10, 0)$	$.1(10) + .2(0) = 1$

The minimum value of the objective function is .7 (i.e., 70¢) occurring at $(3, 2)$. Therefore, she should take 3 red pills and 2 blue pills, for a minimum cost of 70¢ per day.

33. Let $x =$ number of barrels of gasoline and
     $y =$ number of barrels of oil.

The linear programming problem may be stated as follows:

Find $x \geq 0$ and $y \geq 0$ with the following constraints.

$$x \geq 2y$$
$$y \geq 3 \text{ million}$$
$$x \leq 6.4 \text{ million}$$

where Revenue $= 1.9x + 1.5y$ and is to be maximized.

Solve the corresponding system of boundary equations for their points of intersection to find the vertices and sketch the graph representing the intersection of the constraints.

The vertices of the feasible region are

$$(6.4, 3), (6, 3) \text{ and } (6.4, 3.2).$$

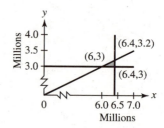

Evaluate the objective function at each vertex.

Point	Revenue $= 1.9x + 1.5y$	(in millions)
$(6.4, 3)$	$1.9(6.4) + 1.5(3) = 16.66$	
$(6, 3)$	$1.9(6) + 1.5(3) = 15.9$	
$(6.4, 3.2)$	$1.9(6.4) + 1.5(3.2) = 16.96$	←maximum

Thus, producing 6.4 million barrels of gasoline and 3.2 million barrels of fuel oil should yield the maximum revenue of $16,960,000.

35. Let $x =$ number of medical kits and
     $y =$ number of containers of water.

Use a table to summarize the given information.

	Volume	Weight
Medical Kits	1	10
Containers of water	1	20
Maximum allowed	6000	80000

Weight constraint:

$$10x + 20y \leq 80000.$$

Volume constraint:

$$(1)x + (1)y \leq 6000.$$

Objective function:

$$6x + 10y$$

The problem may be stated as follows:

For $x \geq 0$ and $y \geq 0$,

$$10x + 20y \leq 80000 \quad \text{(Pounds)}$$
$$x + y \leq 6000. \quad \text{(Cubic feet)}$$

These constraints may be written in the simpler equivalent form

$$x + 2y \leq 8000$$
$$x + y \leq 6000.$$

The number of people aided $= 6x + 10y$ and is to be maximized.

Solve the corresponding equations for their points of intersection to find the vertices and sketch the graph representing the intersection of the constraints.

The vertex points are

$$(0, 4000), (4000, 2000) \text{ and } (6000, 0).$$

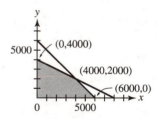

Evaluate the objective function at each vertex.

Point	people aided $= 6x + 10y$
$(0, 4000)$	$6(0) + 10(4000) = 40000$
$(4000, 2000)$	$6(4000) + 10(2000) = 44000 \leftarrow$ maximum
$(6000, 0)$	$6(6000) + 10(0) = 36000$

Thus, ship 4000 medical kits and 2000 containers of water in order to maximize the number of people aided.

CHAPTER 8 TEST

1. $d = \sqrt{(x_2 - x_1)^2 + (y_2 - y_1)^2}$

$\quad = \sqrt{(2 - (-3))^2 + (1 - 5)^2}$

$\quad = \sqrt{(5)^2 + (-4)^2}$

$\quad = \sqrt{25 + 16}$

$\quad = \sqrt{41}$

3. $3x - 2y = 8$

To find the x-intercept, let $y = 0$.

$$3x - 2(0) = 8$$
$$3x = 8$$
$$x = \frac{8}{3}$$

The x-intercept is $\left(\frac{8}{3}, 0\right)$.

To find the y-intercept, let $x = 0$.

$$3(0) - 2y = 8$$
$$-2y = 8$$
$$y = -4$$

The y-intercept is $(0, -4)$.

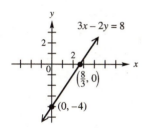

5. The slope-intercept form of the equation of the line
    (a) passing through the point $(-1, 3)$, with slope $-2/5$.

Use the point-slope form of the line to write the equation:

$$y - y_1 = m(x - x_1)$$
$$y - 3 = -\frac{2}{5}[x - (-1)]$$
$$y = -\frac{2}{5}x - \frac{2}{5} + 3$$
$$y = -\frac{2}{5}x + \frac{13}{5}.$$

(b) passing through $(-7, 2)$ and perpendicular to $y = 2x$.

Use the slope of the given line, 2, to find the slope of any perpendicular line, $-\frac{1}{2}$. Use the point-slope form of the line to write the equation:

$$y - y_1 = m(x - x_1), \text{ or}$$
$$y - 2 = -\frac{1}{2}[x - (-7)]$$
$$y = -\frac{1}{2}x - \frac{7}{2} + 2$$
$$y = -\frac{1}{2}x - \frac{3}{2}.$$

(c) the line shown in the textbook displays.

The line displays show two points on the line, $(-2, 3)$ and $(6, -1)$. Use these points to create the slope of the line:

$$m = \frac{y_2 - y_1}{x_2 - x_1}$$
$$= \frac{-1 - 3}{6 - (-2)}$$
$$= \frac{-4}{8}$$
$$= -\frac{1}{2}.$$

Use this slope and either known point to write the equation:

$$y - y_1 = m(x - x_1)$$
$$y - 3 = -\frac{1}{2}[x - (-2)]$$
$$y = -\frac{1}{2}x - 1 + 3$$
$$y = -\frac{1}{2}x + 2.$$

7. (a) Choose the data points represented by the two years, $(3, 21696)$ and $(7, 25050)$ to create slope.

$$m = \frac{y_2 - y_1}{x_2 - x_1}$$
$$= \frac{25050 - 21696}{7 - 3}$$
$$= \frac{3354}{4}$$
$$= 838.5$$

Use the slope, the point–slope form of the line, and either data point to write the equation of the line.

$$y - y_1 = m(x - x_1), \text{ or}$$
$$y - 21696 = 838.5(x - 3)$$
$$y - 21696 = 838.5x - 838.5(3)$$
$$y = 838.5x - 2515.5 + 21696$$
$$y = 838.5x + 19180.5$$

(b) To approximate median income for 1995, let $x = 5$ and solve for y.

$$y = 838.5(5) + 19180.5$$
$$= 4192.5 + 19180.5$$
$$= 23373$$

The predicted value, \$23,373 is a bit less than the actual value.

9. Estimate the values for two points on the line such as $(0, 1)$ and $(3, 3)$. Use these values to create the slope.

$$m = \frac{y_2 - y_1}{x_2 - x_1}$$
$$= \frac{3 - 1}{3 - 0}$$
$$= \frac{2}{3}$$

Because one of our points is the y-intercept, $(0, 1)$, we may choose to use the slope-intercept form of the line to write the equation.

$$y = mx + b$$
$$y = \frac{2}{3}x + 1$$

11. Given the function $f(x) = \frac{2}{x-3}$, then the domain is $(-\infty, 3) \cup (3, \infty)$.

13. $f(x) = -(x + 3)^2 + 4$

By inspection of the given standard form of the equation, the axis is at $x = -3$ and the vertex is at $(-3, 4)$. The domain is $(-\infty, \infty)$ and the range is $(-\infty, 4]$.

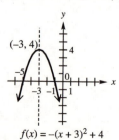

$f(x) = -(x + 3)^2 + 4$

15. (a) $5.1^{4.7} \approx 2116.31264888$

(b) $e^{-1.85} \approx .157237166314$

(c) $\ln 23.56 \approx 3.15955035878$

17. $P = \$12,000$, $n = 4$, $r = 4\% = .04$, and $t = 3$ years.
(a) Use the compounding formula
$A = P\left(1 + \frac{r}{n}\right)^{nt}$:

$$A = 12000\left(1 + \frac{.04}{4}\right)^{4 \cdot 3}$$
$$= 12000(1.01)^{12}$$
$$\approx \$13,521.90.$$

(b) Use the compounding formula
$A = Pe^{rt}$ for continuous compounding:

$$A = 12000e^{(.04)3}$$
$$= 12000e^{.12}$$
$$\approx \$13,529.96.$$

19. $2x + 3y = 2$
$3x - 4y = 20$

Multiply the first equation by 4 and the second by 3 then add.

$$8x + 12y = 8$$
$$\underline{9x - 12y = 60}$$
$$17x = 68$$
$$x = 4$$

Substitute this value for x into the first equation and solve for x.

$$2(4) + 3y = 2$$
$$3y = 2 - 8$$
$$3y = -6$$
$$y = -2$$

Since the ordered pair $(4, -2)$ satisfies both equations, it checks. The solution set is $\{(4, -2)\}$.

21. $2x + 3y - 6z = 11$
$x - y + 2z = -2$
$4x + y - 2z = 7$

Eliminate y and z from the second and third equations by adding.

$$x - y + 2z = -2$$
$$\underline{4x + y - 2z = 7}$$
$$5x = 5$$
$$x = 1$$

It is not possible to find a unique value for y and z corresponding to the value $x = 1$. This means that the system is dependent and has an infinite number of answers.

Therefore, replace $x = 1$ into the third equation (this may be done in any of the original three equations) and solve for y in terms of z.

$$4(1) + y - 2z = 7$$
$$y - 2z = 7 - 4$$
$$y = 2z + 3$$

The solution set may now be expressed as $\{(1, 2z + 3, z)\}$.

23. Let $x =$ the sale price with a 10% commission,
$y =$ the sale price with a 6% commission and
$z =$ the sale price with a 5% commission.

The total property sold is then given by

$$x + y + z = 280,000.$$

The total commission is given by

$$.10x + .06y + .05z = 17,000.$$

In addition there is the following relationship between the sales

$$z = x + y.$$

Simplifying the second equation and writing all in standard form results in the system

$$x + y + z = 28,0000$$
$$10x + 6y + 5z = 1,700,000$$
$$-x - y + z = 0.$$

Adding the first and last equation results in the elimination of x and y terms.

$$x + y + z = 280,000$$
$$\underline{-x - y + z = 280,000}$$
$$2z = 280,000$$
$$z = 140,000$$

Multiply the first equation $x + y + z = 280000$ by -6 and add to the second equation to eliminate the y terms.

$$-6x - 6y - 6z = -1,680,000$$
$$\underline{10x + 6y + 5z = 1,700,000}$$
$$4x - z = 20,000$$

Replace the value for z in this last equation and solve for x.

$$4x - 140,000 = 20,000$$
$$4x = 160,000$$
$$x = 40,000$$

Replace x and z in the equation $z = x + y$ by there values and solve for y.

$$140,000 = 40,000 + y$$
$$140,000 - 40,000 = y$$
$$y = 100,000$$

Thus, Keshon Grant sold a property for $40,000 with a 10% commission, another property for $100,000 with a 6% commission and a third for $140,000 with a 5% commission.

25. Let $x = $ the number of VIP rings produced, and $y = $ the number of SST rings produced.

Since the company can produce no more than 24 rings in a day, the constraint is represented by:

$$x + y \leq 24.$$

Since the company can afford no more than 60 hours of labor in a day, the constraint is represented by

$$3x + 2y \leq 60.$$

The profit function is given by $20x + 40y$.

The linear programming problem may be stated as follows:

Find $x \geq 0$ and $y \geq 0$ such that

$$x + y \leq 24$$
$$3x + 2y \leq 60$$

and Profit $= 20x + 40y$ is maximized.

Solve the corresponding equations for their points of intersection to find the vertices and sketch the graph representing the intersection of the constraints.

To find the vertex points, solve the system of boundary equations.

$$x + y = 24$$
$$3x + 2y = 60$$

Multiply the first equation by -2 and add to the second equation.

$$-2x - 2y = -48$$
$$\underline{3x + 2y = 60}$$
$$x = 12$$

Substitute for this value for x in the first equation $x + y = 24$.

$$12 + y = 24$$
$$y = 12$$

Thus, a corner point (or vertex) is $(12, 12)$.

The vertices of the feasible region are

$$(0, 24), (12, 12) \text{ and } (20, 0).$$

Sketch the graph of the feasible region representing the intersection of all of the constraints.

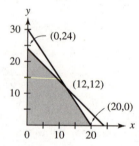

Evaluate the objective function at each vertex.

Point	Profit $= 20x + 40y$	
$(0, 24)$	$20(0) + 40(24) = 960$	←maximum
$(12, 12)$	$20(12) + 40(12) = 720$	
$(20, 0)$	$20(20) + 40(0) = 400$	

Thus, to maximize its profit the Alessic company should design and sell 0 VIP rings and 24 SST rings for a profit of $960.

9 GEOMETRY

Chapter Goals

Upon completion of this chapter, the student should be able to

- Symbolize lines, half-lines, rays, and line segments.
- Classify angles.
- Understand relationships between angles.
- Identify various types of curves and polygons.
- Apply the Pythagorean theorem.
- Find perimeter and area of plane figures.
- Find volume and surface area of three-dimensional figures.
- Understand relationships between similar triangles.
- Understand the differences between Euclidean and Non-Euclidean geometries.
- Determine topological equivalency.
- Determine the genus of an object.
- Determine if a network is traversable.

Chapter Summary

The study of geometry often begins with undefined terms such as point, line, and plane in addition to ways to symbolize them. The chart in Section 9.1 gives a summary of figures and symbols for lines, half-lines, rays and line segments.

An angle is the union of two rays that have a common endpoint. Angles are classified according to their degree measurement. An acute angle measures between 0° and 90°; a right angle measures exactly 90°; an obtuse angle measures between 90° and 180°.

The terms complementary and supplementary refer to relationships between angles. Two angles are complementary if the sum of their degree measures equals 90°; two angles are supplementary if the sum of their degree measures equals 180°. The term vertical is also used to describe angle relationships. When two lines intersect, the angles that are formed that are directly opposite each other are called vertical angles. They have the same degree measure.

Curve is another term that is often undefined; yet we can draw and describe one. In geometry we are concerned with simple curves that can be drawn without lifting the pencil from the paper, without passing through any point twice; and with closed curves, which have their starting and ending points the same and are also drawn without lifting the pencil from the paper.

A simple closed curve made up of only straight lines segments is called a polygon. Several kinds of polygons and their properties can be studied. The chart in the text in Section 9.2 gives the names of several polygons classified according to the number of sides. Also, the types of quadrilaterals, which are four-sided polygons are described in Section 9.2.

A triangle, which is a three-sided polygon, has several interesting characteristics. One of these characteristics is that the sum of the measures of the angles is always 180°. Triangles can be classified according to their angle measures or according to their sides. If a triangle is named according to its angle measures, it is either right, acute, or obtuse. If it is named according to the length of its sides, it is equilateral if all sides are equal; scalene if all sides are unequal; or isosceles if two sides are equal in length.

A circle is a special simple closed curve. It is defined as the set of points in a plane, each of which is the same distance from a fixed point. Several terms that are related to circles, such as radius, diameter, chord, secant line, and arc are also described in Section 2.

Finding perimeter and area is another important topic in the study of geometry. Formulas for the perimeter and area of several figures can be found throughout Section 3 and in the Chapter 9 summary in the text. Here is one example of using a perimeter formula to find the length of one side of a square.

EXAMPLE A   Perimeter of a square

A stained-glass window in a church is in the shape of a square. The perimeter of the square is 7 times the length of a side in meters, decreased by 12. Find the length of a side of the window.

Solution
The formula for perimeter of a square is $P = 4s$. Translating the second sentence and replacing P with $4s$ yields

$$P = 4s = 7s - 12.$$

Solve the equation for s.

$$4s = 7s - 12$$
$$4s + 12 = 7s - 12 + 12$$
$$4s + 12 = 7s$$
$$4s - 4s + 12 = 7s - 4s$$
$$12 = 3s$$
$$\frac{12}{3} = \frac{3s}{3}$$
$$4 = s$$

The length of the side is 4 m.

EXAMPLE B   Area of a triangle

Find the area of a triangle that has a base of 22 mm and a height of 38 mm.

Solution
$$A = \frac{1}{2}bh$$
$$= \frac{1}{2} \cdot 22 \cdot 38$$
$$= 418 \text{ mm}^2$$

Triangles that have exactly the same size and the same shape are called congruent. The congruence properties of triangles are described in Section 4. They are referred to as SAS for Side-Angle-Side; ASA for Angle-Side-Angle; and SSS for Side-Side-Side. In brief, this mean that two triangles are congruent if these particular parts of each triangle are equal in measure. For example, if two sides and the included angle of one triangle are equal to two sides and the included angle of another triangle, then the two triangles are congruent by the SAS congruence.

Similar triangles have the same shape but not necessarily the same size. Corresponding angles have the same measure and corresponding sides are proportional. These relationships can be used to solve problems such as the following.

EXAMPLE C   Similar triangles

Given a triangle with sides 10 cm, 20 cm, and 25 cm, find the lengths of the missing sides of a similar triangle that has longest side 75 cm.

Solution
Let $x =$ length of the shortest side. Computations are easier if the fraction on the right is simplified.

$$\frac{10}{x} = \frac{25}{75}$$
$$\frac{10}{x} = \frac{25 \cdot 1}{25 \cdot 3}$$
$$\frac{10}{x} = \frac{1}{3}$$
$$x \cdot 1 = 10 \cdot 3$$
$$x = 30 \text{ cm}$$

Let $y =$ length of the medium side.

$$\frac{20}{y} = \frac{25}{75}$$
$$\frac{20}{y} = \frac{25 \cdot 1}{25 \cdot 3}$$
$$\frac{20}{y} = \frac{1}{3}$$
$$y \cdot 1 = 20 \cdot 3$$
$$y = 60 \text{ cm}$$

Right triangles have a special property called the Pythagorean property or the Pythagorean theorem. In words it can be stated, the sum of the squares of the two shorter sides is equal to the square of the longest side. The longest side is called the hypotenuse. The formula is $a^2 + b^2 = c^2$; a and b represent the lengths of the two shorter sides, and c is the length of the hypotenuse. Here is an example to determine whether a triangle is a right triangle.

EXAMPLE D Pythagorean property

The lengths of the sides of a triangle are 6, 8, and 10. Is this a right triangle?

Substitute the given values into the Pythagorean theorem to see if it makes a true statement.

$$a^2 + b^2 = c^2$$
$$6^2 + 8^2 = 10^2$$
$$36 + 64 = 100$$

The sum of the squares of the two shorter sides is equal to the square of the hypotenuse. Therefore it is a right triangle.

The Pythagorean theorem can be applied to real-world settings such as the following:

EXAMPLE E Application of Pythagorean property

Robert wants to construct a TV antenna on top of the family room of his house. The family room has a flat roof and the antenna is 8 feet tall. He will use three cables of equal length attached to the top of the antenna and placed 6 feet from the base of the antenna. What is the total length of these three cables?

Solution
To find the length of one cable use the Pythagorean theorem.
Let x = length of one cable.

$$x^2 = 8^2 + 6^2$$
$$x^2 = 64 + 36$$
$$x^2 = 100$$
$$\sqrt{x^2} = \sqrt{100}$$
$$x = 10$$

Then the total length of the three cables is $3 \cdot 10 = 30$ feet.

Formulas for the volume of several three-dimensional figures can be found in Section 5 and also in the Chapter 9 summary. To find the volume of a sphere with a radius of 7.4 cm, consider this example.

EXAMPLE F Volume of a sphere

Sphere: radius = 7.4 cm

Here is the solution using 3.14 for the value of π.

$$V = \frac{4}{3}\pi r^3$$
$$= \frac{4}{3}(3.14)(7.4)^3$$
$$= \frac{4}{3}(3.14)(405.224)$$
$$= \frac{4}{3}(1272.4034)$$
$$\approx 1696.5 \text{ cm}^3$$

EXAMPLE G Volume and surface area of a box

Given a box with the following dimensions, find its (a) volume and (b) surface area.

$l = 6$ in, $w = 5$ in, $h = 3.2$ in

(a) $V = lwh$
$\quad = 6 \cdot 5 \cdot 3.2$
$\quad = 96 \text{ in}^3$

(b) $S = 2lh + 2hw + 2lw$
$\quad = 2(6)(3.2) + 2(3.2)(5) + 2(6)(5)$
$\quad = 38.4 + 32 + 60$
$\quad = 130.4 \text{ in}^2$

In Section 6 three topics are explored: non-Euclidean geometry, topology, and networks. Topology is an interesting branch of geometry that classifies objects according to their genus, which describes the number of holes in an object. Graph theory is another branch of modern geometry in which networks are studied. A network is a diagram showing the various paths between points. Leonard Euler solved the now-famous Koenigsburg Bridge problem, which was an early introduction to networks. A network is said to be traversable if a route can be found that "crosses each bridge" only once in finding a path from a beginning point to an ending point.

Finally, in Section 7, the new studies of chaos and fractals are described. The word fractal means "fractional dimension."

9.1 EXERCISES

1. The sum of the measures of two complementary angles is <u>90</u> degrees.

3. The measures of two vertical angles are <u>equal</u>.

5. It is true that a line segment has two endpoints.

7. It is false that if A and B are distinct points on a line, then ray AB and ray BA represent the same set of points. The initial point of ray AB is point A; the initial point of ray BA is point B.

9. It is true that if two lines are parallel, they lie in the same plane.

11. It is true that segment AB and segment BA represent the same set of points. A and B are the endpoints of the line segment and can be names in either order.

13. It is false that there is no angle that is its own supplement. A $90°$ or right angle is supplementary to a $90°$ angle.

15. (a) $\overset{\leftrightarrow}{AB}$ (b) A •————————• B

17. (a) $\overset{\leftrightarrow}{CB}$ (b) ◄————•——•————•————► A B   C

19. (a) $\overset{\rightarrow}{BC}$ (b) ○————•———•————► B   C   D

21. (a) $\overset{\leftrightarrow}{BA}$ (b) ◄————————•——————• A   B

23. (a) $\overset{\leftrightarrow}{CA}$ (b) •————•————• A   B   C

25. Letter F. Line segment PQ is the same as line segment QP.

27. Letter D. Ray QR names the same set of points as ray QS. The initial point is Q, and the set of points passes through both R and S.

29. Letter B. Half-line RP is the same as half-line RQ. The initial point for both is R.

31. Letter E. Line segment PS is the same as line segment SP.

33. $\overset{\leftrightarrow}{MN} \cup \overset{\leftrightarrow}{NO}$ names the same set of points as $\overset{\leftrightarrow}{MO}$. The union symbol joins the two line segments.

35. $\overset{\leftrightarrow}{MO} \cap \overset{\leftrightarrow}{OM}$ indicates the intersection or overlap of the same line segment. Therefore a simpler way is either $\overset{\leftrightarrow}{MO}$ or $\overset{\leftrightarrow}{OM}$.

37. $\overset{\rightarrow}{OP} \cap O$ have no points in common because point O is not part of the half-line $\overset{\rightarrow}{OP}$. Therefore the intersection is the empty set, symbolized by $\emptyset$.

39. $\overset{\leftrightarrow}{NP} \cap \overset{\leftrightarrow}{OP}$ indicates the same set of points as $\overset{\leftrightarrow}{OP}$.

41. $90 - 28 = 62°$

43. $90 - 89 = 1°$

45. $(90 - x)°$

47. $180 - 132 = 48°$

49. $180 - 26 = 154°$

51. $(180 - y)°$

53. $\angle ABE$ and $\angle CBD$; $\angle ABD$ and $\angle CBE$

55. (a) $52°$. They are vertical angles.

(b) $180 - 52 = 128°$. They are supplementary angles.

57. The designated angles are supplementary; their sum is $180°$.
$$(10x + 7) + (7x + 3) = 180$$
$$17x + 10 = 180$$
$$17x = 170$$
$$x = 10$$

Then one angle measure is $10 \cdot 10 + 7 = 107°$ and the other angle measure is $7 \cdot 10 + 3 = 73°$.

59. The angles are vertical so they have the same measurement. Set the algebraic expressions each to each other.
$$3x + 45 = 7x + 5$$
$$-4x = -40$$
$$x = 10$$

Then one angle measure is $3 \cdot 10 + 45 = 75°$ and the other angle measure is $7 \cdot 10 - 5 = 75°$.

61. The angles are vertical so they have the same measurement. Set the algebraic expressions each to each other.
$$11x - 37 = 7x + 27$$
$$4x = 64$$
$$x = 16$$

Then one angle measure is $11 \cdot 16 - 37 = 139°$ and the other angle measure is $7 \cdot 16 + 27 = 139°$.

63. The designated angles are supplementary; their sum is $180°$.
$$(3x + 5) + (5x + 15) = 180$$
$$8x + 20 = 180$$
$$8x = 160$$
$$x = 20$$

Then one angle measure is $3 \cdot 20 + 5 = 65°$ and the other angle measure is $5 \cdot 20 + 15 = 115°$.

65. The designated angles are complementary; their sum is 90°.

$$(5k + 5) + (3x + 5) = 90$$
$$8k + 10 = 90$$
$$8k = 80$$
$$k = 10$$

Then one angle measure is $5 \cdot 10 + 5 = 55°$ and the other angle measure is $3 \cdot 10 + 5 = 35°$.

67. Alternate interior angles have equal measures.

$$2x - 5 = x + 22$$
$$x = 27$$

Then the measure of each angle is $2 \cdot 27 - 5 = 49°$, and $27 + 29 = 49°$.

69. The angles are supplementary; their sum is 180°.

$$(x + 1) + (4x - 56) = 180$$
$$5x - 55 = 180$$
$$5x = 235$$
$$x = 47$$

Then one angle measure is $47 + 1 = 48°$ and the other angle measure is $4 \cdot 47 - 56 = 132°$.

71. Let $x =$ measure of the angle,
$180 - x =$ measure of its supplement,
$90 - x =$ measure of its complement.

$$180 - x = 25 + 2(90 - x)$$
$$180 - x = 25 + 180 - 2x$$
$$180 - x = 205 - 2x$$
$$180 - x + 2x = 205 - 2x + 2x$$
$$180 + x = 205$$
$$180 - 180 + x = 205 - 180$$
$$x = 25°$$

73. Let $x =$ the measure of the angle,
$180 - x =$ measure of its supplement,
$90 - x =$ measure its complement.

$$(180 - x) + (90 - x) = 210$$
$$270 - 2x = 210$$
$$270 - 270 - 2x = 210 - 270$$
$$-2x = -60$$
$$\frac{-2x}{-2} = \frac{-60}{-2}$$
$$x = 30°$$

75. Writing exercise

77. Some of the unknown angles must be solved before other unknown angles. Here is one order that they can be solved.
$\angle 1 = 55°$; vertical angle to 55°.
$\angle 8 = 180 - 120 = 60°$; supplementary angle to 120°.
$\angle 6 = 180 - 60 = 120°$; supplementary angle to $\angle 8$.
$\angle 7 = 60°$; vertical angle to $\angle 8$.
$\angle 5 = 60°$; alternate interior angle to $\angle 7$.
$\angle 3 = 60°$; vertical angle to $\angle 5$.
$\angle 2 = 180 - (55 + 60) = 180 - 55 - 60 = 65°$.
Angles 2, 3, and 55 all add to 180°.
$\angle 4 = 180 - (55 + 60) = 180 - 55 - 60 = 65°$;
straight angle composed of $\angle 4$, $\angle 5$, and 55°.
$\angle 10 = 55°$; alternate interior angle to the 55° angle.
$\angle 9 = 55°$; vertical angle to $\angle 10$.

79. Refer to the figure for Exercise 66 (in the text). The interior angles are $\angle 3$, $\angle 4$, $\angle 5$, and $\angle 6$. The interior angles on the same side of the transversal are $\angle 4$ and $\angle 6$, and $\angle 3$ and $\angle 5$.

(a) $\angle 3 = \angle 2$; vertical angles.

(b) Measure of $\angle 3 =$ measure of $\angle 6$; alternate interior angles.

(c) Measure of $\angle 2 =$ measure of $\angle 6$; by parts (a) and (b) above.

(d) $\angle 2$ and $\angle 4$ are supplementary, by definition.

(e) $\angle 6$ and $\angle 4$ are supplementary, by parts (c) and (d) above.

The result can be shown similarly for $\angle 3$ and $\angle 5$. Therefore interior angles on the same side of a transversal are supplementary.

9.2 EXERCISES

1. A segment joining two points on a circle is called a(n) <u>chord</u>.

3. A regular triangle is called a(n) <u>equilateral</u> triangle.

5. False. A rhombus does not have equal angle measures.

7. False. The sum of the angle measures of a triangle always equals 180°. If a triangle had two obtuse angles, then the sum of the measure of the angles would exceed 180° which is impossible. Therefore a triangle has at most one obtuse angle.

9. True. A square is a rhombus with four 90° angles.

11. Writing exercise.

13. Both. It is closed and there are no intersecting curves.

15. Closed

17. Closed

19. Neither

21. Convex

23. Convex

25. Not convex

27. Right; scalene

29. Acute; equilateral

31. Right; scalene

33. Right; isosceles

35. Obtuse; scalene

37. Acute; isosceles

39. Writing exercise

41. Writing exercise

43. The sum of the measures of the three angles of any triangle is always 180.

$$x + (x + 20) + (210 - 3x) = 180$$
$$-x + 230 = 180$$
$$-x = -50$$
$$x = 50$$

The measure of angle A is 50°; angle B is $50 + 20 = 70$°; angle C is $210 - 3 \cdot 50 = 60$°.

45.
$$(x - 30) + (2x - 120) + \left(\frac{1}{2}x + 15\right) = 180$$
$$3\frac{1}{2}x - 135 = 180$$
$$\frac{7}{2}x = 315$$
$$\frac{2}{7} \cdot \frac{7}{2}x = 315 \cdot \frac{2}{7}$$
$$x = 90$$

The measure of angle A is $90 - 30 = 60$°; angle B is $2 \cdot 90 - 120 = 60$°; angle C is $\frac{1}{2} \cdot 90 + 15 = 60$°.

47. Let $x =$ the angle measure of A (or B)
$x + 24 =$ the angle measure of C

$$x + x + (x + 24) = 180$$
$$3x + 24 = 180$$
$$3x = 156$$
$$x = 52$$

The measure of angle A (or B) is 52°; the measure of angle C is $52 + 24 = 76$°.

49. The measure of the exterior angle of a triangle is equal to the sum of the measures of the two opposite interior angles.

$$10x + (15x - 10) = 20x + 25$$
$$25x - 10 = 20x + 25$$
$$25x - 20x - 10 = 20x - 20x + 25$$
$$5x - 10 = 25$$
$$5x - 10 + 10 = 25 + 10$$
$$5x = 35$$
$$x = 7$$

Then $(20 \cdot 7) + 25 = 165$°.

51.
$$(2 - 7x) + (100 - 10x) = 90 - 20x$$
$$-17x + 102 = 90 - 20x$$
$$-17x + 102 - 102 = 90 - 102 - 20x$$
$$-17x = -12 - 20x$$
$$3x = -12$$
$$x = -4$$

Then $90 - 20 \cdot (-4) = 90 + 80 = 170$°.

53. Writing exercise

55. According to figure 19, angles 3 and 6 are supplementary; the sum of the measures of their angles is equal to 180°. Also the sum of the interior measures of the angles of a triangle is always equal to 180°. Therefore, the sum of the measures of angles 1 and 2 must be equivalent to the measure of angle 6.

9.3 EXERCISES

1. The perimeter of an equilateral triangle with side length equal to <u>12</u> inches is the same as the perimeter of a rectangle with length 10 inches and width 8 inches. The perimeter of the rectangle is $2 \cdot 10 + 2 \cdot 8 = 36$. The perimeter of the triangle is also 36. If all three sides must have the same length, the one side has length $36 \div 3 = 12$ inches.

3. If the area of a certain triangle is 24 square inches, and the base measures 8 inches, then the height must measure <u>6</u> inches. The formula for the area of a triangle is $A = \frac{1}{2}bh$. Substitute the given values into the formula and solve for h.

$$24 = \frac{1}{2} \cdot 8h$$
$$24 = 4h$$
$$6 = h$$

5. The area of an equilateral triangle with side length 6 inches is $9\sqrt{3}$ square inches. Use Heron's area formula with $a = b = c = 6$. Then $s = \frac{1}{2}(6 + 6 + 6) = 9$:

$$A = \sqrt{s(s-a)(s-b)(s-c)}$$
$$= \sqrt{9(9-6)(9-6)(9-6)}$$
$$= \sqrt{9(3)(3)(3)}$$
$$= \sqrt{243}$$
$$= \sqrt{81 \cdot 3}$$
$$= 9\sqrt{3}$$

7. $A = lw$
$A = 4 \cdot 3$
$A = 12 \, \text{cm}^2$

9. $A = lw$
$A = 2\frac{1}{2} \cdot 2$
$A = 5 \, \text{cm}^2$

11. $A = bh$
$A = 4 \cdot 2$
$A = 8 \, \text{in.}^2$

13. $A = bh$
$A = 3 \cdot 1.5$
$A = 4.5 \, \text{cm}^2$

15. $A = \frac{1}{2}bh$
$A = \frac{1}{2} \cdot 22 \cdot 38$
$A = \frac{1}{2} \cdot \frac{22}{1} \cdot \frac{38}{1}$
$A = 418 \, \text{mm}^2$

17. $A = \frac{1}{2}h(b + B)$
$A = \frac{1}{2} \cdot 2(3 + 5)$
$A = 1(8)$
$A = 8 \, \text{cm}^2$

19. $A = \pi r^2$
$A = (3.14)(1)^2$
$A = 3.14 \, \text{cm}^2$

21. The diameter is 36, so the radius is 18 m.

$$A = \pi r^2$$
$$A = (3.14)(18)^2$$
$$A = (3.14)(324)$$
$$A = 1017.36 \, \text{m}^2$$

23. Let s = length of a side of the window. Use the formula $P = 4s$. Replace P with $7s - 12$ and solve for s.

$$P = 4s$$
$$7s - 12 = 4s$$
$$7s - 7s - 12 = 4s - 7s$$
$$-12 = -3s$$
$$\frac{-12}{-3} = \frac{-3s}{-3}$$
$$4 = s$$

The length of a side of the window is 4 m.

25. The formula for perimeter of a triangle is $P = a + b + c$. Translating the problem, let a be the shortest side. Then $b = 100 + a$ and $c = 200 + a$. Replace a, b, and c in the formula with these expressions; replace P with 1200 and solve for a.

$$P = a + b + c$$
$$1200 = a + (100 + a) + (200 + a)$$
$$1200 = 3a + 300$$
$$900 = 3a$$
$$300 = a$$

Side a is 300 ft.; $b = 100 + 300 = 400$ ft.; side $c = 200 + 300 = 500$ ft.

27. One formula for circumference is $C = 2\pi r$. Translating the second sentence of the problem, $C = 6r + 12.88$. Equate these two expressions for C and solve the equation for r.

$$6r + 12.88 = 2\pi r$$
$$6r + 12.88 = 2(3.14)r$$
$$6r + 12.88 = 6.28r$$
$$6r - 6r + 12.88 = 6.28r - 6r$$
$$12.88 = 0.28r$$
$$\frac{12.88}{0.28} = \frac{0.28r}{0.28}$$
$$46 = r$$

The radius is 46 ft.

29. The formula for the area of a trapezoid is $A = \frac{1}{2}h(b + B)$. Substitute the numerical values given in the problem and compute to find area.

$$A = \frac{1}{2}h(b + B)$$
$$A = \frac{1}{2}(165.97)(115.80 + 171.00)$$
$$A = \frac{1}{2}(165.97)(286.8)$$
$$A = \frac{47600.196}{2}$$
$$A = 23,800.098$$

Rounded to the nearest hundredth, the area is 23,800.10 sq ft.

31. You would need to use perimeter. Fencing is sold by linear measure.

	r	d	C	A
33.	6 in.	12 in.	12π in.	36π in.^{2}
34.	9 in.	18 in.	18π in.	81π in.^{2}
35.	5 ft.	10 ft.	10π ft.	25π ft.^{2}
36.	20 ft.	40 ft.	40π ft.	400π ft.^{2}
37.	6 cm	12 cm	12π cm	36π cm^2
38.	9 cm	18 cm	18π cm	81π cm^2
39.	10 in.	20 in.	20π in.	100π in.^{2}
40.	16 in.	32 in.	32π in.	256π in.^{2}

Table for Exercises 33–40

33. $d = 2r = 2 \cdot 6 = 12$
$C = 2\pi r = 2\pi \cdot 6 = 12\pi$
$A = \pi r^2 = \pi \cdot 6^2 = 36\pi$

35. $r = \dfrac{1}{2} \cdot 10 = 5$
$C = \pi d = \pi \cdot 10 = 10\pi$
$A = \pi r^2 = \pi \cdot 5^2 = 25\pi$

37. $d = \dfrac{C}{\pi} = \dfrac{12\pi}{\pi} = 12$
$r = \dfrac{1}{2} \cdot 12 = 6$
$A = \pi \cdot 6^2 = 36\pi$

39. $r^2 = \dfrac{A}{\pi} = \dfrac{100\pi}{\pi} = 100.$ Then $r = \sqrt{100} = 10.$
$d = 2r = 2 \cdot 10 = 20$
$C = 2\pi r = 2 \cdot \pi \cdot 10 = 20\pi$

41. Use the formula $P = 4s$, replacing s with x and P with 58.
$$P = 4x$$
$$58 = 4x$$
$$\frac{58}{4} = \frac{4x}{4}$$
$$14.5 = x$$

43. Use the formula $P = 2l + 2w$, replacing l and w with the expressions in x and replacing P with 38.
$$38 = 2(2x - 3) + 2(x + 1)$$
$$38 = 4x - 6 + 2x + 2$$
$$38 = 6x - 4$$
$$42 = 6x$$
$$\frac{42}{6} = \frac{6x}{6}$$
$$7 = x$$

45. Use the formula $A = s^2$, replacing s with x and A with 26.01.

$$26.01 = x^2$$
$$\sqrt{26.01} = \sqrt{x^2}$$
$$5.1 = x$$

47. Use the formula $A = \frac{1}{2}bh$, replacing b and h with the expressions in x and replacing A with 15.

$$15 = \frac{1}{2}x(x + 1)$$
$$2 \cdot 15 = \frac{2}{1} \cdot \frac{1}{2}x(x + 1)$$
$$30 = x(x + 1)$$
$$30 = x^2 + x$$
$$0 = x^2 + x - 30$$

Now factor the trinomial and set each factor equal to zero.

$$0 = (x + 6)(x - 5)$$

$$x + 6 = 0 \quad \text{or} \quad x - 5 = 0$$
$$x = -6 \qquad\qquad x = 5$$

The solution -6 is not meaningful because the base of a triangle must be positive a number. The answer then is 5.

49. Use the formula $C = 2\pi r$, replacing C with 37.58 and r with the expression in x.

$$37.68 = 2(3.14)(x + 1)$$
$$37.68 = 6.28(x + 1)$$
$$\frac{37.68}{6.28} = \frac{6.28(x + 1)}{6.28}$$
$$6 = x + 1$$
$$5 = x$$

51. Use the formula $A = \pi r^2$, replacing A with 18.0864 and r with x.

$$18.0864 = 3.14x^2$$
$$\frac{18.0864}{3.14} = \frac{3.14x^2}{3.14}$$
$$5.76 = x^2$$
$$\sqrt{5.76} = \sqrt{x^2}$$
$$2.4 = x$$

53. (a) $A = 4 \cdot 5 = 20$ cm^2

(b) $A = 8 \cdot 10 = 80$ cm^2

(c) $A = 12 \cdot 15 = 180$ cm^2

(d) $A = 16 \cdot 20 = 320$ cm^2

(e) The rectangle in part (b) had sides twice as long as the sides of the rectangle in part (a). Divide the larger area by the smaller ($80 \div 20 = 4$). By doubling the sides, the area increased $\underline{4}$ times.

(f) To get the rectangle in part (c) each side of the rectangle of part (a) was multiplied by $\underline{3}$. This made the larger area $\underline{9}$ times the side of the smaller area ($180 \div 20 = 9$).

(g) To get the rectangle of part (d) each side of the rectangle of part (a) was multiplied by $\underline{4}$. This made the area increase to $\underline{16}$ times what it was originally ($320 \div 20 = 16$).

(h) In general, if the length of each side of a rectangle is multiplied by n, the area is multiplied by $\underline{n^2}$.

55. Because each measurement is multiplied by 2, the area will increase by $2^2 = 4$. Then $4 \cdot 200 = \$800$.

57. If the radius of a circle is multiplied by n, then the area of the circle is multiplied by $\underline{n^2}$.

59. Find the area of the parallelogram and the area of the triangle. Then add the two area values.

Parallelogram $\quad A = 6 \cdot 10 = 60$

Triangle $\quad A = \frac{1}{2}(10)(4) = 20$

Total area $\quad 60 + 20 = 80$

61. There are 2 semicircles or equivalently 1 full circle with radius of 3. Find the area of this circle and of the rectangle.

Rectangle $\quad A = 8 \cdot 6 = 48$
Circle $\quad A = (3.14) \cdot 3^2 = 28.26$
Total area $\quad 48 + 28.26 = 76.26$

63. Find the area of the trapezoid that surrounds the triangle. Subtract the area of the triangle.

Trapezoid $\quad A = \frac{1}{2}(12)(18+11) = 174$

Triangle $\quad A = \frac{1}{2}(12)(7) = 42$

Shaded area $\quad 174 - 42 = 132$ ft^2

65. Find the area of the rectangle that surrounds the triangles. Subtract the areas of the triangles. The length of the rectangle is $48 + 48 = 96$.

Rectangle $\quad A = 74 \cdot 96 = 7104$
One triangle $\quad A = \frac{1}{2}(48)(36) = 864$
Shaded area $\quad 7104 - 2(864) = 5376$ cm^2.

67. Find the area of the square that surrounds the circle. Subtract the area of the circle. The diameter is 26; therefore, the radius is $26 \div 2 = 13$.

Square $\quad A = 26^2 = 676$
Circle $\quad A = (3.14)(13)^2 = 530.66$
Shaded area $\quad 676 - 530.66 = 145.34$ m^2

69. The best buy is the pizza with the lowest cost per square inch or unit price if you have enough money and you can eat all of it!

10" pizza $\quad A = (3.14)(5^2) = 78.5$ in^2
Unit price $\quad \frac{5.99}{78.5} \approx \$.076$
12" pizza $\quad A = (3.14)(6^2) = 113.04$ in^2
Unit price $\quad \frac{7.99}{113.04} \approx \$.071$
14" pizza $\quad A = (3.14)(7^2)$ or 153.86 in^2
Unit price $\quad \frac{8.99}{153.86} \approx \$.058$

The best buy is the 14" pizza.

71. The best buy is the pizza with the lowest cost per square inch or unit price.

10" pizza $\quad A = (3.14)(5^2) = 78.5$ in^2
Unit price $\quad \frac{9.99}{78.5} \approx \$.127$
12" pizza $\quad A = (3.14)\sqrt{6^2} = 113.04$ in^2
Unit price $\quad \frac{11.99}{113.04} \approx \$.106$
14" pizza $\quad A = (3.14)(7^2)$ or 153.86 in^2
Unit price $\quad \frac{12.99}{153.86} \approx \$.084$

The best buy is the 14" pizza.

73. The key is to construct OB and to realize that the diagonals of a rectangle are equal in length. So by inspection $OB = AC = 13$ in. OB is a radius. Therefore the diameter $= 2 \cdot 13 = 26$ in.

75. The key is to construct TV and UW to create more triangles. By inspection, all the small triangles are equal. $PQRS$ has 8 triangles. $TUVW$ has 4 triangles. Therefore $TUVW$ has half the area of $PQRS$, which is 625 ft^2. Otherwise find the area by first solving for the length of one side using the Pythagorean theorem.

77. The key is to construct a perpendicular line from E to side DC. Let the point of intersection of side DC be labeled point F. By inspection there are two sets of equal triangles: $\triangle DAE$ and $\triangle EFD$; $\triangle CBE$ and $\triangle EFC$. Then the area of the shaded region is half the area of the square. Therefore,

$$\text{Area} = \frac{36^2}{2} = \frac{1296}{2} = 648 \text{ in.}^2$$

79. The key is to construct a square using two radii from O and bounding the shaded region. The area of the small square is r^2 and the area of the quarter circle is $\pi r^2/4$. Therefore the area of the shaded region is

$$r^2 - \frac{\pi r^2}{4} = r^2\left(1 - \frac{\pi}{4}\right).$$

9.4 EXERCISES

$SAS = Side\text{-}Angle\text{-}Side$
$ASA = Angle\text{-}Side\text{-}Angle$
$SSS = Side\text{-}Side\text{-}Side$

1. SSS because the correspondence of two sides is given and the third side, AB, is common to both triangles.

3. SAS because the correspondence of one side is given, $\angle ABD = \angle CBD$ (the included angle) because they are both right angles, and side BD is common to both triangles.

5. ASA because the correspondence of two angles is given and the included side, AC, is common to both triangles.

7. If $\angle B$ measures $46°$, then $\angle A$ measures $\underline{67°}$ and $\angle C$ measures $\underline{67°}$. In an isosceles triangle, the angles opposite the equal sides are also equal in measure. Thus $\angle A = \angle C$. The sum of the angles of the triangle is 180. Then $180 - 46 = 134$ and $134 \div 2 = 67$.

9. The length of side $AB = 12$ because it has the same measure as BC. Then $30 - 2 \cdot 12 = 6$ in.

11. Writing exercise

13. Corresponding angles are equal in measure.

$$\angle A \text{ and } \angle P$$
$$\angle B \text{ and } \angle Q$$
$$\angle C \text{ and } \angle R$$

Corresponding sides are equal in length.

$$\overleftrightarrow{AB} \text{ and } \overleftrightarrow{PQ}$$
$$\overleftrightarrow{AC} \text{ and } \overleftrightarrow{PR}$$
$$\overleftrightarrow{CB} \text{ and } \overleftrightarrow{RQ}$$

15. Sometimes it is helpful to sketch the triangles, drawing them side by side. It is easier to determine the corresponding sides.

Corresponding angles are equal in measure.

$$\angle HGK \text{ and } \angle EGF \text{ because they are vertical}$$
$$\angle H \text{ and } \angle F$$
$$\angle K \text{ and } \angle E$$

Corresponding sides are equal in length.

$$\overleftrightarrow{HK} \text{ and } \overleftrightarrow{EF}$$
$$\overleftrightarrow{GH} \text{ and } \overleftrightarrow{GF}$$
$$\overleftrightarrow{GK} \text{ and } \overleftrightarrow{GE}$$

17. $\angle P = \angle C = 78°$
$\angle M = \angle B = 46°$
$\angle N = \angle A = 180 - (78 + 46) = 180 - 124 = 56°$
because the sum of the angle measures must equal $180°$.

19. $\angle T = \angle X = 74°$
$\angle V = \angle Y = 28°$
$\angle W = \angle Z = 180 - (74 + 28) = 180 - 102 = 78°$
because the sum of the angle measures must equal $180°$.

21. $\angle T = \angle P = 20°$
$\angle V = \angle Q = 64°$
$\angle U = \angle R = 180 - (20 + 64) = 180 - 84 = 96°$
because the sum of the angle measures must equal $180°$.

23. Corresponding sides must be proportional.

$$\frac{a}{8} = \frac{25}{10}$$
$$10 \cdot a = 8 \cdot 25$$
$$\frac{10a}{10} = \frac{200}{10}$$
$$a = 20$$
and
$$\frac{b}{6} = \frac{25}{10}$$
$$10 \cdot b = 6 \cdot 25$$
$$\frac{10b}{10} = \frac{150}{10}$$
$$b = 15$$

25. Corresponding sides must be proportional.

$$\frac{a}{12} = \frac{6}{12}$$
$a = 6$ because the denominators are equal

and

$$\frac{b}{15} = \frac{6}{12}$$
$$12 \cdot b = 15 \cdot 6$$
$$\frac{12b}{12} = \frac{90}{12}$$
$$b = \frac{6 \cdot 15}{6 \cdot 2}$$
$$b = \frac{15}{2}$$

27. Corresponding sides must be proportional.

$$\frac{x}{4} = \frac{9}{6}$$
$$6 \cdot x = 4 \cdot 9$$
$$\frac{6x}{6} = \frac{36}{6}$$
$$x = 6$$

29. Corresponding sides must be proportional.

$$\frac{x}{50} = \frac{220}{100}$$
$$100 \cdot x = 50 \cdot 220$$
$$\frac{100x}{100} = \frac{11000}{100}$$
$$x = 110$$

31. Corresponding sides must be proportional. In the third step, reduce the fraction on the right to lowest terms to make computations easier.

$$\frac{c}{100} = \frac{10 + 90}{90}$$
$$\frac{c}{100} = \frac{100}{90}$$
$$\frac{c}{100} = \frac{10}{9}$$
$$9 \cdot c = 100 \cdot 10$$
$$\frac{9c}{9} = \frac{1000}{9}$$
$$c = 111\frac{1}{9}$$

33. Let h = height of the tree.

$$\frac{h}{45} = \frac{2}{3}$$
$$3 \cdot h = 45 \cdot 2$$
$$\frac{3h}{3} = \frac{90}{3}$$
$$h = 30 \text{ m}$$

35. Let x = length of the mid-length side.

$$\frac{5}{x} = \frac{4}{400}$$
$$4 \cdot x = 5 \cdot 400$$
$$\frac{4x}{4} = \frac{2000}{4}$$
$$x = 500 \text{ m}$$

Now let y = the longest side.

$$\frac{7}{y} = \frac{4}{400}$$
$$4 \cdot y = 7 \cdot 400$$
$$\frac{4y}{4} = \frac{2800}{4}$$
$$y = 700 \text{ m}$$

The lengths of the other two sides are 500 meters and 700 meters.

37. Let h = height of the building. In step 2 the fraction on the right is reduced to make further computations easier.

$$\frac{h}{15} = \frac{300}{40}$$
$$\frac{h}{15} = \frac{15}{2}$$
$$2 \cdot h = 15 \cdot 15$$
$$\frac{2h}{2} = \frac{225}{2}$$
$$h = 112.5 \text{ ft.}$$

39. Let x = length of carved body.

$$\frac{x}{\text{actual body height}} = \frac{\text{height of carved head}}{\text{actual head height}}$$

$$\frac{x}{6\frac{1}{3}} = \frac{60}{\frac{3}{4}}$$

$$\frac{3}{4} \cdot x = 6\frac{1}{3} \cdot 60$$

$$\frac{3}{4}x = \frac{19}{3} \cdot \frac{60}{1}$$

$$\frac{3}{4}x = 380$$

$$\frac{4}{3} \cdot \frac{3}{4}x = 380 \cdot \frac{4}{3}$$

$$x = 506\frac{2}{3} \text{ feet}$$

41. Use the Pythagorean theorem $a^2 + b^2 = c^2$ with $a = 8$ and $b = 15$.

$$8^2 + 15^2 = c^2$$
$$64 + 225 = c^2$$
$$289 = c^2$$
$$\sqrt{289} = \sqrt{c^2}$$
$$17 = c$$

43. Use the Pythagorean theorem $a^2 + b^2 = c^2$ with $b = 84$ and $c = 85$.

$$a^2 + 84^2 = 85^2$$
$$a^2 + 7056 = 7225$$
$$a^2 = 169$$
$$\sqrt{a^2} = \sqrt{169}$$
$$a = 13$$

45. Use the Pythagorean theorem $a^2 + b^2 = c^2$ with $a = 14$ and $b = 48$.

$$14^2 + 48^2 = c^2$$
$$196 + 2304 = c^2$$
$$2500 = c^2$$
$$\sqrt{2500} = \sqrt{c^2}$$
$$50 \text{ m} = c$$

47. No. The area of the square attached to side a is $4 \cdot 4 = 16$. The area of the square attached to side b is $11 \cdot 11 = 121$. The sum of these area values is $16 + 121 = 137$, but the area of the square attached to side c is $12 \cdot 12 = 144$. It is not a right triangle

49. The sum of the squares of the two shorter sides of a right triangle is equal to the square of the longest side.

51. Given $r = 2$ and $s = 1$,

$$\begin{aligned}
a &= r^2 - s^2 &&= 2^2 - 1^2 = 4 - 1 &&= 3 \\
b &= 2rs &&= 2 \cdot 2 \cdot 1 &&= 4 \\
c &= r^2 + s^2 &&= 2^2 + 1^2 = 4 + 1 &&= 5
\end{aligned}$$

The Pythagorean triple is $(3, 4, 5)$.

53. Given $r = 4$ and $s = 3$,

$$\begin{aligned}
a &= r^2 - s^2 &&= 4^2 - 3^2 = 16 - 9 &&= 7 \\
b &= 2rs &&= 2 \cdot 4 \cdot 3 &&= 24 \\
c &= r^2 + s^2 &&= 4^2 + 3^2 = 16 + 9 &&= 25
\end{aligned}$$

The Pythagorean triple is $(7, 24, 25)$.

55. Given $r = 4$ and $s = 2$,

$$\begin{aligned}
a &= r^2 - s^2 &&= 4^2 - 2^2 = 16 - 4 &&= 12 \\
b &= 2rs &&= 2 \cdot 4 \cdot 2 &&= 16 \\
c &= r^2 + s^2 &&= 4^2 + 2^2 = 16 + 4 &&= 20
\end{aligned}$$

The Pythagorean triple is $(12, 16, 20)$.

57. Substitute the expressions in r and s for a and b.

$$\begin{aligned}
a^2 + b^2 &= \left(r^2 - s^2\right)^2 + (2rs)^2 \\
&= r^4 - 2r^2s^2 + s^4 + 4r^2s^2 \\
&= r^4 + 2r^2s^2 + s^2 \\
&= \left(r^2 + s^2\right)^2 \\
&= c^2
\end{aligned}$$

59. When $m = 3$,

$$\frac{m^2 - 1}{2} = \frac{3^2 - 1}{2} = \frac{8}{2} = 4$$

$$\frac{m^2 + 1}{2} = \frac{3^2 + 1}{2} = \frac{10}{2} = 5$$

The Pythagorean triple is $(3, 4, 5)$.

61. When $m = 7$,

$$\frac{m^2 - 1}{2} = \frac{7^2 - 1}{2} = \frac{48}{2} = 24$$

$$\frac{m^2 + 1}{2} = \frac{7^2 + 1}{2} = \frac{50}{2} = 25$$

The Pythagorean triple is $(7, 24, 25)$.

63. Replace a^2 with m^2, replace b^2 with $\left(\frac{m^2-1}{2}\right)^2$ and show that their sum simplifies to c^2.

$$a^2 + b^2 = m^2 + \left(\frac{m^2-1}{2}\right)^2$$
$$= m^2 + \frac{m^4 - 2m^2 + 1}{4}$$
$$= \frac{4m^2}{4} + \frac{m^4 - 2m^2 + 1}{4}$$
$$= \frac{m^4 + 2m^2 + 1}{4}$$
$$= \frac{(m^2+1)^2}{4}$$
$$= \left(\frac{m^2+1^2}{2}\right)^2$$
$$= c^2$$

65. For $n = 2$,

$$2n = 2 \cdot 2 \qquad = 4$$
$$n^2 - 1 = 2^2 - 1 \quad = 3$$
$$n^2 + 1 = 2^2 + 1 \quad = 5$$

The Pythagorean triple is $(4, 3, 5)$.

67. For $n = 4$,

$$2n = 2 \cdot 4 \qquad = 8$$
$$n^2 - 1 = 4^2 - 1 \quad = 15$$
$$n^2 + 1 = 4^2 + 1 \quad = 17$$

The Pythagorean triple is $(8, 15, 17)$.

69. Replace a with $2n$ and b with $n^2 - 1$ in the Pythagorean theorem, and show that the expression simplifies to c^2.

$$a^2 + b^2 = (2n)^2 + \left(n^2 - 1\right)^2$$
$$= 4n^2 + n^4 - 2n^2 + 1$$
$$= n^4 + 2n^2 + 1$$
$$= \left(n^2 + 1\right)^2$$
$$= c^2$$

71. Let b = length of longer leg,
$b + 1$ = length of hypotenuse c,
7 = length of shorter leg, a.
Substitute these expressions into the Pythagorean theorem, $a^2 + b^2 = c^2$, and solve for b.

$$7^2 + b^2 = (b+1)^2$$
$$49 + b^2 = b^2 + 2b + 1$$
$$49 + b^2 - b^2 = b^2 - b^2 + 2b + 1$$
$$49 = 2b + 1$$
$$48 = 2b$$
$$24 = b$$

The longer leg is 24 m.

73. Let h = the height of the tower, one of the legs of the triangle,
$2h + 2$ = the hypotenuse, c,
$a = 30$, another leg of the triangle.
Substitute these expressions into the Pythagorean theorem, $a^2 + b^2 = c^2$, and solve for h.

$$30^2 + h^2 = (2h + 2)^2$$
$$900 + h^2 = 4h^2 + 8h + 4$$
$$900 + h^2 - h^2 = 4h^2 - h^2 + 8h + 4$$
$$900 = 3h^2 + 8h + 4$$
$$900 - 900 = 3h^2 + 8h + 4 - 900$$
$$0 = 3h^2 + 8h - 896$$
$$0 = (3h + 56)(h - 16)$$

$$3h + 56 = 0 \qquad \text{or} \quad h - 16 = 0$$
$$h = -\frac{56}{3} \qquad\qquad h = 16$$

A negative height is not meaningful. The height is 16 feet.

75. Let h = the height of the break,
$a = 3$ ft, one leg of the triangle,
$c = 10 - h$, the hypotenuse of the triangle.
Substitute these expressions into the Pythagorean theorem, $a^2 + b^2 = c^2$, and solve for h.

$$3^2 + h^2 = (10 - h)^2$$
$$9 + h^2 = 100 - 20h + h^2$$
$$9 + h^2 - h^2 = 100 - 20h + h^2 - h^2$$
$$9 = 100 - 20h$$
$$9 - 100 = 100 - 100 - 20h$$
$$9 - 100 = 100 - 100 - 20h$$
$$-91 = -20h$$
$$4.55 = h$$

The height of the break is 4.55 ft.

77. Let $c =$ the length of the diagonal.

$$12^2 + 15^2 = c^2$$
$$144 + 225 = c^2$$
$$369 = c^2$$
$$\sqrt{369} = \sqrt{c^2}$$
$$19.21 \approx c$$

Then $.21(12) = 2.52$, which is 3 inches to the nearest inch. The diagonal should be 19 feet, 3 inches.

79. Let $c =$ the length of the diagonal.

$$16^2 + 24^2 = c^2$$
$$256 + 576 = c^2$$
$$832 = c^2$$
$$\sqrt{832} = \sqrt{c^2}$$
$$28.84 \approx c$$

Then $.84(12) = 10.08$, which is 10 inches to the nearest inch. The diagonal should be 28 feet, 10 inches.

81. (a) The formula for the area of a trapezoid is $A = \frac{1}{2}h(b + B)$. If ZY is the base B, it can be expressed as b. Base b is then WX, which can also be expressed as a. The height of the trapezoid is WZ, which can also be expressed as $b + a$. Substitute these expressions for B, b, and h.

$$A = \frac{1}{2}(b + a)(a + b) = \frac{1}{2}(a + b)(a + b).$$

(b) The area of $\triangle PWX$ is $\frac{1}{2}ab$. The area of $\triangle PZY$ is $\frac{1}{2}ab$. The area of $\triangle PXY$ is $\frac{1}{2}c^2$.

(c)
$$\frac{1}{2}(a + b)(a + b) = \frac{1}{2}ab + \frac{1}{2}ab + \frac{1}{2}c^2$$
$$\frac{1}{2}\left(a^2 + 2ab + b^2\right) = \frac{1}{2}ab + \frac{1}{2}ab + \frac{1}{2}c^2$$
$$\frac{1}{2}a^2 + ab + \frac{1}{2}b^2 = ab + \frac{1}{2}c^2$$
$$\frac{1}{2}a^2 + \frac{1}{2}b^2 = \frac{1}{2}c^2$$
$$a^2 + b^2 = c^2$$

83. Draw a line connecting B and D to form two right triangles. The area of $\triangle DAB$ is $\frac{1}{2} \cdot 6 \cdot 8 = 24$. The area of $\triangle BCD$ cannot be found until the length of CD is known. First find the length of the hypotenuse BD for $\triangle DAB$: $6^2 + 8^2 = 36 + 64 = 100$. The hypotenuse is 10. To find the length of side CD, use the Pythagorean Theorem again:

$$2^2 + (CD)^2 = 10^2$$
$$4 + (CD)^2 = 100$$
$$(CD)^2 = 96$$
$$\sqrt{(CD)^2} = \sqrt{96}$$
$$CD = \sqrt{16 \cdot 6}$$
$$CD = 4\sqrt{6}.$$

Then the area of $\triangle BCD$ is $\frac{1}{2} \cdot 2 \cdot 4\sqrt{6} = 4\sqrt{6}$. Add the areas of the two triangles to find the area of the quadrilateral: $24 + 4\sqrt{6}$.

85. Draw a line from the upper vertex that is perpendicular to the base of 24. Two right triangles are created, each with a base of 12. Use the Pythagorean theorem to find the height, x.

$$x^2 + 12^2 = 13^2$$
$$x^2 + 144 = 169$$
$$x^2 = 25$$
$$x = 5$$

Now separate the two triangles and rearrange them so that the height of each, x, lie next to each other to form the base of a triangle. Then $2x = 10$.

87. The area of the pentagon can be found by adding the area of the square and the area of the triangle. If all sides of the pentagon are equal and the perimeter is 80, the length of each side is $80 \div 5 = 16$. Then the area of the square is $16^2 = 256$. Find the height of the triangle in order to use the area formula. One way this can be done is to draw a line from Q to base SR that forms a right angle with the base. This will divide the base in half so that both of the smaller triangles are right triangles. The Pythagorean theorem can be used to find the height.

$$8^2 + h^2 = 16^2$$
$$64 + h^2 = 256$$
$$h^2 = 192$$
$$h = \sqrt{192}$$
$$h = \sqrt{64 \cdot 3}$$
$$h = 8\sqrt{3}$$

Use the formula for the area of a triangle.

$$A = \frac{1}{2}bh$$
$$A = \frac{1}{2}16 \cdot 8\sqrt{3}$$
$$A = 64\sqrt{3}$$

The area of the pentagon is $256 + 64\sqrt{3}$.

89. In section 9.2 it is shown that any angle inscribed in a semicircle must be a right angle; $\angle ACB$ is then a right angle. Use the Pythagorean theorem to find the length of the hypotenuse AB, which is also the diameter of the circle. Finally, divide the diameter by two to obtain the length of the radius.

$$8^2 + 6^2 = (AB)^2$$
$$64 + 36 = (AB)^2$$
$$100 = (AB)^2$$
$$10 = AB$$

Then $10 \div 2 = 5$ in.

EXTENSION: RIGHT TRIANGLE TRIGONOMETRY

Recall the trigonometric ratios.

$$\sin\theta = \frac{opp\,\theta}{hyp}; \cos\theta = \frac{adj\,\theta}{hyp}; \tan\theta = \frac{opp\,\theta}{adj\,\theta}$$

1. $\sin\theta = \dfrac{3}{5}$; $\cos\theta = \dfrac{4}{5}$; $\tan\theta = \dfrac{3}{4}$

3. Use the Pythagorean theorem to find the length of the hypotenuse.

$$6^2 + 8^2 = c^2$$
$$36 + 64 = c^2$$
$$100 = c^2$$
$$10 = c$$

Then $\sin\theta = \dfrac{8}{10} = \dfrac{4}{5}$; $\cos\theta = \dfrac{6}{10} = \dfrac{3}{5}$; $\tan\theta = \dfrac{8}{6} = \dfrac{4}{3}$.

5. Use the Pythagorean theorem to find the length of the adjacent side.

$$10^2 + b^2 = 26^2$$
$$100 + b^2 = 676$$
$$b^2 = 576$$
$$b = 24$$

Then $\sin\theta = \dfrac{10}{26} = \dfrac{5}{13}$; $\cos\theta = \dfrac{24}{26} = \dfrac{12}{13}$; $\tan\theta = \dfrac{10}{24} = \dfrac{5}{12}$.

7. (a) Since the sum of the acute angles of a right triangle is 90°, the two acute angles of a right triangle are <u>complements</u> of each other.

   (b) The measure of this angle is $90° - \theta$.

(c) The sine of angle θ is the ratio of the side opposite the angle to the hypotenuse or a/c. The cosine of the unmarked acute angle is the ratio of the side adjacent to the angle to the hypotenuse or a/c.

(d) The two functions are equal.

9. $\sin 30° = \dfrac{a}{2a} = \dfrac{1}{2}$ $\qquad$ $\sin 60° = \dfrac{\sqrt{3}\,a}{2a} = \dfrac{\sqrt{3}}{2}$

$\cos 30° = \dfrac{\sqrt{3}\,a}{2a} = \dfrac{\sqrt{3}}{2}$ $\qquad$ $\cos 60° = \dfrac{a}{2a} = \dfrac{1}{2}$

$\tan 30° = \dfrac{a}{\sqrt{3}\,a} = \dfrac{1}{\sqrt{3}}$ $\qquad$ $\tan 60° = \dfrac{\sqrt{3}\,a}{a} = \sqrt{3}$

11. One of several ways to use the calculator to find the angle measure is as follows:

$$\boxed{3}\;\boxed{\div}\;\boxed{5}\;\boxed{=}\;\boxed{\text{inverse sine key}}$$

The answer is 36.9°.

13. One of several ways to use the calculator to find the angle measure is as follows:

$$\boxed{21}\;\boxed{\div}\;\boxed{20}\;\boxed{=}\;\boxed{\text{inverse tangent key}}$$

The answer is 46.4°.

15. $\tan 28° = \dfrac{4.8}{x}$

$.5317 = \dfrac{4.8}{x}$

$.5317x \approx 4.8$

$\dfrac{.5317x}{.5317} \approx \dfrac{4.8}{.5317}$

$x \approx 9.0$ ft.

17. $\tan 23° = \dfrac{4.7}{x}$

$.4245 = \dfrac{4.7}{x}$

$.4245x \approx 4.7$

$\dfrac{.4245x}{.4245} \approx \dfrac{4.7}{.4245}$

$x \approx 11.1$ ft.

19. Let $x =$ the height of the flagpole.

$$\tan 38° = \frac{x - 6}{24.8}$$
$$.7813 = \frac{x - 6}{24.8}$$
$$(24.8.6)(.7813) \approx x - 6$$
$$19.376 \approx x - 6$$
$$25.376 \approx x$$

Rounded to the nearest tenth, the height is about 25.4 ft.

21. Let θ = the angle measurement.

$$\sin \theta = \frac{71.3}{77.4}$$
$$\sin \theta = .9212$$
$$\theta \approx 67°$$

23. The surveyor has created a right triangle because the two angles are complementary.
    Let x = the length of RS.

$$\tan 32° = \frac{x}{53.1}$$
$$.6249 = \frac{x}{53.1}$$
$$(53.1)(.6249) \approx x$$
$$33.2 \text{ ft.} \approx x$$

25. Writing exercise

9.5 EXERCISES

1. True. If the volume is 64 cubic inches, one side of the cube is 4 inches because $4 \cdot 4 \cdot 4 = 64$. Then the area of one face is $4 \cdot 4 = 16$. A cube has six faces so that $6 \cdot 16 = 96$ square inches is the total surface area.

3. True. A dodecahedron has 12 faces.

5. False. The new cube will have six times the volume of the original cube.

7. (a) $V = lwh$

$$= 2 \cdot 1\frac{1}{2} \cdot 1\frac{1}{4}$$
$$= 2 \cdot \frac{3}{2} \cdot \frac{5}{4}$$
$$= \frac{15}{4}$$
$$= 3\frac{3}{4} \text{ m}^3$$

   (b) $S = 2lh + 2hw + 2lw$

$$= 2 \cdot 2 \cdot \frac{5}{4} + 2 \cdot \frac{5}{4} \cdot \frac{3}{2} + 2 \cdot 2 \cdot \frac{3}{2}$$
$$= 5 + \frac{15}{4} + 6$$
$$= 5 + 3\frac{3}{4} + 6$$
$$= 14\frac{3}{4} \text{ m}^2$$

9. It may be helpful to use parentheses to indicate multiplication when working with decimal values. Otherwise it is possible to confuse the multiplication dot and the decimal points.

(a) $V = lwh$
$$= (6)(5)(3.2)$$
$$= 96 \text{ in}^3$$

(b) $S = 2lh + 2hw + 2lw$
$$= 2(6)(3.2) + 2(3.2)(5) + 2(6)(5)$$
$$= 38.4 + 32 + 60$$
$$= 130.4 \text{ in}^2$$

11. (a) $V = \frac{4}{3}\pi r^3$
$$= \frac{4}{3}(3.14)(40)^3$$
$$= \frac{4}{3}(3.14)(64000)$$
$$\approx 267,946.67 \text{ ft}^3$$

    (b) $S = 4\pi r^2$
$$= 4(3.14)(40)^2$$
$$= 4(3.14)(1600)$$
$$= 20,096 \text{ ft}^2$$

13. (a) $V = \pi r^2 h$
$$= (3.14)(5)^2(7)$$
$$= (3.14)(25)(7)$$
$$= 549.5 \text{ cm}^3$$

    (b) $S = 2\pi r^2 + 2\pi rh$
$$= 2(3.14)(5)^2 + 2(3.14)(5)(7)$$
$$= 2(3.14)(25) + 2(3.14)(5)(7)$$
$$= 157 + 219.8$$
$$= 376.8 \text{ cm}^2$$

15. (a) $V = \frac{1}{3}\pi r^2 h$
$$= \frac{1}{3}(3.14)(3)^2(7)$$
$$= \frac{1}{3}(3.14)(9)(7)$$
$$= 65.94 \text{ m}^3$$

    (b) $S = \pi r\sqrt{r^2 + h^2}$
$$= (3.14)(3)\sqrt{3^2 + 7^2}$$
$$= (3.14)(3)\sqrt{9 + 49}$$
$$= (3.14)(3)\sqrt{58}$$
$$= 9.42\sqrt{58}$$
$$\approx 71.74 \text{ m}^2$$

17. Remember that B represents the area of the base.

$$V = \frac{1}{3}Bh$$
$$= \frac{1}{3}(8 \cdot 9) \cdot 7$$
$$= \frac{504}{3}$$
$$= 168 \text{ in}^3$$

19. $V = \pi r^2 h$
$$= (3.14)(6.3)^2(15.8)$$
$$= (3.14)(36.69)(15.8)$$
$$= 1969.10 \text{ cm}^3$$

21. First find the radius by taking half of the diameter: $r = \frac{1}{2}(7.2) = 3.6$. Then use the formula for volume of a right circular cylinder.

$$V = \pi r^2 h$$
$$= (3.14)(3.6)^2(10.5)$$
$$= (3.14)(12.96)(10.5)$$
$$\approx 427.29 \text{ cm}^3$$

23. First find the radius by taking half of the diameter: $r = \frac{1}{2}(9) = 4.5$. Then use the formula for volume of a right circular cylinder.

$$V = \pi r^2 h$$
$$= (3.14)(4.5)^2(8)$$
$$= (3.14)(20.25)(8)$$
$$= 508.68 \text{ cm}^3$$

25. Remember that B represents the area of the base.

$$V = \frac{1}{3}Bh$$
$$= \frac{1}{3}(230)^2 \cdot 137$$
$$= \frac{1}{3}(52900) \cdot 137$$
$$= \frac{7,247,300}{3}$$
$$\approx 2,415,766.67 \text{ m}^3$$

27. Change 1/2 to the decimal value .5 for ease of computation.

$$V = \frac{1}{3}\pi r^2 h$$
$$= \frac{1}{3}(3.14)(.5)^2(2)$$
$$= \frac{1}{3}(3.14)(.25)(2)$$
$$= \frac{1.57}{3}$$
$$= .523 \text{ m}^3$$

	r	d	V	S
29.	6 in	12 in	288π in^3	144π in^2
31.	5 ft	10 ft	$\frac{500}{3}\pi$ ft^3	100π ft^2
33.	2 cm	4 cm	$\frac{32}{3}\pi$ cm^3	16π cm^2
35.	1 m	2 m	$\frac{4}{3}\pi$ m^3	4π m^2

Table for Exercises 29–36

29. $d = 2r = 2 \cdot 6 = 12$
$$V = \frac{4}{3}\pi r^3 = \frac{4}{3}\pi(6)^3 = \frac{4}{3}\pi(216) = 288\pi$$
$$S = 4\pi r^2 = 4\pi(6)^2 = 4\pi(36) = 144\pi$$

31. $r = \frac{1}{2}d = \frac{1}{2} \cdot 10 = 5$
$$V = \frac{4}{3}\pi r^3 = \frac{4}{3}\pi(5)^3 = \frac{4}{3}\pi(125) = \frac{500}{3}\pi$$
$$S = 4\pi r^2 = 4\pi(5)^2 = 4\pi(25) = 100\pi$$

33. Use the formula for the volume of a sphere to solve for r, by replacing V with the given value, $\frac{32}{3}\pi$.

$$V = \frac{4}{3}\pi r^3$$
$$\frac{32}{3}\pi = \frac{4}{3}\pi r^3$$
$$\frac{32}{3}\pi \div \frac{4}{3}\pi = \frac{4}{3}\pi r^3 \div \frac{4}{3}\pi$$
$$\frac{32\pi}{3} \cdot \frac{3}{4\pi} = r^3$$
$$8 = r^3$$
$$2 = r$$

$$d = 2r = 2 \cdot 2 = 4$$
$$S = 4\pi r^2 = 4\pi(2)^2 = 4\pi(4) = 16\pi$$

35. Use the formula for the surface area of a sphere to solve for r, by replacing S with the given value, 4π.

$$S = 4\pi r^2$$
$$4\pi = 4\pi r^2$$
$$\frac{4\pi}{4\pi} = r^2$$
$$1 = r^2$$
$$1 = r$$

$$d = 2r = 2 \cdot 1 = 2$$
$$V = \frac{4}{3}\pi r^3 = \frac{4}{3}\pi(1)^3 = \frac{4}{3}\pi(1) = \frac{4}{3}\pi$$

37. Volume is a measure of capacity.

39. The volume of the original cube is x^3. Let the length of the side of the new cube be represented by y. Then $y^3 = 2x^3$. Solve for y by taking the cube root of both sides of the equation.

$$y^3 = 2x^3$$
$$\sqrt[3]{y^3} = \sqrt[3]{2x^3}$$
$$y = \sqrt[3]{2x^3}$$
$$y = x\sqrt[3]{2}$$

41. If the new diameter is 3 times the old diameter, then the new volume will be 3^3 or 27 times greater. Therefore the cost will also be 27 times greater, or $27 \cdot 300 = \$8100$.

43. If the new diameter is 5 times the old diameter, then the new volume will be 5^3 or 125 times greater. Therefore the cost will also be 125 times greater, or $125 \cdot 300 = \$37,500$.

45. $V = lwh$
$$60 = 6 \cdot 4 \cdot x$$
$$60 = 24x$$
$$2.5 = x$$

47. In this exercise x = the diameter of the sphere. Therefore, $r = \frac{x}{2}$.

$$V = \frac{4}{3}\pi r^3$$
$$36\pi = \frac{4}{3}\pi\left(\frac{x}{2}\right)^3$$
$$36\pi = \frac{4}{3}\pi \cdot \frac{x^3}{8}$$
$$36 = \frac{4}{3} \cdot \frac{x^3}{8}$$
$$36 = \frac{x^3}{6}$$
$$216 = x^3$$
$$6 = x$$

49. Look at the figure and try some values for the edges of each side that will create the given areas. One side has edges 6 in. and 5 in.; the adjacent side has edges 5 in. and 7 in.; and the third side has edges 7 in. and 6 in. Write these values on the edges of the rectangular box to verify that it can be done to create the given areas. The three dimensions of the box, then, are 6, 7, and 5. The volume of the box is $6 \cdot 7 \cdot 5 = 210$ in.^{3}.

51. The formula for the volume of a sphere is $V = \frac{4}{3}\pi r^3$, so the radius of the sphere must be found. To find the radius of the circle that is formed by the intersection, set 576π equal to πr^2, the formula for the area of a circle, and solve for r.

$$576\pi = \pi r^2$$
$$\frac{576\pi}{\pi} = \frac{\pi r^2}{\pi}$$
$$576 = r^2$$
$$24 = r$$

Now use the Pythagorean theorem to find the length of the hypotenuse, which is also the radius of the sphere.

$$7^2 + 24^2 = c^2$$
$$49 + 576 = c^2$$
$$625 = c^2$$
$$25 = c$$

Now compute the volume of the sphere.

$$V = \frac{4}{3}\pi r^3$$
$$= \frac{4}{3}\pi(25)^3$$
$$= \frac{4}{3}\pi(15625)$$
$$= \frac{62500}{3}\pi \text{ in.}^3.$$

53. Rotate the inscribed square 45° so that one of its diagonals is horizontal and the other vertical. Notice that the length of the diagonal is the same length as the side of the circumscribed square. This length is $2r$. That means that the area of the circumscribed square is $A = 4r^2$. Returning to the inscribed square, the length $2r$ is the length of the hypotenuse of a right triangle. Use the Pythagorean theorem to find the length of a side of this square.

Let x = the length of each leg.

$$x^2 + x^2 = (2r)^2$$
$$2x^2 = 4r^2$$
$$x^2 = 2r^2$$

Because the area of this square is equal to x^2, the ratio of the two areas can be determined

$$\frac{\text{area of the circumscribed square}}{\text{area of the inscribed square}} = \frac{4r^2}{2r^2} = \frac{2}{1}$$

The ratio is 2 to 1.

55. Draw a line connecting one diameter RT; draw a line connecting another diameter QS. Recall from section 9.2 that any angle inscribed in a semicircle is a right angle, which means that $\angle RPT$ and $\angle QPS$ are both right angles. From the Pythagorean theorem, $PR^2 + PT^2$ equals the square of the diameter, 12^2. Also, $PQ^2 + PS^2$ equals the square of the diameter. Finally,

$$PR^2 + PT^2 + PQ^2 + PS^2 = 12^2 + 12^2 = 288.$$

9.6 EXERCISES

The chart in the text characterizes certain properties of Euclidean and non-Euclidean geometries. Study it and use it to respond to Exercises 1–10.

1.   Euclidean

3.   Lobachevskian

5.   greater than

7.   Riemannian

9.   Euclidean

11. Writing exercise

13. C. Both are of genus 2, meaning they have two holes.

15. A and E. All are of genus 0, having no holes.

17. A and E. All three are of genus 0, having no holes.

19. None of them

21. No, both have no holes. They are of genus 0.

23. Yes, the slice of American cheese is of genus 0 and the slice of Swiss cheese is of genus 1 or more.

25. A compact disc has one hole, so it is of genus 1.

27. A sheet of loose-leaf paper made for a three-ring binder has three holes, so it is of genus 3.

29. A wedding band has one hole, so it is of genus 1.

31. A, C, D, and F are even vertices because each has two paths leading to or from the vertex; B and E are odd because each has three paths leading to or from the vertex.

33. A, B, C, and F are odd because each has three paths leading to or from the vertex; D, E, and G are even. D and E each have two paths leading to or from the vertex; G has four.

35. A, B, C, and D are odd vertices because each has three paths leading to or from the vertex; E is even because it has four.

37. There are two odd vertices at the extremities and the rest are even. Therefore the network is traversable.

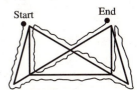

39. Not traversable. It has more than two odd vertices.

41. The network has exactly 2 odd vertices. It is traversable.

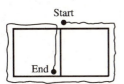

43. Yes. There are exactly two rooms (vertices) with an odd number of doors (paths), the top left room and the bottom left room. The rest of the rooms and the exterior of the house have an even number of doors. Therefore, the house (network) is traversable.

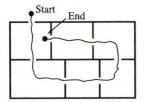

45. No. There are more than two rooms with an odd number of doors (paths).

9.7 EXERCISES

1.   The least number of these squares that can be put together edge to edge to form a larger square is 4.

3.   The length of each edge of the new square is 2.

5. $\dfrac{\text{new size}}{\text{old size}} = \dfrac{4}{1} = 4$

7.   The scale factor is

$$\frac{\text{new length}}{\text{old length}} = \frac{4}{1} = 4.$$

The ratio of

$$\frac{\text{new size}}{\text{old size}} = \frac{16}{1} = 16.$$

9.   Each ratio in the bottom row is the square of the scale factor in the top row.

11.

Scale factor	2	3	4	5	6	10
Ratio of new size to old size	4	9	16	25	36	100

13. Some examples are: $3^d = 9$ and $d = 2$; $4^d = 16$ and $d = 2$; $5^d = 25$ and $d = 2$.

15. The scale factor between these two cubes is

$$\frac{\text{new length}}{\text{old length}} = \frac{2}{1} = 2.$$

The ratio of

$$\frac{\text{new size}}{\text{old size}} = \frac{8}{1} = 8.$$

17. Each ratio in the bottom row is the cube of the scale factor in the top row.

19. The scale factor between stage 1 and stage 2 is <u>3 to 1</u>.

21. $3^d = 4$

Use trial and error:

$$3^{1.5} = 5.196\ldots$$
$$3^{1.25} = 3.948\ldots$$
$$3^{1.26} = 3.992\ldots$$
$$3^{1.27} = 4.036\ldots$$
$$3^{1.261} = 3.996\ldots$$
$$3^{1.262} = 4.001\ldots$$
$$3^{1.263} = 4.005\ldots$$
$$d = 1.262 \text{ to three decimal places,}$$

or solve using logarithms.

$$3^d = 4$$
$$\ln 3^d = \ln 4$$
$$d \ln 3 = \ln 4$$
$$d = \frac{\ln 4}{\ln 3}$$
$$d = 1.262 \text{ to three decimal places.}$$

23. Old size $= 1$, new size $= \underline{3}$

25. $2^d = 3$

Use trial and error:

$$2^{1.5} = 2.828\ldots$$
$$2^{1.6} = 3.031\ldots$$
$$2^{1.55} = 2.928\ldots$$
$$2^{1.58} = 2.990\ldots$$
$$2^{1.59} = 3.010\ldots$$
$$2^{1.584} = 2.998\ldots$$
$$2^{1.585} = 3.000\ldots$$
$$d = 1.585 \text{ to three decimal places}$$

Or solve using logarithms.

$$2^d = 3$$
$$\ln 2^d = \ln 3$$
$$d \ln 2 = \ln 3$$
$$d = \frac{\ln 3}{\ln 2}$$
$$d = 1.585 \text{ to three decimal places}$$

27. Given $k = 3.4$, $x = .8$ and formula $y = kx(1 - x)$

$$y = 3.4(.8)(1 - .8)$$
$$= 3.4(.8)(.2)$$
$$= .544$$

$$y = 3.4(.544)(1 - .544)$$
$$= 3.4(.544)(.456)$$
$$\approx .843$$

$$y = 3.4(.843)(1 - .843)$$
$$= 3.4(.843)(.157)$$
$$\approx .450$$

$$y = 3.4(.450)(1 - .450)$$
$$= 3.4(.450)(.550)$$
$$\approx .842$$

$$y = 3.4(.842)(1 - .842)$$
$$= 3.4(.842)(.158)$$
$$\approx .452$$

$$y = 3.4(.452)(1 - .452)$$
$$= 3.4(.452)(.548)$$
$$\approx .842$$

$$y = 3.4(.842)(1 - .842)$$
$$= 3.4(.842)(.158)$$
$$\approx .452$$

The attractors are evidently .842 and .452.

CHAPTER 9 TEST

1. (a) The measure of its complement is $90 - 38 = 52°$.

    (b) The measure of its supplement is $180 - 38 = 142°$.

    (c) It is an acute angle because it is less than $90°$.

2. The designated angles are supplementary; their sum is $180°$.

$$(2x + 16) + (5x + 80) = 180$$
$$7x + 96 = 180$$
$$7x = 84$$
$$x = 12$$

Then one angle measure is $2 \cdot 12 + 16 = 40°$ and the other angle measure is $5 \cdot 12 + 80 = 140°$. A check is that their sum is indeed $180°$.

3. The angles are vertical so they have the same measurement. Set the algebraic expressions equal to each other.

$$7x - 25 = 4x + 5$$
$$3x = 30$$
$$x = 10$$

Then one angle measure is $7 \cdot 10 - 25 = 45°$ and the other angle measure is $4 \cdot 10 + 5 = 45°$.

4. The designated angles are complementary; their sum is $90°$.
$$(4x + 6) + 10x = 90$$
$$14x + 6 = 90$$
$$14x = 84$$
$$x = 6$$

Then one angle measure is $4 \cdot 6 + 6 = 30°$ and the other angle measure is $10 \cdot 6 = 60°$.

5. The designated angles are supplementary.

$$(7x + 11) + (3x - 1) = 180$$
$$10x + 10 = 180$$
$$10x = 170$$
$$x = 17$$

Then one angle measure is $7 \cdot 17 + 11 = 130°$ and the other angle measure is $3 \cdot 17 - 1 = 50°$. A check is that their sum is indeed $180°$.

6. These are alternate interior angles, which are equal to each other.
$$13y - 26 = 10y + 7$$
$$3y = 33$$
$$y = 11$$

Then one angle measure is $13 \cdot 11 - 26 = 117°$ and the other angle measure is $10 \cdot 11 + 7 = 117°$.

7. Writing exercise

8. Letter C is false because a triangle cannot have both a right angle and an obtuse angle. A right angle measures $90°$ and an obtuse angle measures greater than $90°$; however, the sum of all three angles of any triangle is exactly $180°$.

9. The curve is simple and closed.

10. The curve is neither simple nor closed.

11. The sum of the three angle measures is $180°$.

$$(3x + 9) + (6x + 3) + 21(x - 2) = 180$$
$$(3x + 9) + (6x + 3) + 21x - 42 = 180$$
$$30x - 30 = 180$$
$$30x = 210$$
$$x = 7$$

Then one angle measure is $3 \cdot 7 + 9 = 30°$, a second angle measure is $6 \cdot 7 + 3 = 45°$, and the third angle measure is $21(7 - 2) = 21(5) = 105°$. A check is that their sum is indeed $180°$.

12. $A = lw$
$$= 6 \cdot 12$$
$$= 72 \text{ cm}^2$$

13. $A = bh$
$$= 12 \cdot 5$$
$$= 60 \text{ in}^2$$

14. $A = \frac{1}{2}bh$
$$= \frac{1}{2} \cdot 17 \cdot 8$$
$$= 68 \text{ m}^2$$

15. $A = \frac{1}{2}h(b + B)$
$$= \frac{1}{2} \cdot 9(16 + 24)$$
$$= \frac{9}{2}(40)$$
$$= 180 \text{ m}^2$$

16. Replace A in the formula for the area of a circle and solve for r.

$$A = \pi r^2$$
$$144\pi = \pi r^2$$
$$\frac{144\pi}{\pi} = \frac{\pi r^2}{\pi}$$
$$144 = r^2$$
$$12 = r$$

Now replace r in the formula for the circumference.

$$C = 2\pi r$$
$$= 2\pi \cdot 12$$
$$= 24\pi \text{ in.}$$

17. Use the formula for the circumference $C = \pi d$.

$$C = (3.14) \cdot 630$$
$$\approx 1978 \text{ ft.}$$

18. Subtract the area of the triangle from the area of the semicircle. First, the area of the semicircle is:

$$A = \frac{1}{2}\pi r^2$$
$$= \frac{1}{2}(3.14)(10)^2$$
$$= \frac{1}{2}(3.14)(100)$$
$$= 157$$

Now find the area of the triangle.

$$A = \frac{1}{2}bh$$
$$= \frac{1}{2} \cdot 20 \cdot 10$$
$$= 100$$

Finally $157 - 100 = 57 \, \text{cm}^2$.

19. SAS. Side AB is common to both triangles.

20. Let h = height of the pole.

$$\frac{h}{30} = \frac{30}{45}$$
$$45 \cdot h = 30 \cdot 30$$
$$\frac{45h}{45} = \frac{900}{45}$$
$$h = 20 \text{ feet}$$

21. Use the Pythagorean theorem to find c.

$$a^2 + b^2 = c^2$$
$$20^2 + 21^2 = c^2$$
$$400 + 441 = c^2$$
$$841 = c^2$$
$$\sqrt{841} = \sqrt{c^2}$$
$$29 = c$$

The length of the diagonal is 29 m.

22. (a) $V = \frac{4}{3}\pi r^3$
$$= \frac{4}{3}(3.14)(6)^3$$
$$= \frac{4}{3}(3.14)(216)$$
$$\approx 904.32 \text{ in.}^3$$

(b) $S = 4\pi r^2$
$$= 4(3.14)(6)^2$$
$$= 4(3.14)(36)$$
$$= 452.16 \text{ in.}^2$$

23. (a) $V = lwh$
$$= 12 \cdot 9 \cdot 8$$
$$= 864 \text{ ft}^3$$

(b) $S = 2lh + 2hw + 2lw$
$$= 2 \cdot 12 \cdot 8 + 2 \cdot 8 \cdot 9 + 2 \cdot 12 \cdot 9$$
$$= 192 + 144 + 216$$
$$= 552 \text{ m}^2$$

24. (a) $V = \pi r^2 h$
$$= (3.14)(6)^2(14)$$
$$= (3.14)(36)(14)$$
$$= 1582.56 \text{ m}^3$$

(b) $S = 2\pi r^2 + 2\pi rh$
$$= 2(3.14)(6)^2 + 2(3.14)(6)(14)$$
$$= 2(3.14)(36) + 2(3.14)(6)(14)$$
$$= 226.08 + 572.52$$
$$= 753.60 \text{ m}^2$$

25. Writing exercise

26. (a) A page of a book and the cover of the same book are topologically equivalent because they both have no holes; they are of genus 1.

(b) A pair of glasses with the lenses removed and the Mona Lisa are not topologically equivalent. The glasses have two holes, but the Mona Lisa has none.

27. (a) Yes.

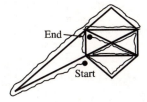

(b) No, because the network has more than two odd vertices.

28. Given $x = .6$ and formula $y = 2.1x(1 - x)$

$$y = 2.1(.6)(1 - .6)$$
$$= 2.1(.6)(.4)$$
$$= .504$$

$$y = 2.1(.504)(1 - .504)$$
$$= 2.1(.504)(.496)$$
$$\approx .525$$

$$y = 2.1(.525)(1 - .525)$$
$$= 2.1(.525)(.475)$$
$$\approx .524$$

$$y = 2.1(.524)(1 - .524)$$
$$= 2.1(.524)(.476)$$
$$\approx .524$$

The only attractor, rounded to three decimal places is .524.

10   COUNTING METHODS

Chapter Goals

The general goal of this chapter is to provide the student with several mathematical tools and procedures as an aid to answering the question—How many? Specifically, the study of this chapter should enable the student to

- List the outcomes of one, two, and three part tasks by systematic methods.

- Evaluate the number of outcomes for a given task by applying the fundamental counting principle and formulas for permutations and combinations.

- Use Pascal's triangle as an aid to generate combination values and coefficients for the binomial theorem.

- Apply methods of indirect counting to evaluate the number of outcomes when appropriate.

Chapter Summary

Counting methods are useful in many areas of mathematics but are especially important in the study of probability. A **cardinal number** answers the question "How many?" The set of cardinal numbers is the same as the set of whole numbers or $\{0, 1, 2, 3, \dots\}$. When asking "How many?" it becomes useful to conceive of the objects or items in question as elements of a set. We then speak of the **cardinality** of the set (or the number of elements in the set). For example, the cardinality of the set $\{a, b, c\}$ is three, or symbolically, if we let $A = \{a, b, c\}$, then $n(A) = 3$. The cardinality, $n(\emptyset)$, of the empty set is 0, and so on.

We can find cardinalities—or answer the question of "How many?"—by listing the elements of a set and then counting them. Two typical listing methods include the use of **tree diagrams** and **product tables**. Product tables may be used when an outcome involves two tasks (or a task may be considered to have two parts). Tree diagrams are useful if two or more tasks are indicated and the number of branches is not excessive. These methods allow us to systematically list occurrences, outcomes, etc. (i.e., elements of sets) and have the advantage of enabling one to see the members, thus categorizing them if needed. See EXAMPLES A and B.

The word "fair" means that each possible outcome has the same likelihood of occurring. Tossing a "fair" coin, die, etc. is often used in examples related to counting methods and probabilities.

EXAMPLE A   Product table–two part task

Julie tosses a fair dime into the air followed by a red die. Construct a table showing all possible outcomes.

Solution

| | | | red die | | | |
coin	1	2	3	4	5	6
Heads	$(H,1)$	$(H,2)$	$(H,3)$	$(H,4)$	$(H,5)$	$(H,6)$
Tails	$(T,1)$	$(T,2)$	$(T,3)$	$(T,4)$	$(T,5)$	$(T,6)$

Tree diagrams are especially useful to organize rather complex conditions as is shown in the next example.

EXAMPLE B   Tree diagram–three part task

Oscar is ordering dinner at his favorite restaurant. He can have either soup (clam chowder or vegetable) or salad (only shrimp salad is available today). The available entrees are beef, ham, and fresh salmon. Cheesecake and apple pie are offered for dessert. Oscar has found that his stomach cannot cope with two kinds of seafood in the same meal. Furthermore, he cannot handle both vegetables and cheesecake, and any time he has both shrimp and ham, he must totally avoid dessert. Use a tree diagram to determine the number of choices Oscar has for tonight's meal.

Solution
Set up a tree diagram which satisfies the above conditions and indicates at each stage the possible results for the decision (choice of soup or salad, entree, and dessert) to be made.

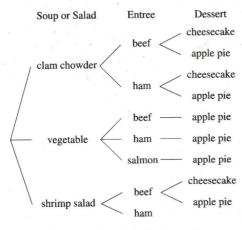

Total number of choices: 10

Often it is not necessary or practical to list all of the outcomes. The **fundamental counting principle** will, in many cases, enable us to compute the total number of outcomes. When a task consists of several parts (or several tasks are indicated to arrive at an outcome), we can multiply the number of ways each part may be accomplished together to arrive at the total number of ways to complete the entire task(s) This method of counting by products generalizes as follows:

Fundamental counting principle   When a task consists of k separate parts, if the first part can be done in n_1 ways, the second part can be done in n_2 ways, and so on through the kth part, which can be done in n_k ways; then the total number of possible results for completing the task is given by the product $n_1 \cdot n_2 \cdot n_3 \cdot \cdot \cdot n_k$.

EXAMPLE C   Fundamental counting principle

Felix is having lunch with his friend Oscar (EXAMPLE B above). But Felix has a much stronger stomach, and combinations of foods do not bother him. He also is trying to decide the number of choices he has for the evening's meal.

Solution
We may consider the problem of deciding how many meals are possible as a three part task—choosing the soup (or salad), choosing the entree, and choosing the dessert. There are three choices for the soup or salad (clam chowder, vegetable soup, or shrimp salad), three choices for the entree (beef, ham, or salmon), and two choices for the dessert (cheesecake or apple pie). Since Felix can eat any combination, we may apply the fundamental counting principle to give $\underline{3} \cdot \underline{3} \cdot \underline{2} = 18$ possible meals.

In building a tree diagram, different stages are developed with each stage consisting of several branches. Each stage corresponds to a task (or part of a task) when using the fundamental counting principle, and the number of branches corresponds to the number of ways the individual task can be performed. Thus, a tree diagram with three stages, where there are two branches in the first stage, three branches in the second stage, and two more branches in the third stage, will have a total of $\underline{2} \cdot \underline{3} \cdot \underline{2} = 12$ separate branches (routes or outcomes) by the fundamental counting principle. A product table with four columns and five rows represents a two part task—choosing the element in a row (4 possibilities) followed by choosing an element in a column (5 possibilities), and the fundamental counting principle indicates that there will be $\underline{4} \cdot \underline{5} = 20$ outcomes or entries in the table.

If a problem boils down to choosing r items from a set of n items, and repetitions are not allowed, it can usually be done with either the use of **permutations** or **combinations** as appropriate. The key as to which to use is to decide whether or not the order of the r items makes a difference. If the order in which items are selected makes a difference (<u>order is important</u>), then each outcome or result is a **permutation** (arrangement), and a permutations formula may be used. See EXAMPLE D.

If the <u>order</u> in choosing items from a set is <u>not important</u>, then the result is considered a **combination**, and the combinations formula may be used. See EXAMPLE E.

The following are very useful formulas for the calculation of permutations and/or combinations. (REMEMBER: Repetitions are not allowed.) See **Guidelines for Choosing a Counting Method** in the text.

Permutations formulas

Order of elements is important as in arrangements

$$P(n, r) = n(n-1)(n-2) \cdots (n-r+1)$$

- *Useful for hand calculation*
- *Observe that there are "r" factors in the product*

$$P(n, r) = \frac{n!}{(n-r)!}$$

- *Useful when using a calculator with "!" button*

$$P(n, n) = n! = n(n-1)(n-2) \cdots 2 \cdot 1$$

- *Special case of above, where $r = n$*

Combinations formula

Order of elements is unimportant

$$C(n, r) = \frac{P(n, r)}{r!} = \frac{n!}{r!(n-r)!}$$

- *Useful for counting subsets of r-elements*

EXAMPLE D   Permutations

(a) Five children in Mrs. Blixt's first grade class raise their hands simultaneously to be excused to use the restrooms. But school policy dictates that only one student may leave the classroom at a time. How many ways can Mrs. Blixt choose the 1st, 2nd, 3rd, 4th, and 5th child to leave the classroom?

Solution

Since the order of leaving is important, we may consider this as a permutation problem. The number of arrangements of the five children taken five at a time equals $P(5, 5) = 5! = 5 \cdot 4 \cdot 3 \cdot 2 \cdot 1 = 120$. Thus, there are 120 different orders in which Mrs. Blixt could choose the children to leave the classroom.

(b) If the bell for recess rings just after the third child returns to the classroom, how many possible orders could Mrs. Blixt have chosen from?

Solution

Since she could order any of the five students only three at a time before the bell rang, we can compute
$P(5, 3) = \underline{5} \cdot \underline{4} \cdot \underline{3} = 60$ possible orders.

Note: This same problem (a or b) could easily be worked by applying the fundamental counting principle. In the case of (b), the first task is to choose the first student to leave the room (5 possibilities), followed by the second task of choosing the second student to leave (4 possibilities), followed by the third task of choosing the third student to leave (3 possibilities). Multiplying these values together ($5 \cdot 4 \cdot 3 = 60$), we arrive at the same product as above.

EXAMPLE E   Combinations

How many different four-member committees could be formed from the 100 U.S. senators?

Solution

Since a committee can be selected without regard to order, we may consider the problem as that of combinations—in the language of sets, the number of four-member subsets that can be formed from a set consisting of one hundred elements or

$$\begin{aligned}
C(100, 4) &= \frac{100!}{4!(100-4)!} \\
&= \frac{100 \cdot 99 \cdot 98 \cdot 97 \cdot 96!}{4! \cdot 96!} \\
&= \frac{100 \cdot 99 \cdot 98 \cdot 97}{4 \cdot 3 \cdot 2 \cdot 1} \\
&= 3,921,225
\end{aligned}$$

Be sure to divide out like factors before actual multiplication.

Pascal's triangle (see text for examples) offers another way to generate combinations. Combinations are used extensively in mathematics. Coefficients, for example, of the binomial expansion of $(a + b)^n$ can be generated by the formula for combinations, or by Pascal's triangle. See the text for a discussion of the Binomial Theorem [expansion of $(a + b)^n$].

In order to ease the computation (especially useful for more complex problems) indirect counting is often used. The number of ways, for instance, that a certain condition can be satisfied is the total number of possible results minus the number of ways the condition would not be satisfied. Using set notation, this **complements principle of counting** can be stated as follows:

$$n(A) + n(A') = n(U), \text{ or}$$
$$n(A) = n(U) - n(A')$$

- *Used if one knows, or can readily calculate, the total number of outcomes, $n(U)$, and the number of ways that the condition for set A is not satisfied, $n(A')$.*

EXAMPLE F Complements principle

If you toss eight fair coins, how many ways can you obtain at least one head?

Solution

By the fundamental counting principle, there are $2^8 = 256$ possible ways (outcomes) if you toss eight fair coins (two outcomes for the first {H or T} multiplied by two for the second, ... times two for the eighth). Since it is easier to calculate A' (no heads)—there is only one, all tails, than A (at least one head), use the complements formula

$$n(A) = n(U) - n(A') \text{ or}$$
$$n(A) = 256 - 1$$
$$= 255$$

ways to get at least one head.

Another useful indirect counting method used is indicated by the **special additive counting principle**. If two sets (or conditions) have no common elements (conditions cannot occur simultaneously), then the number of elements in the union (one set of conditions "or" the other) of the two sets is given by the sum of the individual cardinalities or

$$n(A \cup B) = n(A) + n(B)$$

- *Used if we are "or"-ing conditions or sets and the sets are disjoint; i.e., $(A \cap B) = \emptyset$.*

The **general additive counting principle** is used especially if the two sets which we are "or"-ing are not disjoint; that is,

they have common elements. But it may be used whether the sets are disjoint or not since the last term, $n(A \cap B)$, reduces to 0 if the sets are disjoint. It says that the number of elements in the union of two sets is given by the sum of the number of elements in each individual set minus the number of elements in their intersection or

$$n(A \cup B) = n(A) + n(B) - n(A \cap B)$$

- *Used if there are some common elements in A and in B.*

Note: Since $(A \cap B) \subseteq A$ and $(A \cap B) \subseteq B$, counting the elements in A, i.e., $n(A)$, and in B, i.e., $n(B)$, will count the elements of $(A \cap B)$ twice. This is corrected by simply subtracting one counting, $n(A \cap B)$, from the sum.

EXAMPLE G Disjoint—Special Additive Principle

Verify that there are four ways to get a royal flush.

Solution

There is only one royal flush to be obtained for each suit. We can use the special additive principle (since each suit is disjoint) arriving at

$$1 + 1 + 1 + 1 = 4.$$

EXAMPLE H Common elements–General Additive Principle

If a single card is drawn from a standard 52-card deck, in how many ways could it be a club or a black card?

Solution

Using the <u>general</u> additive principle because some cards are both black and clubs (i.e., the sets are not disjoint), we count the number of black cards (26) + the number of clubs (13)—the number of cards that are both black and clubs (13) = 13.

Note: We could have reasoned that the only black cards that were not clubs are the spades (13) and subtracted this number from the total number of black cards.

10.1 EXERCISES

In Exercises 1–8 consider the set N = {A, B, C, D, E} for {Andy (a man), Bill (a man), Cathy (a woman), David (a man), and Evelyn (a woman)}. List and count the different ways of electing each of the following slates of officers.

1. A president and a treasurer.

   Agreeing that the first letter represents the president and that the second represents the treasurer, we can generate systematically the following symbolic list and count the resulting possibilities: AB, AC, AD, AE; BA, BC, BD, BE; CA, CB, CD, CE; DA, DB, DC, DE; EA, EB, EC, ED. By counting, there are 20 ways to elect a president and treasurer.

3. A president and a treasurer if the two officers must not be the same sex.

   Since the men include A, B, and D, and the women are C and E, we must also eliminate those doubles from the same list that represent all men (AB, BA, AD, DA, BD, and DB) and those representing all women (CE and EC). We have lost a total of 8 from the original list of 20 (Exercise 1), leaving 12 ways to elect a president and treasurer. The results include: AC, AE, BC, BE, CA, CB, CD, DC, DE, EA, EB, ED. Note that a new list with the above restrictions could be generated rather than subtracting from the original list.

5. A president, a secretary, and a treasurer, if the president must be a man and the other two must be women.

   Generating a new symbolic list where the first member must be a man and the second and third, women, we get

   ACE, AEC, BCE, BEC, DCE, and DEC.

   The officers may be elected in 6 different ways.

List and count the ways club N could appoint a committee of three members under the following conditions.

7. There are no restrictions.

   One method would be to list all triples. Remembering, however, that ABC is the same committee as BAC or CAB, cross out all triples with the same three letters. We are left with: ABC, ABD, ABE, ACD, ACE, ADE, BCD, BCE, BDE, CDE. Therefore, there are 10 ways to select the 3-member committees with no restrictions.

For Exercises 9–25, refer to Table 2 (the product table for rolling two dice) in the text.

9. Only 1 member of the product table (1, 1) represents an outcome where the sum of the dice is two.

11. Only 3 members of the product table (3, 1), (2, 2), (1, 3) represent outcomes where the sum of the dice is four.

13. There are 5 members of the product table, (5, 1), (4, 2), (3, 3), (2, 4), and (1, 5), which represent outcomes where the sum is six.

15. There are 5 members of the product table, (6, 2), (5, 3), (4, 4), (3, 5), and (2, 6), which represent outcomes where the sum is 8.

17. There are only 3 members of the product table, (6, 4), (5, 5), and (4, 6), which represent outcomes where the sum of the dice is ten.

19. Only 1 member, (6, 6), of the product table yields an outcome where the sum is twelve.

21. Half of all 36 outcomes suggested by the product table should represent a sum which is even; the other half, odd. Thus, there are 18 outcomes which will be even. They are:

    (1, 1), (3, 1), (2, 2), (1, 3), (5, 1), (4, 2), (3, 3), (2, 4), (1, 5), (6, 2), (5, 3), (4, 4), (3, 5), (2, 6), (6, 4), (5, 5), (4, 6), (6, 6).

23. To find the sums between 6 and 10, we must count pairs in which the sum is 7, 8, or 9.

    Sum is 7: (1, 6), (2, 5), (3, 4), (4, 3), (5, 2), (6, 1).

    Sum is 8: (2, 6), (3, 5), (4, 4), (5, 3), (6, 2).

    Sum is 9: (6, 3), (5, 4), (4, 5), (3, 6).

    Since there are six pairs with a sum of 7, five pairs with a sum of 8, and 4 pairs with a sum of 9 there are:

    $$6 + 5 + 4 = 15$$

    pairs with a sum of 6 through 8 inclusive.

25. Construct a product table showing all possible two-digit numbers using digits from the set {1, 2, 3, 4, 5, 6}.

		2nd digit				
		1	2	3	4	5	6
	1	11	12	13	14	15	16
	2	21	22	23	24	25	26
1st	3	31	32	33	34	35	36
digit	4	41	42	43	44	45	46
	5	51	52	53	54	55	56
	6	61	62	63	64	65	66

27. The following numbers in the table are numbers with repeating digits:

    11, 22, 33, 44, 55, 66.

29. A counting number larger than 1 is prime if it is divisible by itself and 1 only. The following numbers in the table are *prime numbers*:

$$11, 13, 23, 31, 41, 43, 53, 61.$$

31. The following are (perfect) *square* numbers:

$$16, 25, 36, 64.$$

33. The numbers 2^4, 2^3, and 2^6 are powers of 2 found in the product table from Exercise 25. That is,

$$16, 32, \text{ and } 64.$$

35. Extend the tree diagram of Exercise 34 for four fair coins. Then list the ways of getting the following results.

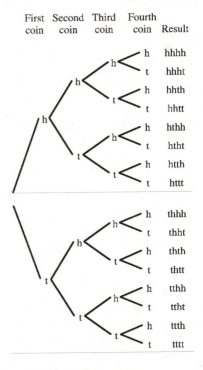

First Second Third Fourth
coin   coin   coin   coin   Result

(a) More that three tails

There is only one such outcome:

tttt.

(b) Fewer than three tails

List those outcomes with 0, 1 or 2 tails:

hhhh, hhht, hhth, hhtt,
hthh, htht, htth, thhh
thht, thth, tthh.

(c) At least three tails

List those outcomes with 3 or 4 tails:

httt, thtt, ttht, ttth, tttt.

(d) No more than three tails

List those outcomes with 0, 1, 2, or 3 tails:

hhhh, hhht, hhth, hhtt, hthh,
htht, htth, httt, thhh, thht,
thth, thtt, tthh, ttht, ttth.

37. Begin with the largest triangles which have the long diagonals as their bases. There is 1 triangle on each side of the (2) diagonals. This gives 4 large triangles. Count the intermediate size triangles, each with a base along the outside edge of the large square. There are 4 of these. Furthermore, each of these intermediate sized triangles contain two right triangles within. There are a total of 8 of these. Thus, the total number of triangles is

$$4 + 4 + 8 = 16.$$

39. Begin with the larger right triangle at the center square. There are 4. Pairing two of these triangles with each other forms 4 isosceles triangles within the center box. Within each of the four right triangles in the square are two smaller right triangles for a total of 8. Associated with each exterior side of the octagon are 8 triangles each containing two other right triangles (one of which has already been counted) for a total of 16. There are 4 more isosceles triangles which have their two equal side lengths as exterior edges of the octagon.

Thus, the number of triangles contained in the figure is

$$4 + 4 + 8 + 16 + 4 = 44.$$

41. Label the figure as shown below, so that we can refer to the small squares by number.

		1	2	
	3	4	5	6
	7	8	9	10
		11	12	

Find the number of squares of each size and add the results.

There are twelve 1×1 squares, which are labeled 1 through 12.

Name the 2×2 squares by listing the small squares they contain:

1, 2, 4, 5	5, 6, 9, 10
8, 9, 11, 12	3, 4, 7, 8
4, 5, 8, 9	

There are five 2×2 squares. There are no squares larger than 2×2.

Thus, there are a total of

$$12 + 5 = 17$$

squares contained in the figure.

43. Examine carefully the figure in the text.

There are sixteen 1×1 squares with horizontal bases.

There are three 2×2 squares in the each of the first and second rows, the second and third rows, as well as the third and fourth rows with horizontal bases. Thus, there are a total of nine 2×2 squares with horizontal bases.

There are two 3×3 squares with horizontal bases found in the first, second, and third rows as well as two 3×3 squares with horizontal base found in the second, third and fourth rows. Thus, there are a total of four 3×3 squares with horizontal bases.

There is one 4×4 square (the large square itself).

Visualize the squares along the diagonals (at a slant).

There are twenty-four 1×1 squares with bases along diagonals.

There are thirteen 2×2 squares with bases along diagonals.

There are four 3×3 squares with bases along diagonals.

There is only one 4×4 square with bases along diagonals.

Add the results:

Size	Number of squares
1×1 (horizontal)	16
1×1 (slant)	24
2×2 (horizontal)	9
2×2 (slant)	13
3×3 (horizontal)	4
3×3 (slant)	4
4×4 (horizontal)	1
4×4 (slant)	1
	72

There are 72 squares in the figure.

45. There are $3 \times 3 = 9$ cubes in each of the bottom two layers. This gives a total of 18 in the bottom two layers.

There are $3 \times 2 = 6$ cubes in each of the middle to layers. This gives a total of 12 in the bottom two layers

There are $3 \times 1 = 3$ cubes in the top two layers. This gives a total of 6 the bottom two layers. All together there are

$$18 + 12 + 6 = 36$$

$(1 \times 1 \times 1)$ cubes.

The visible cubes are:

Location	Number of cubes
Top two layers	6
Middle two layers	8
Bottom two layers	10
(exclude cubes in corners which have already been counted)	24

Thus, the number of cubes in the stack that are not visible is: $36 - 24 = 12$. One could ignore the top two levels since each cube is visible.

47. There are $4 \times 4 = 16$ small cubes on the bottom layer; $3 \times 3 = 9$ on the second layer; $2 \times 2 = 4$ on the third layer, and 1 on the top layer. This gives a total of

$$16 + 9 + 4 + 1 = 30$$

$(1 \times 1 \times 1)$ cubes. Of these, the following are visible.

There are 4 cubes along each edge of the bottom layer for a total of 10 cubes. There are 3 cubes along each edge of the second layer for a total of 6 cubes. There are 2 cubes along each edge of the third layer for a total of 3 cubes. Remember not to count the back corner cube twice. The top layer cube is visible, so ignore it. Thus, there are a total of

$$10 + 6 + 3 = 19$$

$(1 \times 1 \times 1)$ cubes in the bottom three layers. Of these, the following are visible.

Location	Number of cubes
Bottom layer	4
Second layer	3
Third layer	2
	9

Thus, the number of cubes in the stack that are not visible is: $19 - 9 = 10$.

49. Label the figure as shown below.

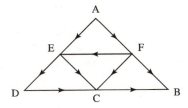

List all the paths in a systematic way.

AFB, AEFB, AECB, AEDCB, AFECB, AFEDCB

represent all paths with the indicated restrictions. Thus, there are 6 paths.

51. To determine the number of ways in which 40 can be written as the sum of two primes, use trial and error in a systematic manner. Test each prime, starting with 2, as a possibility for the smaller prime.

(Since $30 - 2 = 28$, and 28 is not a prime, 2 will not work.)

We obtain the following list:

$$40 = 3 + 37$$
$$40 = 11 + 29$$
$$40 = 17 + 23.$$

Thus, 40 can be written as the sum of two primes in 3 different ways.

53. A group of twelve strangers sat in a circle, and each one got acquainted only with the person to the left and the person to the right. Then all twelve people stood up and each one shook hands (once) with each of the others who was still a stranger. How many handshakes occurred?

One strategy is to place and number each person as on a 12-hour clock. Beginning with person #1: he will shake hands with 9 people. This is also true for #2. But person #3 will only shake hands with 8 people, having already shaken hands with #1. Similarly, #4 will shake hands with 7 people, #5 with 6 people, #6 with 5 people, #7 with 4 people, #8 with 3 people, #9 with 2 people, #10 with 1 person.

Adding, we get 54 handshakes.

55. How many of the numbers from 10 through 100 have the sum of their digits equal to a perfect square?

One strategy would be to list the numbers from 10 to 100. Then check each by adding digits: e.g., 10, yes (since $1 + 0 = 1$, a perfect square number). Similarly, 13, yes (since $1 + 3 = 4$) and so on ... until 100, yes (since $1 + 0 + 0 = 1$). In total there are 18 such numbers: {10, 13, 18, 22, 27, 31, 36, 40, 45, 54, 63, 72, 81, 90, 79, 88, 97, 100}.

57. Make a systematic list, or table, of numbers between 100 and 400 which contain the digit 2. As you are listing the numbers look for patterns.

	102	112					
120–129							
	132	142	152	162	172	182	192
200–299							
	302	312					
320–329							
	332	342	352	362	372	382	392

This table shows that there are

$$2 + 10 + 7 + 100 + 2 + 10 + 7 = 138 \text{ numbers}$$

between 100 and 400 which contain the digit 2.

59. Draw a tree diagram showing all possible switch settings.

First Switch	Second Switch	Third Switch	Fourth Switch	Switch Settings
			0	0 0 0 0
		0	1	0 0 0 1
	0		0	0 0 1 0
		1	1	0 0 1 1
0		0	0	0 1 0 0
			1	0 1 0 1
	1	1	0	0 1 1 0
			1	0 1 1 1
		0	0	1 0 0 0
	0		1	1 0 0 1
		1	0	1 0 1 0
1			1	1 0 1 1
		0	0	1 1 0 0
	1		1	1 1 0 1
		1	0	1 1 1 0
			1	1 1 1 1

Thus, Michelle can choose 16 different switch settings.

61. There are five switches rather than four, and no two adjacent switches can be on. If no two adjacent switches can be on, the tree diagram that is constructed will not have two "1"s in succession.

First Switch	Second Switch	Third Switch	Fourth Switch	Fifth Switch	Switch Settings
				0	0 0 0 0 0
			0	1	0 0 0 0 1
		0	1	0	0 0 0 1 0
	0	1	0	0	0 0 1 0 0
				1	0 0 1 0 1
0			0	0	0 1 0 0 0
	1	0		1	0 1 0 0 1
			1	0	0 1 0 1 0
		0	0	0	1 0 0 0 0
	0			1	1 0 0 0 1
1	0	1	0	0	1 0 0 1 0
		1	0	0	1 0 1 0 0
				1	1 0 1 0 1

Thus, Michelle can choose 13 different switch settings.

63. A line segment joins the points $(8, 12)$ and $(53, 234)$ in the Cartesian plane. Including its endpoints, how many lattice points does this line segment contain? (A lattice point is a point with integer coordinates.)

Any point (x, y) on the line segment, when used with either endpoint, must yield the same slope as that found of the segment using both endpoints. Therefore, find the slope of the segment.

$$m = \frac{y_2 - y_1}{x_2 - x_1}$$
$$= \frac{234 - 12}{53 - 8}$$
$$= \frac{222}{35}$$
$$= \frac{24}{15}$$

Set up the slope using the unknown point (x, y) and the known endpoint $(8, 12)$.

$$m = \frac{y_2 - y_1}{x_2 - x_1}$$
$$= \frac{y - 12}{x - 8}$$

Since the slope is the same for all points on the line segment, set these equal to each other and solve for y (in terms of x).

$$\frac{y - 12}{x - 8} = \frac{24}{15}$$
$$y - 12 = \frac{24}{15}(x - 8)$$
$$y = \frac{24}{15}(x - 8) + 12$$

All points on the line segment must be solutions for this equation. For the solutions to be integers (with x between 8 and 53), the denominator 15 will have to divide the value $(x - 8)$ evenly so that y remains an integer. That is, the number "$x - 8$" must be a multiple of 15 for values of x. Systematically trying integers for x from and including 8 to 53 will yield the following results. All other values for x between 8 and 53 would not.

x	$x - 8$	Divisible by 15?
8	$8 - 8 = 0$	Yes
23	$23 - 8 = 15$	Yes
38	$38 - 8 = 30$	Yes
53	$53 - 8 = 45$	Yes

Thus, including the endpoints, there are 4 lattice points.

Note that using a graphing calculator with a table feature would provide a quicker solution. Set up the function

$$y = \frac{24}{15}(x - 8) + 12$$

in the calculator. Adjust the "table set" feature to begin at $x = 8$ with $\triangle x$ set to increase by 1 and create the table (set of ordered pairs) for x and y. Scanning the table for those y-values which are whole numbers will yield the same answers as above.

65. Uniform length matchsticks are used to build a rectangular grid as shown in the figure (in the text). If the grid is 15 matchsticks high and 28 matchsticks wide, how many matchsticks are used?

Each row will contain 28 matchsticks. If the grid is 15 matchsticks high there will be 16 rows of matchsticks (including the top and bottom rows). Therefore, there are $16 \times 28 = 448$ matchsticks in all of the rows. Each column contains 15 matchsticks. Thus, there are $15 \times 29 = 435$ matchsticks in the rows. If the grid is 28 matchsticks wide, then there will be 29 columns of matchsticks counting the first and last columns. Thus, there are $29 \times 15 = 435$ matchsticks in the columns. Altogether, the number of matchsticks are

$$448 + 435 = 883.$$

67. Writing exercise

69. (a) Determine the number of two-digit numbers that can be formed using digits from the set $\{1, 2, 3\}$ if repetition of digits is allowed.

    (b) Determine the number of two-digits numbers that can be formed using digits from the set $\{1, 2, 3\}$ if the selection is done with replacement.

71. (a) Find the number of ways to select an ordered pair of letters from the set $\{A, B, C, D, E\}$ if repetition of letters is not allowed.

    (b) Find the number of ways to select an ordered pair of letters from the set $\{A, B, C, D, E\}$ if the selection is done without replacement.

Answers may vary for Exercises 73–77.

73. Examples 2, 3, 4, 5 make use of the strategy of "create a table or chart."

75. Examples 6, 7, 8, 10 make use of the strategy "draw a sketch."

77. Examples 1 through 10 make use of the strategy "use common sense."

10.2 EXERCISES

1. Writing exercise

3. (a) No, $(n + m)! \neq n! + m!$

    (b) Writing exercise

Evaluate each expression without using a calculator.

5. $4! = 4 \cdot 3 \cdot 2 \cdot 1 = 24$

7. $\dfrac{8!}{5!} = \dfrac{8 \cdot 7 \cdot 6 \cdot \cancel{5} \cdot \cancel{4} \cdot \cancel{3} \cdot \cancel{2} \cdot \cancel{1}}{\cancel{5} \cdot \cancel{4} \cdot \cancel{3} \cdot \cancel{2} \cdot \cancel{1}} = 336$

9. $\dfrac{5!}{(5-2)!} = \dfrac{5!}{(3)!} = \dfrac{5 \cdot 4 \cdot \cancel{3} \cdot \cancel{2} \cdot \cancel{1}}{\cancel{3} \cdot \cancel{2} \cdot \cancel{1}} = 20$

11. $\dfrac{9!}{6!(6-3)!} = \dfrac{9!}{6!(3)!}$

    $= \dfrac{9 \cdot 8 \cdot 7 \cdot \cancel{6} \cdot \cancel{5} \cdot \cancel{4} \cdot \cancel{3} \cdot \cancel{2} \cdot \cancel{1}}{\cancel{6} \cdot \cancel{5} \cdot \cancel{4} \cdot \cancel{3} \cdot \cancel{2} \cdot \cancel{1} \cdot (3 \cdot 2 \cdot 1)}$

    $= \dfrac{\cancel{9}^{3} \cdot \cancel{8}^{4} \cdot 7}{\cancel{3} \cdot \cancel{2}} = \dfrac{3 \cdot 4 \cdot 7}{1}$

    $= 84$

13. Evaluate
    $$\dfrac{n!}{(n-r)!}, \text{ where } n = 7 \text{ and } r = 4.$$
    $$\dfrac{7!}{(7-4)!} = \dfrac{7!}{(3)!}$$
    $$= \dfrac{7 \cdot 6 \cdot 5 \cdot 4 \cdot \cancel{3} \cdot \cancel{2} \cdot \cancel{1}}{\cancel{3} \cdot \cancel{2} \cdot \cancel{1}}$$
    $$= 840$$

Evaluate each expression using a calculator. (Some answers may not be exact.) For Exercises 15–24, use the factorial key on a calculator, which is labeled $\boxed{x!}$ or $\boxed{n!}$ or, if using a graphing, calculator find "!" in the "Math" menu.

15. $11! = 39{,}916{,}800$

17. $\dfrac{12!}{7!} = 95{,}040$

19. $\dfrac{13!}{(13-3)!} = \dfrac{13!}{(10)!} = 1716$

21. $\dfrac{20!}{10! \cdot 10!} = 184{,}756$

23. Evaluate
    $$\dfrac{n!}{(n-r)!}, \text{ where } n = 23 \text{ and } r = 10.$$
    $$\dfrac{23!}{(23-10)!} = \dfrac{23!}{(13)!}$$
    $$= 4.151586701 \times 10^{12}$$

25. Since there are two possible outcomes for each switch (on/off), we have $2 \cdot 2 \cdot 2 = 2^3 = 8$.

27. Writing exercise

29. Using the fundamental counting principle, this may be considered as a three–part task. There would be
    $$\underline{6} \cdot \underline{6} \cdot \underline{6} = 6^3 = 216.$$

Recall the club $N = \{Andy, Bill, Cathy, David, Evelyn\}$.

31. This is a 5-part task. Use the fundamental counting principle to count the number of ways of lining up all five members for a photograph. There would be
    $$\underline{5} \cdot \underline{4} \cdot \underline{3} \cdot \underline{2} \cdot \underline{1} = 120 \text{ possibilities.}$$

    Similarly, one could use $n!$, where $n = 5$, to arrive at the total number or arrangements of a set of n objects.

33. This is a 2-part task. Since there are three males to choose from and two females to choose from, the number of possibilities is
    $$\underline{3} \cdot \underline{2} = 6.$$

In the following exercises, counting numbers are to be formed using only digits from the set {3, 4, 5}.

35. Choosing two-digit numbers may be considered a 2-part task. Since we can use any of the three given digits for each choice, there are

$$\underline{3} \cdot \underline{3} = 9$$

different numbers that can be obtained.

37. Using the textbook hint, this may be considered a 3-part task. (1) Since there are only 3 positions that the two adjacent 4's can take (1st and 2nd, 2nd and 3rd, and 3rd and 4th positions), (2) two remaining positions that the 3 can take, and (3) the one last remaining digit must filled by the 5, there are

$$\underline{3} \cdot \underline{2} \cdot \underline{1} = 6$$

different numbers that may be created.

39. Choosing from each of the three food categories is a 3 part task. There are four choices from the soup and salad category, two from the bread category, and three from the entree category. Applying the fundamental counting principle gives

$$4 \cdot 2 \cdot 3 = 24$$

different dinners that may be selected.

41. Since there are 2 choices (T or F) for each question, we have

$$2^6 = 64$$

possible ways.

For each situation in Exercises 43–48, use the table in the text to determine the number of different sets of classes Tiffany can take.

43. All classes shown are available. Choose the number of possible courses from each category and apply the fundamental theorem:

$$\underline{3} \cdot \underline{3} \cdot \underline{4} \cdot \underline{5} = 180.$$

45. All sections of Minorities in America and Women in American Culture are filled already. The filled classes reduce the options in the sociology category by 2. Thus, there are

$$\underline{3} \cdot \underline{3} \cdot \underline{4} \cdot \underline{3} = 108$$

possible class schedules.

47. Funding has been withdrawn for three of the computer courses and for two of the sociology courses. The reductions to the computer and sociology categories leave only 1 computer course and 3 sociology courses to choose from. Thus, there are

$$\underline{3} \cdot \underline{3} \cdot \underline{1} \cdot \underline{3} = 27$$

possible class schedules.

49. This is a 3-part task. Applying the fundamental counting principle, there are

$$\underline{2} \cdot \underline{5} \cdot \underline{6} = 60$$

different outfits that Sean may wear.

51. The number of different zip codes that can be formed using all of those same five digits, 86726, would be the number of arrangements of the 5 digits. This is given by

$$\underline{5} \cdot \underline{4} \cdot \underline{3} \cdot \underline{2} \cdot \underline{1} = 5! = 120.$$

Andy, Betty, Clyde, Dawn, Evan, and Felicia have reserved six seats in a row at the theater, starting at an aisle seat. (Refer to Example 7 in this section.)

Aaron (A), Bobbette (B), Chuck (C), Deirdre (D), Ed (E), and Fran (F) have reserved six seats in a row at the theater, starting at an aisle seat.

53. Using the textbook hint, divide the task into the series of six parts shown below, performed in order.

(a) If A is seated first, how many seats are available for him?

Six seats are available.

(b) Now, how many are available for B?

Five are available since A is already seated.

(c) Now, how many are available for C?

Four are available since A and B are already seated.

(d) Now, how many for D?

Three, since A, B, and C are already seated.

(e) Now, how many for E?

Two, since A, B, C, and D are already seated.

(f) Now, how many for F?

One, since all other seats have been taken.

Thus, by the fundamental counting principle, there are

$$6 \cdot 5 \cdot 4 \cdot 3 \cdot 2 \cdot 1 = 720$$

arrangements.

55. In how many ways can they arrange themselves if the men and women are to alternate seats and a man must sit on the aisle? Using the textbook hint, first answer the following series of questions:

(a) How many choices are there for the person to occupy the first seat, next to the aisle? (It must be a man.)

Three, since there are only 3 men.

(b) Now, how many choices of people may occupy the second seat from the aisle? (It must be a woman.)

Three, since there are only 3 women.

(c) Now, how many for the third seat (one of the remaining men)?

Two, since one of the men has already been seated.

(d) Now, how many for the fourth seat (a woman)?

Two, since one of the women has already been seated.

(e) Now, how many for the fifth seat (a man)?

One since only one man remains.

(f) Now, how many for the sixth seat (a woman)?

One, since only one woman remains.

Thus, there are

$$3 \cdot 3 \cdot 2 \cdot 2 \cdot 1 \cdot 1 = 36$$

arrangements.

57. Writing exercise

59. Repeat Example 4. This time, allow repeated digits. Does the order in which digits are considered matter in this case?

There are 9 choices for the first digit since it can't be 0. There are 10 choices for the second digit. Finally, there are only 5 choices for the third (unit's) digit since it must be odd. Thus, by the fundamental counting principle there are

$$9 \cdot 10 \cdot 5 = 450.$$

10.3 EXERCISES

Evaluate each of the following expressions.

1. Begin at 6, use four factors.

$$P(6,4) = 6 \cdot 5 \cdot 4 \cdot 3 = 360$$

Alternatively, use the factorial formula for permutations

$$P(n,r) = \frac{n!}{(n-r)!}$$

where $n = 6$ and $r = 4$.

$$\begin{aligned}
P(6,4) &= \frac{6!}{(6-4)!} \\
&= \frac{6!}{(6-4)!} \\
&= \frac{6!}{2!} \\
&= 360
\end{aligned}$$

3. $$\begin{aligned}
C(15,3) &= \frac{P(15,3)}{3!} \\
&= \frac{P(15,3)}{3!} \\
&= \frac{15 \cdot 14 \cdot 13}{3 \cdot 2 \cdot 1} \\
&= \frac{5 \cdot 7 \cdot 13}{1} \\
&= 455
\end{aligned}$$

Alternatively, use the factorial formula for combinations

$$C(n,r) = \frac{n!}{r!(n-r)!}$$

where $n = 15$ and $r = 3$.

$$\begin{aligned}
C(15,3) &= \frac{15!}{3!(15-3)!} \\
&= \frac{15 \cdot 14 \cdot 13 \cdot 12!}{(3 \cdot 2 \cdot 1) \cdot 12!} \\
&= \frac{15 \cdot 14 \cdot 13}{3 \cdot 2} \\
&= 5 \cdot 7 \cdot 13 \\
&= 455
\end{aligned}$$

5. To find the number of permutations of 21 things taken 4 at a time, evaluate $P(21,4)$. Begin at 21, use four factors.

$$P(21,4) = 21 \cdot 20 \cdot 19 \cdot 18 = 143640$$

Thus, the answer is $143,640$.

Determine the number of combinations (subsets) of each of the following.

7.   To find the number of combinations of 10 things taken 5 at a time, evaluate $C(10,5)$.

$$C(10,5) = \frac{P(10,5)}{5!}$$
$$= \frac{10 \cdot 9 \cdot 8 \cdot 7 \cdot 6}{5 \cdot 4 \cdot 3 \cdot 2 \cdot 1}$$
$$= 252$$

Alternatively, use the factorial formula for combinations

$$C(n,r) = \frac{n!}{r!(n-r)!}$$

where $n = 10$ and $r = 5$.

$$C(10,5) = \frac{10!}{5!(10-5)!}$$
$$= \frac{10 \cdot 9 \cdot 8 \cdot 7 \cdot 6 \cdot 5!}{(5 \cdot 4 \cdot 3 \cdot 2 \cdot 1) \cdot 5!}$$
$$= \frac{10 \cdot 9 \cdot 8 \cdot 7 \cdot 6}{5 \cdot 4 \cdot 3 \cdot 2 \cdot 1}$$
$$= 252$$

Use a calculator to evaluate each expression.

9.   Use the nPr or $P(n,r)$ button on a scientific calculator in the following order 25 $\boxed{nPr}$ 8. Or, with a graphing calculator, find nPr. It is usually found in the probability menu. Insert, in the same order, 25 $\boxed{nPr}$ 8.

$$P(25,8) = 4.3609104 \times 10^{10}$$

11.   Writing exercise

13.   Writing exercise

15.   Since a first place winner can not also be a second or third place winner in the same race, repetitions are not allowed. Also order is important. Therefore, we are able to use permutations. The number of ways in which 1st, 2nd, and 3rd place winners can occur in a race with six runners competing is given by

$$P(6,3) = 6 \cdot 5 \cdot 4 = 120.$$

17.   With no repeated digits, there would be

$$P(10,4) = 10 \cdot 9 \cdot 8 \cdot 7 = 5040$$

possible PIN numbers.

19.   Since no repetitions are allowed and the order of selection is important, the number of ways to choose a the three prize winners is

$$P(10,3) = 10 \cdot 9 \cdot 8 = 720.$$

21.   Since no repetitions are allowed and the order of selection is important, the number of ways for the security team to visit the twelve offices is

$$P(12,12) = 12! = 479,001,600.$$

23.   Since the order of testing each sample is not important (samples are subsets), use combinations to arrive at

$$C(24,5) = 42,504$$

ways.

25.   Consider the five-member committees as subsets of the 100 U.S. Senators.

$$C(100,5) = \frac{100!}{5! \cdot (95)!} = 75,287,520.$$

Alternatively, we could use

$$C(100,5) = \frac{P(100,5)}{5!} = 75,287,520.$$

27.   Since order is not important and repetition is not allowed, use combinations. The number of different ways to select the lottery numbers is

$$C(39,7) = 15,380,937.$$

29.   Use the hint furnished in the text. Calculate the number of ways of selecting the downward paths. For any selection of the downward steps the horizontal paths are fixed. Thus, there are

$$C(10,3) = 120$$

different paths that may be followed to get from A to B.

31.   (a) Since repetitions are not allowed and the order of selection is unimportant, the number of different sets is

$$C(51,6) = \frac{51!}{6!45!} = 18,009,460.$$

(b) Since two of Diane's lottery numbers are already determined, she only needs to select 4 numbers from the remaining 49 numbers. (Notice that the rules require six *distinct* numbers, that is, repetitions are not allowed.) The number of ways in which Diane can complete her list is

$$C(49,4) = 211,876.$$

33. Use the fundamental counting principle, where the first task is to choose which suit. There are four possible suits to choose from. Once the choice of suit is made, the 2nd task is to choose how many ways one can select 5 cards from the 13 in any particular suit. Thus, the number of ways to choose five card hands of the same suit is

$$4 \cdot C(13, 5) = 5148.$$

35. Consider this to be a two-part problem: The number of choices for the first call letter, 2, followed by the number of arrangements of the following two letters can represent, $P(25, 2)$. Observe that permutations are used since no letter may be repeated and that we are looking for arrangements (order is important). Choose the two letters from the remaining 25 that have not been selected. Apply the fundamental counting principle to arrive at

$$2 \cdot P(25, 2) = 1200$$

sets of call letters.

37. There are $C(8, 2)$ different games to be played if each team plays another only once. Since the teams are scheduled to play each other three times we have

$$3 \cdot C(8, 2) = 3 \cdot 28 = 84$$

games altogether.

39. (a) Use the fundamental counting principle where the first task is to choose the girls and the 2nd is to choose the boys. Since the order is important and there are no repetitions, use permutations to calculate each. The number of different programs is

$$P(8, 8) \cdot P(7, 7) = 8! \cdot 7! = 203, 212, 800.$$

(b) Consider the problem as a three-part task: Choosing the girl to perform first, the boy to perform last, followed by choosing the remaining 13 performers. There are 8 possibilities for the girl, 7 possibilities for the boy, and $P(13, 13) = 13!$ possibilities for the other 13 performers. Thus, by the fundamental counting principle the number of programs is given by

$$8 \cdot 7 \cdot 13! = 348, 713, 164, 800.$$

(c) Since the first and the last are fixed, we have only 1 choice for the first and last performer. Using the fundamental counting principle, the number of programs is given by

$$1 \cdot 13! \cdot 1 = 6, 227, 020, 800.$$

(d) The first must be a girl–8 possibilities, followed by a boy–7 possibilities, followed by a girl–7 possibilities, etc. By use of the fundamental counting principle, there are

$$(8 \cdot 7) \cdot (7 \cdot 6) \cdot (6 \cdot 5) \cdot (5 \cdot 4) \cdot (4 \cdot 3) \cdot (3 \cdot 2) \cdot (2 \cdot 1)$$

possibilities. This is equivalent, by reordering the factors, to

$$8! \cdot 7! = 203, 212, 800.$$

(e) Choosing the first, eighth, and fifteenth positions first, we have $8 \cdot 7 \cdot 6$ ways. Follow this by choosing the remaining positions, 12! Altogether, the number of different program arrangements is given by

$$(8 \cdot 7 \cdot 6) \cdot 12! = 160, 944, 537, 600.$$

41. (a) Since there are 13 clubs and 39 non-clubs in the deck, the four clubs can be chosen in $C(13, 4)$ ways, and the one non-club can be chosen in 39 ways. Use the fundamental counting principle to combine the two tasks. The number of possible hands is

$$\begin{aligned} C(13, 4) \cdot 39 &= 715 \cdot 39 \\ &= 27, 885. \end{aligned}$$

(b) The two face cards can be chosen from the twelve face cards $C(12, 2)$ ways. The three non-face cards can be chosen from the forty non-face cards in $C(40, 3)$ ways. Using the fundamental counting principle to combine both tasks, the number of hands is

$$C(12, 2) \cdot C(40, 3) = 652, 080.$$

(c) The two red cards can be chosen in $C(26, 2)$ ways, the two clubs can be chosen in $C(13, 2)$ ways, and the spade can be chosen in 13 ways. Applying the fundamental counting principle, the number of possible hands is

$$C(26, 2) \cdot C(13, 2) \cdot 13 = \ = 329, 550.$$

43. Since each pair of points determines a different line, we are looking for the number of 2-element subsets of a set of 7 elements. Subsets are combinations, so the number of lines determined by seven points in a plane, no three of which are collinear, is

$$C(7, 2) = \frac{7!}{2!5!} = 21.$$

45. Use the fundamental counting principle since repetitions are allowed. We have 40 choices for the first, second, and third number. The total number of combinations is

$$40 \cdot 40 \cdot 40 = 40^3 = 64, 000.$$

47. Use the fundamental counting principle. The number of ways to select a winner from the first and from the second race is

$$6 \cdot 8 = 48.$$

Thus, buying 48 different tickets will assure a winner.

49. Follow the textbook hint and answer the 6 questions to solve the exercise.

   (a) There are $C(9, 2)$ ways to select two-person committees.

   (b) There are $C(7, 3)$ ways to select three-person committee.

   Note that there is only one way to choose the 4 members of the four-person committee. Since there are only 4 left after the other committees are selected, we must choose these 4.

   (c) There are 2 ways to select the chair of the two-person committee.

   (d) There are 3 ways to select the chair of the three-person committee.

   (e) There are 4 ways to select the chair of the four-person committee.

   Thus, the total number of ways to make these selections is

   $$C(9, 2) \cdot C(7, 3) \cdot 2 \cdot 3 \cdot 4 = 30,240.$$

51. (a) How many numbers can be formed using all six digits 2, 3, 4, 5, 6, and 7?

   Since the order or arrangement of these digits is important (each being a different number), we can consider the answer to the question to be

   $$P(6, 6) = 6! = 720$$

   different numbers.

   (b) The first number is 456,789; the second number is 456,798; the third number is 456,879 and so forth. The number of arrangements (permutations) of the last five digits is given by $5! = 120$. Thus, the 121st number is 546,789 (where we have moved to the sixth digit from right and interchanged the fifth and sixth digit–4 and 5). There are another 120 permutations (with this change on the fifth and sixth digits) bringing us to the 241st number: 645,789. In a similar manner, numbers beginning with the new first digit–6, we have another 120 permutations using the new set of numbers. For the 361st number, we change the first digit to seven, giving us 745,689. The 362nd number is 745,698. The 363rd number is 745,869 and the 364th number is 745,896

53. Five percent of the 60 students is $(.05)60$ or 3 students. Five percent of the 40 students is $(.05)40$ or 2 students. The number of ways that he can assign A-grades to the first class is $C(60, 3)$. The number of ways that he can

assign A-grades to the second class is $C(40, 2)$. Apply the fundamental counting principle to find the total number of ways the professor may assign A-grades to his students.

$$C(60, 3) \cdot C(40, 2) = 26,691,600$$

55. This question is the same as asking how many 5-element subsets can be formed from a set with 30 elements. The number of different samples is thus,

$$C(30, 5) = 142,506.$$

57. $C(12, 9) = \dfrac{12!}{9!(12 - 9)!}$

   $= \dfrac{12!}{9!3!}$

   $= 220$

   $C(12, 3) = \dfrac{12!}{3!(12 - 3)!}$

   $= \dfrac{12!}{3!9!}$

   $= 220$

   Thus, $C(12, 9) = C(12, 3)$.

59. (a) Since $P(n, r) = \dfrac{n!}{(n - r)!}$,

   $$P(n, 0) = \dfrac{n!}{(n - 0)!}$$

   $= \dfrac{n!}{n!}$

   $= 1.$

   (b) Writing exercise

10.4 EXERCISES

Read the following combination values directly from Pascal's triangle.

1. To find from the value of $C(4, 3)$ from Pascal's triangle read entry number 3 in row 4 (remember that the top row is row "0" and that in row 4 the "1" is entry "0").

$$C(4, 3) = 4$$

For exercises 2–8, refer to Table 5 in the text.

3. To find the value of $C(6, 4)$ from Pascal's triangle, read entry number 4 in row 6.

$$C(6, 4) = 15$$

5. To find the value of $C(8, 2)$ from Pascal's triangle, read entry number 2 in row 8.

$$C(7, 5) = 28$$

7. To find the value of $C(9, 7)$ from Pascal's triangle, read entry number 7 in row 9.

$$C(9, 7) = 36$$

9. Selecting the committee is a two-part task. There are $C(7, 1)$ ways of choosing the one Democrat and $C(3, 3)$ way of choosing the remaining 3 Republicans. The combination values can be read from Pascal's triangle. By the fundamental counting principle, the total number of ways is

$$C(7, 1) \cdot C(3, 3) = 7 \cdot 1 = 7.$$

11. A committee with exactly three Democrats will consist of three Democrats and one Republican. Selecting the committee is a two-part task. There are $C(7, 3)$ ways of choosing three Democrats and $C(3, 1)$ ways to choose the one remaining Republican. Hence, there are

$$C(7, 3) \cdot C(3, 1) = 35 \cdot 3 = 105$$

ways in total.

13. There are

$$C(8, 3) = 56$$

ways to choose three different positions for heads. Using Pascal's triangle, find row 8 entry 3. Remember to count first row and first entry as 0. The remaining positions will automatically be tails.

15. There are

$$C(8, 5) = 56$$

ways to choose exactly five different positions for heads. Using Pascal's triangle, this would be found in row 8, entry 5.

17. The number of selections for four rooms is given by

$$C(9, 4) = 126.$$

Using Pascal's triangle, this would be found in row 9, entry 4.

19. The number of selections that succeed in locating the class is given by total number of selections (Exercise 17) minus the number of ways which will fail to locate the classroom (Exercise 18), or

$$C(9, 4) - C(8, 4) = 126 - 70 = 56$$

ways.

21. The number of 0-element subsets for a set of five elements is entry 0 (the first entry) in row 5 of Pascal's triangle. This number is 1.

23. The number of 2-element subsets for a set of five elements is entry 2 (the third entry) in row 5 of Pascal's triangle. This number is 10.

25. The number of 4-element subsets for a set of five elements is entry 4 (the fifth entry) in row 5. This number is 5.

27. The total number of subsets is given by

$$C(5, 0) + C(5, 1) + C(5, 2) + C(5, 3) + C(5, 4) + C(5, 5)$$
$$= 1 + 5 + 10 + 10 + 5 + 1 = 32.$$

This is the sum of elements in the fifth row of Pascal's triangle.

29. (a) All are multiples of the row number.

(b) The same pattern holds.

(c) Row 11:

1 11 55 165 330 462 462 330 165 55 11 1

All are multiples of 11. Thus, the same pattern holds.

31. Following the indicated sums

$$1, 1, 2, 3, 5,$$

the sequence continues

$$8, 13, 21, 34, \ldots .$$

A number in this sequence comes from the sum of the two preceding terms. This is the Fibonacci sequence.

33. Row 8 would be the next row to begin and end with 1 with all other entries 0 (each internal entry in row 8 of Pascal's triangle is even).

35. The sum of the squares of the entries across the top row equals the entry at the bottom vertex. Choose, for example, the second triangle from the bottom.

$$1^2 + 3^2 + 3^2 + 1^2$$
$$= 1 + 9 + 9 + 1$$
$$= 20 \text{ (the vertex value)}$$

37. The sum $= N$; Any entry in the array equals the sum of the two entries immediately above it and immediately to its left.

39. The sum $= N$; any entry in the array equals the sum of the row of entries from the cell immediately above it to the left boundary of the array.

41. Reading the coefficients from row 6 of Pascal's triangle and applying the binomial theorem, we obtain

$$(x+y)^6 = x^6 + 6x^5y + 15x^4y^2 + 20x^3y^3 + 15x^2y^4$$
$$+ 6xy^5 + y^6.$$

43. Reading the coefficients from row 3 of Pascal's triangle and applying the binomial theorem, we obtain

$$(z+2)^3 = z^3 + 3z^2(2) + 3z(2^2) + 2^3$$
$$= z^3 + 6z^2 + 12z + 8.$$

45. Reading the coefficients from row 4 of Pascal's triangle and applying the binomial theorem, we obtain

$$(2a+5b)^4 = (2a)^4 + 4(2a)^3(5b) + 6(2a)^2(5b)^2$$
$$+ 4(2a)(5b)^3 + (5b)^4$$
$$= 16a^4 + 160a^3b + 600a^2b^2 + 1000ab^3$$
$$+ 625b^4.$$

47. Reading the coefficients from row 6 of Pascal's triangle and applying the binomial theorem, we obtain

$$(b-h)^6 = [b+(-h)]^6$$
$$= b^6 + 6b^5(-h) + 15b^4(-h)^2 + 20b^3(-h)^3$$
$$+ 15b^2(-h)^4 + 6b(-h)^5 + (-h)^6$$
$$= b^6 - 6b^5 + 15b^4h^2 - 20b^3h^3 + 15b^2h^4$$
$$- 6bh^5 + h^6.$$

49. For the expansion $(x+y)^n$, there will be $n+1$ terms.

51. Here $n = 14$ and $r = 5$. Then $r - 1 = 4$ and $n - r + 1 = 10$. Substituting these values into the result of Exercise 44, we find that the 5th term of $(x+y)^{14}$ is

$$\frac{14!}{10!4!} x^{10}y^4 = 1001x^{10}y^4.$$

53. Prove $C(n,r) = C(n-r, r-1) + C(n-1, r)$.

$$C(n-r, r-1) + C(n-1, r)$$
$$= \frac{(n-1)!}{(r-1)![(n-1)-(r-1)]!} + \frac{(n-1)!}{r![(n-1)-r]!}$$
$$= \frac{(n-1)!}{(r-1)!(n-r)!} + \frac{(n-1)!}{r!(n-r-1)!}$$
$$= \frac{n}{n} \cdot \frac{(n-1)!}{(r-1)!(n-r)!} \cdot \frac{r}{r} + \frac{n}{n} \cdot \frac{(n-1)!}{r!(n-r-1)!} \cdot \frac{(n-r)}{(n-r)}$$
$$= \frac{n! \cdot r}{n \cdot r! \cdot (n-r)!} + \frac{n! \cdot (n-r)}{n \cdot r! \cdot (n-r)!}$$
$$= \frac{n! \cdot r + n! \cdot (n-r)}{n \cdot r! \cdot (n-r)!}$$
$$= \frac{n! \cdot [r + (n-r)]}{n \cdot r! \cdot (n-r)!}$$
$$= \frac{n! \cdot \cancel{n}}{\cancel{n} \cdot r! \cdot (n-r)!}$$
$$= \frac{n!}{r! \cdot (n-r)!}$$
$$= C(n,r)$$

10.5 EXERCISES

1. Writing exercise

3. The total number of subsets is 2^4 for a set with 4 elements. The only subset which is not a proper subset is the given set itself. Thus, by the complements principle, the number of proper subsets is

$$2^4 - 1 = 16 - 1 = 15.$$

5. By the fundamental counting principle, there are 2^7 different outcomes if seven coins are tossed. There is only one way to get no heads (all tails); thus, by the complements principle, there are

$$2^7 - 1 = 128 - 1 = 127$$

outcomes with at least one head.

7. "At least two" tails is the complement of "one or none." There are $2^7 = 128$ different outcomes if seven coins are tossed, 1 way of getting no tails (all heads), and 7 ways of getting one tail (tail on 1st coin, or tail on 2nd coin, etc.). Thus, by the complements principle, the number of ways to get at least two tails is

$$2^7 - (1 + 7) = 128 - 8 = 120.$$

Refer to Table 2 in the first section of the textbook chapter.

9. Counting outcomes with a 2 on the red die (row 2), there are a total of 6.

11. Counting the number of outcomes in row 4 (4 on red die) + those in column 4 (4 on green die) and subtracting the outcome counted twice, the number of outcomes with "a 4 on at least one of the dice" is

$$6 + 6 - 1 = 11.$$

13. There are nine two digit multiples of 10 (10, 20, 30, ..., 90). Altogether there are $9 \cdot 10 = 90$ two-digit numbers by the fundamental counting principle. Thus, by the complements principle, the number of "two-digit numbers which are not multiples of ten" is

$$90 - 9 = 81.$$

15. (a) The number of different sets of three albums she could choose is

$$C(8,3) = 56.$$

(b) The number which would not include *Unchained* is

$$C(7,3) = 35.$$

(c) The number that would contain *Unchained* is

$$56 - 35 = 21$$

17. The total number of ways of choosing any three days of the week is $C(7,3)$. The number of ways of choosing three days of the week that do not begin with S is $C(5,3)$. Thus, the number of ways of choosing any three days of the week such that "at least one of them begin with S" is

$$C(7,3) - C(5,3) = 25.$$

19. If the order of selection is important, the number of choices of restaurants is $P(8,3)$. The number of choices of restaurants that would not include seafood is $P(6,3)$. Thus, the number of choices such that at least one of the three will serve seafood is

$$P(8,3) - P(6,3) = 216.$$

21. The total number of ways to arrange 3 people among ten seats is $P(10,3)$. The number of ways to arrange three people among the seven (non-aisle) seats is $P(7,3)$. Therefore, by the complements principle, the number of arrangements with at least one aisle seat is

$$P(10,3) - P(7,3) = 510.$$

23. "At least one of these faculty members" is the complement of "none of these faculty members." There is a total of $C(25,4)$ ways of choosing 4-person committees and $C(23,4)$ ways to choose 4-person committees, excluding the two professors. Thus, applying the complements principle, the number of ways of choosing four faculty committees "with at least one of these two professors" is

$$C(25,4) - C(23,4) = 3795.$$

25. Let $C =$ the set of clubs and $J =$ the set of jacks. Then, $C \cup J$ is the set of cards which are face cards or jacks, and $C \cap J$ is the set of cards which are both clubs and jacks, that is, the jack of clubs. Using the general additive counting principle, we obtain

$$n(C \cup J) = n(C) + n(J) - n(C \cap J)$$
$$= 13 + 4 - 1$$
$$= 16.$$

27. Let $M =$ the set of students who enjoy music and $L =$ the set of students who enjoy literature. Then $M \cup L$ is the set students who enjoy music or literature, and $M \cap L$ is the set of students who enjoy both music and literature. Using the general additive counting principle, we obtain

$$n(M \cup L) = n(M) + n(L) - n(M \cap L)$$
$$= 30 + 15 - 10$$
$$= 35.$$

29. There are $C(13,5)$ 5-card hands of clubs. Thus, by the complements principle, the number of hands containing "at least one card that is not a club" is

$$2,598,960 - C(13,5) = 2,597,673.$$

31. The number of 5-card hands drawn from the 40 non-face cards in the deck is given by $C(40,5)$. Thus, by the complements principle the number of 5-card hands with "at least one face card" is

$$2,598,960 - C(40,5) = 1,940,952.$$

For Exercises 33–36, the given original set has 12 elements.

33. "At most two elements" is the same as 0, 1, or 2 elements. Thus, the number of subsets is
$$C(12,0) + C(12,1) + C(12,2) = 1 + 12 + 66 = 79.$$

35. "More than two elements" is the complement of "at most two elements." Find the number of subsets with "at most two elements" by adding the number of 0-member subsets, 1-member subsets and 2-member subsets.
$$C(12,0) + C(12,1) + C(12,2)$$
$$= 1 + 12 + 66$$
$$= 79.$$

There are 2^{12} subsets altogether. Thus, by the complements principle, the number of subsets of more than two elements is

$$2^{12} - 79 = 4096 - 79$$
$$= 4017.$$

37. The complement of "at least one letter or digit repeated" is "no letters or digits repeated." There are $P(26,3) \cdot P(10,3)$ license plates with no digits repeated, and using the fundamental counting principle we have $26 \cdot 26 \cdot 26 \cdot 10 \cdot 10 \cdot 10 = 26^3 \cdot 10^3$ license plates where any digit can be repeated. By the complements principle, the number of different license numbers with at least one letter or digit repeated is

$$26^3 \cdot 10^3 - P(26,3) \cdot P(10,3) = 6,344,000.$$

39. To choose sites in only one state works as follows: The number of ways to choose three monuments in New Mexico is $C(4,3)$, the number of ways to chose three monuments in Arizona is $C(3,3)$, and the number of ways to pick three monuments in California is $C(5,3)$. Since these components are disjoint, we may use the special additive principle. The number of ways to choose the monuments is

$$C(4,3) + C(3,3) + C(5,3) = 4 + 1 + 10$$
$$= 15.$$

41. "Sites in fewer than all three states" is the complement of choosing "sites in all three states." Since there are 12 monuments altogether, the total number of ways to select three monuments (with no restrictions) is $C(12,3)$. Choosing sites in all three states requires choosing one site in each state, which can be done in $4 \cdot 3 \cdot 5$ ways. Using the complements principles, the number of ways to choose sites in fewer than all three states is
$$C(12,3) - 4 \cdot 3 \cdot 5 = 220 - 60$$
$$= 160.$$

43. Writing exercise

45. Writing exercise

47. Writing exercise

49. Writing exercise

CHAPTER 10 TEST

1. To find three-digit numbers from the set $\{0,1,2,3,4,5,6\}$ use the fundamental counting principle:
$$\underline{6} \cdot \underline{7} \cdot \underline{7} = 294.$$

2. To find even three-digit numbers from the set $\{0,1,2,3,4,5,6\}$ use the fundamental counting principle:
$$\underline{6} \cdot \underline{7} \cdot \underline{4} = 168.$$

3. To find three-digit numbers without repeated digits from the set $\{0,1,2,3,4,5,6\}$ use the fundamental counting principle:
$$\underline{6} \cdot \underline{6} \cdot \underline{5} = 180.$$

4. To find three-digit multiples of five without repeated digits from the set $\{0,1,2,3,4,5,6\}$ use the fundamental counting principle and the special additive principle: Multiples of 5 end in "0" or "5." There are
$$\underline{6} \cdot \underline{5} \cdot \underline{1} = 30$$
multiples that end in 0 and
$$\underline{5} \cdot \underline{5} \cdot \underline{1} = 25$$
that end in 5. Thus, the number of three-digit multiples of five without repeated are
$$30 + 25 = 55.$$

5. Make a systematic listing of triangles. Beginning with the smaller inside triangle there are 4 right triangles off the horizontal bisector of the larger triangle. These triangles may be combined to create 4 larger isosceles triangles. There are 2 isosceles triangles – inside the upper left and lower left corners of the largest triangle and 1 larger right triangles above and 1 below the horizontal bisector of the larger triangle. Of course, count the largest isosceles triangle itself. The total number of triangles is
$$4 + 4 + 2 + 1 + 1 + 1 = 13$$

6.

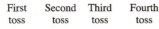

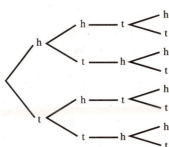

7.   There is only one set of 4 digits that add to 30
$(9 + 8 + 7 + 6 = 30)$. The number of arrangements of
the digits 9876 is given by

$$P(4, 4) = 4! = 24.$$

8.   To find the number of 3-digit numbers with no repeated
odd digits from $\{0, 1, 2\}$, use the fundamental counting
principle. The only digit which can not be repeated is
the odd digit "1" so look (a) at the cases where 1 is
positioned in each of the three possible places and (b)
those numbers that do not contain the odd digit 1.

(a) There are

$$1 \cdot 2 \cdot 2 = 4$$

3-digit numbers that begin with 1 and

$$1 \cdot 1 \cdot 2 = 2$$

3-digit numbers whose middle digit is 1, and

$$1 \cdot 2 \cdot 1 = 2$$

3-digit numbers whose last digit is 1. Thus, using the
special additive principle, the number of 3-digit numbers
without which contain the odd digit 1 is

$$4 + 2 + 2 = 8.$$

(b) The condition "no repeated odds digit" is
automatically satisfied by any number that contains "no
odd digits." Using the only the digits $\{0, 2\}$ and the
fundamental counting principle, there are

$$1 \cdot 2 \cdot 2 = 4$$

such numbers.

There are, therefore

$$8 + 4 = 12$$

numbers that satisfy the condition "no repeated odds
digit." The numbers are

$$100, 102, 120, 122, 200, 201,$$
$$202, 210, 212, 220, 221, 222.$$

9. $5! = 5 \cdot 4 \cdot 3 \cdot 2 \cdot 1 = 120$

10. $\dfrac{8!}{5!} = \dfrac{8 \cdot 7 \cdot 6 \cdot 5!}{5!}$

$= \dfrac{8 \cdot 7 \cdot 6}{1}$

$= 336$

11. $P(12, 4) = 12 \cdot 11 \cdot 10 \cdot 9$
$= 11,880$

12. $C(7, 3) = \dfrac{P(7, 3)}{3!}$

$= \dfrac{7 \cdot \cancel{6} \cdot 5}{\cancel{3} \cdot \cancel{2} \cdot 1}$

$= 35$

13.   Since the arrangement of the letters is important and no
repetitions are allowed, use $P(26, 5)$.

$$P(26, 5) = 26 \cdot 25 \cdot 24 \cdot 23 \cdot 22 = 7,893,600$$

14.   Since repetitions are allowed, use the fundamental
counting principle.

$$32^5 = 33,554,432$$

15.   Since order is not important, use $C(12, 4)$.

$$C(12, 4) = \dfrac{P(12, 4)}{4!}$$

$$= \dfrac{12 \cdot 11 \cdot 10 \cdot 9}{4 \cdot 3 \cdot 2 \cdot 1}$$

$$= 495$$

16.   Use the fundamental counting principle along with
$C(12, 2)$ and $C(10, 2)$.

$$C(12, 2) \cdot C(10, 2) = 66 \cdot 45$$
$$= 2970$$

17.   Use the fundamental counting principle along with
$C(12, 5)$ and $C(7, 5)$.

$$C(12, 5) \cdot C(7, 5) = 792 \cdot 21$$
$$= 16,632$$

18.   The complement of "a set of three or more of the
players" is "zero, one, or two of the players." The total
number of subsets of the 12 players is

$$2^{12} = 4096.$$

The total number of 0-member subsets and 1-member
subsets and 2-member subsets is

$$C(12,0) + C(12,1) + C(12,2)$$
$$= 1 + 12 + 66$$
$$= 79.$$

By the complements principle, the number of ways that the coach has in selecting a set of three or more of the players is

$$4096 - 79 = 4017.$$

19. With no restrictions, use the fundamental counting principle to determine the number of positions that a row of four switches may be set.

$$2 \cdot 2 \cdot 2 \cdot 2 = 2^4 = 16$$

21. Use the fundamental counting principle to determine the number of positions that a row of four switches may be set if the first and fourth position must be set the same.

$$2 \cdot 2 \cdot 2 \cdot 1 = 2^3 = 8$$

Complete a tree diagram like Exercise 59 in 10.1 to answer questions 22–23.

First Switch	Second Switch	Third Switch	Fourth Switch	Switch Settings
			0	0000
		0	1	0001
	0	1	0	0010
			1	0011
0		0	0	0100
			1	0101
	1	1	0	0110
			1	0111
		0	0	1000
	0		1	1001
		1	0	1010
1			1	1011
		0	0	1100
	1		1	1101
		1	0	1110
			1	1111

Alternatively, you may also choose to create separate tree diagrams instead, according to the restrictions imposed by each exercise.

22. The following represent switch settings where "no two adjacent switches can be off. Remember that "0" typically represents an "off" switch.

0101	0110	0111	1010
1011	1101	1110	1111

There are 8 switch settings that satisfy the restriction.

23. The following represent switch settings where "no two adjacent switches can be set the same."

0101	1010

There are 2 switch settings that satisfy the restriction.

24. The complement of "at least two switches must be on" is "zero or 1 switch(es) is/are on." Without restrictions there are $2^4 = 16$ different switch settings. There is only 1 way for zero switches to be on and 4 ways for only 1 switch to be on. Thus, by the complements principle, the number of ways for at least two switches to be on is

$$16 - (1 + 4) = 11.$$

Alternatively, you may count those settings in the above tree diagram which reflect the given restriction.

Use the set $\{A, B, C, D, E, F, G\}$ to answer Exercises 25–29.

25. Since the letter "D" must be a member, all that is necessary is to choose three members from the six remaining members. The number of 4-member subsets is then

$$C(6,3) = 20.$$

26. Since the letters "A" and "E" must be members, all that is necessary is to choose two members from the five remaining members. The number of 4-member subsets is then

$$C(5,2) = 10.$$

27. $C(5,3)$ represents the number of 4-member subsets that contain exactly one of the letters but not the other. There will be twice this number if we do the same for the 2nd letter. Thus, the number of ways to select 4-member subsets that contain exactly one of the letters but not the other is

$$2 \cdot C(5,3) = 20.$$

29. "More consonants then vowels" can happen two ways. One, with 4 consonants and no vowels, can be found by $C(5, 4)$. The second, with 3 consonants and 1 vowel, can be found by choosing one of the two vowels first, $C(2, 1)$ ways, followed by choosing the 3 consonants $C(5, 3)$ ways. Since one "or" the other will satisfy the restrictions (and they cannot both happen at the same time), apply the special addition principle to find the total number of subset choices for "more consonants then vowels."

$$C(5, 4) + C(2, 1) \cdot C(5, 3)$$
$$= 5 + 2 \cdot 10$$
$$= 25$$

31. Because $C(n, r)$ and $C(n, r + 1)$ are the two entries just above $C(n + 1, r + 1)$, evaluate $C(n + 1, r + 1)$ by adding their values.

$$C(n + 1, r + 1) = C(n, r) + C(n, r + 1)$$
$$= 495 + 220$$
$$= 715$$

33. Writing exercise

11   PROBABILITY

Chapter Goals

The general goal of this chapter is to provide an overview of basic probability concepts. Upon completion of this chapter, the reader should be able to

- Explain three definitions of probability.
- Calculate probabilities using addition and multiplication rules.
- Calculate odds in favor of (or against) the occurrence of an event.
- Calculate binomial probabilities.
- Calculate conditional probabilities.
- Calculate expected value (mathematical expectation).
- Use simulation techniques in the calculation of probabilities.
- Use characteristics of Pascal's triangle as an aid in providing computational shortcuts.

Chapter Summary

A **probability** is a measure of the degree of belief that a particular outcome will happen. A probability may assume values from 0 to 1.0 inclusive, or expressed as a percent, from 0% to 100% inclusive. A value close to 0 indicates that the event is unlikely to happen. A value close to 1.0 indicates the event is quite likely to happen. For example, a probability of 0.9 (highly likely) might express your degree of belief that you will get an "A" in your mathematics class. A probability of 0.25 (considerable uncertainty) might be a business manager's best estimate of the likelihood of increased sales next month.

Making decisions is a part of all our lives. Since there is often considerable uncertainty in decision making, it is usually important that all the known risks involved be scientifically evaluated. This is especially true if the decisions involve business, scientific research, management, or other responsibilities. **Probability theory**, referred to by many as the science of uncertainty, is very helpful in this evaluation. Probability also plays an important role, for example, in the application of inferential statistics (the process of drawing conclusions about a population based on a sample) by providing the bridge between the population and the sample taken from it. Indeed the theory of probability is an important tool in the solution of many problems (scientific, social, business, and technological) of the modern world.

Before examining the various definitions of probability and the associated rules for calculating probabilities, several key words need to be defined. An **experiment** is either an activity which is observed or is the act of taking some type of measurement. Examples of an experiment include tossing one or more dice, drawing cards from a bridge deck, weighing the contents from a sample of cereal boxes to determine how many produced are underweight, or selecting club members to act as officers.

A particular result of an experiment is called an **outcome**. One outcome in the experiment of tossing a die might be three dots representing the number 3.   The set of all possible outcomes is called the **sample space**. The sample space in the above experiment is the set of outcomes $\{1, 2, 3, 4, 5, 6\}$. An **event** of interest is some subset of the sample space. For example, we might specify the event E to be all outcomes in the above experiment which are divisible by 2; that is $E = \{2, 4, 6\}$.

Three approaches to (or definitions of) probability encountered include those referred to as theoretical, empirical, and subjective. The **theoretical** (or classical) definition of probability is dependent on the fact that each outcome in the sample space must be equally likely to occur, and if this is the case, then the probability of a particular event happening is computed by dividing the number of favorable outcomes by the total number of possible outcomes. That is,

$$P(\text{Event}) = \frac{\text{number of favorable outcomes}}{\text{total number of outcomes}},$$

or

$$P(E) = \frac{n(E)}{n(S)}.$$

where $n(E)$ is the number of elements in the event, E, of interest and $n(S)$ is the number of elements in the sample space, S. A favorable outcome is considered a success, and thus we may interpret the classical probability ratio alternatively as

$$P(\text{Event}) = \frac{\text{number of successes}}{\text{total number of outcomes}}.$$

EXAMPLE A   Theoretical probability

Find the probability of getting each of the following events when a single die is rolled.

(a) A number less than 3.

Solution
The outcomes 1 or 2 would be considered successes and belong to the event of interest. Thus,

$$P(\text{a number less than 3}) = \frac{\text{number of successes}}{\text{total number of outcomes}}$$

$$= \frac{2}{6} = \frac{1}{3}.$$

(b) A number different from 4.

Solution
The outcomes 1, 2, 3, 5, or 6 would be considered successes. Thus,

$$P(\text{a number different than 4}) = \frac{\text{number of successes}}{\text{total number of outcomes}}$$

$$= \frac{5}{6}.$$

EXAMPLE B   Theoretical probability

In 5-card poker, find the probability of being dealt "four of a kind."

Solution
Use the table of poker hands (found in the text). According to the table, there are 624 ways to arrive at "four of a kind" and a total of 2,598,960 different 5-card hands that could be dealt. Thus,

$$P(\text{four of a kind}) = \frac{\text{number of successes}}{\text{total number of outcomes}}$$

$$= \frac{624}{2,598,960}$$

$$= \frac{1}{4165} \approx 0.00024.$$

Probabilities determined (or estimated) by observation are called **empirical** (experimental or statistical). These probabilities typically use a relative frequency interpretation; that is, they are found by dividing the number of times the event has occurred in the past by the total number of observations. This would be done, for example, if the experiment involved tossing a loaded die where one or more sides were more likely to come up than the others. One may toss the die a number of times recording the number of times each side came up and divide that total by the total number of tosses (observations) to get an empirical probability associated with each side.

A **subjective** probability is based on personal judgment, intuition, or whatever information is available. The chances of rain tomorrow, or the probability that Raindance will win next week's derby, are based on the personal or subjective view of the weather observer, or the racetrack oddsmaker.

If one knows (or can calculate) the probability that a particular event "will not" occur, then the probability that the event "will" occur is given by the **complements rule** for probability: for any event E, $P(E) = 1 - P(\text{not } E)$. This may be written symbolically as $\boldsymbol{P(E) = 1 - P(E')}$ or, if the event E is considered a success,

$$\boldsymbol{P(\text{success}) = 1 - P(\text{failure})}.$$

EXAMPLE C   Complements rule

According to one study, if you are an American, the probability of being overweight is 63%. What is the probability of an American not being overweight?

Solution
The complements formula says $P(E) = 1 - P(E')$. Or equivalently,

$$P(E) + P(E') = 1 \text{ or}$$
$$P(E') = 1 - P(E).$$

Thus,

$$P(E') = 1 - 63\% = 37\% \text{ or } .37,$$

where E' represents the event of not being overweight.

Closely related to the idea of probability is that of **odds**. If the probability of an event is described as the ratio of favorable outcomes to total outcomes, then the "odds in favor of an event" would be the ratio of favorable outcomes to unfavorable outcomes with the "odds against an event" defined similarly as the reciprocal of the "odds in favor."

The **odds in favor** of an event E are given by

$$\frac{\text{number of outcomes favorable to the event}}{\text{number of unfavorable outcomes}},$$

or

$$\frac{n(E)}{n(E')}.$$

The odds against an event E are given by

$$\frac{\text{number of outcomes unfavorable to the event}}{\text{number of favorable outcomes}},$$

or

$$\frac{n(E)'}{n(E)}.$$

EXAMPLE D   Odds in favor

Find the odds in favor of drawing a three of hearts from a thoroughly mixed deck of 52 cards.

Solution

$$\text{Odds in favor of three of hearts} = \frac{\text{number of successes}}{\text{number of failures}}$$
$$= \frac{1}{51}, \text{ or 1 to 51.}$$

If the related probabilities are known, it is often easier to express the odds in terms of these probabilities. That is,

---

The **odds in favor** of an event E are given by

$$\frac{\text{probability of event occurring}}{\text{probability of event not occurring}}, \text{ or}$$

$$\frac{P(E)}{P(E')}, \text{ or } \frac{P(E)}{1 - P(E)},$$

using the complements formula.

The **odds against** an event are given by the reciprocal of the above, that is

$$\frac{\text{probability of event not occurring}}{\text{probability of event occurring}}, \text{ or}$$

$$\frac{P(E')}{P(E)}, \text{ or } \frac{1 - P(E)}{P(E)}, \text{ or } \frac{P(E')}{1 - P(E')},$$

using the complements formula.

EXAMPLE E   Changing from probability to odds

If the probability of a wife outliving her spouse is .54, what are the odds that her husband will outlive her?

Solution
By the complements formula, the probability that her husband will outlive her is $1 - .54 = .46$.

The odds that her husband will outlive her is given by

$$\frac{P(\text{husband outlives wife})}{P(\text{husband does not outlive wife})} = \frac{0.46}{0.54} = \frac{46}{54} \approx 0.85.$$

The above allows one to move easily from known probabilities to the corresponding odds in favor or against. To find, however, the probability of an event occurring when one knows only the odds in favor of an event occurring, the following formula is used:

If the **odds in favor** of event E are a/b (usually expressed as a to b or $a : b$), then

$P(E) = a/(a + b)$, where a represents the number of successes (favorable outcomes), b represents the number of failures (unfavorable outcomes), and $a + b$ is the total number of outcomes.

EXAMPLE F   Changing from odds to probability

If the odds against having a motorcycle accident are 3 to 5, then what is the probability of having a motorcycle accident?

Solution
Since the odds against having a motorcycle accident are $3/5$, the odds in favor of having such an accident must be $5/3$ (one might think of this ratio as representing the number of successes divided by the number of failures). The probability of having a motorcycle accident is then given by

$$\frac{5}{5 + 3} = \frac{5}{8}$$

(which one might consider as the number of successes divided by the total number of outcomes, successes + failures).

Events of interest may be combined using the rules of addition and/or the rules of multiplication. These are sometimes referred to as **composite events** and the associated probabilities as **joint probabilities**. There are two rules of addition, the **general addition rule** and the **special addition rule**. The general addition rule states that if A and B are any two events for the same experiment, then

$P(A \text{ or } B) = P(A) + P(B) - P(A \text{ and } B),$

or symbolically as

$P(A \cup B) = P(A) + P(B) - P(A \cap B).$

By this rule, the probability of observing one or the other or both of the two events equals the sum of their individual probabilities minus the probability that they both will occur. The key word here is "or." When we are "or–ing" in events or probabilities, we are using addition rules.

EXAMPLE G   General addition rule

A local community has two newspapers. The *Morning News* is read by 45 percent of the local households. The *Evening Times* is read by 60 percent of the local households. Twenty percent of the households read both papers.

(a) What is the probability that a particular household in the city reads at least one paper?

Solution
Letting N represent the event of reading the *Morning News* and T represent the event of reading the *Evening Times* and using the general rule of addition, we have

$$P(N \text{ or } T) = P(N) + P(T) - P(N \text{ and } T)$$
$$= 0.45 + 0.60 - 0.20 = 0.85.$$

Observe that "at least one paper" is equivalent to "reading the *News* or the *Times* or both."

(b) What is the probability that a particular household in the community does not read either paper?

Solution
In the previous example, it was found that the probability that a household reads at least one newspaper is 0.85. Using the complements formula, we have

$$P(\text{neither } N \text{ or } T) = 1 - P(N \text{ or } T)$$
$$= 1 - 0.85 = 0.15,$$

indicating that 15 percent of the households in the community read neither the *News* nor the *Times*.

The **special addition rule** states that if the events of interest are "mutually exclusive," then the probability of event A or event B occurring is given by the sum of the two individual probabilities or symbolically as

$P(A \cup B) = P(A) + P(B).$

The condition that A and B are **mutually exclusive** means that the two events cannot both happen or symbolically, that $A \cap B = \emptyset$. Tossing a die with event A representing the set of outcomes where an even number comes up and event B representing the set of outcomes where an odd number comes up is an example of mutually exclusive events where both cannot occur at the same time.

The special addition rule can be applied to more than two events.

> If $E_1, E_2, \ldots, E_n$ are pairwise mutually exclusive events, then
>
> $$P(E_1 \cup E_2 \cup \ldots \cup E_n) = P(E_1) + P(E_2) + \ldots + P(E_n).$$

EXAMPLE H   Special addition rule

If a single card is drawn from a standard 52-card deck, find the probability that it will be black or a king of hearts?

Solution

Since a card cannot be both black and a king of hearts (they are mutually exclusive), we can use the special addition rule:

$$P(\text{black or king of hearts}) = P(\text{black}) + P(\text{king of hearts})$$

$$= \frac{26}{52} + \frac{1}{52}$$

$$= \frac{27}{52}.$$

The **special multiplication rule** (often called the multiplication rule of probability for independent events) is used in combining two events when the probability of the second event does not depend on the outcome of the first event. Such events are called **independent**. Finding the probability that two cards drawn from a deck are both queens when the first card is replaced in the deck before the second card is drawn is an example of two events which are independent of each other. Note that if the first card is not replaced after it is drawn, this will affect the probability that the second card is a queen, and, thus, the events are considered **not independent** or **dependent** events.

If there are two independent events A and B, the probability of A and B occurring is found by multiplying the two probabilities.

That is, for independent events,

> $$P(A \text{ and } B) = P(A \cap B) = P(A) \times P(B).$$

The special multiplication rule can be applied to more than two events also. If $E_1, E_2, \ldots, E_n$ are n events, each one independent of each of the others, then

> $$P(E_1 \cap E_2 \cap \ldots \cap E_n) = P(E_1) \cdot P(E_2) \ldots \cdot P(E_n).$$

EXAMPLE I   Special multiplication rule

Three cards are drawn from a standard 52-card deck. After each card is drawn from the deck, it is replaced. What is the probability of drawing three hearts?

Solution

Because each card is replaced before the next drawing, the probability associated with one drawing in no way affects the probability of the next drawing and thus the events are independent which means that we can apply the special multiplication rule:

$P(\text{all three are hearts}) =$
$P(\text{first is a heart and second is a heart and third is a heart})$

$$= P(H_1 \cap H_2 \cap H_3) = P(H_1) \cdot P(H_2) \cdot P(H_3)$$

$$= \frac{13}{52} \cdot \frac{13}{52} \cdot \frac{13}{52} = \frac{1}{4} \cdot \frac{1}{4} \cdot \frac{1}{4} = \frac{1}{64}.$$

The **general multiplication rule** of probability is used to combine events when the probability of the second event depends on the outcome of the first event. That is, the two events are not independent. The probability of A and B occurring under these circumstances is written

> $$P(A \text{ and } B) = P(A \cap B) = P(A) \cdot P(B \mid A).$$

where $P(B \mid A)$ is read as "the probability that event B occurs given that A has occurred" and is thus called a **conditional probability**. Stating the general multiplication rule in terms of the two events E_1 and E_2, we have the equivalent formula

> $$P\left(E_1 \text{ and } E_2\right) = P(E_1 \cap E_2)$$
> $$= P(E_1) \cdot P(E_2 \mid E_1).$$

This formula may also be generalized as shown in the next example.

EXAMPLE J General multiplication rule of probability

If three cards are drawn without replacement from a standard 52-card deck, find the probability that the first is a spade and the second and third are hearts.

Solution
If we symbolize the first card drawn is a spade by the event S_1, the second is a heart by H_2, and the third is a heart by H_3, then we can indicate the probability we are trying to find by

$$
\begin{aligned}
P(S_1 &\cap H_2 \cap H_3) \\
&= P(S_1) \cdot P(H_2 \mid S_2) \cdot P(H_3 \mid S_1 \cap H)_2. \\
&= \frac{13}{52} \cdot \frac{13}{51} \cdot \frac{12}{50} \\
&= \frac{1}{4} \cdot \frac{13}{51} \cdot \frac{6}{25} \quad \text{canceling like factors} \\
&= \frac{1}{2} \cdot \frac{13}{17} \cdot \frac{1}{25} \quad \text{canceling like factors} \\
&= \frac{13}{850}.
\end{aligned}
$$

Dividing both sides of the general multiplication rule, $P(E_1 \cap E_2) = P(E_1) \cdot P(E_2 \mid E_1)$, by $P(E_1)$ gives a formula (alternate conditional probability formula) commonly used for finding the value of **conditional probabilities**:

$$
P(E_2 \mid E_1) = \frac{P(E_1 \cap E_2)}{P(E_1)}.
$$

Used if associated probabilities are known or can readily be computed.

When the number of elements in E_1 and E_2 are known and the associated probabilities are not known, use the similar formula:

$$
P(E_2 \mid E_1) = \frac{n(E_1 \cap E_2)}{n(E_1)}.
$$

Used when the number of elements (cardinalities) of the events of interest can more readily be obtained than the associated probabilities.

In many problems involving conditional probabilities, the easiest approach to their solution is to reduce the size of the original sample space using the given information and then apply the classic definition for evaluating the particular probability; that is,

$$
P(\text{event}) = \frac{n(\text{event})}{n(\text{reduced sample space})}.
$$

EXAMPLE K Conditional probability using reduced sample space

Two cards are drawn successively from a standard 52-card deck, without replacement. Find the probability that the second will be a queen of hearts given that the first was a spade.

Solution
The original sample space (complete deck) can be reduced by the spade that was selected on the first drawing to one of 51 cards. Thus,

$$
P(\text{second is queen of hearts} \mid \text{first is spade}) = \frac{1}{51}.
$$

EXAMPLE L Conditional probability using formula

A coin is thrown; then a die is tossed. Find the probability of obtaining a 4, given that heads came up.

Solution
Method 1. Let F be the event in which a 4 turns up, and let H be the event in which heads come up. Thus,

$$
\begin{aligned}
P(F \mid H) &= \frac{P(F \cap H)}{P(H)} = \frac{P(F) \cdot P(H)}{P(H)}. \\
&= \frac{\frac{1}{6} \cdot \frac{1}{2}}{\frac{1}{2}} = \frac{\frac{1}{12}}{\frac{1}{2}} \\
&= \frac{1}{12} \cdot \frac{2}{1} = \frac{1}{6}.
\end{aligned}
$$

Method 2 (without formula). We know that heads came up, so our reduced sample space is given by
$$S = \{(H, 1), (H, 2), (H, 3), (H, 4), (H, 5), (H, 6)\}.$$

Only one outcome is favorable (considered a success), $(H, 4)$. Thus,

$$
P(F \mid H) = \frac{1}{6}.
$$

EXAMPLE M   **Conditional   probability   using formula**

Two dice are thrown, and a friend tells you that the first die shows a 6. Find the probability that the sum of the numbers showing on the two dice is 7.

Solution

Method 1. Let S_1 be the event in which the first die shows a 6, and let S_2 be the event in which the sum is 7. There is only one member of the sample space where the first die is a 6 and the second is a 1, $n(S_2 \cap S_1)$, and there are 6 members where we have a 6 on the first and 1, 2, ..., or 6 on the second, $n(S_1)$. Thus,

$$P(S_1 \mid S_2) = \frac{n(S_2 \cap S_1)}{n(S_1)} = \frac{1}{6}.$$

Note that we could have just as easily calculated the probabilities $P(S_2 \cap S_1)$ and $P(S_1)$ and used the alternative conditional probability formula.

Method 2 (without formula). We know that a 6 came up on the first die, so our new (reduced) sample space is $\{(6, 1), (6, 2), (6, 3), (6, 4), (6, 5), (6, 6)\}$. Hence,

$$P(S_1 \mid S_2) = \frac{1}{6},$$

because there is only one favorable outcome (6, 1).

When applying probability formulas, each repetition of an experiment is called a trial. A trial of any experiment where the outcomes are classified as either success or failure is referred to as a Bernoulli trial. If our experiment is to toss a coin in the air and if we are interested in the event of getting tails (t), then each trial would yield a success (getting a tail) or a failure (getting a head)—each trial is thus a Bernoulli trial. If we repeat the experiment, let's say twice, we can create a probability distribution for the likelihood of getting 0 tails, 1 tail, or 2 tails as in the following table where "x" is called a random variable and stands for the number of successes.

x	$P(x)$	
0	1/4	$P(0 \text{ tails}) = P(\text{1st is a head and 2nd is a head})$
		$= P(h) \cdot P(h) = (1/2) \cdot (1/2).$
1	1/2	$P(\text{1st is a } t \text{ and 2nd is not or the 2nd is a } t \text{ and 1st is not})$
		$= P(t_1) \cdot P(h_2) + P(t_2) \cdot P(h_1)$
		$= (1/2) \cdot (1/2) + (1/2) \cdot (1/2)$
		$= (1/4) + (1/4) = 1/2$
2	1/4	$P(\text{1st is a tail and 2nd is a tail})$
		$= P(t_1) \cdot P(t_2)$
		$= (1/2) \cdot (1/2) = 1/4$

Sum = 4/4 = 1

Such a distribution for repeated Bernoulli trials is commonly called a binomial probability distribution. The associated probabilities are referred to as **binomial probabilities**. As the number of trials increases, we can continue to compute the probabilities as "joint" probabilities (as in the above table) using repeated addition and multiplication rules. Or we can simplify the task by using the **binomial probability formula**:

For n repeated trials, where
  p = probability of success and
  q = probability of failure on each trial, the probability of exactly x successes is given by

$$P(x) = C(n, x)p^x\, q^{n-x}$$

where $p + q = 1$.

Remember that $C(n, x)$ represents the number of combinations that can be drawn from a set of n objects, x at a time, and is computed by

$$C(n, x) = \frac{n!}{x!\,(n - x)!}.$$

EXAMPLE N   Binomial probability

A fair die is rolled five times. Find the probability that an odd number will occur three times.

Solution

Let odds be success. Then this is a binomial experiment with $n = 5, p = \frac{3}{6} = \frac{1}{2}$, $q = \frac{1}{2}$, and $x = 3$. Thus,

$$P(\text{odd numbers occur 3 times}) =$$

$$\begin{aligned}
P(3) &= C(5,3) \cdot \left(\frac{1}{2}\right)^3 \cdot \left(\frac{1}{2}\right)^{5-3} \\
&= \frac{5!}{3!(5-3)!} \cdot \left(\frac{1}{2}\right)^3 \cdot \left(\frac{1}{2}\right)^2 \\
&= \frac{5 \cdot 4 \cdot 3!}{3! \cdot 2! \cdot 2^3 \cdot 2^2} \\
&= \frac{5 \cdot 4}{2 \cdot 2^5} = \frac{5}{16}.
\end{aligned}$$

Many decisions in business and science are made on the basis of what the outcomes of specific decisions will be. An important tool of this analysis which can be developed using probability is that of **expected value**. Expected value is very useful in making decisions in business and the sciences. It was originally developed for and is used frequently in gambling, so let us introduce it in this context.

Suppose that you toss a coin three times and you receive $1 for each head that results. How much would you expect to win each time you played if you played a large number of times?

The probabilities of 0, 1, 2, and 3 heads are 1/8, 3/8, 3/8, and 1/8, respectively (these can, remember, be calculated as binomial probabilities). Thus, if you played this game a large number of times, you would average

$$\$0 \cdot \frac{1}{8} + \$1 \cdot \frac{3}{8} + \$2 \cdot \frac{3}{8} + \$3 \cdot \frac{1}{8} = \$1.50$$

each time you played the game. Therefore, a fair price for you to pay in order to play this game is $1.50. This is called the **expected value** for this game. If the cost to you for playing the game one time is set by a gambling casino at $1.50, then we would consider the game to be a **fair game**. If the casino, however, offers to let people play this game for $2 (a more likely event), the casino could expect to make an average of

$.50 for every game that is played. If the cost of playing the game is higher than the expected value, we say that the game favors the casino.

The casino would have to pay $3 to some people, $2 to some, $1 to some, and $0 to others, but in the long run it could expect to make on average $.50 for each game that is played.

The following represents the definition of **mathematical expectation** or **expected value**:

If a given quantity x can have any of the values x_1, x_2, x_3, ... , x_n, and the corresponding probabilities of these values occurring are $P(x_1)$, $P(x_2)$, $P(x_3)$, ... , $P(x_n)$, then the expected value of the quantity x is given by

$$x_1 \cdot P(x_1) + x_2 \cdot P(x_2) + x_3 \cdot P(x_3), \dots ,$$
$$x_n \cdot P(x_n).$$

EXAMPLE O   Expected value

The Homeowner's Insurance Company insures 100,000 residences. Their records indicate that during a year they will pay out the following for insurance claims:

$100,000 with probability 0.0001
$ 50,000 with probability 0.001
$ 25,000 with probability 0.002
$ 5,000 with probability 0.008
$ 1,000 with probability 0.02

What amount of money would the company expect to pay on average per residence for claims made by its clients?

Solution

The expected value of the company's payments is

$$\begin{aligned}
&\$100{,}000(0.0001) + \$50{,}000(0.001) \\
&\quad + \$25{,}000(0.002) + \$5000(0.008) \\
&\quad + \$1000(0.02) \\
&= \$10 + \$50 + \$50 + \$40 + \$20 = \$170.
\end{aligned}$$

Useful Formulas

Classical definition of probability

Each outcome in the sample space must be equally likely to occur

$$P(\text{event}) = \frac{\text{number of favorable outcomes}}{\text{total number of outcomes}} = \frac{n(E)}{n(S)}$$

Complements formula

$$P(E) = 1 - P(E') \quad \text{or} \quad P(\text{success}) = 1 - P(\text{failure})$$

Odds

Using cardinalities (i.e. counting elements in event of interest and sample space)

The **odds in favor** of an event E are given by

$$\frac{\text{number of outcomes favorable to the event}}{\text{number of unfavorable outcomes}}, \text{ or } \frac{n(E)}{n(E')}.$$

The **odds against** an event E are given by

$$\frac{\text{number of outcomes unfavorable to the event}}{\text{number of favorable outcomes}}, \text{ or } \frac{n(E')}{n(E)}.$$

Odds

Using probabilities

The **odds in favor** of an event E are given by

$$\frac{\text{probability of event occurring}}{\text{probability of event not occurring}}, \text{ or } \frac{P(E)}{P(E')}, \text{ or } \frac{P(E)}{1 - P(E')}.$$

The **odds against** an event are given by the reciprocal of the odds in favor, or

$$\frac{\text{probability of event not occurring}}{\text{probability of event occurring}}, \text{ or } \frac{P(E')}{P(E)}, \text{ or } \frac{1 - P(E)}{P(E)}, \text{ or } \frac{P(E')}{1 - P(E')}.$$

General addition rule of probability

Events need not be mutually exclusive

$$P(A \cup B) = P(A) + P(B) - P(A \cap B)$$

Special addition rule of probability

Events must be mutually exclusive

$$P(A \cup B) = P(A) + P(B)$$

or more generally

$$P(E_1 \cup E_2 \cup \ldots \cup E_n) = P(E_1) + P(E_2) + \ldots + P(E_n).$$

General multiplication rule

Events may be dependent

$$P(A \text{ and } B) = P(A \cap B) = P(A) \cdot P(B \mid A)$$

Special multiplication rule of probability

Events must be independent

$$P(A \cap B) = P(A) \cdot P(B)$$

Or more generally

$$P(E_1 \cap E_2 \cap \ldots \cap E_n) = P(E_1) \cdot P(E_2) \ldots \cdot P(E_n).$$

Used when the number of elements in the events of interest can more readily be obtained than the associated probabilities.

Conditional probability formula

$$P(E_2 \mid E_1) = \frac{n(E_1 \cap E_2)}{n(E_1)}$$

Alternate conditional probability formula

Used if associated probabilities are known or can readily be computed.

$$P(E_2 \mid E_1) = \frac{P(E_1 \cap E_2)}{P(E_1)}$$

Binomial probability formula

$$P(x) = C(n, x) \, p^x \, q^{n-x}$$
For n repeated trials, where
p = probability of success and
q = probability of failure on each trial

Used for repeated Bernoulli trials—each outcome is classified as either a success or a failure.

Mathematical expectation (expected value)

If a given quantity x can have any of the values $x_1, x_2, x_3, \ldots, x_n,$
and the corresponding probabilities of these values occurring are
$P(x_1), P(x_2,) P(x_3), \ldots, P(x_n)$, then the **expected value** of the quantity s given by

$$x_1 \cdot P(x_1) + x_2 \cdot P(x_2) + x_3 \cdot P(x_3), \ldots, + x_n \cdot P(x_n).$$

11.1 EXERCISES

The sample space is $\{1, 2, 3\}$.

1. The number of regions in the sample space is $n(S) = 3$. Each region has the same area and thus, has the same likelihood of occurring. Therefore,

    (a) $P(\text{red}) = \dfrac{n(\text{red region})}{n(S)}$
    $$= \frac{1}{3}.$$

    (b) $P(\text{yellow}) = \dfrac{n(\text{yellow region})}{n(S)}$
    $$= \frac{1}{3}.$$

    (c) $P(\text{blue}) = \dfrac{n(\text{blue region})}{n(S)}$
    $$= \frac{1}{3}.$$

3. The number of regions in the sample space is $n(S) = 6$. Each region (piece of the pie) has the same area and thus, has the same likelihood of occurring. The probability of landing on any one of the six regions is 1/6, but we must account for the fact that some colors shade more than one region. Therefore,

    (a) $P(\text{red}) = \dfrac{n(\text{red regions})}{n(S)}$
    $$= \frac{3}{6} = \frac{1}{2}.$$

    (b) $P(\text{yellow}) = \dfrac{n(\text{yellow regions})}{n(S)}$
    $$= \frac{2}{6} = \frac{1}{3}.$$

    (c) $P(E) = \dfrac{n(\text{blue regions})}{n(S)}$
    $$= \frac{1}{6}.$$

5. (a) The sample space is $\{1, 2, 3\}$.

    (b) The number of favorable outcomes is 2.

    (c) The number of unfavorable outcomes 1.

(d) The total number of possible outcomes is 3.

(e) The probability of an odd number is given by

$$P(\text{odd number}) = P(E)$$
$$= \frac{\text{number of favorable outcomes}}{\text{total number of outcomes}}$$
$$= \frac{2}{3}.$$

(f) The odds in favor of an odd number is given by

$$\text{Odds in favor} = \frac{\text{number of favorable outcomes}}{\text{number of unfavorable outcomes}}$$
$$= \frac{2}{1}, \text{ or 2 to 1.}$$

7. (a) The sample space is
    $$\{11, 12, 13, 21, 22, 23, 31, 32, 33\}.$$

    (b) The probability of an odd number is given by

    $$P(\text{odd number}) = P(E)$$
    $$= \frac{\text{number of favorable outcomes}}{\text{total number of outcomes}}$$
    $$= \frac{6}{9} = \frac{2}{3}.$$

    (c) The probability of a number with repeated digit is given by

    $$P(\text{number with repeated digits }) = \frac{3}{9} = \frac{1}{3}.$$

    (d) the probability of a number greater than 30 is given by
    $$P(\text{number greater than 30}) = \frac{3}{9} = \frac{1}{3}.$$

    (e) The primes are $\{11, 13, 23, 31\}$. Thus, the probability for a prime number is given by

    $$P(\text{prime number}) = \frac{4}{9}.$$

9. (a) The odds in favor of selecting a red ball is given by

    $$\text{Odds in favor} = \frac{\text{number of favorable outcomes}}{\text{number of unfavorable outcomes}}$$
    $$= \frac{4}{7}, \text{ or 4 to 7.}$$

(b) The odds in favor of selecting a yellow ball is given by

$$\text{Odds in favor} = \frac{\text{number of favorable outcomes}}{\text{number of unfavorable outcomes}}$$
$$= \frac{5}{6}, \text{ or 5 to 6.}$$

(c) The odds in favor of selecting a blue ball is given by

$$\text{Odds in favor} = \frac{\text{number of favorable outcomes}}{\text{number of unfavorable outcomes}}$$
$$= \frac{2}{9}, \text{ or 2 to 9.}$$

11. (a) $P(\text{Buddy Holly}) = \dfrac{1}{50}$

(b) $P(\text{The Drifters}) = \dfrac{2}{50} = \dfrac{1}{25}$

(c) $P(\text{Bobby Darin}) = \dfrac{3}{50}$

(d) $P(\text{The Coasters}) = \dfrac{4}{50} = \dfrac{2}{25}$

(e) $P(\text{Fats Domino}) = \dfrac{5}{50} = \dfrac{1}{10}$

The sample space for three fair coin tosses is {*hhh, hht, hth, htt, thh, tht, tth, ttt*}.

13. Product table for "sum"

		2nd die					
	+	1	2	3	4	5	6
	1	2	3	4	5	6	7
	2	3	4	5	6	7	8
1st	3	4	5	6	7	8	9
die	4	5	6	7	8	9	10
	5	6	7	8	9	10	11
	6	7	8	9	10	11	12

(a) Of the 36 possible outcomes, one 1 gives a sum of 2, so

$$P(\text{sum is 2}) = \frac{1}{36}.$$

(b) The sum of 3 appears 2 times in the body of the table. Thus,

$$P(\text{sum of 3}) = \frac{2}{36} = \frac{1}{18}.$$

(c) The sum of 4 appears 3 times in the body of the table. Thus,

$$P(\text{sum of 4}) = \frac{3}{36} = \frac{1}{12}.$$

(d) The sum of 5 appears 4 times in the table. Thus,

$$P(\text{sum of 4}) = \frac{4}{36} = \frac{1}{9}.$$

(e) The sum of 6 appears 5 times in the table. Thus,

$$P(\text{sum of 6}) = \frac{5}{36}.$$

(f) The sum of 7 appears 6 times in the table. Thus,

$$P(\text{sum of 7}) = \frac{6}{36} = \frac{1}{6}.$$

(g) The sum of 8 appears 5 times in the table. Thus,

$$P(\text{sum of 8}) = \frac{5}{36}.$$

(h) The sum of 9 appears 4 times in the table. Thus,

$$P(\text{sum of 7}) = \frac{4}{36} = \frac{1}{9}.$$

(i) The sum of 10 appears 3 times in the table. Thus,

$$P(\text{sum of 10}) = \frac{3}{36} = \frac{1}{12}.$$

(j) The sum of 11 appears 2 times in the table. Thus,

$$P(\text{sum of 11}) = \frac{2}{36} = \frac{1}{18}.$$

(k) The sum of 12 appears 1 time in the table. Thus,

$$P(\text{sum of 12}) = \frac{1}{36}.$$

In Exercises 14–15, answers are computed to three decimal places.

15. The total number of children born was

$$37{,}052 + 35{,}192 = 72{,}244.$$

(a) The empirical probability that one of these births chosen at random is a boy is given by

$$P(\text{boy}) = \frac{37{,}052}{72{,}244} \approx .513.$$

(b) The empirical probability that one of these births chosen at random is a girl is given by

$$P(\text{girl}) = \frac{35{,}192}{72{,}244} \approx .487.$$

17. Writing exercise

19. Since there is no dominance, only RR will result in red flowers. Thus,

$$P(\text{red}) = \frac{1}{4}.$$

21. Since only rr will result in white flowers

$$P(\text{white}) = \frac{1}{4}.$$

23. (a) Since round peas are dominant over wrinkled peas, the combinations RR, Rr, and rR will all result in round peas. Thus,

$$P(\text{round}) = \frac{3}{4}.$$

(b) Since wrinkled peas are recessive, only rr will result in wrinkled peas. Thus,

$$P(\text{wrinkled}) = \frac{1}{4}.$$

25. Cystic Fibrosis occurs in 1 of every 250,000 non-Caucasian births, so the empirical probability that cystic fibrosis will occur in a randomly selected non-Caucasian birth is

$$P = \frac{1}{250,000} = .000004.$$

Construct a chart similar to Table 2 in the textbook and determine the probability of each of the following events.

		Second parent	
+		C	c
First	C	CC	Cc
Parent	c	cC	cc

27. C represents the normal (disease-free gene) and c represents the cystic fibrosis gene. Since c is a recessive gene, only the combination cc results in a child with the disease. Thus, the probability that their first child will have the disease is given by

$$P = \frac{1}{4}.$$

29. Only the combination CC results in a child who neither has nor carries the disease, so the required probability is give by

$$P = \frac{1}{4}.$$

31. The combination cC results in a child who is a healthy cystic fibrosis carrier. This combination occurs twice in

the table (while the other combination that gives a carrier, Cc, does not occur). Thus, the required probability is given by

$$P = \frac{2}{4} = \frac{1}{2}.$$

33. Sickle-cell anemia occurs in about 1 of every 500 black baby births, so the empirical probability that a randomly selected black baby will have sickle-cell anemia is

$$P = \frac{1}{500} = .002.$$

35. From Exercise 33, the probability that a particular black baby will have sickle-cell anemia is .002. Therefore, among 80,000 black baby births, about

$$.002(80,000) = 160$$

occurrences of sickle-cell anemia would be expected.

For Exercises 36–38 let S represent the normal gene and s represent th sickle-cell gene. The possibilities for a child with parents who both have sickle-cell trait are given in the following table that gives the possibilities when one parent is a carrier and the other is a non-carrier.

		Second parent	
+		S	s
First	S	SS	Ss
Parent	s	sS	ss

37. Since the combinations Ss and sS result in sickle-cell trait, the probability that the child will have sickle-cell trait is given by

$$P = \frac{1}{4} = \frac{1}{2}.$$

Use the table corresponding to Exercise 39, where "CC" corresponds to being healthy.

39. (a) The empirical probability formula is

$$P(E) = \frac{\text{number of times event occurred}}{\text{number of times experiment performed}}.$$

Since the number of times that the event described in the exercise has occurred is 0, the probability fraction has a numerator of 0. The denominator is some natural number n. Thus,

$$P(E) = \frac{0}{n} = 0.$$

(b) There is no basis for establishing a theoretical probability for this event.

(c) A woman may break the 10 second barrier at any time in the future, so it is possible that this event will occur.

41. Writing exercise

One approach to Exercises 42–44 is to consider the following: Odds in favor = $\frac{a}{b}$; Odds against = $\frac{b}{a}$; Probability of same event = $\frac{a}{a+b}$

43. The odds in favor of event E are 12 to 19, where $a = 12$ and $b = 19$. Since

$$P(E) = \frac{a}{a+b}$$

we have

$$P(E) = \frac{12}{12+19} = \frac{12}{31}.$$

45. From Table 1 in the text there are 2,598,960 5-card poker hands. Of these 36 are straight flushes. Thus,

$$P(\text{straight flush}) = \frac{36}{2,598,960} \approx .00001385.$$

Refer to Table 1 in the text. Answers are given to eight decimal places.

47. Since there are 13 different possible 4 of a kind hands, the probability of being dealt four queens is 1/13 of the total number of 4 of a kind hands Thus, the probability is

$$\left(\frac{1}{13}\right) \cdot \left(\frac{624}{2,598,960}\right) \approx .00001847.$$

49. If a dart hits the square target shown in the text at random, what is the probability that it will hit in a colored region? (Hint: Compare the area of the colored regions to the total area of the target.)

The area of the colored regions is given by is given by

$$(6cm)^2 - (4cm)^2 + (2cm)^2 = 24\ cm^2$$

The total area is given by

$$(8cm)^2 = 64\ cm^2.$$

Thus, the probability of hitting a colored region is given by

$$P = \frac{24}{64} = \frac{3}{8} = .375.$$

51. Use the fundamental counting principle to determine the number of favorable seating arrangements where each man will sit immediately to the left of his wife.

The first seat can be occupied by one of the three men, the second by his wife, the third by one of the two remaining men, etc. or, $3 \cdot 1 \cdot 2 \cdot 1 \cdot 1 \cdot 1 = 6$; since there are $6! = 720$ possible arrangements of the six people in the six seats we have

$$P = \frac{6}{720} = \frac{1}{120} \approx .0083.$$

53. Use the fundamental counting principle to determine the number of ways the women can sit in three adjacent seats.

The first task is to decide in which seat the first woman is to sit. There are 4 choices (seats 1, 2, 3 or 4). Once this is decided, a second task would be to decide how many arrangements the three women can make sitting together (3!). The last task is to decide how many arrangements the three men could make sitting in the remaining three seats (3!). Thus, there are

$$4 \cdot 3! \cdot 3! = 144$$

ways to accommodate the three women sitting together.

The probability of this occurring is given by

$$P = \frac{144}{720} = \frac{1}{5} = .2,$$

where $6! = 720$ is the total number of seat arrangements one can make.

55. Two distinct numbers are chosen randomly from the set

$$\{-2, -4/3, -1/2, 0, 1/2, 3/4, 3\}.$$

To evaluate the probability that they will be the slopes of two perpendicular lines, find the size of the sample space and the size of the event of interest. The size of the sample space, $n(S)$ is given by $C(7,2) = 21$ since we are choosing 2 items from a set of 7. The size of the event of interest, $n(E)$ is 2 (remember that perpendicular lines must have slopes that are the negative reciprocals of each other). These are either -2 and $1/2$ (either order) or $-4/3$ and $3/4$. Thus,

$$P = \frac{n(E)}{n(S)} = \frac{2}{21} \approx .095.$$

57. Since repetitions are not allowed and order is not important in selecting courses, use combinations. The number of ways of choosing any four courses from the list of ten is $C(6,4)$. Let F be the event of interest "all four courses selected are science courses." Then

$$P(F) = \frac{\text{number of favorable outcomes}}{\text{total number of outcomes}}$$
$$= \frac{C(6,4)}{C(10,4)} = \frac{15}{210}$$
$$= \frac{1}{14} \approx .071.$$

59. The total number of ways to make the three selections, in order, is given by

$$P(36, 3) = 42840.$$

Only 1 of these ways represents a success. Thus,

$$P = \frac{1}{42840} \approx 0.000023.$$

61. The first eight primes are $3, 5, 7, 11, 13, 17, 19$. The sample space consists of all combinations of the set of 8 elements taken 2 at a time, the size of which is given by $C(8,2) = 28$. Thus,

$$E = \{19 + 5, 17 + 7, 13 + 11\} \text{ and}$$

$$P = \frac{n(E)}{n(S)} = \frac{3}{28} \approx .107.$$

63. Two integers are randomly selected from the set $\{1, 2, 3, 4, 5, 6, 7, 8, 9\}$ and are added together. Find the probability that their sum is 11 if they are selected as follows:

(a) With replacement, the event of interest is $E = \{2 + 9, 9 + 2, 3 + 8, 8 + 3, 4 + 7, 7 + 4, 5 + 6, 6 + 5.\}$ Thus, $n(E) = 8$. Since there are $9 \cdot 9 = 9^2 = 81$ ways of selecting the two digits to add together, we have $n(S) = 81$ and

$$P(\text{sum is eleven}) = \frac{8}{81} \approx .049.$$

(b) Without replacement, the event of interest is

$$E = \{2 + 9, 3 + 8, 4 + 7, 5 + 6\}$$

and $n(E) = 4$. However (without replacement),

$$n(S) = C(9, 2) = 36.$$

Thus,

$$P(\text{sum is eleven}) = \frac{4}{36} = \frac{1}{9} \approx .111.$$

65. Let S be the sample space, which is the set of all three-digit numbers. The total number of three-digit numbers is given by the fundamental counting principle,

$$n(S) = 9 \cdot 10 \cdot 10 = 900.$$

Let E be the event "a palindromic three-digit number is chosen." The number of three-digit palindromic numbers is given by

$$n(E) = 9 \cdot 10 \cdot 1,$$

again using the fundamental counting principle. Remember that there are only 9 ways $\{1, 2, \ldots 9\}$ to choose the first digit and 10 ways $\{0, 1, \ldots 9\}$ to choose the second. The third is fixed since it must be the same as the first. Thus,

$$P(E) = \frac{n(R)}{n(S)} = \frac{9 \cdot 10}{9 \cdot 10^2} = \frac{9}{9 \cdot 10} = \frac{1}{10}.$$

11.2 EXERCISES

1. Yes, since event A and event B cannot happen at the same time.

3. Writing exercise

Use the sample space $S = \{1, 2, 3, 4, 5, 6\}$ for Exercises 5–7.

5. Let E be the event "not prime." Then $E = \{1, 4, 6\}$, the non-prime numbers in S. Thus,

$$P(E) = \frac{3}{6} = \frac{1}{2}.$$

7. Let $E = \{2, 4, 6\}$ and $F = \{2, 3, 5\}$.

$$P(E \text{ or } F) = P(E) + P(F) - P(E \text{ and } F)$$
$$= \frac{3}{6} + \frac{3}{6} - \frac{1}{6}$$
$$= \frac{5}{6},$$

by the general addition rule of probability.

9. Let A be the event "less than 3" and B be the event "greater than 4." Thus, $A = \{1, 2\}$ and $B = \{5, 6\}$.

Since A and B are mutually exclusive events, use the special addition rule of probability:

$$P(A \text{ or } B) = P(A) + P(B)$$
$$= \frac{2}{6} + \frac{2}{6}$$
$$= \frac{4}{6} = \frac{2}{3}.$$

11. (a) Since the two events, drawing a king (K) and drawing a queen (Q) are mutually exclusive, use the special addition rule:

$$P(K \text{ or } Q) = P(K) + P(Q)$$
$$= \frac{4}{52} + \frac{4}{52} = \frac{8}{52} = \frac{2}{13}.$$

(b) Use the formula for finding odds in favor:

$$\text{Odds in favor of } E = \frac{P(E)}{P(E')}. \text{ Thus,}$$

$$\frac{P(E)}{P(E')} = \frac{P(K \text{ or } Q)}{P(\text{not } (K \text{ or } Q))}$$
$$= \frac{P(K \text{ or } Q)}{1 - P(K \text{ or } Q)}$$
$$= \frac{2/13}{1 - (2/13)}$$
$$= \frac{2/13}{11/13}$$
$$= \frac{2}{13} \cdot \frac{13}{11}$$
$$= \frac{2}{11}, \text{ or } 2 \text{ to } 11.$$

13. (a) Let S be the event of "drawing a spade" and F be the event of "drawing a face card." Then, $n(S) = 13$ and $n(F) = 12$. (There are 3 face cards in each suite and 4 suites.) There are 3 face cards that are also spades so that $n(S \text{ and } F) = 3$. Thus, by the general additive rule for probability,

$$P(S \text{ or } F) = P(S) + P(F) - P(S \text{ and } F)$$
$$= \frac{13}{52} + \frac{12}{52} - \frac{3}{52}$$
$$= \frac{11}{26}.$$

(b) Use the formula for finding odds in favor of an event E:

$$\text{Odds in favor of } E = \frac{P(E)}{P(E')}. \text{ Thus,}$$

$$\frac{P(E)}{P(E')} = \frac{P(K \text{ or } Q)}{P(\text{not } (K \text{ or } Q))}$$
$$= \frac{P(K \text{ or } Q)}{1 - P(K \text{ or } Q)}$$
$$= \frac{11/26}{1 - (11/26)}$$
$$= \frac{11/26}{15/26}$$
$$= \frac{11}{26} \cdot \frac{26}{15}$$
$$= \frac{11}{15}, \text{ or } 11 \text{ to } 15.$$

15. (a) Let H be the event "a heart is drawn" and S be the event "a seven is drawn." We want to find $P(H \text{ or } S)'$, or $P(H \cup S)'$. Since there are 13 hearts in the deck and 3 other sevens which are not hearts, $n(H \cup N) = 16$ and

$$P(H \cup S) = \frac{16}{52} = \frac{4}{13}.$$

Thus, $P(H \cup S)' = 1 - P(H \cup S)$

$$= 1 - \frac{4}{13}$$
$$= \frac{9}{13}.$$

(b) The number of cards which are hearts or are sevens totals 16. Thus the number of cards which are not hearts or not sevens is $52 - 16 = 36$. The odds in favor of not hearts or not sevens are 36 to 16, or 9 to 4.

Construct a table showing the sum for each of the 36 equally likely outcomes.

		2nd die					
	+	1	2	3	4	5	6
	1	2	3	4	5	6	7
	2	3	4	5	6	7	8
1st	3	4	5	6	7	8	9
die	4	5	6	7	8	9	10
	5	6	7	8	9	10	11
	6	7	8	9	10	11	12

17. Let E be the event of getting a sum which is an even number. Counting the number of occurrences in the sum table for these even outcomes represents the numerator of the probability fraction.

Then,

$$P(E) = \frac{18}{36}.$$

Let M be the event of getting sums which are multiples of three, $\{3, 6, 9, \text{ or } 12\}$. Counting the number of occurrences in the sum table for these outcomes represents the numerator of the probability fraction.

Then,

$$P(M) = \frac{12}{36}$$

and

$$P(E \text{ and } M) = \frac{6}{36}.$$

Thus, by the general addition law,

$$P(E \cup M) = P(E) + P(M) - P(E \text{ and } M)$$
$$= \frac{18}{36} + \frac{12}{36} - \frac{6}{36}$$
$$= \frac{24}{36} = \frac{2}{3}.$$

19. Since these are mutually exclusive events, use the special additive rule. Since there is only one sum less than 3 (the sum of 2),

$$P(\text{sum less than 3}) = \frac{1}{36},$$

and since there are six sums greater than 9,

$$P(\text{sum greater than 9}) = \frac{6}{36}.$$

Thus,

$$P(\text{sum less than 3 or greater than 9}) = \frac{1}{36} + \frac{6}{36} = \frac{7}{36}.$$

21. $P(S) = P(A \cup B \cup C \cup D)$
$= P(A) + P(B) + P(C) + P(D)$
$= 1$

Refer to Table 1 (11.1 in the textbook) and give answers to six decimal places.

23. Let F be the event of drawing a full house and S be the event of drawing a straight. Using Table 1 in the text to determine

$$n(F) = 3744$$

and

$$n(S) = 10,200,$$

we have

$$P(F) = \frac{3744}{2,598,960} \text{ and } P(S) = \frac{10,200}{2,598,960}.$$

Since the events are mutually exclusive,

$$P(F \text{ or } S) = P(F) + P(S)$$
$$= \frac{3,744}{2,598,960} + \frac{10,200}{2,598,960}$$
$$= \frac{5108}{2,598,960} + \frac{54,912}{2,598,960}$$
$$= \frac{13,944}{2,598,960} \approx .005365.$$

25. The events are mutually exclusive, so use the special addition rule:

$P(\text{nothing any better than two pairs})$
$= P(\text{no pair or one pair or two pairs})$
$= P(\text{no pair}) + P(\text{one pair}) + P(\text{two pairs})$
$$= \frac{1,302,540}{2,598,960} + \frac{1,098,240}{2,598,960} + \frac{123,522}{2,598,960}$$
$$= \frac{2,524,332}{2,598,960} \approx .971285.$$

27. "Par or above" is represented by all categories from 70 up. Since these are mutually exclusive, use the special addition rule:

$P(\text{Par or above})$
$= 0.30 + 0.23 + 0.09 + 0.06 + 0.04 + 0.03 + 0.01$
$= 0.76.$

29. "Less than 90" is represented by all categories under the 90–94 category. Since these are mutually exclusive, use the special addition rule:

$P(\text{Less than 90})$
$= 0.04 + 0.06 + 0.14 + 0.30 + 0.23 + 0.09 + 0.06$
$= 0.92.$

31. Odds of Sue's shooting below par $= \dfrac{P(\text{below par})}{P(\text{par or above})}$

$$P(\text{par or above}) = .76 \text{ (Exercise 27), and}$$
$$P(\text{below par}) = 1 - P(\text{par or above})$$
$$= 1 - .76 = .24,$$

by the complements rule of probability.

Thus,

$$\text{odds in favor} = \frac{.24}{.76} = \frac{6}{19}, \text{ or 6 to 19.}$$

33. Let x denote the sum of two distinct numbers selected randomly from the set $\{1, 2, 3, 4, 5\}$. Construct the probability distribution for the random variable x.

Create a "product table" to list the elements in the sample space. Note: can't use $1 + 1 = 2$, etc. Why?

+	1	2	3	4	5
1	–	3	4	5	6
2	–	–	5	6	7
3	–	–	–	7	8
4	–	–	–	–	9
5	–	–	–	–	–

Thus,

x	$P(x)$
3	$1/10 = .1$
4	$1/10 = .1$
5	$2/10 = .2$
6	$2/10 = .2$
7	$2/10 = .2$
8	$1/10 = .1$
9	$1/10 = .1.$

For Exercises 35–38, let A be an event within the sample space S, and let $n(A) = a$ and $n(S) = s$.

35. $n(A') + n(A) = n(S)$
    $n(A') + a = s$
    Thus, $n(A') = s - a$

37. $P(A) + P(A') = \dfrac{a}{s} + \dfrac{s-a}{s}$
    $= \dfrac{a+s-a}{s}$
    $= \dfrac{s}{s} = 1$

We want to form three-digit numbers using the set of digits $\{0, 1, 2, 3, 4, 5\}$. For example, 501 and 224 are such numbers but 035 is not.

39. The number of three-digit numbers is, by the fundamental counting principle,

    $$5 \cdot 6 \cdot 6 = 180.$$

    Remember that we can't choose "0" for the first digit.

41. If one three-digit number is chosen at random from all those that can be made from the above set of digits, find the probability that the one chosen is not a multiple of 5.

    The number of three-digit numbers is, by the fundamental counting principle,

    $$5 \cdot 6 \cdot 6 = 180 \text{ (Exercise 39)}.$$

    There are

    $$5 \cdot 6 \cdot 2 = 60$$

    three-digit numbers that are multiples of 5 (Exercise 40). Thus,

    $$P(\text{multiple of 5}) = \frac{60}{180} = \frac{1}{3}.$$

By the complements principle,

$$P(\text{not a multiple of 5}) = 1 - P(\text{multiple of 5})$$
$$= 1 - \frac{60}{180}$$
$$= 1 - \frac{1}{3}$$
$$= \frac{2}{3}.$$

43. Since box C contains the same number of green marbles as blue and the number of green and blue marbles was the same to begin with, then box A and box B must contain exactly the same number of marbles after all the marbles are drawn. Because this is certain, the probability is 1.

11.3 EXERCISES

1. The events are independent since the outcome on the first toss has no effect on the outcome of the second toss.

3. The two planets are selected, with replacement, from the list in Table 5. The events "first is closer than Earth" and "second is farther than Uranus" are independent since the first selection has no affect on the second. This is because the first selection was replaced before the second was made.

5. The answers are all guessed on a twenty-question multiple choice test. Let A be the event "first answer correct" and let B be the event "last answer correct." The events A and B are independent since the first answer choice does not affect the last answer choice.

7. The probability that the student selected is "female" is

    $$\frac{42}{100} = \frac{21}{100}.$$

9. Since a total of 68 of the 100 students are "not motivated primarily by money," the probability is

    $$\frac{68}{100} = \frac{17}{25}.$$

11. Given that the student selected is motivated primarily by "sense of giving to society," the sample space is reduced to 34 students. Of these 19 are male, so the probability is

    $$\frac{19}{34}.$$

In Exercises 13–16 the first puppy chosen is replaced before the second is chosen. Note that "with replacement" means that the events may be considered independent and we can apply the special multiplication rule of probability.

13. The probability that both select a "poodle" is

$$P(P_1 \text{ and } P_2) = P(P_1) \cdot P(P_2)$$
$$= \left(\frac{4}{7}\right) \cdot \left(\frac{4}{7}\right)$$
$$= \frac{16}{49}.$$

15. The probability that "Rebecka selects a terrier and Aaron selects a retriever" is given by

$$P(T_1 \text{ and } R_2) = P(T_1) \cdot P(R_2)$$
$$= \left(\frac{2}{7}\right) \cdot \left(\frac{1}{7}\right)$$
$$= \frac{2}{49}.$$

In Exercises 17–22 the first puppy chosen is not replaced before the second is chosen. Thus, the events are not independent. Therefore, apply the general multiplication rule of probability.

17. The probability that "both select a poodle" is given by

$$(P_1 \text{ and } P_2) = P(P_1) \cdot P(P_2 \mid P_1)$$
$$= \left(\frac{4}{7}\right) \cdot \left(\frac{3}{6}\right)$$
$$= \frac{2}{7}.$$

Remember that the second probability is conditional to the first event as having occurred and hence, the sample space is reduced.

19. The probability that "Aaron select a retriever," given "Rebecka selects a poodle" is found by

$$P(R_2 \mid P_1) = \frac{1}{6}.$$

Note that Rebecka's choice decreased the sample space by one dog.

21. The probability that "Aaron selects a retriever," given "Rebecka selects a retriever" is found by

$$P(R_2 \mid R_1) = \frac{0}{6} = 0.$$

Note that after Rebecka's selection there are no remaining retriever's for for Aaron to retrieve.

23. Since the cards are dealt without replacement, when the second card is drawn, there will be 51 cards left, of which 12 are spades. Thus,

$$P(S_2 \mid S_1) = \frac{12}{51} = \frac{4}{17}.$$

Note that both event of interest (numerator) and the sample space (denominator) was reduced by the selection of the first card, a spade.

25. Since the cards are dealt without replacement, the events "first is face card" and "second is face card" are not independent, so be sure to use the general multiplication rule of probability. Let F_1 be the event "first is a face card" and F_2 be the event "second is a face card." Then,

$$P(\text{two face cards}) = P(F_1 \text{ and } F_2)$$
$$= P(F_1) \cdot P(F_2 \mid F_1)$$
$$= \frac{12}{52} \cdot \frac{11}{51}$$
$$= \frac{3}{13} \cdot \frac{11}{51}$$
$$= \frac{11}{221}.$$

27. The probability that the "first card dealt is a jack and the second is a face card" is found by

$$P(J_1 \text{ and } F_2) = P(J_1) \cdot (F_2 \mid J_1)$$
$$= \frac{4}{52} \cdot \frac{11}{51}$$
$$= \frac{11}{663}.$$

Remember that there are only 11 face cards left once the Jack is drawn.

Use the results of Exercise 28 to find each of the following probabilities when a single card is drawn from a standard 52-card deck.

29. $P(\text{queen} \mid \text{face card}) = \dfrac{n(F \text{ and } Q)}{n(F)}$
$$= \frac{4}{12} = \frac{1}{3}$$

31. $P(\text{red} \mid \text{diamond}) = \dfrac{n(D \text{ and } R)}{n(D)}$
$$= \frac{13}{13} = 1$$

33. From the integers 1 through 10, the set of primes are $\{2, 3, 5, 7\}$. The set of odds are $\{1, 3, 5, 7.9\}$. Since there only three primes in the set of 4 odd numbers, the second probability fraction become 2/4 and

$$P(\text{prime}) \cdot P(\text{odd} \mid \text{prime}) = \frac{4}{10} \cdot \frac{3}{4}$$
$$= \frac{3}{10}.$$

This is the same value as computed in the text for $P(\text{odd}) \cdot P(\text{prime} \mid \text{odd})$, that is, the probability of selecting an integer from the set which is "odd and prime."

35. Since the birth of a boy (or girl) is independent of previous births the probability of having three boys successively is

$$\left(\frac{1}{2}\right) \cdot \left(\frac{1}{2}\right) \cdot \left(\frac{1}{2}\right) = \frac{1}{8}.$$

37. Let S represent a sale purchase for more than $100. Then,

$$P(\text{both sales more than } \$100) = P(S_1 \text{ and } S_2)$$
$$= P(S_1) \cdot P(S_2)$$
$$= (.80) \cdot (.80)$$
$$= .640.$$

39. Let S represent a sale purchase for more than $100. Then, S' represents a sale purchase that is not more than $100. Since $P(S) = .80$,

$$P(S') = 1 - .80$$
$$= .20,$$

by the complements principle.

$$P(\text{none of the first 3 sales more than } \$100)$$
$$= P(\text{not } S_1 \text{ and not } S_2 \text{ and not } S_3)$$
$$= P(S'_1) \cdot P(S'_2) \cdot P(S'_3)$$
$$= (.20) \cdot (.20) \cdot (.20)$$
$$= .008.$$

41. Since the probability the cloud will move in the critical direction is .05, the probability that it will not move in the critical direction is

$$1 - .05 = .95,$$

by the complements formula.

43. The probability that the cloud would not move in the critical direction for each launch is $1 - .05 = .95$. The probability that the cloud would not move in the critical direction for any 5 launches is $(.95)^5$ by the special multiplication rule. The probability that any 5 launches will result in at least one cloud movement in the critical direction is the complement of the probability that a cloud would not move in the critical direction for any 5 launches, or

$$1 - (.95)^5 \approx .23.$$

Four men and three women are waiting to be interviewed for jobs. If they are all selected in random order, find the probability of each of the following events.

45. $\quad P(\text{all women first})$
$$= P(W_1) \cdot P(W_2 \mid W_1) \cdot P(W_3 \mid W_1 \text{ and } W_2)$$
$$= \frac{3}{7} \cdot \frac{2}{6} \cdot \frac{1}{5}$$
$$= \frac{1}{35}$$

47. The probability that the first person interviewed will be a woman is

$$P(W_1) = \frac{3}{7}.$$

49. Consider choosing the last to be interviewed first. Then the probability that the "last person interviewed will be a woman" is

$$P(W_7) = \frac{3}{7}.$$

51. (a) "At least three" is the complement of "one or two" and we can't get two girls from one birth. Thus, find only the probability of having 2 girls with two births,
$$P(gg) = P(g) \cdot P(g)$$
$$= \frac{1}{2} \cdot \frac{1}{2}$$
$$= \frac{1}{4}.$$

Therefore, by the complements principle,

$$P(\text{at least three births}) = 1 - \frac{1}{4} = \frac{3}{4}.$$

(b) "At least four births" is the complement of "two or three births." From (a) above

$$P(\text{two births to get } gg) = \frac{1}{4}.$$

To calculate $P(\text{three births to get } gg)$ examine Figure 2 in the text. The three success are ggb, gbg, and bgg. Don't count, however, the outcome ggb as a success since this has already been counted when computing the probability associated with two births.

$$P(\text{three births to get } gg) = \frac{2}{8} = \frac{1}{4}.$$

P(two or three births to get gg)

$\quad = P$(two births to get gg) $+ P$(three births to get gg)

$\quad = \dfrac{1}{4} + \dfrac{1}{4}$

$\quad = \dfrac{2}{4} = \dfrac{1}{2}$

Finally, the probability of "at least four births" may be calculated by the complements principle:

$$P(\text{at least four births to get gg}) = 1 - \frac{1}{2} = \frac{1}{2}.$$

(c) "At least five births" is the complement of " two or three or four births."

P(two or three or four births to obtain gg)

$\quad = P$(two or three births to get gg) $+ P$(four births to get gg)

P(two or three births to get gg) $= 1/2$ was calculated in (b) above. To calculate P(four births to get gg) extend the tree diagram (Figure 2 in the text). Count all of the outcomes with gg (two girls) as successes except bggb and ggbb since they have already been used in calculating the earlier probability. There are then 3 outcomes which may be considered as successes. Thus, P(four births to get gg)) $= 3/16$.

P(two or three or four births to obtain gg)

$\quad = P$(two or three births to get gg) $+ P$(four births to get gg)

$\quad = \dfrac{1}{2} + \dfrac{3}{16} = \dfrac{8}{16} + \dfrac{3}{16}$

$\quad = \dfrac{11}{16}.$

Finally, the probability of "at least five births to obtain gg" may be calculated by the complements principle:

$$P(\text{at least five births to obtain gg}) = 1 - \frac{11}{16} = \frac{5}{16}.$$

53. Since the two tosses are independent events, use the special multiplication rule:

$$\begin{aligned} P(hh) &= P(h) \cdot P(h) \\ &= (.5200) \cdot (.5200) \\ &= .2704. \end{aligned}$$

A coin, biased so that $P(h) = .5200$ and $P(t) = .4800$, is tossed twice. Give answers to four decimal places.

55. Since the two tosses are independent events, use the special multiplication rule:

$$\begin{aligned} P(th) &= P(t) \cdot P(h) \\ &= (.4800) \cdot (.5200) \\ &= .2496. \end{aligned}$$

57. Writing exercise

59. Using the fundamental counting principle, there are

$$2 \cdot 2 \cdot 2 \cdot 2 \cdot 2 \cdot 2 = 2^6 = 64$$

different switch settings. The probability of randomly getting 1 of the 64 possible settings is

$$\frac{1}{64} \approx .0156.$$

61. (Review: For Further Thought in this section.)

P(at least one duplication of switch settings)

$\quad = 1 - P$(no duplication)

$\quad = 1 - \dfrac{63}{64}$

$\quad \approx .016$ for two neighbors;

$$1 - \frac{63}{64} \cdot \frac{62}{64} = 1 - \frac{63 \cdot 62}{(64^2)}$$

$\approx .046$ for three neighbors;

$\cdots$

$$1 - \frac{63 \cdot 62 \cdot 61 \cdot 60 \cdot 59 \cdot 58 \cdot 57 \cdot 56}{(64)^8}$$

$\approx .445$ for nine neighbors;

and

$$1 - \frac{63 \cdot 62 \cdot 61 \cdot 60 \cdot 59 \cdot 58 \cdot 57 \cdot 56 \cdot 55}{(64)^9}$$

$\approx .523 > \dfrac{1}{2}$ for ten neighbors.

Thus, the minimum number of neighbors who must use this brand of opener before the probability of at least one duplication of settings is greater than 1/2 is ten.

63. Since the events are not independent, use the general multiplication rule. Let R_1 represent rain on the first day, R_2 represent rain on the second day.

P(rain on two consecutive days in November)

$\quad = P(R_1 \cap R_2)$

$\quad = P(R_1) \cdot P(R_2 \mid R_1)$

$\quad = (.500) \cdot (.800)$

$\quad = .400$

65. Since the events are not independent, use the general multiplication rule. Let R_1 represent rain on November 1, R_2 represent rain November 2, and R_3 represent rain on November 3.

P(rain on November 1st and 2nd, but not on the 3rd)
$P(R_1 \cap R_2 \cap \text{not } R_3)$
$= P(R_1) \cdot P(R_2 \mid R_1) \cdot P(\text{not } R_3 \mid R_2)$
$= (.500) \cdot (.800) \cdot (1 - .800)$
$= .080$

Note that the probability of not raining after a rainy day is the complement of the probability that it does rain.

67. To find the probability of "no engine failures" begin by letting F represent a failed engine. The probability that a given engine will fail, $P(F)$, is .10. This means that, by the complements principle, the probability that an engine will not fail is given by

$$P(\text{not } F) = 1 - .10 = .90.$$

Since engines "not failing" are independent events, use the special product rule.

P(no engine failures)
$= P(\text{not } F_1 \cap \text{ not } F_2 \cap \text{not } F_3 \cap \text{ not } F_4)$
$= P(\text{not } F_1) \cdot P(\text{not } F_2) \cdot P(\text{not } F_3) \cdot P(\text{not } F_4)$
$= (.90)^4 = .6561$

69. The probability of "exactly two engine failures" can be found by applying the fundamental counting principle, where the first task is to decide which two engines fail $[C(4, 2)]$; the second task, find the probability that one of these engines fail (.10); followed by the second engine failing (.20); followed by finding the probability the the third engine does not fail $(1 - .30 = .70)$; followed by the probability that last engine does not fail $(1 - .30 = .70)$. Thus,

$$P = C(4, 2) \cdot (.10) \cdot (.20) \cdot (.70)^2$$
$$= .0588.$$

Refer to text discussion of the rules for "one-and-one" basketball. Karin Wagner, a basketball player, has a 60% foul shot record. (She makes 60% of her foul shots.) Find the probability that, on a given one-and-one foul shooting opportunity, Karin will score the following number of points.

71. A one-and-one foul shooting opportunity means that Susan gets a second shot only if she makes her first shot. The probability of scoring "no points" means that she missed her first shot. Thus,

P(scoring no points)
$= 1 - P$(scoring at least one point)
$= 1 - .70$
$= .30.$

73. The Probability of "scoring two points" is given by

P(scoring two points)
$= P$(scoring the 1st shot and scoring the 2nd shot)
$= P$(scoring on 1st shot) $\cdot P$(scoring on 2nd shot)
$= (.70) \cdot (.70)$
$= .49.$

75. Writing exercise

11.4 EXERCISES

1. Let heads be "success."
Then $n = 3, p = q = \frac{1}{2}$, and $x = 0$.

By the binomial probability formula,

$$P(0) = C(3, 0) \left(\frac{1}{2}\right)^0 \left(\frac{1}{2}\right)^{3-0}$$
$$= \frac{3!}{0! \, (3 - 0)!} \cdot 1 \cdot \left(\frac{1}{2}\right)^3$$
$$= \frac{3!}{1 \cdot 3!} \cdot \frac{1}{2^3}$$
$$= 1 \cdot \frac{1}{8} = \frac{1}{8}.$$

Note $C(3, 0) = 1$, and we could easily reason this result without using the combination formula since there is only one way to choose 0 things from a set of 3 things–take none out.

3. Let heads be "success."
Then $n = 3, p = q = \frac{1}{2}$, and $x = 2$.

By the binomial probability formula,

$$P(2 \text{ heads}) = C(3, 2) \cdot \left(\frac{1}{2}\right)^2 \cdot \left(\frac{1}{2}\right)^{3-2}$$
$$= \frac{3!}{2!(3 - 2)!} \cdot \frac{1}{4} \cdot \frac{1}{2}$$
$$= \frac{3 \cdot 2 \cdot 1}{2 \cdot 1 \cdot 4 \cdot 2} = \frac{3}{8}.$$

5. Use the special addition rule for calculating the probability of "1 or 2 heads."

$$P(1 \text{ or } 2 \text{ heads}) = P(1) + P(2)$$
$$= C(3,1) \cdot \left(\tfrac{1}{2}\right)^1 \cdot \left(\tfrac{1}{2}\right)^{3-1}$$
$$+ C(3,2) \cdot \left(\tfrac{1}{2}\right)^2 \cdot \left(\tfrac{1}{2}\right)^{3-2}$$
$$= 3 \cdot \frac{1}{2} \cdot \frac{1}{4} + 3 \cdot \frac{1}{4} \cdot \frac{1}{2}$$
$$= \frac{6}{8} = \frac{3}{4}$$

7. "No more than 1" is the same as "0 or 1."

$$P(0 \text{ or } 1) = P(0) + P(1)$$
$$= C(3,0)\left(\frac{1}{2}\right)^0 \left(\frac{1}{2}\right)^3$$
$$+ C(3,1)\left(\frac{1}{2}\right)^1 \left(\frac{1}{2}\right)^2$$
$$= 1 \cdot 1 \cdot \frac{1}{8} + 3 \cdot \frac{1}{2} \cdot \frac{1}{4}$$
$$= \frac{1}{8} + \frac{3}{8}$$
$$= \frac{1}{2}$$

9. Assuming boy and girl babies are equally likely, find the probability that a family with three children will have exactly two boys.

$$P(2 \text{ boys}) = C(3,2)\left(\frac{1}{2}\right)^2 \left(\frac{1}{2}\right)^1$$
$$= 3 \cdot \frac{1}{4} \cdot \frac{1}{2}$$
$$= \frac{3}{8}$$

11. If n fair coins are tossed, the probability of exactly x heads is the fraction whose numerator is entry number $\underline{x}$ of row number $\underline{n}$ in Pascal's triangle, and whose denominator is the sum of the entries in row number $\underline{n}$. That is, $x; n; n$.

For Exercises 12–18, refer to Pascal's triangle. Since seven coins are tossed, we will use row number 7 of the triangle. (Recall that the first row is row number 0 and that the first entry in each row is entry number 0.) The sum of the numbers in row 7 is $2^7 = 128$, which will be the denominator in each of the probability fractions.

```
              1
           1     1
        1     2     1
     1     3     3     1
  1     4     6     4     1
1     5    10    10     5     1
1   6    15    20    15     6     1
1   7   21    35    35    21    7     1
```

13. For the probability "1 head," the numerator of the probability fraction is entry 1 of row 7 of Pascal's triangle, or 7, and the denominator is the sum of the elements in row 7, or

$$1 + 7 + 21 + 35 + 35 + 21 + 7 + 1 = 128.$$

Thus,

$$P(1 \text{ head}) = \frac{7}{128}.$$

15. For the probability "3 heads," the numerator of the probability fraction is entry number 3 of row 7 of Pascal's triangle, or 35, and the denominator is the sum of the elements in row 7, or 128. Thus,

$$P(3 \text{ heads}) = \frac{35}{128}.$$

17. For the probability "5 heads," the numerator of the probability fraction is entry number 5 of row 7 of Pascal's triangle, or 21, and the denominator is the sum of the elements in row 7, or 128. Thus,

$$P(5 \text{ heads}) = \frac{21}{128}.$$

19. For the probability "7 heads," the numerator of the probability fraction is entry 7 of row 7 of Pascal's triangle, or 1, and the denominator is the sum of the elements in row 7, or 128. Thus,

$$P(7 \text{ heads}) = \frac{1}{128}.$$

For Exercises 21–23, a fair die is rolled three times and a 4 is considered "success," while all other outcomes are "failures"

21. Here $n = 3$, $p = \frac{1}{6}$, $q = \frac{5}{6}$, and $x = 1$.

$$P(1) = C(3,1)\left(\frac{1}{6}\right)^1\left(\frac{5}{6}\right)^2$$
$$= 3 \cdot \frac{1}{6} \cdot \frac{25}{36}$$
$$= \frac{25}{72}$$

23. Here $n = 3$, $p = \frac{1}{6}$, $q = \frac{5}{6}$, and $x = 3$.

$$P(3) = C(3,3)\left(\frac{1}{6}\right)^3\left(\frac{5}{6}\right)^0$$
$$= 1 \cdot \frac{1}{216} \cdot 1$$
$$= \frac{1}{216}$$

Answers are rounded to three decimal places.

25. Here $n = 5$, $p = \frac{1}{3}$, and $x = 4$.
Since $p = \frac{1}{3}$,

$$q = 1 - p$$
$$= 1 - \frac{1}{3}$$
$$= \frac{2}{3}.$$

Substitute these values into the binomial probability formula:

$$P(4) = C(5,4)\left(\frac{1}{3}\right)^4\left(\frac{2}{3}\right)^1$$
$$= 5 \cdot \frac{2}{3^5}$$
$$\approx .041.$$

27. Here $n = 20$, $p = \frac{1}{8}$, and $x = 2$.
Since $p = \frac{1}{8}$,

$$q = 1 - p$$
$$= 1 - \frac{1}{8}$$
$$= \frac{7}{8}.$$

Substitute these values into the binomial probability formula:

$$P(2) = C(20,2)\left(\frac{1}{8}\right)^2\left(\frac{7}{8}\right)^{18}$$
$$= \frac{20!}{2!18!} \cdot \frac{7^{18}}{8^2 8^{18}}$$
$$= 190 \cdot \frac{7^{18}}{8^{20}}$$
$$\approx .268.$$

29. Writing exercise

31. Writing exercise

For Exercises 32–35, let a correct answer be a "success." Then $n = 10$, $p = 2/6 = 1/3$, and $q = 1 - p = 2/3$.

33. The probability of getting "exactly 7 correct answers" is given by

$$P(7 \text{ correct answers}) = C(10,7)\left(\frac{1}{3}\right)^7\left(\frac{2}{3}\right)^3$$
$$= 120 \cdot \frac{2^3}{3^{10}}$$
$$\approx .016.$$

35. "At least seven" means seven, eight, nine, or ten correct answers.

$$P(7 \text{ or } 8 \text{ or } 9 \text{ or } 10)$$
$$= P(7) + P(8) + P(9) + P(10)$$
$$= C(10,7)\left(\frac{1}{3}\right)^7\left(\frac{2}{3}\right)^3 + C(10,8)\left(\frac{1}{3}\right)^8\left(\frac{2}{3}\right)^2$$
$$+ C(10,9)\left(\frac{1}{3}\right)^9\left(\frac{2}{3}\right)^1 + C(10,10)\left(\frac{1}{3}\right)^{10}\left(\frac{2}{3}\right)^0$$
$$\approx .01626 + .00305 + .00034 + .00002$$
$$\approx .01967 \approx .020$$

37. For "exactly 1 to have undesirable side effects," $x = 1$ and $n = 8$, $p = .30$, $q = 1 - p = .70$. Thus,

$$P(1) = C(8,1)(.30)^1(1 - .30)^{8-1}$$
$$= 8(.30)(.70)^7$$
$$\approx .198.$$

39. "More than two" is the complement of 0, 1, or 2. Thus,

$$
\begin{aligned}
P(\text{more than two}) &= 1 - P(0, 1 \text{ or } 2) \\
&= 1 - [P(0) + P(1) + P(2)] \\
&= 1 - [C(8,0)(.30)^0(.70)^8 \\
&\quad \cdot C(8,1)(.30)^1(.70)^7 \\
&\quad + C(8,2)(.30)^2(.70)^6] \\
&\approx 1 - [1 \cdot 1 \cdot (.057648) \\
&\quad + 8(.30)(.08235) \\
&\quad + (28)(.09)(.11765)] \\
&\approx .448.
\end{aligned}
$$

41. For the probability that "from 4 through 6" will attend college, $n = 9, p = .50,$ and $q = 1 - p = .50.$ Thus,

$$
\begin{aligned}
P(4 \text{ or } 5 \text{ or } 6) &= P(4) + P(5) + P(6) \\
&= C(9,4)(.50)^4(.50)^5 \\
&\quad + C(9,5)(.50)^5(.50)^4 \\
&\quad + C(9,6)(.50)^6(.50)^3 \\
&\approx .246 + .246 + .164 \\
&\approx .656.
\end{aligned}
$$

43. For the probability that "all 9 enroll in college," $x = 9, n = 9, p = .50,$ and $q = 1 - p = .50.$ Thus,

$$
\begin{aligned}
P(\text{all } 9) &= P(9) \\
&= C(9,9)(.50)^9(.50)^0 \\
&= 1 \cdot (.001953125) \cdot 1 \\
&\approx .002.
\end{aligned}
$$

45. "At least half of the 6 trees" is the complement of "0, 1, or 2 trees." Here $p = .65, q = 1 - .65 = .35$ and $n = 6.$ Using the complements rule the probability is found by

$$
\begin{aligned}
P(\text{at least half}) &= 1 - P(0 \text{ or } 1 \text{ or } 2) \\
&= 1 - [P(0) + P(1) + P(2)] \\
&= 1 - [C(6,0)(.65)^0(.35)^6 \\
&\quad + C(6,1)(.65)^1(.35)^5 \\
&\quad + C(6,2)(.65)^2(.35)^4] \\
&\approx 1 - [.0018 + .0205 + .0951] \\
&\approx 1 - .1174 \approx .883.
\end{aligned}
$$

See discussion in text concerning the probability of a first success on the xth trial which follows x − 1 trials which are failures. The formula is as follows:

$$P(F_1 \text{ and } F_2 \text{ and } \ldots \text{ and } F_{x-1} \text{ and } S_x) = q^{x-1} \cdot p.$$

47. Writing exercise

49. Let $p = .038, q = 1 - 0.38 = 0.62,$ and $x = 4.$ Thus,

$$
\begin{aligned}
P(F_1 \text{ and } F_2 \text{ and } \ldots \text{ and } F_{5-1} \text{ and } S_5) \\
= q^{4-1} \cdot p = q^3 \cdot p \\
= (.62)^3(.38) \\
= (.238328)(.38) \\
\approx .091.
\end{aligned}
$$

51. Let an aborted launching be a "success."

Then $p = .04, q = 1 - .04 = .96,$ and $x = 20.$ Thus,

$$
\begin{aligned}
P(\text{20th launch is first launch aborted}) &= q^{20} \cdot p \\
&= (.96)^{20} \cdot (.04) \\
&\approx (.442) \cdot (.04) \\
&\approx .018.
\end{aligned}
$$

Heads means walk one block north and tails means walk one block south. In each case, ask how many successes, say heads, would be required and use the binomial formula. See discussion in text about a random walk.

53. To end up 6 blocks north of his corner, Harvey must go 8 blocks north and 2 blocks south, so he must toss 8 heads and 2 tails. Use the binomial probability formula with

$$n = 10, p = \frac{1}{2}, q = \frac{1}{2}, \text{and } x = 8.$$

Then,

$$
\begin{aligned}
P(8 \text{ heads, } 2 \text{ tails}) &= (8) \\
&= C(10,8)\left(\frac{1}{2}\right)^8\left(\frac{1}{2}\right)^2 \\
&= 45 \cdot \left(\frac{1}{2}\right)^{10} \\
&= \frac{45}{1024} \approx .044.
\end{aligned}
$$

55. It will be Impossible for Harvey to end up 5 blocks south of his corner since the number of heads + the number of tails must $= 10$ and at the same time the number of tails − the number of heads must $= 5.$

To show this, solve the system of equations:

$$
\begin{aligned}
&\quad\; h + t = 10 \\
&\quad\; t - h = 5 \\
\text{or} \quad &\; 2t = 15 \text{ (adding equations)} \\
&\quad\; t = 7.5 \text{ and } h = 2.5.
\end{aligned}
$$

Thus, $P(-5) = 0.$

57. P(at least 2 blocks north)

$= P(2 \text{ or } 3 \text{ or } \ldots \text{ or } 10 \text{ blocks north})$

$= P(2 \text{ blocks north}) + P(3 \text{ blocks north}) \ldots$
  $+ P(10 \text{ blocks north})$

$= P(6 \text{ heads and } 4 \text{ tails})$
  $+ 0[\text{since 3 blocks north is impossible} -$
  similar to Exercise 55]
  $+ P(7 \text{ heads and } 3 \text{ tails})$
  $+ 0 \text{ [5 blocks north is impossible]}$
  $+ P(8 \text{ heads and } 2 \text{ tails})$
  $+ 0[7 \text{ blocks north is impossible}]$
  $+ P(9 \text{ heads and } 1 \text{ tail})$
  $+ 0 \text{ [9 blocks north is impossible]}$
  $+ P(10 \text{ heads})$

$= P(6 \text{ heads}) + 0 + P(7 \text{ heads}) + 0$
  $+ P(8 \text{ heads}) + P(9 \text{ heads}) + 0 + P(10 \text{ heads})$

$= C(10,6)\left(\dfrac{1}{2}\right)^6 \left(\dfrac{1}{2}\right)^4$

$\quad + C(10,7)\left(\dfrac{1}{2}\right)^7 \left(\dfrac{1}{2}\right)^3$

$\quad + C(10,8)\left(\dfrac{1}{2}\right)^8 \left(\dfrac{1}{2}\right)^2$

$\quad + C(10,9)\left(\dfrac{1}{2}\right)^9 \left(\dfrac{1}{2}\right)^1$

$\quad + C(10,10)\left(\dfrac{1}{2}\right)^{10} \left(\dfrac{1}{2}\right)^0$

$= 210 \cdot \left(\dfrac{1}{2}\right)^{10} + 120 \cdot \left(\dfrac{1}{2}\right)^{10} + 45 \cdot \left(\dfrac{1}{2}\right)^{10}$

$\quad + 10 \cdot \left(\dfrac{1}{2}\right)^{10} + 1 \cdot \left(\dfrac{1}{2}\right)^{10}$

$= (210 + 120 + 45 + 10 + 1) \cdot \left(\dfrac{1}{2}\right)^{10}$

$= \dfrac{210 + 120 + 45 + 10 + 1}{1024}$

$= \dfrac{193}{512} \approx .377$

59.   In order to end up *on* his corner, Harvey must go 5 blocks north and 5 blocks south. Let $p = \frac{1}{2}, q = \frac{1}{2}$, and $x = 5$:

$P(5 \text{ heads, } 5 \text{ tails}) = P(5)$

$= C(10,5)\left(\dfrac{1}{2}\right)^5 \left(\dfrac{1}{2}\right)^5$

$= 252\left(\dfrac{1}{2}\right)^{10}$

$= 252\left(\dfrac{1}{1024}\right)$

$= \dfrac{252}{1024} = \dfrac{63}{256} \approx .246.$

11.5 EXERCISES

1.   Writing exercise

3.   Five fair coins are tossed. A tree diagram may be helpful to create the following sample space. Use the following sample space to create the individual probabilities.

hhhhh	*hhhht*	*hhhth*	*hhhtt*
hhthh	*hhtht*	*hhtth*	*hhttt*
hthhh	*hthht*	*hthth*	*hthtt*
htthh	*httht*	*httth*	*htttt*
thhhh	*thhht*	*thhth*	*thhtt*
ththh	*ththt*	*thtth*	*thttt*
tthhh	*tthht*	*tthth*	*tthtt*
ttthh	*tttht*	*tttth*	*ttttt*

number of heads, x	Probability $P(x)$	Product $x \cdot P(x)$
0	1/32	0
1	5/32	5/32
2	10/32	20/32
3	10/32	30/32
4	5/32	20/32
5	1/32	5/32

Thus, the expected value is given by:

Expected value

$= 0 + \dfrac{5}{32} + \dfrac{20}{32} + \dfrac{30}{32} + \dfrac{20}{32} + \dfrac{5}{32}$

$= \dfrac{80}{32} = \dfrac{5}{2}.$

5. List the given information in a table. Then calculate $P(x)$, the product $x \cdot P(x)$, and their total.

Number Rolled	Payoff	Probability	Product
6	$3	1/6	$(3/6)
5	$2	1/6	$(2/6)
4	$1	1/6	$(1/6)
0–3	$0	3/6	$0

Expected value: $(6/6) = \$1$

7. List the given information in a table. Then complete the table as follows.

Number Rolled	Payoff	Probability	Product
1	−$1	1/6	−$(1/6)
2	$2	1/6	$(2/6)
3	−$3	1/6	−$(3/6)
4	$4	1/6	$(4/6)
5	−$5	1/6	−$(5/6)
6	$6	1/6	$(6/6)

Expected value: $(3/6) = 50¢$

The expected net winnings for this game are 50¢.

9. List the given information in a table, and complete the probability and product columns Remember that the expected value is the sum of the product column.

Number of heads	Payoff	Probability $P(x)$	Product $x \cdot P(x)$
3	10¢	1/8	(10/8)¢
2	5¢	3/8	(15/8)¢
1	3¢	3/8	(9/8)¢
0	0¢	1/8	0¢

Expected value: $(34/8)¢ = (17/4)¢$

Since it costs 5¢ to play, the expected net winnings are

$$\frac{17}{4}¢ - 5¢ = \frac{17}{4}¢ - \frac{20}{4}¢ = -\frac{3}{4}¢.$$

Because the expected net winnings are not zero, 5¢ is not a fair price to pay to play this game.

If two cards are drawn from a standard 52-card deck, find the expected number of spades in each of the following cases.

11. The expected number of absences on a given day is

$$x_1 \cdot P(x_1) + x_2 \cdot P(x_2) + x_3 \cdot P(x_3) + x_4 \cdot P(x_4)$$
$$+ x_5 \cdot P(x_5)$$
$$= 0(.12) + 1(.32) + 2(.35) + 3(.14) + 4(.07)$$
$$= 1.72.$$

A college foundation raises funds by selling raffle tickets for a new car worth $36,000.

13. (a) Since 600 tickets are sold, a person who buys one ticket will have a probability of $1/600 \approx .00167$ of $1 - .0017 = .9983$ of not winning anything. For this person, the expected value is

$$\$36,000(.00167) + \$0(.9983) \approx \$60,$$

and the expected *net* winnings (since the ticket costs $120) are

$$\$60 - \$120 = -\$60.$$

(b) By selling 600 tickets at $120 each, the foundation takes in

$$500(\$120) = \$72,000.$$

Since they had to spend $36,000 for the car, the total profit for the foundation is

$$\text{revenue} - \text{cost} = \text{profit}$$
$$\$72,000 - \$36,000 = \$36,000.$$

(c) Without having to pay for the car, the foundation's total profit will be all of the revenue from the ticket sales, which is $72,000.

Five thousand raffle tickets are sold. One first prize of $1,000, two second prizes of $500 each, and five third prizes of $100 each are to be awarded, with all winners selected randomly.

15. The associated probabilities are

$$P(\text{1st prize}) = \frac{1}{5000},$$
$$P(\text{2nd prize}) = \frac{2}{5000}, \text{ and}$$
$$P(\text{3rd prize}) = \frac{5}{5000}.$$

The expected winnings, ignoring the cost of the raffle ticket, are given by

$$\$1000\left(\frac{1}{5000}\right) + \$500\left(\frac{2}{5000}\right) + \$100\left(\frac{5}{5000}\right)$$
$$= \$.20 + \$.20 + \$.10$$
$$= \$.50, \text{ or } 50¢.$$

17. Since 5000 tickets were sold for $1 each, the sponsor's revenue was 5000 ($1) = $5000. The sponsor's cost was the sum of all the prizes:

$$1(\$1000) + 2(\$500) + 5(\$100)$$
$$= \$2500.$$

Therefore, the sponsor's profit is

$$\text{revenue} - \text{cost} = \text{profit}$$
$$\$5000 - \$2500 = \$2500.$$

19. List the given information in a table, and complete the probability and product columns. Remember that the expected value is the sum of the product column.

Number of families	Probability $P(x)$	Product $x \cdot P(x)$
1020	1020/10000	$1 \cdot (1020/10000) = 1020/10000$
3370	3370/10000	$2 \cdot (3370/10000) = 6740/10000$
3510	3510/10000	$3 \cdot (3510/10000) = 10530/10000$
1340	1340/10000	$4 \cdot (1340/10000) = 5360/10000$
510	510/10000	$5 \cdot (510/10000) = 2550/10000$
80	80/10000	$6 \cdot (80/10000) = 480/10000$
170	170/10000	$0 \cdot (170/10000) = 0$

Expected value: $26684/10000 \approx 2.7$

21. The expected value is

$$1200(.3) + 500(.5) + (-800)(.2)$$
$$+ 360 + 250 - 160$$
$$= 450.$$

Since this expected value is positive, the expected change in the number of electronics jobs is an increase of 450.

23. Writing exercise

25. The optimist viewpoint would ignore the probabilities and hope for the best possible outcome which is Project C since it may return up to $400,000.

27. If the contestant takes a chance on the other two prizes, the expected winnings will be

$$\$5000(.20) + \$8000(.15)$$
$$= \$1000 + \$1200$$
$$= \$2200.$$

29. Compute the remaining values in Column 5 (Expected Value).

Row 5: $(50,000)(.5) = 25,000$
Row 6: $(100,000)(.6) = 60,000$
Row 7: $(20,000)(.8) = 16,000$

31. Account classification in Column 7. Let x = account value in Column 6. Then, the classifications are:

$$A \qquad x \geq \$55,000$$
$$B \quad \$45,000 \leq x < \$55,000$$
$$C \qquad x < \$45,000.$$

Row 1: C
Row 2: C
Row 3: C
Row 4: B
Row 5: C
Row 6: A
Row 7: B

33.

$P(\text{matching 3 numbers})$
$$= \frac{C(20,3)C(60,3)}{C(80,6)} = .1298$$

$P(\text{matching 4 numbers})$
$$= \frac{C(20,4) \cdot C(60,2)}{C(80,6)} = .0285$$

$P(\text{matching 5 numbers})$
$$= \frac{C(20,5) \cdot C(60,1)}{C(80,6)} = .0031$$

$P(\text{matching 6 numbers})$
$$= \frac{C(20,6) \cdot C(60,0)}{C(80,6)} = .000129$$

Expected value
$$= (\$.35)(.1298) + (\$2.00)(.0285)$$
$$+ (\$60.00)(.0031)$$
$$+ (\$1250.00)(.000129)$$
$$= \$.44968$$
$$\approx 45¢$$

Since the player pays 60¢ for his ticket, his expected net winnings are about

$$45¢ - 60¢ = -15¢.$$

11.6 EXERCISES

1.   Writing exercise

3.   No, since the probability of an individual girl's birth is (nearly) the same as that for a boy.

5.   Let each of the 50 numbers correspond to one family. For example, the first number, 51592, with middle digits—1(boy), 5(girl), 9(girl)—represents a family with 2 girls and 1 boy. The last number whose middle digits are 800 represents the 50th family which has 1 girl and 2 boys—a success, and so on. Examining each number, we count (tally) 18 successes. Therefore,

$$P(2 \text{ boys and 1 girl}) = \frac{18}{50} = .36.$$

Observe that this is quite close to the .375, predicted by the theoretical value.

Refer to discussion in text regarding foul shooting in basketball. After completing the indicated tally, find the empirical probability that, on a given opportunity, Karin will score as follows.

To construct the tally for Exercises 6–8, begin as follows: Since the first number in the table of random digits is 5, which represents a hit, Susan will get a second shot. The second digit is 7, representing a miss on the second shot. Record the results of the first two shots (the first one-and-one opportunity) as "one point." The second and third opportunities correspond to the pairs 3, 4 and 0, 5. Record each of these results as "two points." For the fourth opportunity, the digit 9 indicates that the first shot was missed, so Susan does not get a second shot. In this case only one digit is used. Record this result as "zero points." Continue in this manner until 50 one-and-one opportunities are obtained. The results of the tally is as follows. Note that this, in effect, is a frequency distribution as discussed in Chapter 12.

Number of Points	Tally - frequency
0	15
1	11
2	24
Total	50

7.   From the tally, we see that 1 point shots occur 11 times. Thus,

$$P(1 \text{ point}) = \frac{11}{50} = .22.$$

9.   Answers will vary.

11.   Writing exercise

CHAPTER 11 TEST

1.   Writing exercise

2.   Writing exercise

3.   There are 39 non-hearts and 13 hearts so the odds against getting a heart are 39 to 13, or 3 to 1 (when reduced).

4.   There are 50 non-red queens and 2 red queens, so the odds against getting a red queen are 50 to 2, or 25 to 1.

5.   There are 12 face cards altogether. Of these, there are 6 black face cards and 2 more non-black kings for a total of 8 cards. There are $52 - 8 = 44$ other cards in the deck. Thus, the odds against getting a black face card or king are 44 to 8, or 11 to 2.

6.

		Second parent	
		C	c
First	C	CC	Cc
Parent	c	cC	cc

7.   There are two outcomes (cC) indicating that the next child will be a carrier. Thus,

$$P(\text{carrier}) = \frac{2}{4} = \frac{1}{2}.$$

8.   There are 3 favorable outcomes $(CC, Cc, \text{and } cC)$. Thus, the odds against a child getting the disease (cc) are 3 to 1.

9.   Use the fundamental counting principle where the first task is to calculate the probability of the initial employee choosing any day of the week $(7/7)$, the second task is the probability for the second employee to choose any other days of the week $(6/7)$. In a similar manner, the third employee's probability must involve a choice from one of the five remaining days with a resulting probability of $(5/7)$. Thus,

$$P = \frac{7}{7} \cdot \frac{6}{7} \cdot \frac{5}{7} = \frac{30}{49}.$$

10.   Use the fundamental counting principle where the first task is to calculate the probability of the initial employee choosing any day of the week $(7/7)$. For the second and third task the employees must choose the same day. The probability in each case is $(1/7)$. Thus,

$$P = \frac{7}{7} \cdot \frac{1}{7} \cdot \frac{1}{7} = \frac{1}{49}.$$

11. The complement of "exactly two choose the same day" is "all three choose different days or all three choose the same day." These complement probabilities were calculated in Exercises 9 and 10 above. Thus, the probability of "exactly two choosing the same day" is given by

$$P = 1 - \left(\frac{30}{49} + \frac{1}{49} \right) = \frac{18}{49}.$$

Observe that the calculation involves both the complements rule and the special addition rule.

Two numbers are randomly selected without replacement from the set $\{1, 2, 3, 4, 5\}$.

12. To find the probability that "both numbers are even" use combinations to select the number of successes – ways of selecting the two even numbers, $C(2,2)$ and to calculate the total number of ways of selecting two of the numbers from the 5, $C(5,2)$. Thus,

$$P(\text{selecting two even numbers}) = \frac{C(2,2)}{C(5,2)} = \frac{1}{10}.$$

As in many of the exercises an alternate solution may be considered here: Let E_1 represent the event of selecting an even number as the first selection and E_2, selecting and even number as the second selection. Using the general multiplication rule the probability is given by

$$P(E_1 \text{ and } E_2) = P(E_1) \cdot P(E_2 | E_1)$$
$$= \frac{2}{5} \cdot \frac{1}{4}$$
$$= \frac{1}{10}.$$

13. To find the probability that "both numbers are prime" use combinations. Since there are three prime numbers $\{2, 3, 5\}$, use $C(3,2)$ to calculate the number of successes. To calculate the total number of ways of selecting two of the numbers from the 5, use $C(5,2)$. Thus,

$$P(\text{selecting two even numbers}) = \frac{C(3,2)}{C(5,2)} = \frac{3}{10}.$$

14. Create a "product table" to list the elements in the sample space and the successes (event of interest). Note that "without replacement" one can only use a selected number once. Hence, there are no diagonal values to the table.

			2nd number			
	+	1	2	3	4	5
	1	–	3	4	5	6
1st	2	3	–	5	6	7
number	3	4	5	–	7	8
	4	5	6	7	–	9
	5	6	7	8	9	–

There are 12 odd sums in the table and a total of 20 sums in the sample space. Thus,

$$P(\text{sum is odd}) = \frac{12}{20} = \frac{3}{5}.$$

15. Similar to Exercise 14, create a "product table" to list the elements in the sample space and the successes (event of interest). Note that "without replacement" one can only use a selected number once. Hence, there are no diagonal values to the table.

			2nd number			
	×	1	2	3	4	5
	1	–	2	3	4	5
	2	2	–	6	8	10
1st	3	3	6	–	12	15
number	4	4	8	12	–	20
	5	5	10	15	20	–

There are 6 odd products in the table and a total of 20 products in the sample space. Thus,

$$P(\text{product is odd}) = \frac{6}{20} = \frac{3}{10}.$$

A three-member committee is selected randomly from a group consisting of three men and two women.

16. Let x represent the number of men on the committee. Then,

x	$P(x)$
0	0
1	3/10
2	6/10
3	1/10

Where,

$$P(1) = \frac{C(3,1) \cdot C(2,2)}{C(5,3)}$$
$$P(2) = \frac{C(3,2) \cdot C(2,1)}{C(5,3)}$$
$$P(3) = \frac{C(3,3) \cdot C(2,0)}{C(5,3)}.$$

Why is $P(0) = 0$?

17. The probability that the "committee members are not all men" is the complement of the "committee are all men," $P(3)$. Hence, use the complements rule,

$$P(\text{committee members are not all men})$$
$$= 1 - P(3)$$
$$= 1 - \frac{1}{10}$$
$$= \frac{9}{10}.$$

18. Complete the table begun in Exercise 16 as an aid to calculating the expected number of men (sum of product column).

x	$P(x)$	$x \cdot P(x)$
0	0	0
1	3/10	3/10
2	6/10	12/10
3	1/10	3/10

Expected number: $18/10 = 9/5$

Create a "product table" such as below for the "sum" from rolling two dice.

		2nd die					
	+	1	2	3	4	5	6
	1	2	3	4	5	6	7
	2	3	4	5	6	7	8
1st	3	4	5	6	7	8	9
die	4	5	6	7	8	9	10
	5	6	7	8	9	10	11
	6	7	8	9	10	11	12

Use for Exercises 19–22.

19. There are 30 non-doubles values and 6 doubles values. The odds against doubles are therefore, 30 to 6, or 5 to 1.

20. A "sum greater than 2" is the complement of a "sum equal to or smaller than 2." There is only one sum which satisfies this condition ("snake eyes"). Thus, by the complements rule

$$P(\text{sum greater than 2}) = 1 - \frac{1}{36} = \frac{35}{36}.$$

21. To find the odds against a "sum of 7 or 11" count the sums that satisfy the condition "sum of 7 or 11." There such 8 such sums. It follows that there are $36 - 8 = 28$ sums that are not 7 or 11. Thus, the odds against a "sum of 7 or 11" are 28 to 8, or 7 to 2.

22. Since there are 4 sums that are even and less than 5,

$$P(\text{sum that is even and less than 5}) = \frac{4}{36} = \frac{1}{9}.$$

For Exercises 23–26 the chance of making par on any one hole is .78.

23. By the special multiplication rule,

$$P(\text{making par on all three holes}) = .78^3 \approx .475.$$

24. Use the fundamental counting principle where the first task is to find the number of ways to choose the two holes he scores par on followed by the tasks of assigning a probability for each hole. Note that since .78 is the probability of scoring par, $1 - .78 = .22$ is the probability of not scoring par on a hole.

$$P(\text{makes par on exactly 2 holes}) = C(3,2)(.78)^2(.22)$$
$$= (3)(.6084)(.22)$$
$$\approx .402$$

25. "At least one of the three holes" is the complement of "none of the three holes." Since the probability of not making par on any of the three holes is $.22^3$,

$$P(\text{at least one of the three holes}) = 1 - (.22)^3$$
$$= 1 - .010648$$
$$\approx .989.$$

26. Use the special multiplication rule since these probabilities are independent. The probability that he makes par on the first and third holes but not on the second is found by

$$P = (.78)(.22)(.78)$$
$$\approx .134.$$

Two cards are drawn, without replacement, from a standard 52-card deck for Exercises 27–30.

27. Let R_1 and R_2 represent the two red cards. Since the cards are not replaced, the events are not independent. Use the general multiplication rule:

$$P(R_1 \text{ and } R_2) = P(R_1) \cdot P(R_2|R_1)$$
$$= \frac{26}{52} \cdot \frac{25}{51}$$
$$= \frac{1}{2} \cdot \frac{25}{51}$$
$$= \frac{25}{102}.$$

28. Let C_1 and C_2 represent two cards of the same color. Since the cards are not replaced, the events are not independent. Use the general multiplication rule. Note that since it doesn't matter what color the first card is, it's probability is 1.

Thus,
$$
\begin{aligned}
P(C_1 \text{ and } C_2) &= P(C_1) \cdot P(C_2|C_1) \\
&= \frac{52}{52} \cdot \frac{25}{51} \\
&= \frac{25}{51}.
\end{aligned}
$$

29. The first card drawn limits the sample space to 51 cards where all 4 queens are still in the deck. The probability then is given by

$$
\begin{aligned}
P(\text{queen given the first card is an ace}) &= P(Q_2|A_1) \\
&= \frac{4}{51}.
\end{aligned}
$$

30. The probability that "the first card is a face card and the second is black" The first (face) card may be red or black. Since this affects the probability associated with the second card, look at both cases and use the special addition rule to add (since we are "or–ing") the results.

Case 1 (first card is a red face card):

$$
\begin{aligned}
P(F_1 \text{ and } B_2) &= P(F_1) \cdot P(B_2|F_1) \\
&= \frac{6}{52} \cdot \frac{26}{51} \\
&= \frac{3}{51} = \frac{1}{17}.
\end{aligned}
$$

Case 2 (first card is a black face card):

$$
\begin{aligned}
P(F_1 \text{ and } B_2) &= P(F_1) \cdot P(B_2|F_1) \\
&= \frac{6}{52} \cdot \frac{25}{51} \\
&= \frac{3}{26} \cdot \frac{25}{51} \\
&= \frac{1}{26} \cdot \frac{25}{17} = \frac{25}{26 \cdot 17}.
\end{aligned}
$$

The probability is

$$
\begin{aligned}
P(\text{Case 1 or Case 2}) &= P(\text{Case 1}) + P(\text{Case 2}) \\
&= \frac{1}{17} + \frac{25}{26 \cdot 17} \\
&= \frac{26 \cdot 1}{26 \cdot 17} + \frac{25}{26 \cdot 17} \\
&= \frac{26 + 25}{26 \cdot 17} = \frac{51}{26 \cdot 17} \\
&= \frac{3 \cdot 17}{26 \cdot 17} = \frac{3}{26}.
\end{aligned}
$$

Exercises 31–33 refer to coin sequence in Example 2 of Section 11.6.

31. The number of 3 successive births is 38. It is helpful to set up a tally to count these.

32. Of the 38 only 1 represents a triple consisting of all girls.

33. The empirical probability that 3 successive births will be all girls is found, using the results of Exercises 31 and 32 above, as

$$
\frac{1}{38} \approx .026.
$$

12 STATISTICS

Chapter Goals

The general goal of this chapter is to provide an overview of some of the basic tools of statistics. Upon completion of this chapter the student should be able to:

- Construct frequency distributions and graphs of sets of data.
- Calculate measures of central tendency including the mean, median, and mode.
- Calculate measures of dispersion including the range and standard deviation.
- Understand properties of the normal curve.
- Solve problems related to a normal distribution.
- Use a normal distribution to approximate a binomial distribution.
- Calculate a linear regression curve and correlation coefficient from a set of data values.

Chapter Summary

Statistics is a branch of mathematics that deals with the collection, analysis, interpretation, and presentation of data. Statistics can be an important tool in decision making especially in areas of uncertainty. There are two main branches of statistics: descriptive statistics and inferential statistics. Descriptive statistics includes collecting, organizing, summarizing, and presenting information. Inferential statistics involves drawing conclusions or making inferences about populations based on information from samples.

Frequency distributions and graphs are convenient ways to organize data in order to recognize patterns and other information at a glance. (See Section 12.1.) The text gives examples of how to construct frequency distributions and common statistical graphs.

When a large amount of data must be analyzed, it is desirable to have a single number that best represents the data. Such a number is referred to as a measure of central tendency. These are described in the following paragraphs.

The mean ($\overline{x}$) of a set of data values or scores is found by adding up the values and dividing by the number of items in the group.

$$\overline{x} = \frac{x_1 + x_2 + x_3 + \ldots + x_n}{n}.$$

The median is the middle number; it divides the group of scores in half. To find the median of a set of values:

1. Arrange the items in numerical order.
2. If the number of items is odd, the median is the middle item in the list.
3. If the number of items is even, the median is the mean of the two middle items.

The median is not as sensitive to extreme values as the mean.

The mode is the value that occurs most often. Some sets of numbers have two values that occur most frequently and are called bimodal. It is also possible for a set to have no mode or more than two modes. For a large number of items, a frequency distribution is a convenient way to find the modal value(s).

EXAMPLE A   Mean, median, and mode

Find the mean, median, and mode of the following set of data: 8, 5, 17, 2, 4, 5, 8, 10, 6, 5.

Mean

$$\overline{x} = \frac{8 + 5 + 17 + 2 + 4 + 5 + 8 + 10 + 6 + 5}{10}$$
$$= \frac{70}{10}$$
$$= 7$$

Median

First arrange the numbers in numerical order.
2, 4, 5, 5, 5, 6, 8, 8, 10, 17

The median is the mean of 5 and 6 because there is an even number of values.

$$\frac{5 + 6}{2} = \frac{11}{2} = 5.5$$

Mode

The mode is 5 because it occurs most often.

Sometimes the calculation of a mean value will involve weighting factors for each of the scores.

The weighted mean of n numbers $x_1, x_2, \ldots, x_n$ that are weighted by the respective factors $f_1, f_2, \ldots, f_n$ is

$$\overline{w} = \frac{x_1 f_1 + x_2 f_2 + \ldots + x_n f_n}{f_1 + f_2 + \ldots + f_n}$$

Calculating a grade point average is an example of a weighted mean. Each course is weighted according to the number of units of the course.

EXAMPLE B   Weighted mean

A student received the indicated grades for the given units of credit. Find the grade point average for the student. Assume A = 4, B = 3, C = 2, D = 1, F = 0.

Units	Grade
7	B
2	A
5	C
1	F

Solution

A table is a useful way to organize the information.

Units	Grade	Grade Points	Units × Grade Points
7	B	3	21
2	A	4	8
5	C	2	10
1	F	0	0
Totals		15	39

Grade point average: $\dfrac{39}{15} = 2.6$

The weighted mean formula is commonly used to find the mean for a frequency distribution. In this case, the weighting factors are the frequencies. EXAMPLE A above can be calculated by this method.

EXAMPLE C   Weighted mean

Complete a frequency distribution and use weighting factors (frequencies) to calculate the mean of the following set of data: 8, 5, 17, 2, 4, 5, 8, 10, 6, 5.

Solution
Set up the work as follows as an aid to use the weighted mean formula.

Value	Frequency	Value × Frequency
2	1	2
4	1	4
5	3	15
6	1	6
8	2	16
10	1	10
17	1	17
Totals	10	70

Then $\dfrac{70}{10} = 7$.

It is usually necessary to know more about a set of numbers than we can learn from a measure of central tendency. For instance, the two sets of numbers $\{6, 4, 5\}$ and $\{5, 0, 10\}$ have the same mean and the same median, 5, but the two sets of numbers are quite different. In fact, the second set of numbers appears to vary considerably, while the first set is grouped closely together.

A number that describes how the numbers of a set are spread out is called a measure of dispersion. An example of such a measure is the range. The range for a set of data is defined as the difference between the highest value and the lowest value in the set.

The two sets $\{6, 4, 5\}$ and $\{5, 0, 10\}$ have ranges $6 - 4 = 2$ and $10 - 0 = 10$, respectively. Because the range is determined by only two numbers of the set, it gives us very little information about the other numbers of the set. In fact, if either the highest or the lowest value happens to be an extreme score, the range will be misleading. The range gives us only a general idea of the spread of the data.

One of the most useful measures of dispersion, the standard deviation, is based on deviations from the mean of the data values; it indicates how widely spread out the individual data values are from the mean.

To calculate s, the sample standard deviation of n numbers:

1. Calculate $\overline{x}$, the mean of the numbers.
2. Find the difference between each data value and the mean: $x - \overline{x}$.
3. Square each deviation, $(x - \overline{x})^2$.
4. Sum the squared deviations.
5. Divide the sum in step 4 by $n - 1$.
6. Take the square root of the result of step 5.

The above method can be summarized by the common formula for standard deviation

$$s = \sqrt{\frac{\sum(x - \overline{x})^2}{n - 1}}$$

This formula is used to find the standard deviation for a sample of a population. To calculate the standard deviation σ, for a population the formula is

$$\sigma = \sqrt{\frac{\sum(x - \overline{x})^2}{n}}$$

If a statistical calculator is used, it will often have a choice of two keys, marked σ_n for a population and σ_{n-1} for a sample.

EXAMPLE D   Range and standard deviation

Find the range and standard deviation of the following sample: 7, 6, 12, 14, 18, 15.

The range is $18 - 6 = 12$.

To find the sample standard deviation:

1. First $\overline{x} = \dfrac{7 + 6 + 12 + 14 + 18 + 15}{6} = \dfrac{72}{6} = 12$
2. For steps 2 and 3, set up a table.

Value	Deviation	Squared Deviation
x	$x - \overline{x}$	$(x - \overline{x})^2$
6	-6	36
7	-5	25
12	0	0
14	2	4
15	3	9
18	6	36

3. Find the sum of the squared deviations.

$$\sum(x - \overline{x})^2 = 36 + 25 + 0 + 4 + 9 + 36 = 110$$

4. Divide this sum by $n - 1$. $\dfrac{110}{5} = 22$

5. Take the square root of this value. $\sqrt{22} \approx 4.7$

Normally distributed populations are predictable. Tables have been established for the normal or bell-shaped distributions which allow these predictions to be made. Any random variable whose graph has this shape is said to have a normal distribution. A normal distribution has the following properties.

1. The total area under the curve is exactly 1 square unit.
2. The graph is bell-shaped and symmetric about a vertical line through its center.
3. The mean, median, and mode are all equal and occur at the center of the distribution.
4. The following three items are numerically equivalent:
   - Area under the curve within an interval
   - Percentage of values that lie within the interval,
   - Probability of a randomly chosen item lying within the same interval.
5. The relative standing of any value in a normal distribution can be found by computing the z-score corresponding to the value.

The Empirical Rule states: About 68% of the data values of a normal distribution lie within 1 standard deviation of the mean, about 95% of the values lie within 2 standard deviations of the mean, and about 99.7% lie within 3 standard deviations.

Most questions about normal distributions require the use of a z-table. The z-scores represent the number of standard deviations a given value is from the mean of the distribution. A z-score of 1, for example, indicates that a specific value in the distribution lies exactly one standard deviation above the mean. A z-score of -2.5 indicates that the value lies exactly 2.5 standard deviations below the mean. The z-table gives the area under the normal curve between the mean and z standard deviations above the mean. Because of symmetry, the z-table may be used for values above the mean or below the mean.

EXAMPLE E   Normal curve using z-scores

Use a z-table to find the percent of all scores that lie between $z = -.82$ and $z = .27$.

Solution
It is helpful to sketch a normal curve with vertical lines showing the approximate boundaries of $z = -.82$ and $z = +.27$.

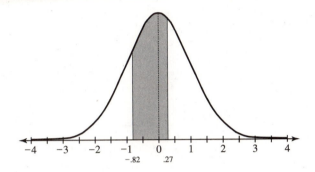

Use the z-table in Section 12.5 or the z-table in the appendix of the text to find the area from the mean to $z = .27$ standard deviations above the mean; this area is .106. To find the area from $z = -.82$ standard deviations to the mean, use the symmetry of the normal curve and find the area between the mean and $z = .82$; this area is .294. The sketch will help to show that these areas must be added. Then the sum of these areas is 0.400 or 40%. The probability that a randomly chosen score will lie within this interval is also 0.4.

When z-scores are unknown, they may be calculated by the z-score formula:

$$z = \frac{\text{value} - \text{mean}}{\text{standard deviation}} = \frac{x - \overline{x}}{s}$$

EXAMPLE F   Application of the Normal curve

In any given year, the U.S. Marines have a very large sample of American youth on active military duty. The Marines know from their records that the mean height of a soldier is 69 inches. This distribution is approximately normal with a standard deviation of 3 inches. What is the percentage of Marines with heights from 66 to 75 inches, including these values? Use the z-table to calculate the answer.

Solution
Use the z-score formula for $x = 66$ and for $x = 75$.

$$z_{66} = \frac{66 - 69}{3} = -1$$

$$z_{75} = \frac{75 - 69}{3} = 2$$

Make a sketch of the normal curve showing the indicated region. Read the z-table for $z = 1$ to see that the area between the mean and 1 is .341. For $z = 2$, the area between the mean and 2 is .477. Add the area values: $.341 + .477 = .818$. The marines can expect 81.8% of their soldiers to have heights between 69 and 75 inches.

It is also possible to answer these questions by applying the empirical rule.

See the text and SOLUTIONS to 12.5 Exercises for examples of how to apply properties of the normal curve to binomial distributions and 12.6 Exercises to calculate linear regression and correlation coefficients.

12.1 EXERCISES

1. (a) Remember that f represents the frequency of each data value, and f/n is a comparison of each frequency to the overall number of data values.

x	f	f/n
0	10	$10/30 \approx 33\%$
1	7	$7/30 \approx 23\%$
2	6	$6/30 \approx 20\%$
3	4	$4/30 \approx 13\%$
4	2	$2/30 \approx 7\%$
5	1	$1/30 \approx 3\%$

(b)

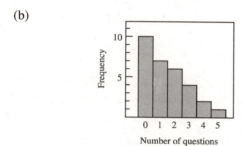

(c)

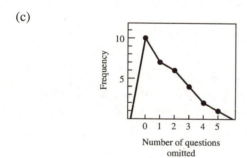

3. (a) In this Exercise, the tally column has been omitted; however, it is a useful tool to use when creating a frequency distribution by hand.

Class Limits	Frequency f	Relative frequency f/n
45–49	3	$3/54 \approx 5.6\%$
50–54	14	$14/54 \approx 25.9\%$
55–59	16	$16/54 \approx 29.6\%$
60–64	17	$17/54 \approx 13.5\%$
65–69	4	$4/54 \approx 7.4\%$

(b)

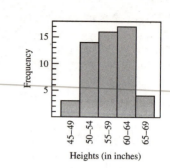

Heights (in inches)

(c)

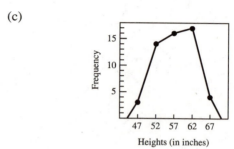

Heights (in inches)

5. (a) In this Exercise, the tally column has been omitted; however, it is a useful tool to use when creating a frequency distribution by hand.

Class Limits	Frequency f	Relative frequency f/n
70–74	2	$2/30 \approx 6.7\%$
75–79	1	$1/30 \approx 3.3\%$
80–84	3	$3/30 = 10\%$
85–89	2	$2/30 \approx 6.7\%$
90–94	5	$5/30 \approx 16.7\%$
95–99	5	$5/30 \approx 16.7\%$
100–104	6	$6/30 = 20\%$
105–109	4	$4/30 \approx 13.3\%$
110–114	2	$2/30 \approx 6.7\%$

(b)

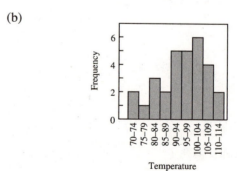

Temperature

(c)

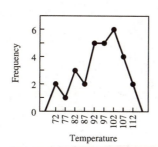

7.
0	7 9 8
1	1 1 2 8 9 4 3 1 0 5 0 5 5
2	0 9 6 6 2 5 2 3 4 4
3	1

9.
0	8 5 4 9 6 9 4 8
1	2 0 1 8 8 2 4 0 8 8 6 3
2	6 6 2 5 1 3
3	0 4 6
4	4

11. Read the vertical axis and total the values:
$10.1 + 10.6 + 10.7 = 31.4$, or about 31%.

13. CBS dropped the most, from a high of about 10.7% at the beginning of the time interval to a low of about 7.8%, which is about a 2.9% drop.

15. Examine the graph for the 1994–95 time period in which NBC and ABC seem to be farthest apart; the difference is $10 - 8.4 = 1.6\%$, and ABC attained the highest ratings.

17. Examine the graph for Consumer Price Index. The tallest bars at the the 3% level in the years 1993 and 1996.

19. Writing exercise

21. Find the sum of the given percentages and subtract this sum from 100%. This gives the "Other Members" percentage.

$$100 - (18.25 + 5.67 + 5.67 + 5.10 + 5.10) = 60.21\%.$$

Now multiply the decimal form of this number by 360°, the number of degrees in the circle to find the number of degrees in the central angle.

$$.6021(360) \approx 217°$$

23. To calculate the number of degrees in each sector of the circle, multiply each percentage in decimal form times 360°. Here are a few examples:

$$.33(360) \approx 119°$$
$$.25(360) = 90°$$
$$.12(360) \approx 43°$$

25. Examine the graph to see that he would be about 79.

27. (a) Examine the 6% curve to see that Dave's money will run out at age 76. If he reaches age 70, his money would last for about

$$76 - 70 = 6 \text{ years.}$$

(b) Writing exercise

29. Writing exercise

31. Writing exercise

33. Writing exercise

35.
Letter	Probability
A	$\dfrac{.08}{.385} \approx .208$
E	$\dfrac{.13}{.385} \approx .338$
I	$\dfrac{.065}{.385} \approx .169$
O	$\dfrac{.08}{.385} \approx .208$
U	$\dfrac{.03}{.385} \approx .078$

37.
Letter	Probability
A	$\dfrac{31}{118} \approx .263$
E	$\dfrac{34}{118} \approx .288$
I	$\dfrac{20}{118} \approx .169$
O	$\dfrac{23}{118} \approx .195$
U	$\dfrac{10}{118} \approx .085$

39.

Class Limits	Probability
10–19	$\frac{6}{40} = .150$
20–29	$\frac{11}{40} = .275$
30–39	$\frac{9}{40} = .225$
40–49	$\frac{7}{40} = .175$
50–59	$\frac{4}{40} = .100$
60–69	$\frac{2}{40} = .050$
70–79	$\frac{1}{40} = .025$

41. The probability is $\frac{7}{40}$.

43. (a) Empirical

(b) Writing exercise

12.2 EXERCISES

1. (a) $\overline{x} = \dfrac{3 + 7 + 12 + 16 + 23}{5} = \dfrac{61}{5} = 12.2$

(b) The data is given in order from smallest to largest; the middle number is 12.

(c) No mode.

3. (a) $\overline{x} = \dfrac{128 + 230 + 196 + 224 + 196 + 233}{6}$

$= \dfrac{1307}{6}$

≈ 201.2

(b) Arrange the values from smallest to largest or vice versa: 128, 196, 196, 224, 230, 233. Find the mean of the two middle numbers: $\dfrac{196 + 224}{2} = 210$.

(c) The value that occurs most frequently is 196.

5. (a) $\overline{x} = \dfrac{3.1 + 4.5 + 6.2 + 7.1 + 4.5 + 3.8 + 6.2 + 6.8}{8}$

$= \dfrac{41.7}{8}$

≈ 5.2

(b) Arrange the values from smallest to largest or vice versa: 3.1, 3.8, 4.5, 4.5, 6.2, 6.2, 6.3, 7.1. Find the mean of the two middle numbers: $\dfrac{4.5 + 6.2}{2} = 5.35$.

(c) The set of values is bimodal: 4.5 and 6.2.

7. (a) $\overline{x} = \dfrac{.78 + .93 + .66 + .94 + .87 + .62 + .74 + .81}{8}$

$= \dfrac{6.35}{8}$

$\approx .8$

(b) Arrange the values from smallest to largest or vice versa: .62, .66, .74, .78, .81, .87, .93, .94. Find the mean of the two middle numbers: $\dfrac{.78 + .81}{2} = .795$.

(c) There is no mode.

9. (a) The sum of the data is 330.4. $\overline{x} = \dfrac{330.4}{7} = 47.2$.

(b) Arrange the values from smallest to largest or vice versa: 1.2, 12.3, 34.5, 45.6, 67.8, 78.9, 90.1. The median is 45.6.

(c) No mode.

11. (a) The sum of the data is 1032. $\overline{x} = \dfrac{1032}{8} = 129$.

(b) Arrange the values from smallest to largest or vice versa: 125, 125, 127, 128, 128, 131, 132, 136. The two middle numbers are both 128. The median is

$\dfrac{128 + 128}{2} = 128$.

(c) There are two modes: 125 and 128.

13. (a) The sum of the numbers is 132,008.

$\overline{x} = \dfrac{132008}{5} \approx 26,402$

(b) The median or middle value is 28,079.

15. (a) The sum of the numbers is 101,977.

$\overline{x} = \dfrac{101977}{10} \approx 10,198$

(b) The median or middle value is

$\dfrac{9185 + 9168}{2} \approx 9177$.

17. The sum of the yearly losses is 19875. Then $\overline{x} = \dfrac{19875}{5} = 3975$. Rounded to the nearest 100 square miles, this is 4000 square miles.

19. The sum of the data is 36.92. Then $\overline{x} = \dfrac{36.92}{7} = 5.27$.

21. The sum for the new list is 16.94. Then $\overline{x} = \dfrac{16.94}{7} = 2.42$.

23. The mean was affected more.

25. $\overline{x} = \dfrac{79 + 81 + 44 + 89 + 79 + 90}{6} = \dfrac{462}{6} \approx 77.$

Arrange in order from smallest to largest; 44, 79, 79, 81, 89, 90. The median is $\dfrac{79 + 81}{2} = 80.$

(c) The mode is 79.

27. Let $x =$ the score he must make.
Replace the score of 44 with x.

$$\frac{x + 79 + 79 + 81 + 89 + 90}{6} = 85$$
$$\frac{6}{1} \cdot \frac{x + 79 + 79 + 81 + 89 + 90}{6} = 85 \cdot 6$$
$$x + 418 = 510$$
$$x = 92.$$

29. (a) Calculate a value · frequency column in order to evaluate the mean.

Value	Frequency	Value · Frequency
603	13	7839
597	8	4776
589	9	5301
598	12	7176
601	6	3606
592	4	2368
Totals	52	31066

Then $\overline{x} = \dfrac{31066}{52} \approx 597.42.$

(b) From part (a), there are 52 items. The formula for the position of the median is

$$\frac{\Sigma f + 1}{2} = \frac{52 + 1}{2} = 26.5.$$

This means that the median is halfway between the 26th and 27th item. A chart showing cumulative frequency shows the value.

Value	Frequency	Cumulative Frequency
603	13	13
597	8	21
589	9	30
598	12	42
601	6	48
592	4	52

The value 589 is the median.

(c) Examine the frequency column to see that the mode is also 603.

31.

Units	Grade	Units · Grade Value
4	C	$4 \cdot 2 = 8$
7	B	$7 \cdot 3 = 21$
3	A	$3 \cdot 4 = 12$
3	F	$3 \cdot 0 = 0$
17		41

Then $\overline{x} = \dfrac{41}{17} \approx 2.41$, to the nearest hundredth.

33. The sum of the area values is 8,506,300. Then

$$\overline{x} = \frac{8506300}{11} \approx 773,300 \text{ square miles.}$$

35. The sum of the populations values is 278,327,413. Then

$$\overline{x} = \frac{278327413}{11} \approx 25,302,000 \text{ to the nearest 1000.}$$

37. Remember that negative values are deducted from the sum. The sum of the values is 7,164,888. Then

$$\overline{x} = \frac{7164888}{10} \approx 716,489.$$

39. Find the sum of the three negative values: $-486,942$. Then

$$\overline{x} = \frac{-486,942}{3} = -162,314.$$

41. The sum of the dollar values is $1,988,800. Then

$$\overline{x} = \frac{1988900}{3} = \$662,933.33.$$

43. The sum of the "Silver" column is 53. Then the position of the median is

$$\frac{\Sigma f + 1}{2} = \frac{53 + 1}{2} = 27.$$

This means that the data entry located at the 27th position is the median. Here is a table with the cumulative frequencies.

Nation	Silver	Cumulative Frequency
Germany	9	9
Norway	10	19
Russia	6	25
Austria	5	30
Canada	5	35
United States	3	38
Finland	4	42
Netherlands	4	46
Japan	1	47
Italy	6	53

Look at the cumulative frequency row for Austria. Compare it to the row above (Russia) to see that it contains the data in positions 26 through 30. Therefore, this is the location of the median and the number of silver medals is 5.

Value	Frequency	Cumulative Frequency
0	1	1
1	1	2
3	1	3
14	2	5
15	1	6
16	2	8
17	2	10
18	3	13
19	1	14
20	1	15

45. (a) The sum of the "Totals" column is 150. Then

$$\overline{x} = \frac{150}{10} = 16.$$

(b) The median is the mean of the two middle numbers:

$$\frac{15 + 13}{2} = 14.$$

(c) The mode is 10, which occurs twice.

Examine the table to see that the median is located in the row for the value of 16.

The mode is 18 with a frequency of 3.

47. (a) $\overline{x} = \dfrac{47 + 51 + 53 + 56 + \ldots + 96}{34} = \dfrac{2544}{34} \approx 74.8$

(b) The scores are listed from smallest to the largest. Because there is an even number of scores, the median is the mean of the two middle numbers, the 17th and 18th scores: $\dfrac{77 + 78}{2} = 77.5.$

(c) The most frequently occurring score is 78.

(b) The median, 16, is most representative of the data.

(c) Writing exercise

53. 2, 3, 5, 7. The median is $\dfrac{3 + 5}{2} = 4.$

(a) There are 4 numbers listed.

(b) $\overline{x} = \dfrac{2 + 3 + 5 + 7}{4} = \dfrac{17}{4} = 4.25$

49. Writing exercise

51. (a)

Value	Frequency	Value · Frequency
0	1	0
1	1	1
3	1	3
14	2	28
15	1	15
16	2	32
17	2	34
18	3	54
19	1	19
20	1	20
Totals	15	206

$$\overline{x} = \frac{206}{15} \approx 13.7.$$

Position of the median is $\dfrac{15 + 1}{2} = 8.$ The position is the 8th piece of data,

55. 1, 3, 6, 10, 15, 21. The median is $\dfrac{6 + 10}{2} = 8.$

(a) There are 6 numbers listed.

(b) $\overline{x} = \dfrac{1 + 3 + 6 + 10 + 15 + 21}{6} = \dfrac{56}{6} \approx 9.33$

57. Arrange the numbers from smallest to largest. At this point it is uncertain where x will lie. However, if a single number must be the mean, median, and mode, then one of the given numbers must be that number. Because the median is the middle number and five values are given, the median must be 70 or 80. Try each of these as a mean to see which one works.

$$\frac{60 + 70 + 80 + 110 + x}{5} = 70$$

$$\frac{5}{1} \cdot \frac{320 + x}{5} = 70 \cdot 5$$

$$320 + x = 350$$

$$x = 30$$

This value does not work.

$$\frac{60 + 70 + 80 + 110 + x}{5} = 80$$

$$\frac{5}{1} \cdot \frac{320 + x}{5} = 80 \cdot 5$$

$$320 + x = 400$$

$$x = 80$$

This value works. The set of numbers is $\{60, 70, 80, 80, 110\}$. The value 80 is the mean, median, and mode.

59. Writing exercise

61. No

12.3 EXERCISES

1. The sample standard deviation will be larger because the denominator is $n - 1$ instead of n.

3. (a) The range is $19 - 2 = 17$.

(b) To find the standard deviation:

1. First find the mean. $\bar{x} = \dfrac{75}{8} = 9.375$

2 and 3. Find each deviation from the mean $(x - \bar{x})$ and square each deviation, $(x - \bar{x})^2$. These steps are shown in the table.

Data	Deviations	Squared Deviations
2	$2 - 9.375 = -7.375$	$(-7.375)^2 = 54.3906$
5	$5 - 9.375 = -4.375$	$(-4.375)^2 = 19.1046$
6	$6 - 9.375 = -3.375$	$(-3.375)^2 = 11.3906$
8	$8 - 9.375 = -1.375$	$(-1.375)^2 = 1.8906$
9	$9 - 9.375 = -0.375$	$(-0.375)^2 = 0.1406$
11	$11 - 9.375 = 1.625$	$(1.625)^2 = 2.6406$
15	$15 - 9.375 = 5.625$	$(5.625)^2 = 31.6406$
19	$19 - 9.375 = 9.625$	$(9.625)^2 = 92.6406$
Total		213.8388

4. Sum the squared deviations. The sum is 213.8388.
5. Divide by $n - 1$.

$$\frac{213.8388}{8 - 1} \approx 30.5484$$

6. Take the square root. $\sqrt{30.5484} \approx 5.53$

5. (a) The range is $41 - 22 = 19$.

(b) To find the standard deviation:

1. First find the mean. $\bar{x} = \dfrac{210}{7} = 30$.

2 and 3. Find each deviation from the mean $(x - \bar{x})$ and square each deviation, $(x - \bar{x})^2$. These steps are shown in the table.

Data	Deviations	Squared Deviations
25	$25 - 30 = -5$	$(-5)^2 = 25$
34	$34 - 30 = 4$	$(4)^2 = 16$
22	$22 - 30 = -8$	$(-8)^2 = 64$
41	$41 - 30 = 11$	$(11)^2 = 121$
30	$30 - 30 = 0$	$(0)^2 = 0$
27	$27 - 30 = -3$	$(-3)^2 = 9$
31	$31 - 30 = 1$	$(1)^2 = 1$
Total		236

4. Sum the squared deviations. The sum is 236.
5. Divide by $n - 1$.

$$\frac{236}{7 - 1} \approx 39.3333$$

6. Take the square root. $\sqrt{39.3333} \approx 6.27$

Some of the details are omitted in the following exercises. See Exercises 3–6 for details of computing standard deviation. A spreadsheet is a very useful tool in obtaining the intermediate calculations.

7. (a) The range is $331 - 308 = 23$.

(b) To find the standard deviation:

1. First find the mean. $\bar{x} = \dfrac{3204}{10} = 320.4$

2 and 3. Find each deviation from the mean $(x - \bar{x})$ and square each deviation, $(x - \bar{x})^2$.
4. Sum the squared deviations. The sum is 540.4.
5. Divide by $n - 1$.

$$\frac{540.4}{10 - 1} \approx 60.004$$

6. Take the square root. $\sqrt{60.004} \approx 7.75$

9. (a) The range is $85.62 - 84.48 = 1.14$.

(b) To find the standard deviation:

1. First find the mean. $\bar{x} = \dfrac{763.62}{9} = 84.84\overline{6}$

2 and 3. Find each deviation from the mean $(x - \bar{x})$ and square each deviation, $(x - \bar{x})^2$.
4. Sum the squared deviations. The sum is 1.091401.
5. Divide by $n - 1$.

$$\frac{1.091401}{9 - 1} \approx .136425$$

6. Take the square root. $\sqrt{.136425} \approx .37$

11. (a) The range is $9 - 1 = 8$.

(b) To find the standard deviation:
1. First find the mean.

Value	Frequency	Value · Frequency
9	3	27
7	4	28
5	7	35
3	5	15
1	2	2
Totals	21	107

Then $\bar{x} = \dfrac{107}{21} \approx 5.095$.

2 and 3. Find each deviation from the mean $(x - \bar{x})$ and square each deviation, $(x - \bar{x})^2$. These steps are shown in the table.

Value	Deviations	Squared Deviations	Freq · $(x - \bar{x})^2$
9	3.905	15.249	3(15.249)
7	1.905	3.629	7(3.629)
5	−.095	.009	5(.009)
3	−2.095	4.389	3(4.389)
1	−4.095	16.769	1(16.769)
Total			115.809

4. The fourth column shows the frequency of each value multiplied by the squared deviation. After multiplying each of these, find the sum: 115.809.

5. Divide by $n - 1$. Remember that the total number of values is 21.

$$\frac{115.809}{21 - 1} \approx 5.790$$

6. Take the square root. $\sqrt{5.790} \approx 2.41$

13. According to Chebyshev's theorem, the fraction of scores that lie within 2 standard deviations of the mean is at least

$$1 - \frac{1}{2^2} = 1 - \frac{1}{4} = \frac{3}{4}.$$

15. According to Chebyshev's theorem, the fraction of scores that lie within 7/2 standard deviations of the mean is at least

$$1 - \frac{1}{\left(\frac{7}{2}\right)^2} = 1 - \frac{1}{\frac{49}{4}} = 1 - \frac{4}{49} = \frac{45}{49}.$$

17. According to Chebyshev's theorem, the fraction of scores that lie within 3 standard deviations of the mean is at least

$$1 - \frac{1}{(3)^2} = 1 - \frac{1}{9} = \frac{8}{9}.$$

Divide 8 by 9 and change the decimal $.\overline{8}$ to 88.9%.

19. According to Chebyshev's theorem, the fraction of scores that lie within 5/3 standard deviations of the mean is at least

$$1 - \frac{1}{\left(\frac{5}{3}\right)^2} = 1 - \frac{1}{\frac{25}{9}} = 1 - \frac{9}{25} = \frac{16}{25}.$$

Divide 16 by 25 and change the decimal .64 to 64%.

21. Since 54 is 2 standard deviations below the mean $(70 - 2 \cdot 8 = 54)$ and 86 is 2 standard deviations above the mean $(70 + 2 \cdot 8 = 86)$, find the minimum fraction of values that lie within 2 standard deviations of the mean. See Exercise 13 for the answer 3/4.

23. Since 38 is 4 standard deviations below the mean $(70 - 4 \cdot 8 = 38)$ and 102 is 4 standard deviations above the mean $(70 + 4 \cdot 8 = 102)$, find the minimum fraction of values that lie within 4 standard deviations of the mean. See Exercise 14 for the answer 15/16.

25. This is equivalent to finding the largest fraction of values that lie outside 2 standard deviations from the mean. There are at least $1 - \dfrac{1}{2^2} = 1 - \dfrac{1}{4} = \dfrac{3}{4}$ of the values within 2 standard deviations of the mean. Thus, the largest fraction of values that lie outside this range would be: $1 - \dfrac{3}{4} = \dfrac{1}{4}$.

27. To find how many standard deviations below the mean is the value of 42:

$$z = \frac{42 - 70}{8} = -3.5$$

Also the value 98 is 3.5 standard deviations above the mean. Then we must find the largest fraction of values that lie outside 3.5 or 3 1/2 standard deviations from the mean. There are at least

$$1 - \frac{1}{\left(\frac{7}{2}\right)^2} = 1 - \frac{1}{\left(\frac{49}{4}\right)} = 1 - \frac{4}{49} = \frac{45}{49}$$

of the values within 3 1/2 standard deviations of the mean. Thus, the largest fraction of values that lie outside this range of values would be: $1 - \dfrac{45}{49} = \dfrac{4}{49}$.

29. The sum of the values is $2430. Then
$$x = \frac{2430}{12} = \$202.50$$

31. The standard deviation is $80.38. Then
$202.50 - 80.38 = \$122.12$, and
$202.50 + 80.38 = \$282.88$. There are six bonus amounts that fall within these two boundaries: $175, $185, $190, $205, $210, and $215.

33. $1 - \dfrac{1}{2^2} = 1 - \dfrac{1}{4} = \dfrac{3}{4}$. Then 3/4 of the data is
$$\frac{3}{4} \cdot 12 = 9.$$

There should be at least 9 amounts.

35. To find the mean average life of Brand B, find the sum of the amounts given in the text: 963,510. Then
$$\overline{x} = \frac{963510}{20} \approx 48,167$$

This mean is greater than the mean for Brand A, which is 43,560; Brand B has the longer average life.

37. Writing exercise

39. $\overline{x} = \dfrac{18 + 19 + 21 + 23 + 25 + 27 + 30}{7} \approx 23.29$

To find the standard deviation:

1. The mean is 23.29.

2 and 3. Find each deviation from the mean $(x - \overline{x})$ and square each deviation, $(x - \overline{x})^2$. These steps are shown in the table.

Data	Deviations	Squared Deviations
18	$18 - 23.29 = -5.29$	$(-5.29)^2 = 27.9841$
19	$19 - 23.29 = -4.29$	$(-4.29)^2 = 18.4041$
21	$21 - 23.29 = -2.29$	$(-2.29)^2 = 5.2441$
$\vdots$	$\vdots$	$\vdots$
30	$30 - 23.29 = 6.71$	$(6.71)^2 = 45.0241$
Total		113.4281

4. Sum the squared deviations. The sum is 113.4281.
5. Divide by $n - 1$.
$$\frac{113.4281}{7 - 1} \approx 18.905$$

6. Take the square root. $\sqrt{18.905} \approx 4.35$

41. Writing exercise

43. Writing exercise

45. (a) Test cities: $\overline{x} = \dfrac{18 + 15 + 7 + 10}{4} = \dfrac{50}{4} = 12.5$.

(b) Control cities:
$$\overline{x} = \frac{+1 + (-8) + (-5) + 0}{4} = \frac{-12}{4} = -3.0.$$

(c) To find the standard deviation for the test cities:

Steps 1, 2, and 3. Find each deviation from the mean for the test cities, $(x - \overline{x})$ and square each deviation, $(x - \overline{x})^2$. These steps are shown in the table.

Data	Deviations	Squared Deviations
18	$18 - 12.5 = 5.5$	$(5.5)^2 = 30.25$
15	$15 - 12.5 = 2.5$	$(2.5)^2 = 6.25$
7	$7 - 12.5 = -5.5$	$(-5.5)^2 = 30.25$
10	$10 - 12.5 = -2.5$	$(-2.5)^2 = 6.25$
Total		73

4. Sum the squared deviations. The sum is 73.
5. Divide by $n - 1$.
$$\frac{73}{4 - 1} \approx 24.333$$

6. Take the square root. $\sqrt{24.333} \approx 4.9$

(d) To find the standard deviation for the control cities:

Steps 1, 2, and 3. Find each deviation from the mean for the control cities, $(x - \overline{x})$ and square each deviation, $(x - \overline{x})^2$. These steps are shown in the table.

Data	Deviations	Squared Deviations
+1	$1 - (-3.0) = 4.0$	$(4.0)^2 = 16$
−8	$-8 - (-3.0) = -5.0$	$(-5.0)^2 = 25$
−5	$-5 - (-3.0) = -2.0$	$(-2.0)^2 = 4$
0	$0 - (-3.0) = 3.0$	$(3.0)^2 = 9$
Total		54

4. Sum the squared deviations. The sum is 54.
5. Divide by $n - 1$.
$$\frac{54}{4 - 1} \approx 18$$

6. Take the square root. $\sqrt{18} \approx 4.2$

(e) $12.5 - (-3.0) = 12.5 + 3.0 = 15.5$

(f) $15.5 - 7.95 = 7.55$; $15.5 + 7.95 = 23.45$

47. Writing exercise

49. No, because the table gives only how many pieces of data occur within a given interval.

51. Writing exercise

53. (a) The smallest item would be the mode because the highest peak of the curve is to the left.

    (b) The largest item would be the mode because the highest peak of the curve is to the right.

    (c) Writing exercise

12.4 EXERCISES

For each of Exercises 1–4, make use of z-scores.

1. The z-score for Chris:

    $$z = \frac{5 - 4.6}{2.1} = 0.19$$

    The z-score for Lynn:

    $$z = \frac{6 - 4.6}{2.3} = 0.48$$

    Lynn's score is farther from the mean in a positive direction, so her score is better.

3. The z-score for Jutta's Brand A tires:

    $$z = \frac{37,000 - 45,000}{4500} \approx -1.78$$

    The z-score for Arvind's Brand B tires:

    $$z = \frac{35,000 - 38,000}{2080} \approx -1.44$$

    Arvind's score is closer to the mean in a negative direction, so that score is better.

5. Find 15% of 40 items: $.15(40) = 6$. Select the 7th item in the data set, which is 58.

7. The third decile is the same as 30%. Find 30% of 40 items: $.30(40) = 12$. Select the 13th item in the data set, which is 62.

9. Find the mean by dividing the sum of the Games Played column by 10, the number of players.

    $$\overline{x} = \frac{476}{10} = 47.6$$

    Find s by using the six step process described in Section 12.3 or by entering the data into a calculator or spreadsheet that the contains the function to calculate s. The standard deviation is 3.27. Use the formula

$z = \frac{x - \overline{x}}{s}$ where x is Charles Barkley's games played.

$$z = \frac{42 - 47.6}{3.27} \approx -1.71$$

11. The sum of the data in the Average Rebounds Per Game column is 113.3 and there are 10 players. Then

    $$\overline{x} = \frac{113.3}{10} = 11.33.$$

    Find s by using the six step process described in Section 12.3 or by entering the data into a calculator or spreadsheet that the contains the function to calculate s. The standard deviation is 0.949. Use the formula $z = \frac{x - \overline{x}}{s}$ where x is Chris Webber's average rebounds per game.

    $$z = \frac{13.0 - 11.33}{0.949} \approx 1.76$$

13. Find 55% of 10 items: $.55(10) = 5.5$. Select the 6th item in the data set for defensive rebounds after they have been put into rank order: 323, 338, 341, 349, 361, 369, 371, 396, 412, 418. The 6th value is 369 defensive rebounds made by Antonio McDyess.

15. The second decile is the same as the 20th percentile. Find 20% of 10 items: $.20(10) = 2$. Select the 3rd item in the data set for games played after they have been put into rank order: 42, 42, 46, 47, 49, 50, 50, 50, 50, 50. The 3rd value is 46 games played made by Alonzo Mourning.

17. See Exercise 9 for the values of $\overline{x}$ and s. The z-score for Kevin Garnett is

    $$z = \frac{47 - 47.6}{3.27} \approx -0.18.$$

    Find the mean for Defensive Rebounds by summing the data and dividing by 10. The value of s is 32.1.

    $$\overline{x} = \frac{3678}{10} = 367.8.$$

    The z-score for Vlade Divac is

    $$z = \frac{361 - 367.8}{32.1} \approx -0.21.$$

    Kevin Garnett was relatively higher because his z-score of -0.18 is greater than Vlade Divac's z-score of -0.21.

19. (a) The median is 11.2.

    (b) The range is $13.0 - 10.0 = 3$.

    (c) The distribution is skewed to the right.

21. Writing exercise

23. Writing exercise

25. Writing exercise

27. The "skewness coefficient" is a measure of the overall distribution.

29. (a) No; this would only be true if Q_1 and Q_3 are symmetric about Q_2.

    (b) Writing exercise

31. Writing exercise

33. Writing exercise

35. Warner's season rating consists of the following:

Numerator terms	Substitute values	Simplify
$+\left(250 \times \dfrac{C}{A}\right)$	$=\left(250 \times \dfrac{325}{499}\right)$	≈ 162.83
$+\left(1000 \times \dfrac{T}{A}\right)$	$=\left(1000 \times \dfrac{41}{499}\right)$	≈ 82.16
$+\left(12.5 \times \dfrac{Y}{A}\right)$	$=\left(12.5 \times \dfrac{4353}{499}\right)$	≈ 109.04
$+\,6.25$	$=\,+\,6.25$	$=\,+\,6.25$
$-\left(1250 \times \dfrac{I}{A}\right)$	$=\left(1250 \times \dfrac{13}{499}\right)$	≈ -32.57
TOTAL		≈ 327.71

$$\text{Rating} = \frac{\text{TOTAL}}{3} = \frac{327.71}{3} = 109.2$$

37. Q_2 is the median: $\dfrac{84.1 + 83.6}{2} \approx 83.9$.

    Q_1 is the median of the lower half, which is 81.1.

    Q_3 is the median of the upper half, which is 94.2.

39. The eighth decile is the 80th percentile. First find 80% of 20 items: $.80(20) = 16$. Then locate the 17th item after arranging all items in order from smallest to largest: 75.8, 77.6, 77.7, 77.9, 78.6, 80.6, 81.1, 81.2, 81.7, 82.0, 83.5, 83.6, 84.1, 85.1, 86.5, 90.0, 90.7, 94.2, 94.6, 109.2. The eighth decile is 90.7.

41.

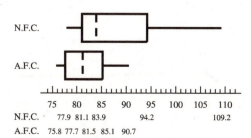

43. The midquartile is found by dividing the sum of Q_1 and Q_3 by 2. For the N.F.C.

$$\frac{Q_1 + Q_3}{2} = \frac{81.1 + 94.2}{2} = 87.65$$

For the A.F.C.

$$\frac{Q_1 + Q_3}{2} = \frac{77.7 + 81.5}{2} = 79.6$$

The midquartile for the N.F.C. is greater by about 6 points.

45. The interquartile range for the N.F.C. is

$$Q_3 - Q_1 = 94.2 - 81.1 = 13.1.$$

The interquartile range for the A.F.C. is

$$Q_3 - Q_1 = 85.1 - 77.7 = 7.4.$$

The interquartile range for the N.F.C. is greater by about 6 units.

47. Repeat the process from Exercise 46, using 28/202 for T/A instead of replacing it with the lower value of 0.11875.

Numerator terms	Substitute values	Simplify
$\left(250 \times \dfrac{C}{A}\right)$	$=\left(250 \times \dfrac{110}{202}\right)$	≈ 136.14
$\left(1000 \times \dfrac{T}{A}\right)$	$=\left(1000 \times \dfrac{28}{202}\right)$	≈ 138.61
$\left(12.5 \times \dfrac{Y}{A}\right)$	$=\left(12.5 \times \dfrac{2194}{202}\right)$	≈ 135.77
SUM		410.52
		$+\,6.25$
$\left(1250 \times \dfrac{I}{A}\right)$	$=\left(1250 \times \dfrac{12}{202}\right)$	$-\,74.26$
TOTAL		$=342.51$

$$\text{Rating} = \frac{\text{TOTAL}}{3} = \frac{342.51}{3} \approx 114.2$$

49. Writing exercise

12.5 EXERCISES

1. Discrete because the variable can take on only fixed number values such as 1, 2, 3, etc., up to and including 50.

3. Continuous because the variable is not limited to fixed values. It is measurable rather than countable.

5. Discrete because the variable is limited to fixed values.

7. This represents all values to the right of the mean, which is 50% of the total number of values or 50% of 100 students: $.50(100) = 50$ students.

9. These values are 1 standard deviation above and 1 standard deviation below the mean, respectively: (86 ± 1). By the empirical rule, 68% of all scores lie within 1 standard deviation of the mean. Then $.68(100) = 68$.

11. Less than 100 represents all values below the mean. This is 50% of the area under the curve or 50% of all the data.

13. The score 70 lies 2 standard deviations below the mean, and 130 lies 2 standard deviations above the mean. According to the empirical rule, 95% of the data lies within 2 standard deviations of the mean.

15. To find the percent of area between the mean and 1.5 standard deviations, use Table 10 to locate a z-score of 1.5. Read the value of A in the next column. This is the area from the mean to the corresponding value of z. The area is .433 which is 43.3%.

17. Because of the symmetry of the normal curve, the area between the mean and a z value of -1.08 is the same as that from the mean to a z value of $+1.08$. Use Table 10 to locate a z-score of 1.08. Read the value of A in the next column. This is the area from the mean to the corresponding value of z. The area is .360 which is 36.0%.

19. It is helpful to sketch the area under the normal curve.

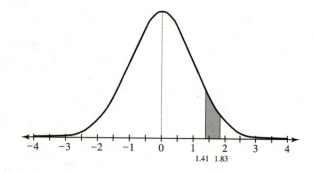

To find the area between $z = 1.41$ and $z = 1.83$, use the z-table to find the area between the mean and $z = 1.83$. The area is .466. Find the area between the mean and $z = 1.41$, which is .421. Subtracting, $.466 - .421 = .045$ which is 4.5%.

21. It is helpful to sketch the area under the normal curve.

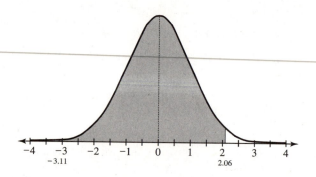

Add the areas under the curve from the mean to $z = -3.11$ to that from the mean to $z = 2.06$. The first area is the same as that from the mean to $z = +3.11$, which is .499. The area from the mean to $z = 2.06$ is .480. Find the total area: $.499 + .480 = .979$ or 97.9%.

23. If 10% of the total area is to the right of the z-score, there is $50\% - 10\% = 40\%$ of the area between the mean and the value of z. From the z-table, find .40 in the A column. This area under the curve of 40% or .40 yields a z-score of 1.28.

25. If 9% of the total area is to the left of z, the z-score is below the mean, so the answer will be negative. (A sketch helps to see this.) Subtract 9% from 50% to find the amount of area between the mean and the value of z: $50\% - 9\% = 41\%$. Now find .41 in the A column to locate the appropriate value for z: 1.34. The z-score is -1.34.

27. Since the mean is 600 hr, we can expect half of the bulbs or 5000 bulbs to last at least 600 hrs.

29. Find the z-score for 675:

$$z = \frac{675 - 600}{50} = 1.5.$$

Find the z-score for 740:

$$z = \frac{740 - 600}{50} = 2.8.$$

It is helpful to sketch the area under the normal curve.

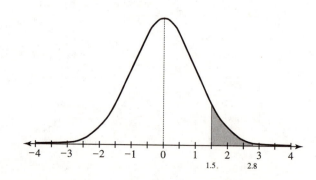

Find the amount of area under the normal curve between 1.5 and 2.8 by finding the corresponding values for A and then subtracting the smaller from the larger:

$$.497 - .433 = .064.$$

Finally find 64% of 10,000:

$$.064(10000) = 640.$$

31. Find the z-score for 740:

$$z = \frac{740 - 600}{50} = 2.8.$$

It is helpful to sketch the area under the normal curve.

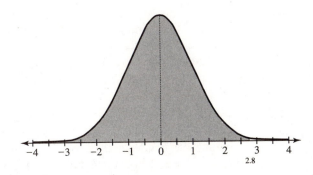

Find the amount of area under the normal curve between the mean and 2.8 by finding the corresponding value for A: .497. Add this area value to .5:

$$.497 + .5 = .997.$$

Finally find 99.7% of 10,000:

$$.997(10000) = 9970.$$

33. Because the mean is 1850 and the standard deviation is 150, the value 1700 corresponds to $z = -1$. Use the empirical rule to evaluate the corresponding area under the curve. The area between $z = -1$ and the mean is half of 68% or 34%. The area to the right of the mean is 50%. The total area to the right of $z = -1$ is $34 + 50 = 84\%$. If the z-table is used, the answer is 84.1%, because the area value in the table for $z = 1$ is .341.

35. Find the z-score for 1750:

$$z = \frac{1750 - 1850}{150} \approx -0.67.$$

Find the z-score for 1900:

$$z = \frac{1900 - 1850}{150} = 0.33.$$

It is helpful to sketch the area under the normal curve.

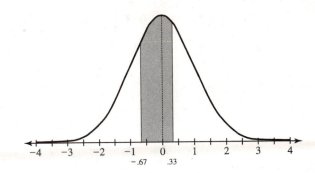

Find the amount of area under the normal curve between -0.66 and 0.33 by finding the corresponding values for $Area$ and then adding the values:

$$.249 + .129 = .378.$$

This is 37.8%.

37. The z-score corresponding to 24 oz when $s = .5$ is found by

$$z = \frac{24 - 24.5}{.5} = -1.$$

The fraction of boxes that are underweight is equivalent to the area under the curve to the left of -1.
It is helpful to sketch the area under the normal curve.

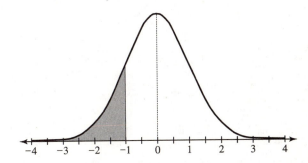

From the z-table, the area under the curve between the mean and $+1$ is .341. Subtract .341 from .5 to obtain the area under the curve to the right of $+1$: $.5 - .341 = .159$. Because of symmetry this is also the amount of area under the curve to the left of -1. The answer is .159 or 15.9%.

39. The z-score corresponding to 24 oz when $s = .3$ is found by

$$z = \frac{24 - 24.5}{.3} \approx -1.67.$$

The fraction of boxes that are underweight is equivalent to the area under the curve to the left of -1.67.
It is helpful to sketch the area under the normal curve.

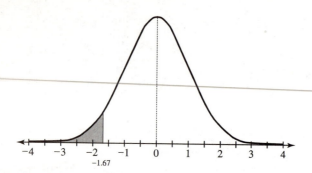

From the z-table, the area under the curve between the mean and $+1.67$ is .453. Subtract .453 from .5 to obtain the area under the curve to the right of $+1.67$: $.5 - .453 = .047$. Because of symmetry this is also the amount of area under the curve to the left of -1.67. The answer is .047 or 4.7%.

41. The mean plus 2.5 times the standard deviation, $\overline{x} + 2.5s$, corresponds to $z = 2.5$ no matter what the values of $\overline{x}$ and s are:

$$z = \frac{\text{value} - \text{mean}}{\text{standard deviation}}$$
$$= \frac{(\overline{x} + 2.5s) - \overline{x}}{s}$$
$$= \frac{2.5s}{s}$$
$$= 2.5$$

The fraction of the population between the mean and $z = 2.5$ is .494, by the z-table. The fraction below the mean is .5. The sum of these is $.5 + .494 = .994$ or 99.4%.

43. The RDA is the value corresponding to $z = 2.5$. Use the z-score formula to find x:

$$z = \frac{x - \overline{x}}{s}$$
$$2.5 = \frac{x - 159}{12}$$
$$(12) \cdot 2.5 = \frac{x - 159}{12} \cdot \frac{12}{1}$$
$$30 = x - 159$$
$$189 = x$$

The RDA is 189 units.

45. Find the z-score for 7:

$$z = \frac{7 - 7.45}{3.6} \approx -0.13.$$

Find the z-score for 8:

$$z = \frac{8 - 7.45}{3.6} \approx 0.15.$$

The area under the curve to the left of $-.13$ and the area under the curve to the right of .15 must be added to answer the question. Use the z-table to obtain .052 for the area between the mean and $+.13$. The area to the right of .13 is then $.5 - .052 = .448$. Because of symmetry, this is the amount of area to the left of -1.3. Use the z-table to obtain .060 for the area between the mean and .15. The area to the right of .15 is then $.5 - .060 = .440$. The sum of these two areas is $.448 + .440 = 0.888$.

47. Find the z-score for 2.2:

$$z = \frac{2.2 - 1.5}{4} = 1.75.$$

Find the area under the normal curve between the mean and 1.75 which is .460. Subtract .460 from .5 to obtain the area under the curve to the right of $z = 1.75$: $.5 - .460 = .040$ or 4.0%. Find 4.0% of five dozen: $(.04)(5 \cdot 12) = 2.4$. The answer is between 2 and 3 eggs.

49. Find the area as a percent between $\overline{x} + (1/2)s$ and $\overline{x} + (3/2)s$. This is the area between $z = 1/2$ or .5 and $z = 3/2$ or 1.5. The area under the curve from the mean to $z = .5$ is .191; the area under the curve from the mean to $z = 1.5$ is .433. Subtract: $.433 - .191 = .242$ or 24.2%.

51. Writing exercise

53. To find the bottom cutoff grade, $.15 + .08 = .23$ must be the amount of area under the normal curve to the right of the grade. That means that the amount of area between the mean and this cutoff grade is $.5 - .23 = .27$. Locate this area value in the A column; it corresponds to a z-score of .74. Then use the z-score formula to find x:

$$z = \frac{x - \overline{x}}{s}$$
$$.74 = \frac{x - 75}{5}$$
$$(5) \cdot (.74) = \frac{x - 75}{5} \cdot \frac{5}{1}$$
$$3.7 = x - 75$$
$$78.7 = x$$

Rounded to the nearest whole number, the cutoff score should be 79.

55. To find the bottom cutoff grade .08 must be the amount of area under the normal curve to the left of the grade. That means that the amount of area between the mean and this cutoff grade is $.5 - .08 = .42$. Locate this area value in the A column; it corresponds to a z-score of 1.40 or 1.41. Use 1.405. However, the negative z-score must be used because the grade is below the mean.

$$z = \frac{x - \overline{x}}{s}$$
$$-1.405 = \frac{x - 75}{5}$$
$$(5) \cdot (-1.405) = \frac{x - 75}{5} \cdot \frac{5}{1}$$
$$-7.025 = x - 75$$
$$68 \approx x$$

The cutoff grade should be 68.

57. Replace $\overline{x}$, s and z in the z-score formula and solve for x:

$$z = \frac{x - \overline{x}}{s}$$
$$1.44 = \frac{x - 76.8}{9.42}$$
$$(9.42) \cdot (1.44) = \frac{x - 76.8}{9.42} \cdot \frac{9.42}{1}$$
$$13.5648 = x - 76.8$$
$$90.4 \approx x$$

59. Replace $\overline{x}$, s and z in the z-score formula and solve for x:

$$z = \frac{x - \overline{x}}{s}$$
$$-3.87 = \frac{x - 76.8}{9.42}$$
$$(9.42) \cdot (-3.87) = \frac{x - 76.8}{9.42} \cdot \frac{9.42}{1}$$
$$-36.4554 = x - 76.8$$
$$40.3 \approx x.$$

61. Writing exercise

EXTENSION: HOW TO LIE WITH STATISTICS

1. There is no explanation of what the solid line represents.

3. The dashed line rises and then falls.

5. Percent of decrease $= \dfrac{\text{old value} - \text{new value}}{\text{old value}}$

$$\frac{2.40 - 1.72}{2.40} \approx .28 \text{ or } 28\%.$$

7. How long have Toyotas been sold in the United States? How do other makes compare?

9. The dentists preferred Trident Sugarless Gum to what? Which and how many dentists were surveyed? What percentage responded?

11. How quiet is a glider?

13. The maps convey their impressions in terms of *area* distributions, whereas personal income distributions may be quite different. The map on the left probably implies too high a level of government spending, while that on the right implies too low a level.

15. By the time the figures were used, circumstances may have changed greatly. (The Navy was much larger.) Also, New York City was most likely not typical of the nation as a whole.

17. (b) Change the fractions to percents and compare:

$$\frac{6}{50} = 12\% \qquad \frac{15}{50} = 30\% \qquad \frac{29}{50} = 58\%.$$

These values are very close to the overall population values.

19. (c) Change the fractions to percents and compare:

$$\frac{50}{120} \approx 42\% \qquad \frac{31}{120} \approx 26\% \qquad \frac{39}{120} \approx 33\%.$$

These values are very close to the overall population values.

21. (a) The first equation implies that $m = c - 2$. Replace m and a in the following equation and solve for c:

$$10 = m + a + c$$
$$10 = (c - 2) + 2c + c$$
$$10 = 4c - 2$$
$$12 = 4c$$
$$c = 3.$$

Then $m = 3 - 2 = 1$ and $a = 2 \cdot 3 = 6$.

(b) The total number of staff is $7 + 25 + 18 = 50$. The fractions of the entire office staff are: managers, 7/50; agents, 25/50; and clerical, 18/50. Now use the number in the sample as the denominator to solve for each variable using the technique of solving proportions from Section 7.3.

$$\frac{m}{10} = \frac{7}{50}$$

$$50 \cdot m = 7 \cdot 10$$

$$m = \frac{70}{50} \text{ or } 1.4 \text{ managers}$$

$$\frac{a}{10} = \frac{25}{50}$$

$$50 \cdot a = 25 \cdot 10$$

$$a = \frac{250}{50} \text{ or } 5 \text{ agents}$$

$$\frac{c}{10} = \frac{18}{50}$$

$$50 \cdot c = 18 \cdot 10$$

$$c = \frac{180}{50} \text{ or } 3.6 \text{ clerical}$$

12.6 EXERCISES

1. The equation of the least squares line is $y' = ax + b$ where a and b are found as follows:

$$a = \frac{n(\Sigma xy) - (\Sigma x)(\Sigma y)}{n(\Sigma x^2) - (\Sigma x)^2}$$

$$= \frac{10(75) - (30)(24)}{10(100) - (30)^2}$$

$$= \frac{750 - 720}{1000 - 900}$$

$$= \frac{30}{100}$$

$$= .3$$

$$b = \frac{\Sigma y - a(\Sigma x)}{n}$$

$$= \frac{24 - .3(30)}{10}$$

$$= \frac{24 - 9}{10}$$

$$= 1.5$$

Then the equation for the least squares line is

$$y' = .3x + 1.5.$$

3. The regression equation is $y' = .3x + 1.5$. Find y' when $x = 3$ tons.

$$y' = .3(3) + 1.5$$

$$= .9 + 1.5$$

$$= 2.4 \text{ decimeters}$$

5. The regression equation is $y' = .556x - 17.8$. Find y' when $x = 120°$.

$$y' = .556(120) - 17.8$$

$$= 66.72 - 17.8$$

$$= 48.92°$$

7. The table shows how to calculate all the sums that are needed in the formula for the least squares line.

x	y	x^2	y^2	xy
62	120	3844	14400	7440
62	140	3844	19600	8680
63	130	3969	16900	8190
65	150	4225	22500	9750
66	142	4356	20164	9372
67	130	4489	16900	8710
68	135	4624	18225	9180
68	175	4624	30625	11900
70	149	4900	22201	10430
72	168	5184	28224	12096
$\Sigma x = 663$	$\Sigma y = 1439$	$\Sigma x^2 = 44059$	$\Sigma y^2 = 209739$	$\Sigma xy = 95748$

The equation of the least squares line is $y' = ax + b$ where a and b are found as follows:

$$a = \frac{n(\Sigma xy) - (\Sigma x)(\Sigma y)}{n(\Sigma x^2) - (\Sigma x)^2}$$

$$= \frac{10(95748) - (663)(1439)}{10(44059) - (663)^2}$$

$$= \frac{957480 - 954057}{440590 - 439569}$$

$$= \frac{3423}{1021}$$

$$\approx 3.35.$$

$$b = \frac{\Sigma y - a(\Sigma x)}{n}$$

$$= \frac{1439 - 3.35(663)}{10}$$

$$= \frac{1439 - 2221.05}{10}$$

$$= -78.2.$$

The value for b calculated here differs slightly from the one in the text because of rounding error. It is highly recommended that a scientific calculator or spreadsheet be used, because then rounding is only done at the very end of all the calculations. Here the value of a was rounded before using it in the calculation for b. Then the equation for the least squares line is

$$y' = 3.35x - 78.2.$$

The value for b in the text is -78.4.

9. The regression equation is $y' = 3.35x - 78.4$. Find y' when $x = 70$ in.

$$y' = 3.35(70) - 78.4$$
$$= 234.5 - 78.4$$
$$\approx 156 \text{ pounds}$$

11.

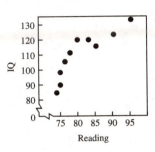

13. The regression equation is $y' = 2x - 51$. Find y' when $x = 65$.

$$y' = 2(65) - 51$$
$$= 130 - 51$$
$$= 79$$

15. Use a scientific calculator or a spreadsheet to calculate the various sums needed in the formula, or see Exercise 7 for the detailed process. Actually, a calculator or spreadsheet can calculate the value of r.

$\Sigma x = 15$; $\Sigma y = 418$; $\Sigma x^2 = 55$; $\Sigma y^2 = 30266$; $\Sigma xy = 1186$; and $n = 6$.

Substitute the sums into the formula for the coefficient of correlation.

$$r = \frac{n(\Sigma xy) - (\Sigma x)(\Sigma y)}{\sqrt{n(\Sigma x^2) - (\Sigma x)^2} \cdot \sqrt{n(\Sigma y^2) - (\Sigma y)^2}}$$
$$= \frac{6(1186) - (15)(418)}{\sqrt{6(55) - (15)^2} \cdot \sqrt{6(30266) - (418)^2}}$$
$$= \frac{7116 - 6270}{\sqrt{330 - 225} \cdot \sqrt{181596 - 174724}}$$
$$= \frac{846}{\sqrt{105} \cdot \sqrt{6872}}$$
$$\approx \frac{846}{10.25 \cdot 82.90}$$
$$\approx .996$$

17.

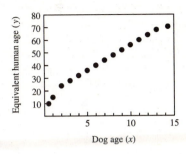

19. Writing exercise

21. Use a scientific calculator or a spreadsheet to calculate the various sums needed in the formula. Actually, a calculator or spreadsheet can calculate the appropriate values for a and b needed for the regression line equation. See Exercise 7 for the details of calculating the following sums.

$\Sigma x = 56$; $\Sigma y = 77.7$; $\Sigma x^2 = 560$; $\Sigma y^2 = 1110.43$; $\Sigma xy = 786.4$ and with $n = 8$.

The equation of the least squares line is $y' = ax + b$ where

$$a = \frac{n(\Sigma xy) - (\Sigma x)(\Sigma y)}{n(\Sigma x^2) - (\Sigma x)^2}$$
$$= \frac{8(786.4) - (56)(77.7)}{8(560) - (56)^2}$$
$$= \frac{6291.2 - 4351.2}{4480 - 3136}$$
$$= \frac{1940}{1344}$$
$$\approx 1.44.$$

$$b = \frac{\Sigma y - a(\Sigma x)}{n}$$
$$= \frac{77.7 - 1.44(56)}{8}$$
$$= \frac{77.7 - 80.64}{8}$$
$$\approx -.37.$$

Again, it is recommended that a scientific calculator or spread sheet be used to calculate these values in order to avoid rounding error. (The value of a has been rounded here.) The value for b that is calculated in the text is $-.39$. Then the equation for the least squares line

$$y' = 1.44x - .39.$$

23. A correlation of .99 is very strong.

25. Use a scientific calculator or a spreadsheet to calculate the various sums needed in the formulas. Actually, a calculator or spreadsheet can calculate the appropriate values for a and b needed for the regression line equation. See Exercise 7 for the details of calculating the following sums.

$\Sigma x = 134267$; $\Sigma y = 1297$; $\Sigma x^2 = 2228673709$;
$\Sigma y^2 = 173547$; $\Sigma xy = 18429024$ and with $n = 10$.

The equation of the least squares line is $y' = ax + b$ where

$$a = \frac{n(\Sigma xy) - (\Sigma x)(\Sigma y)}{n(\Sigma x^2) - (\Sigma x)^2}$$
$$= \frac{10(18429024) - (134267)(1297)}{10(2228673709) - (134267)^2}$$
$$= \vdots$$
$$\approx .00238.$$

$$b = \frac{\Sigma y - a(\Sigma x)}{n}$$
$$= \frac{1297 - .000238(134267)}{10}$$
$$= \vdots$$
$$\approx 97.7.$$

Again, it is recommended that a scientific calculator or spread sheet be used to calculate these values in order to avoid rounding error. Then the equation for the regression line

$$y' = .00238x + 97.7.$$

27. The linear correlation is moderate.

CHAPTER 12 TEST

1. Examine the first graph to see that in 1880 about 57.1% of total U.S. workers were in farm occupations. In 1900 about 37.5% were in farm occupations. Somewhere between 1880 and 1900, the percentage would have been 50%.

2. No, the percentage was 4 times as great, not the number of workers.

3. Examine the line graph. Find 1960 along the horizontal axis and the value of about 4 million along the vertical axis.

4. Use the bottom two graphs to answer this question. Find the number of farms (in millions) in the graph on the left for each year: about 6.3 in 1940 and about 2.1 in 1998. Multiply each of these values by the size of the average farm; the sizes of the average farm are found in the bar

graph on the right: 174 for 1940 and 435 for 1998.

$$6.3(174) = 1096.2 \text{ and } 2.1(435) = 913.5$$

Find the percentage:

$$\frac{1096.2 - 913.5}{913.5} \times 100 = 20\%.$$

5. A tally column is not shown here, but it is useful when creating a frequency distribution by hand.

Class Limits	Frequency f	Relative frequency f/n
6–10	3	$3/22 \approx .14$
11–15	6	$6/22 \approx .27$
16–20	7	$7/22 \approx .32$
21–25	4	$4/22 \approx .18$
26–30	2	$4/22 \approx .09$

6. (a)

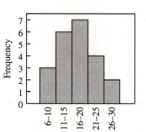

Number of client contacts

(b)

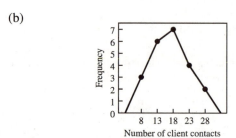

Number of client contacts

7. If the first class had limits of 7–9, this would be a width of 3. The smallest value is 8, and the largest value is 30. All the data must be included within the classes that are constructed. The classes would be:

$$7–9$$
$$10–12$$
$$13–15$$
$$16–18$$
$$19–21$$
$$22–24$$
$$25–27$$
$$28–30.$$

This is a total of 8 classes.

8.

Value	Frequency	Value · Frequency
8	3	24
10	8	80
12	10	120
14	8	112
16	5	80
18	1	18
Totals	35	434

Then $\bar{x} = \dfrac{434}{35} = 12.4$.

9. There are 35 items. The formula for the position of the median is

$$\frac{\Sigma f + 1}{2} = \frac{35 + 1}{2} = 18.$$

This means that the median is 18th item when the values are arranged from smallest to largest. A chart showing cumulative frequency shows that the value is in the row for a cumulative frequency of 21.

Value	Frequency	Cumulative Frequency
8	3	3
10	8	11
12	10	21
14	8	29
16	5	34
18	1	35

The value 12 is the median.

10. Examine the frequency column to see that the mode is 12, with a frequency of 10.

11. The range of the data is $18 - 8 = 10$.

12. The leaves are arranged in rank order (from smallest to largest).

```
3 | 3 8
4 | 3 5 8 9
5 | 0 2 5
6 | 1 1 4 5 6 7 7 8
7 | 0 1 2 3 7 7 8 9 9
8 | 0 4 4
9 | 1
```

13. There are 33 items. The formula for the position of the median is

$$\frac{\Sigma f + 1}{2} = \frac{33 + 1}{2} = 17.$$

This means that the median is the 17th item when the values are arranged from smallest to largest. Count the values in the stem-and-leaf display in the text to see that the 17th value is 35.

14. The mode is 33; it occurs 3 times.

15. The range of the data is $60 - 23 = 37$.

16. The third decile is the same as the 30th percentile. First

$$.30(33) = 9.9.$$

Then the 10th value is the location of the third decile. This value is 31.

17. The eighty-fifth percentile is located by first taking 85% of the 33 data items:

$$.85(33) = 28.0.$$

Then the 29th value is located at the 85th percentile. This value is 49.

18. The values shown in the boxplot are: Minimum value, 23; Maximum value, 60; Q_1, 29.5; Q_2, 35, and Q_3, 43.

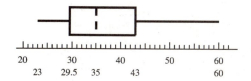

19. The scores 70 and 90 are each two standard deviations away from the mean because the mean is 80, and the value of one standard deviation is 5. According to the empirical rule, approximately 95% of the data lies within two standard deviations of the mean.

20. The score of 95 is three standard deviations above the mean; the score of 65 is three standard deviations below the mean. According to the empirical rule, approximately 99.7% of the data lie within this interval. Therefore, $100 - 99.7 = .3\%$ lie outside this interval.

21. The score of 75 is one standard deviation below the mean. Using the empirical rule, if 68% of the scores lie within one standard deviation of the mean, then 34% of the scores lie between 75 and 80. Subtract 34% from 50% to find the percentage of scores that are less than 75: $50 - 34 = 16\%$.

22. The score of 85 is one standard deviation above the mean; the score of 90 is two standard deviations above the mean. From Exercise 21 we know that about 34% of the scores lie between 80 and 85. Also from the empirical rule, approximately 95% of the scores lie within two standard deviations. Half of 95% is 47.5%. Subtract 34% from 47.5% to find the percentage of scores between 85 and 90: $47.5 - 34 = 13.5\%$.

23. First find the z-score for 6.5:

$$z = \frac{6.5 - 5.5}{2.1} \approx .48.$$

It is always helpful to sketch the area under the normal

curve that is being sought. Then find the amount of area between the mean and $z = .48$ from Table 10 in Section 12.5. The amount is .184. This area must be added to .5, the amount of area under the normal curve below the mean:

$$.184 + .5 = .684$$

24. First find the z-scores for 6.2 and 9.4:

$$z = \frac{6.2 - 5.5}{2.1} \approx .33$$
$$z = \frac{9.4 - 5.5}{2.1} \approx 1.86.$$

It is always helpful to sketch the area under the normal curve that is being sought.

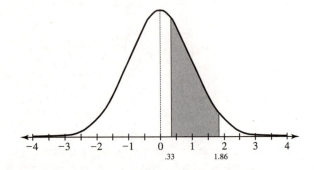

We are trying to find the amount of area under the curve between a z-score of .33 and a z-score of 1.86. Find the amount of area between the mean and $z = 1.86$ from Table 10 in Section 12.5. The amount is .469. Find the amount of area between the mean and $z = .33$; it is .129. The smaller value must be subtracted from the larger.

$$.469 - .129 = .340$$

25. Find the means of the three groups.

Eastern teams $\quad \dfrac{84.6 + 81.8}{2} = 83.2$

Central teams $\quad \dfrac{73.6 + 81.2}{2} = 77.4$

Western teams $\quad \dfrac{82.8 + 81.8}{2} = 82.3$

The Eastern teams had the highest winning average.

26. Find the means of the standard deviations.

Eastern teams $\quad \dfrac{11.8 + 17.4}{2} = 14.6$

Central teams $\quad \dfrac{13.9 + 12.4}{2} = 13.15$

Western teams $\quad \dfrac{10.7 + 11.5}{2} = 11.1$

The Western teams were the most "consistent" because their standard deviation (variation) is the smallest.

27. The average number of games won for all 30 teams is

$$\frac{5(84.6) + 5(73.6) + 4(82.8) + 5(81.8) + 6(81.2) + 5(81.8)}{5 + 5 + 4 + 5 + 6 + 5}$$
$$= \frac{2427.4}{30}$$
$$\approx 80.9$$

28.

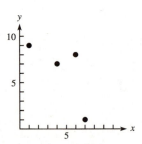

29. It is more efficient to use a calculator or a spreadsheet to find the various sums that are needed in the formulas for a and b. However, if these calculations are done by hand, make a table as follows:

x	y	x^2	y^2	xy
1	9	1	81	9
4	7	16	49	28
6	8	36	64	48
7	1	49	1	7
$\Sigma x = 18$	$\Sigma y = 25$	$\Sigma x^2 = 102$	$\Sigma y^2 = 195$	$\Sigma xy = 92.$

The equation of the least squares line is $y' = ax + b$ where a and b are found as follows:

$$a = \frac{n(\Sigma xy) - (\Sigma x)(\Sigma y)}{n(\Sigma x^2) - (\Sigma x)^2}$$
$$= \frac{4(92) - (18)(25)}{4(102) - (18)^2}$$
$$= \frac{368 - 450}{408 - 324}$$
$$= \frac{-82}{84}$$
$$\approx -.97619$$

$$b = \frac{\Sigma y - a(\Sigma x)}{n}$$
$$= \frac{25 - (-.97619)(18)}{4}$$
$$\approx \frac{25 + 17.5714}{4}$$
$$= 10.64$$

Then the equation for the least squares line, with a and b rounded to the nearest hundredth, is

$$y' = -.98x + 10.64.$$

30. The regression equation is $y' = -.98x + 10.64$. Find y' when $x = 3$.

$$y' = -.98(3) + 10.64$$
$$= -2.94 + 10.64$$
$$= 7.70.$$

31. $$r = \frac{n(\Sigma xy) - (\Sigma x)(\Sigma y)}{\sqrt{n(\Sigma x^2) - (\Sigma x)^2} \cdot \sqrt{n(\Sigma y^2) - (\Sigma y)^2}}$$

$$= \frac{4(92) - (18)(25)}{\sqrt{4(102) - (18)^2} \cdot \sqrt{4(195) - (25)^2}}$$

$$= \frac{368 - 450}{\sqrt{408 - 324} \cdot \sqrt{780 - 625}}$$

$$= \frac{-82}{\sqrt{84} \cdot \sqrt{155}}$$

$$\approx \frac{-84}{(9.165) \cdot (12.4499)}$$

$$\approx -.72$$

32. Writing exercise

33. Writing exercise

13 | CONSUMER MATHEMATICS

Chapter Goals

After completing this chapter the student should be able to

- Calculate simple interest and compound interest.
- Calculate finance charges on charge accounts.
- Understand and calculate true annual interest.
- Understand the basic concepts of buying a house, including calculating mortgage payments.
- Understand the basic concepts of buying and selling stocks.

Chapter Summary

It is necessary to understand the difference between simple interest and compound interest. When interest is calculated only on principal, it is called simple interest and is found by the formula:

Interest = Principal × rate × time, or $I = Prt$.

EXAMPLE A   Simple Interest

Find the simple interest on $650 at 6% for 9 months.

Solution
Use the formula $I = Prt$ with $P = \$650$; $r = .06$; and $t = 9/12$. Remember that t must be expressed in years.

$$
\begin{aligned}
I &= Prt \\
&= (\$650)(.06)\left(\frac{9}{12}\right) \\
&= \$29.25
\end{aligned}
$$

When interest is calculated on the accumulated interest in addition to the principal, it is called compound interest. The following formula can be derived for compound interest.

Compound amount formula: If P dollars are deposited at a rate of interest, r, compounded m times per year, and the money is left on deposit for a total of n periods, then the future value, A, the final amount on deposit, is:

$$
A = P\left(1 + \frac{r}{m}\right)^n
$$

EXAMPLE B   Future Value

Find the future value or final amount on deposit when $2370 is deposited at 10%, compounded quarterly for 5 years.

Solution
$P = \$2370$, $r = .10$, $m = 4$ and $n = 4 \cdot 5 = 20$.

$$
\begin{aligned}
A &= P\left(1 + \frac{r}{m}\right)^n \\
&= \$2370\left(1 + \frac{.10}{4}\right)^{20} \\
&= \$2370(1.025)^{20} \\
&= \$2370(1.63861) \\
&= \$3883.52
\end{aligned}
$$

Two common ways of buying items on credit are closed-end and open-end. An example of closed-end credit would be paying a fixed amount for a fixed time period:

EXAMPLE C   Closed-end credit

The Giordanos buy $8500 worth of furniture for their new home. They pay $3000 down. The store charges 10% add-on interest. The Giordanos will pay off the furniture in 30 monthly payments (2 1/2 years). Find the monthly payment.

Solution

First subtract the down payment from the purchase price of the furniture to find the value of P.

$$\$8500 - \$3000 = \$5500$$

Now use the formula for simple interest with $P = \$5500$, $r = .10$, and $t = \frac{30}{12} = 2.5$.

$$\begin{aligned} I &= Prt \\ &= (\$5500)(.10)(2.5) \\ &= \$1375 \end{aligned}$$

The total amount owed is

$$\begin{aligned} P + I &= \$5500 + \$1375 \\ &= \$6875 \end{aligned}$$

There are $2.5(12) = 30$ monthly payments, so the amount of each monthly payment is

$$\frac{\$6875}{30} = \$229.17$$

There is not a fixed number of payments with open-end credit. Department store charge accounts and bank charge cards such as VISA are examples of this type of credit. To calculate finance charges on an open-end credit account, use the unpaid balance method. The following table assumes an interest rate of 1.5% on the unpaid balance.

EXAMPLE D   Unpaid balance method

Month	Unpaid Balance at Beginning of Month	Finance Charge	Purchases During Month	Returns	Payments	Unpaid. Balance at End of Month
October	$828.63	$12.43	$128.72	$23.15	$125	$821.63
November	821.63	12.32	291.64	0	170	955.59
December	955.59	14.33	147.11	17.15	150	949.88
January	949.88	14.25	27.84	139.82	200	652.15

Here are explanations of how each underlined value was found:

1. Finance charge = Unpaid balance at beginning of month · Interest rate as decimal
2. Unpaid balance at end of month = Unpaid balance at beginning of month + Finance charge + Purchases − Returns − Payments.
3. Unpaid balance at beginning of month = Unpaid balance at end of previous month.

October

Unpaid balance at beginning of month: $828.63
Finance charge: ($828.63)(.015) = $12.43
Unpaid balance at end of month: $828.63 + 12.43 + 128.72 − 23.15 − 125 = $821.63

November

Unpaid balance at beginning of month: $821.63
Finance charge: ($821.63)(.015) = $12.32
Unpaid balance at end of month: $821.63 + 12.32 + 291.64 − 170 = $955.59

December

Unpaid balance at beginning of month: $955.59
Finance charge: ($955.59)(.015) = $14.33
Unpaid balance at end of month: $955.59 + 14.33 + 147.11 − 17.15 − 150 = $949.88

January

Unpaid balance at beginning of month: $949.88
Finance charge: ($949.88)(.015) = $14.25
Unpaid balance at end of month: $949.88 + 14.25 + 27.84 − 139.82 − 200 = $625.15

Another method of calculating finance charges is by the average daily balance method. The next example uses this method of calculation, using an interest rate of 1.5% per month on the average daily balance.

EXAMPLE E   Average daily balance method

First, make a table showing all the transactions, and compute the balance due on each date.

Date	Balance due
August 17	$983.25
August 21	$983.25 + $14.92 = $998.17
August 23	$998.17 − $25.41 = $972.76
August 27	$972.76 + $31.82 = $1004.58
August 31	$1004.58 − $108.00 = $896.58
September 9	$896.58 − $71.14 = $825.44
September 11	$825.44 + $110 = $935.44
September 14	$935.44 + $100 = $1035.44

Next, tabulate the balance due figures, along with the number of days until the balance changed. Use this data to calculate the sum of the daily balances.

Date	Balance due	Number of days until balance changed	$\left(\begin{array}{c}\text{Balance}\\\text{due}\end{array}\right) \times \left(\begin{array}{c}\text{Number}\\\text{of days}\end{array}\right)$
August 17	$983.25	4	$983.25 × 4 = $3933.00
August 21	$998.17	2	$998.17 × 2 = $1996.34
August 23	$972.76	4	$972.76 × 4 = $3891.04
August 27	$1004.58	4	$1004.58 × 4 = $4018.32
August 31	$896.58	9	$896.58 × 9 = $8069.22
September 9	$825.44	2	$825.44 × 2 = $1650.88
September 11	$935.44	3	$935.44 × 3 = $2806.32
September 14	$1035.44	3	$1035.44 × 3 = $3106.32
Totals		31	$29,471.44

$$\text{Average daily balance} = \frac{\text{Sum of daily balances}}{\text{Days in billing period}} = \frac{\$29471.44}{31} = \$950.69$$

and

$$\text{Finance charge} = (.015)(\$950.69) = \$14.26$$

Table 3 in the text is a portion of a table that gives the true annual interest rate. The number of payments column should be labeled as the number of monthly payments. The following example demonstrates how to use the table:

EXAMPLE F   True annual interest

If the amount financed is $1700, the interest charge is $202, and the number of (monthly) payments is 24, the true annual interest rate is found as follows:

Solution
Finance charge per $100 of the amount financed.

$$\frac{\$202}{\$1700} \times \$100 = \$11.88$$

In Table 3, find the "24 payments" row and read across to find the number closest to 11.88, which is 11.86. Read up to find the APR, which is 11.0%.

The Rule of 78 method of calculating the unearned interest when a loan is paid off in full before the due date, is often used, as in the following example:

EXAMPLE G   Rule of 78

If the original finance charge is $323.30, use the rule of 78 to find the unearned interest. The original loan was to be paid off in 18 months, but now it will be paid off with 6 months remaining. In the formula, k represents the number of payments remaining, n is the original number of payments, and F is the original finance charge. Then $k = 6$, $n = 18$, and $F = \$323.30$.

$$U = \frac{k(k+1)}{n(n+1)} \times F$$
$$= \frac{6(6+1)}{18(18+1)} \times \$323.30$$
$$= \$39.70$$

A more accurate method for calculating unearned interest is the actuarial method. It is shown in the following example.

EXAMPLE H   Actuarial Computation of Unearned Interest

Using the same information from Example G, compute the unearned interest by the actuarial method.

Solution
The original loan was to be paid off in 18 months with a regular payment of $201.85, but now it will be paid off with 6 months remaining.

Use the formula $U = kR\left(\dfrac{h}{100+h}\right)$ to find unearned interest, with $k =$ remaining number of payments,

$R =$ regular monthly payment, and $h =$ finance charge per $100, which is $3.53.

$$U = (6)(\$201.85)\left(\frac{\$3.53}{\$100 + \$3.53}\right)$$
$$= (6)(\$201.85)\left(\frac{\$3.53}{\$103.53}\right)$$
$$= \$41.29$$

When buying a home a useful formula is the regular monthly payment formula which gives the regular monthly payment required to repay a loan of P dollars, together with an interest rate r, over a term of t years. This formula is

$$R = \frac{P\left(\frac{r}{12}\right)\left(1 + \frac{r}{12}\right)^{12t}}{\left(1 + \frac{r}{12}\right)^{12t} - 1}$$

Table 4 in the text is a portion of a table that can be used as a quick reference to find the regular monthly payments for a $1000 loan.

EXAMPLE I   Monthly payment on home loan

Loan amount $= \$95,450$
Interest rate $= 15.5\%$
Term of loan $= 5$ years

Solution
In Table 4, find the 15.5% row and read across to the column for 5 years to find entry 24.05319. Since this is the monthly payment amount needed to amortize a loan of $1000, and this loan is for $95,450, the required monthly payment is

$$\frac{\$95450}{\$1000} \times \$24.05319 = \$2295.88$$

If Table 4 cannot be used because either the annual percentage rate or the term of the loan is not in the table, use the formula for regular monthly payment.

Example J   Monthly payment on home loan

Because the value of 17 years is not in Table 4, use the formula for regular monthly payment with $P = \$105,000$, $r = .055$, and $t = 17$.

$$R = \frac{P\left(\frac{r}{12}\right)\left(1 + \frac{r}{12}\right)^{12t}}{\left(1 + \frac{r}{12}\right)^{12t} - 1}$$
$$= \frac{\$105000\left(\frac{.055}{12}\right)\left(1 + \frac{.055}{12}\right)^{12(17)}}{\left(1 + \frac{.055}{12}\right)^{12(17)} - 1}$$
$$= \$793.39$$

13.1 EXERCISES

Note: Except where it is stated that a table has been used, exercises in this chapter have been completed with a calculator or spreadsheet. To ensure as much accuracy as possible, rounded values have been avoided in intermediate steps as much as possible. When rounded values have been necessary, several decimal places have been carried throughout the exercise; only the final answer has been rounded to 1 or 2 decimal places. In most cases, answers involving money are rounded to the nearest cent.

1. Use the formula $I = Prt$ with $P = \$1400$; $r = .08$; and $t = 1$.

$$I = Prt$$
$$= (\$1400)(.08)(1)$$
$$= \$112$$

3. Use the formula $I = Prt$ with $P = \$650$; $r = .06$; and $t = 9/12$. Remember that t must be expressed in years.

$$I = Prt$$
$$= (\$650)(.06)\left(\frac{9}{12}\right)$$
$$= \$29.25$$

5. Use the formula $I = Prt$ with $P = \$2675$; $r = .082$; and $t = 2\,1/2$ or 2.5. Remember that t must be expressed in years.

$$I = Prt$$
$$= (\$2675)(.082)(2.5)$$
$$= \$548.38$$

7. (a) Use the formula $A = P(1 + rt)$, with $P = \$700$; $r = .06$; and $t = 3$.

$$A = P(1 + rt)$$
$$= \$700(1 + .06 \cdot 3)$$
$$= \$700(1 + .18)$$
$$= \$700(1.18)$$
$$= \$826$$

(b) Use the formula $A = P\left(1 + \frac{r}{m}\right)^n$, with $m = $ number of periods per year and $n = $ total number of periods.

$$A = P\left(1 + \frac{r}{m}\right)^n$$
$$= \$700\left(1 + \frac{.06}{1}\right)^3$$
$$= \$700(1.06)^3$$
$$= \$700(1.191016)$$
$$= \$833.71$$

9. (a) Use the formula $A = P(1 + rt)$, with $P = \$2500$; $r = .05$; and $t = 3$.

$$A = P(1 + rt)$$
$$= \$2500(1 + .05 \cdot 3)$$
$$= \$2500(1 + .15)$$
$$= \$2500(1.15)$$
$$= \$2875$$

(b) Use the formula $A = P\left(1 + \frac{r}{m}\right)^n$, with $m = $ number of periods per year and $n = $ total number of periods.

$$A = P\left(1 + \frac{r}{m}\right)^n$$
$$= \$2500\left(1 + \frac{.05}{1}\right)^3$$
$$= \$2500(1.05)^3$$
$$= \$2500(1.157625)$$
$$= \$2894.06$$

11. Use the formula $I = Prt$ with $P = \$7500$; $r = .10$; and $t = 4/12$. Remember that t must be expressed in years.

$$I = Prt$$
$$= (\$7500)(.10)\left(\frac{4}{12}\right)$$
$$= \$250$$

13. First use the formula $I = Prt$ with $P = \$14,800$; $r = .09$; and $t = 10/12$. Remember that t must be expressed in years.

$$I = Prt$$
$$= (\$14800)(.09)\left(\frac{10}{12}\right)$$
$$= \$1110$$

Now the interest amount must be added to the loan amount to find the total amount she must repay.

$$\$1110 + \$14800 = \$15,910$$

15. To find the compound interest that was earned, subtract the principal from the final amount.

$$\$1338.47 - \$975 = \$363.47$$

17. To find the final amount add the principal to the compound interest.

$$\$480 + \$337.17 = \$817.17.$$

19. To find the final amount, use the formula for future value $A = P\left(1 + \frac{r}{m}\right)^n$, with $P = \$7500$, $r = .055$, $m = 1$ and $n = 1 \cdot 25 = 25$.

$$A = P\left(1 + \frac{r}{m}\right)^n$$
$$= \$7500\left(1 + \frac{.055}{1}\right)^{25}$$
$$= \$7500(1.055)^{25}$$
$$= \$7500(3.81339)$$
$$= \$28,600.44$$

Then to find the compound interest subtract the principal from this final amount.

$$\$28600.44 - \$7500 = \$21,100.44$$

21. To find the final amount, use the formula for future value $A = P\left(1 + \frac{r}{m}\right)^n$.

(a) $A = P\left(1 + \frac{r}{m}\right)^n$
$$= \$1000\left(1 + \frac{.10}{1}\right)^3$$
$$= \$1000(1.10)^3$$
$$= \$1000(1.331)$$
$$= \$1331$$

(b) $A = P\left(1 + \frac{r}{m}\right)^n$
$$= \$1000\left(1 + \frac{.10}{2}\right)^{2\cdot3}$$
$$= \$1000(1.05)^6$$
$$= \$1000(1.340095641)$$
$$= \$1340.10$$

(c) $A = P\left(1 + \frac{r}{m}\right)^n$
$$= \$1000\left(1 + \frac{.10}{4}\right)^{4\cdot3}$$
$$= \$1000(1.025)^{12}$$
$$= \$1000(1.34488824)$$
$$= \$1344.89$$

23. To find the final amount, use the formula for future value $A = P\left(1 + \frac{r}{m}\right)^n$.

(a) $A = P\left(1 + \frac{r}{m}\right)^n$
$$= \$12000\left(1 + \frac{.08}{1}\right)^5$$
$$= \$12000(1.08)^5$$
$$= \$12000(1.469328)$$
$$= \$17,631.94$$

(b) $A = P\left(1 + \frac{r}{m}\right)^n$
$$= \$12000\left(1 + \frac{.08}{2}\right)^{2\cdot5}$$
$$= \$12000(1.04)^{10}$$
$$= \$12000(1.480244)$$
$$= \$17,762.93$$

(c) $A = P\left(1 + \frac{r}{m}\right)^n$
$$= \$12000\left(1 + \frac{.08}{4}\right)^{4\cdot5}$$
$$= \$12000(1.02)^{20}$$
$$= \$12000(1.485947)$$
$$= \$17,831.37$$

25. (a) First find the future value by using the formula $A = P\left(1 + \frac{r}{m}\right)^n$ with $P = \$1040$, $r = .076$, $m = 2$, and $n = 2 \cdot 4 = 8$.

$$A = P\left(1 + \frac{r}{m}\right)^n$$
$$= \$1040\left(1 + \frac{.076}{2}\right)^8$$
$$= \$1040(1.038)^8$$
$$= \$1040(1.347655)$$
$$= \$1401.56$$

Then subtract the principal from this amount.

$$\$1401.56 - \$1040 = \$361.56$$

(b) Find the future value, changing to $m = 4$, and $n = 4 \cdot 4 = 16$.

$$A = P\left(1 + \frac{r}{m}\right)^n$$
$$= \$1040\left(1 + \frac{.076}{4}\right)^{16}$$
$$= \$1040(1.019)^{16}$$
$$= \$1040(1.351409)$$
$$= \$1405.47$$

Then subtract the principal from this amount

$$\$1405.47 - \$1040 = \$365.47.$$

(c) Find the future, changing to $m = 12$, and $n = 12 \cdot 4 = 48$.

$$A = P\left(1 + \frac{r}{m}\right)^n$$
$$= \$1040\left(1 + \frac{.076}{12}\right)^{48}$$
$$= \$1040(1.006\overline{3})^{48}$$
$$= \$1040(1.35397)$$
$$= \$1408.13$$

Then subtract the principal from this amount.

$$\$1408.13 - \$1040 = \$368.13$$

(d) Find the future, changing to $m = 365$, and $n = 365 \cdot 4 = 1460$.

$$A = P\left(1 + \frac{r}{m}\right)^n$$
$$= \$1040\left(1 + \frac{.076}{365}\right)^{1460}$$
$$= \$1040(1.000208219)^{1460}$$
$$= \$1040(1.355226)$$
$$= \$1409.44$$

Then subtract the principal from this amount.

$$\$1409.44 - \$1040 = \$369.44$$

(e) Use the formula for continuous growth $A = Pe^{rt}$ with $P = \$1040$; $r = .076$; and $t = 4$

$$A = Pe^{rt}$$
$$= \$1040 \cdot e^{.076 \cdot 4}$$
$$= \$1040 \cdot e^{.304}$$
$$= \$1040(1.3551269056)$$
$$= \$1409.48$$

Then subtract the principal from this amount.

$$\$1409.48 - \$1040 = \$369.48$$

27. Writing exercise

29. Use the present value formula for simple interest with $A = \$1500$, $r = .10$, and $t = 4/12 = .\overline{3}$.

$$P = \frac{A}{1 + rt}$$
$$= \frac{\$1500}{1 + (.10)(.\overline{3})}$$
$$= \$1451.61$$

31. Use the present value formula for compound interest with $A = \$1000$, $r = .06$, $m = 1$, and $n = 1 \cdot 5$.

$$P = \frac{A}{\left(1 + \frac{r}{m}\right)^n}$$
$$= \frac{\$1000}{\left(1 + \frac{.06}{1}\right)^5}$$
$$= \frac{\$1000}{(1.06)^5}$$
$$= \frac{\$1000}{1.3382256}$$
$$= \$747.26$$

33. Use the present value formula for compound interest with $A = \$9860$, $r = .08$, $m = 2$, and $n = 2 \cdot 10 = 20$.

$$P = \frac{A}{\left(1 + \frac{r}{m}\right)^n}$$
$$= \frac{\$9860}{\left(1 + \frac{.08}{2}\right)^{20}}$$
$$= \frac{\$9860}{(1.04)^{20}}$$
$$= \frac{\$9860}{2.1911231}$$
$$= \$4499.98$$

35. Use the present value formula for compound interest with $A = \$500,000$, $r = .08$, $m = 4$, and $n = 4 \cdot 30 = 120$.

$$P = \frac{A}{\left(1 + \frac{r}{m}\right)^n}$$
$$= \frac{\$500000}{\left(1 + \frac{.08}{4}\right)^{120}}$$
$$= \frac{\$500000}{(1.02)^{120}}$$
$$= \frac{\$500000}{10.76516303}$$
$$= \$46,446.11$$

37. Use the present value formula for compound interest with $A = \$500,000$, $r = .08$, $m = 365$, and $n = 365 \cdot 30 = 10950$.

$$P = \frac{A}{\left(1 + \frac{r}{m}\right)^n}$$
$$= \frac{500000}{\left(1 + \frac{.08}{365}\right)^{10950}}$$
$$= \$45,370.91$$

39. Use the formula $Y = \left(1 + \frac{r}{m}\right)^m - 1$, with $r = .05$ and $m = 1$.

$$Y = \left(1 + \frac{.05}{1}\right)^1 - 1$$
$$= (1.05) - 1$$
$$= .05, \text{ which is } 5\%$$

41. Use the formula $Y = \left(1 + \frac{r}{m}\right)^m - 1$, with $r = .05$ and $m = 4$.

$$Y = \left(1 + \frac{.05}{4}\right)^4 - 1$$
$$= (1.0125)^4 - 1$$
$$= 1.0509453 - 1$$
$$= .05095, \text{ which is } 5.095\%$$

43. Use the formula $Y = \left(1 + \frac{r}{m}\right)^m - 1$, with $r = .05$ and $m = 365$.

$$Y = \left(1 + \frac{.05}{365}\right)^{365} - 1$$
$$= (1.000136986)^{365} - 1$$
$$= 1.05127 - 1$$
$$= .05127, \text{ which is } 5.127\%$$

45. Use the formula $Y = \left(1 + \frac{r}{m}\right)^m - 1$, with $r = .05$ and $m = 10,000$.

$$Y = \left(1 + \frac{.05}{10000}\right)^{10000} - 1$$
$$= (1.000005)^{10000} - 1$$
$$= 1.05127 - 1$$
$$= .05127, \text{ which is } 5.127\%$$

47. Because the deposit is more than $1000, but less than $2500, use the yield of 5.56% as the rate in the formula $A = P + I$.

$$A = P + I$$
$$= \$2000 + (.0556)(\$2000)$$
$$= \$2111.20$$

49. Use the formula $A = P\left(1 + \frac{r}{m}\right)^n$ with A having the value $2P$ because we want the amount to double the principle.

$$2P = P\left(1 + \frac{.08}{4}\right)^{4n}$$
$$2 = (1 + .02)^{4n}$$
$$\ln 2 = 4n \cdot \ln(1.02)$$
$$\frac{\ln 2}{\ln(1.02)} = 4n$$
$$\frac{35.00278878}{4} = n$$
$$8.75 = n$$

This is 8 years and .75 of a year: $.75(365) = 274$. The answer then is about 8 years and 274 days.

51. Use the formula $Y = \left(1 + \frac{r}{m}\right)^m - 1$ where $r =$ nominal interest rate; $m =$ number of times interest is compounded per year; $Y =$ effective annual yield. Substitute values and solve for r.

$$Y = \left(1 + \frac{r}{m}\right)^m - 1$$
$$.0615 = \left(1 + \frac{r}{12}\right)^{12} - 1$$
$$1.0615 = \left(1 + \frac{r}{12}\right)^{12}$$
$$\ln(1.0615) = 12 \cdot \ln\left(1 + \frac{r}{12}\right)$$
$$\frac{\ln(1.0615)}{12} = \ln\left(1 + \frac{r}{12}\right)$$
$$.00497358 \approx \ln\left(1 + \frac{r}{12}\right)$$
$$e^{.00497358} \approx 1 + \frac{r}{12}$$
$$1.004985972 \approx 1 + \frac{r}{12}$$
$$.004985972 \approx \frac{r}{12}$$
$$.0598 \approx r$$

The nominal rate is 5.98%.

53.
$$Y = \left(1 + \frac{r}{m}\right)^m - 1$$
$$Y + 1 = \left(1 + \frac{r}{m}\right)^m$$
$$(Y+1)^{\frac{1}{m}} = \left[\left(1 + \frac{r}{m}\right)^m\right]^{\frac{1}{m}}$$
$$(Y+1)^{\frac{1}{m}} = \left(1 + \frac{r}{m}\right)$$
$$(Y+1)^{\frac{1}{m}} - 1 = \frac{r}{m}$$
$$m\left[(Y+1)^{\frac{1}{m}} - 1\right] = r$$

55. Years to double $= \dfrac{70}{\text{Annual inflation rate}}$

Years to double $= \dfrac{70}{2}$

Years to double $= 35$ years

57. Years to double $= \dfrac{70}{\text{Annual inflation rate}}$

Years to double $= \dfrac{70}{9}$

Years to double $= 7.8$ or about 8 years

59. Let r = the inflation rate as a percent.

Years to double $= \dfrac{70}{\text{Annual inflation rate}}$

$7 = \dfrac{70}{r}$

$7r = 70$

$r = 10\%$

61. Let r = the inflation rate as a percent.

Years to double $= \dfrac{70}{\text{Annual inflation rate}}$

$22 = \dfrac{70}{r}$

$22r = 70$

$r \approx 3.2\%$

63. Writing exercise

Item	2005 2%	2020 2%	2005 10%	2020 10%
65. Meal	$4.85	$6.55	$7.24	$32.44
67. Car	$13,815	$18,648	$20,609	$92,363

Table for Exercises 65 and 67

65. Use the formula for continuous compounding $A = Pe^{rt}$, changing r and t as appropriate.

$$A = \$4.39e^{.02(5)}$$
$$A = \$4.85$$

$$A = \$4.39e^{.02(20)}$$
$$A = \$6.55$$

$$A = \$4.39e^{.10(5)}$$
$$A = \$7.24$$

$$A = \$4.39e^{.10(20)}$$
$$A = \$32.44$$

67. Use the formula for continuous compounding $A = Pe^{rt}$, changing r and t as appropriate.

$$A = \$12500e^{.02(5)}$$
$$A = \$13,815$$

$$A = \$12500e^{.02(20)}$$
$$A = \$18,648$$

$$A = \$12500e^{.10(5)}$$
$$A = \$20,609$$
$$A = \$12500e^{.10(20)}$$
$$A = \$92,363$$

	Item	Price	Date	Price in Dec. 1999
69.	Roses	$78	Dec. 96	$83
71.	Motorcycle	$5299	Dec. 94	$5950

Table for Exercises 69 and 71

69. Find the percent change in Table 1 for each of the three consecutive years from 1996 to 1999. These are 2.3, 1.6, and 2.2; as decimals they are .023, .016, and .022. Using the compounding interest concept, the problem can be done efficiently by multiplying the beginning amount, $78, by 1 plus the percent increase for each year:

$$78 \times 1.023 \times 1.016 \times 1.022 \approx \$83.$$

71. Find the percent change in Table 1 for each of the five consecutive years from 1994 to 1999. These are 2.8, 3.0, 2.3, 1.6, and 2.2; as decimals they are .028, .030, .023, .016, and .022. Using the compounding interest concept, the problem can be done efficiently by multiplying the beginning amount, $5299, by 1 plus the percent increase for each year:

$$\$5299 \times 1.028 \times 1.03 \times 1.023 \times 1.016 \times 1.022 \approx \$5960.$$

73. Use the formula $P = \frac{A}{(1+\frac{r}{m})^n}$ with
$A = 300 \cdot \$300,000 = \$90,000,000;\ r = .07;\ m = 4;$
$n = 4 \cdot \frac{18}{12} = 6.$

$$P = \frac{A}{\left(1 + \frac{r}{m}\right)^n}$$
$$P = \frac{\$90000000}{\left(1 + \frac{.07}{4}\right)^6}$$
$$= \$81,102,828.75$$

EXTENSION: ANNUITIES

In Exercises 1–6, for part (a) use the future value formula, replacing R, r, m, and n with the appropriate values.

1. (a)
$$V = \frac{R\left[\left(1 + \frac{r}{m}\right)^n - 1\right]}{\frac{r}{m}}$$
$$= \frac{\$1000\left[\left(1 + \frac{.085}{1}\right)^{10} - 1\right]}{\frac{.085}{1}}$$
$$= \$14,835.10$$

(b) To find the total of all deposits multiply the amount of the regular deposit by the number of years in the Accumulation Period.

$$10(\$1000) = \$10,000.$$

(c) To find the total interest earned, subtract the total deposits from the total accumulation.

$$\$14835.10 - \$10000 = \$4835.10.$$

3. (a)
$$V = \frac{R\left[\left(1 + \frac{r}{m}\right)^n - 1\right]}{\frac{r}{m}}$$
$$= \frac{\$50\left[\left(1 + \frac{.06}{12}\right)^{5 \cdot 12} - 1\right]}{\frac{.06}{12}}$$
$$= \$3488.50$$

(b) To find the total of all deposits multiply the amount of the regular deposit by the total number of deposits, $5 \cdot 12 = 60.$

$$60(\$50) = \$3000.$$

(c) To find the total interest earned, subtract the total deposits from the total accumulation

$$\$3488.50 - \$3000 = \$488.50.$$

5. (a)
$$V = \frac{R\left[\left(1 + \frac{r}{m}\right)^n - 1\right]}{\frac{r}{m}}$$
$$= \frac{\$20\left[\left(1 + \frac{.052}{52}\right)^{3 \cdot 52} - 1\right]}{\frac{.052}{52}}$$
$$= \$3374.70$$

(b) To find the total of all deposits multiply the amount of the regular deposit by the total number of deposits, $3 \cdot 52 = 156.$

$$156(\$20) = \$3120$$

(c) To find the total interest earned, subtract the total deposits from the total accumulation.

$$3374.70 - 3120 = \$254.70$$

7. Writing exercise

For Exercises 9–12, use the formula

$$V = \frac{R\left[\left(1 + \frac{r}{m}\right)\right]^{n+1} - 1]}{\frac{r}{m}}$$

9. $$V = \frac{R\left[\left(1 + \frac{r}{m}\right)^{n+1} - 1\right]}{\frac{r}{m}}$$
$$= \frac{\$50\left[\left(1 + \frac{.06}{12}\right)^{5 \cdot 12 + 1} - 1\right]}{\frac{.06}{12}}$$
$$= \$3555.94$$

11. $$V = \frac{R\left[\left(1 + \frac{r}{m}\right)^{n+1} - 1\right]}{\frac{r}{m}}$$
$$= \frac{\$20\left[\left(1 + \frac{.052}{52}\right)^{3 \cdot 52 + 1} - 1\right]}{\frac{.052}{52}}$$
$$= \$3398.08$$

13. Writing exercise

15. Writing exercise

17. Writing exercise

13.2 EXERCISES

1. Subtract the down payment amount from $2150.

$$\$2150 - \$500 = \$1650$$

3. Referring to Exercise 2, the payment period is 2 years. The total interest is $(\$1650)(.12)(2) = \$396$. Find the sum of the total amount financed and the total interest owed.

$$\$1650 + \$396 = \$2046$$

5. Subtract the value of the old car from the cost of the new car.

$$\$16,500 - \$3000 = \$13,500$$

7. The total interest from Exercise 6 is $(13,500)(.09)(3) = \$3645$. Then the total amount owed is the sum of the principal and the interest.

$$\$13,500 + \$3645 = \$17,145$$

9. The interest charge is:

$$I = Prt$$
$$= (\$4500)(.09)(3)$$
$$= \$1215.$$

The total amount owed is:

$$P + I = \$4500 + \$1215$$
$$= \$5715.$$

There are $3 \cdot 12 = 36$ monthly payments, so the amount of each monthly payment is:

$$\frac{\$5715}{36} = \$158.75.$$

11. The interest charge is:

$$I = Prt$$
$$= (\$750)(.114)\left(\frac{18}{12}\right)$$
$$= \$128.25.$$

The total amount owed is:

$$P + I = \$750 + \$128.25$$
$$= \$878.25.$$

There are 18 monthly payments, so the amount of each monthly payment is:

$$\frac{\$878.25}{18} = \$48.79.$$

13. The interest charge is:

$$I = Prt$$
$$= (\$1580)(.13)\left(\frac{10}{12}\right)$$
$$= \$171.17.$$

The total amount owed is:

$$P + I = \$1580 + \$171.17$$
$$= \$1751.17.$$

There are 10 monthly payments, so the amount of each monthly payment is

$$\frac{\$1751.17}{10} = \$175.12.$$

15. The interest charge is:

$$I = Prt$$
$$= (\$535)(.111)\left(\frac{16}{12}\right)$$
$$= \$79.18.$$

The total amount owed is:

$$P + I = \$535 + \$79.18$$
$$= \$614.18.$$

There are 16 monthly payments, so the amount of each monthly payment is:

$$\frac{\$614.18}{16} = \$38.39.$$

17. First subtract the down payment from the purchase price of the furniture to find the value of P.

$$\$8500 - \$3000 = \$5500.$$

Now use the formula for simple interest with $P = \$5500$, $r = .10$, and $t = 2.5$.

$$I = Prt$$
$$= (\$5500)(.10)(2.5)$$
$$= \$1375.$$

The total amount owed is:

$$P + I = \$5500 + \$1375$$
$$= \$6875.$$

There are $2.5(12) = 30$ monthly payments, so the amount of each monthly payment is:

$$\frac{\$6875}{30} = \$229.17.$$

19. First subtract the down payment from the purchase price of the furniture to find the value of P.

$$\$14240 - \$2900 = \$11340.$$

Now use the formula for simple interest with $P = \$11340$, $r = .10$, and $t = \frac{48}{12} = 4$.

$$I = Prt$$
$$= (\$11340)(.10)(4)$$
$$= \$4536$$

The total amount owed is:

$$P + I = \$11340 + \$4536$$
$$= \$15876.$$

There are 48 monthly payments, so the amount of each monthly payment is:

$$\frac{\$15876}{48} = \$330.75.$$

21. First find the total amount to be paid.

$$(48)(314.65) = \$15,103.20$$

Then the rate is .098 and the time is 4 years to correspond to the 48 months.
Let $x =$ amount borrowed.

Simple interest + Amount borrowed = \$15103.20
$$(.098)(4)x + x = \$15103.20$$
$$.392x + x = \$15103.20$$
$$1.392x = \$15103.20$$
$$x = \$10,850$$

23. The total number of payments will be $12 \cdot t$, where $t =$ the time in years. The amount to be paid per year is equivalent to multiplying the monthly payment by 12. This yields:

$$(\$172.44) \cdot 12 = \$2069.28.$$

Substitute $P = \$8000$, $r = .092$, into the formula $P + Prt = A$, and solve for t.

$$\$8000 + (\$8000)(.092)t = \$2069.28t$$
$$\$8000 + \$736t = \$2069.28t$$
$$\$8000 + \$736t - \$736t = \$2069.28t - \$736t$$
$$\$8000 = \$1333.28t$$
$$\frac{\$8000}{\$1333.28} = \frac{\$1333.28t}{\$1333.28}$$
$$6 \text{ years} = t$$

25. $(\$325.50)(.018) = \$5.86$

27. $(\$242.88)(.0168) = \$4.08$

29.

Month	Unpaid Balance at Beginning of Month	Finance Charge	Purchases During Month	Returns	Payments	Unpaid Balance at End of Month
February	$319.10	$4.79	$86.14	0	$50	$360.03
March	360.03	5.40	109.83	$15.75	60	399.51
April	399.51	5.99	39.74	0	72	373.24
May	373.24	5.60	56.29	18.09	50	367.04

Here are explanations of how each underlined value was found:

1. Finance charge = Unpaid balance at beginning of month $\cdot$ Interest rate as decimal.
2. Unpaid balance at end of month = Unpaid balance at beginning of month + Finance charge + Purchases − Returns − Payments.
3. Unpaid balance at beginning of month = Unpaid balance at end of previous month.

February

Unpaid balance at beginning of month: $319.10.
Finance charge: ($319.10)(.015) = $4.79.
Unpaid balance at end of month: $319.10 + 4.79 + 86.14 − 50 = $360.03.

March

Unpaid balance at beginning of month: $360.03.
Finance charge: ($360.03)(.015) = $5.40.
Unpaid balance at end of month: $360.03 + 5.40 + 109.83 − 15.75 − 60 = $399.51

April

Unpaid balance at beginning of month: $399.51.
Finance charge: ($399.51)(.015) = $5.99.
Unpaid balance at end of month: $399.51 + 5.99 + 39.74 − 72 = $373.24.

May

Unpaid balance at beginning of month: $373.24.
Finance charge: ($373.24)(.015) = $5.60.
Unpaid balance at end of month: $373.24 + 5.60 + 56.29 − 18.09 − 50 = $367.04.

31.

Month	Unpaid Balance at Beginning of Month	Finance Charge	Purchases During Month	Returns	Payment	Unpaid Balance at End of Month
August	$684.17	$10.26	$155.01	$38.11	$100	$711.33
September	711.33	10.67	208.75	0	75	855.75
October	855.75	12.84	56.30	0	90	834.89
November	834.89	12.52	190.00	83.57	150	803.84

Here are explanations of how each underlined value was found:

1. Finance charge = Unpaid balance at beginning of month · Interest rate as decimal.
2. Unpaid balance at end of month = Unpaid balance at beginning of month + Finance charge + Purchases − Returns − Payments.
3. Unpaid balance at beginning of month = Unpaid balance at end of previous month.

August

Unpaid balance at beginning of month: $684.17.
Finance charge: ($684.17)(.015) = $10.26.
Unpaid balance at end of month: $684.17 + 10.26 + 155.01 − 38.11 − 100 = $711.33.

September

Unpaid balance at beginning of month: $711.33.
Finance charge: ($711.33)(.015) = $10.67.
Unpaid balance at end of month: $711.33 + 10.67 + 208.75 − 75 = $855.75

October

Unpaid balance at beginning of month: $855.75.
Finance charge: $(\$855.75)(.015) = \$12.84$.
Unpaid balance at end of month: $\$855.75 + 12.84 + 56.30 - 90 = \$834.89$.

November

Unpaid balance at beginning of month: $834.89.
Finance charge: $(\$834.89)(.015) = \$12.52$.
Unpaid balance at end of month: $\$834.89 + 12.52 + 190.00 - 83.57 - 150 = \$803.84$.

33. $(\$249.94)(.015) = \$3.75$

35. $(\$1073.40)(.01613) = \$17.31$

37. First, make a table showing all the transactions, and compute the balance due on each date.

Date	Balance due
May 9	$728.36
May 17	$728.36 - \$200 = \$528.36$
May 30	$528.36 + \$46.11 = \$574.47$
June 3	$574.47 + \$64.50 = \$638.97$

Next, tabulate the balance due figures, along with the number of days until the balance changed. Use this data to calculate the sum of the daily balances.

Date	Balance due	Number of days until balance changed	$\left(\begin{array}{c}\text{Balance}\\\text{due}\end{array}\right) \times \left(\begin{array}{c}\text{Number}\\\text{of days}\end{array}\right)$
May 9	$728.36	8	$728.36 \times 8 = \$5826.88$
May 17	$528.36	13	$528.36 \times 13 = \$6868.68$
May 30	$574.47	4	$574.47 \times 4 = \$2297.88$
June 3	$638.97	6	$638.97 \times 6 = \$3833.82$
Totals		31	$18,827.26

$$\text{Average daily balance} = \frac{\text{Sum of daily balances}}{\text{Days in billing period}} = \frac{\$18827.26}{31} = \$607.33$$

and

$$\text{Finance charge} = (.015)(\$607.33) = \$9.11.$$

39. First, make a table showing all the transactions, and compute the balance due on each date.

Date	Balance due
June 11	$462.42
June 15	$462.42 − $106.45 = $355.97
June 20	$355.97 + $115.73 = $471.70
June 24	$471.70 + $74.19 = $545.89
July 3	$545.89 − $115.00 = $430.89
July 6	$430.89 + $68.49 = 499.38

Next, tabulate the balance due figures, along with the number of days until the balance changed. Use this data to calculate the sum of the daily balances.

Date	Balance due	Number of days until balance changed	(Balance due) × (Number of days)
June 11	$462.42	4	$462.42 × 4 = $1849.68
June 15	$355.97	5	$355.97 × 5 = $1779.85
June 20	$471.70	4	$471.70 × 4 = $1886.80
June 24	$545.89	9	$545.89 × 9 = $4913.01
July 3	$430.89	3	$430.89 × 3 = $1292.67
July 6	$499.38	5	$499.38 × 5 = $2496.9
Totals		30	$14,218.91

$$\text{Average daily balance} = \frac{\text{Sum of daily balances}}{\text{Days in billing period}} = \frac{\$14,218.91}{30} = \$473.96$$

and

$$\text{Finance charge} = (.015)(\$473.96) = \$7.11.$$

41. (a) Using the unpaid balance method to find the finance charge, multiply the previous unpaid balance by the interest rate.

$$(.015)(\$720) = \$10.80$$

(b) Using the average daily balance method, she has a balance of $720 for the first 28 days of the billing period. For the last 3 days of the billing period the balance drops to $120 because of her $600 payment. Then,

$$\text{Average daily balance} = \frac{\$720 \cdot 28 + \$120 \cdot 3}{31} = \$661.93$$

and her finance charge is

$$(.015)(\$661.93) = \$9.93.$$

43. (a) ($2400)(.018583) = $44.60

(b) At the end of the second month, the interest is first added to the balance before being multiplied by the interest rate.

$$(\$2400 + \$44.60)(.018583) = \$45.43.$$

(c) At the end of the third month, both interest amounts are added to the balance before being multiplied by the interest rate.

$$(\$2400 + \$44.60 + \$45.53)(.018583) = \$46.27.$$

45. From Exercises 43 and 44, the total interest is $136.30. Use the formula $I = Prt$, with $I = \$136.30$, $P = \$2400$, and $t = 1/4$ or .25.

$$\$136.30 = (\$2400)(r)(.25)$$
$$136.30 = 600r$$
$$.227 = r, \text{ which is } 22.7\%.$$

47. There is a 2% charge for each cash advance of $100. Because she had six cash advances of $100 each, the total charge is

$$6 \times .02(100) = \$12.$$

Add this amount to the late payment fee of $15 and the over-the-credit-limit fee of $15.

$$\$12 + \$15 + \$15 = \$42.$$

49. Writing exercise

51. Writing exercise

53. Writing exercise

55. (a) If they choose Bank A, their estimated total yearly cost will be

$$(\$900)(.016)(12) = \$172.80.$$

(b) If they choose Bank A, their estimated total yearly cost will be

$$\$40 + (\$900)(.013)(12) = \$180.40.$$

57. Let x = the average monthly unpaid balance (in dollars).

The annual charges for Bank A are

$$(.016)(12)(x) = .192x,$$

while the annual charges for Bank B are

$$40 + (.013)(12)(x) = 40 + .156x.$$

The charges for Bank B will be less than those for Bank A if

$$40 + .156x < .192x$$
$$40 < .036x$$
$$\frac{40}{.036} = x$$
$$\$1111.11 = x.$$

Since $x = \$1111.11$, the least average monthly unpaid balance that makes Bank B a better deal is $1111.12.

13.3 EXERCISES

1. Find the finance charge per $100 of the amount financed.

$$\frac{75}{1000} \times 100 = \$7.50$$

In Table 3, find the "12 payments" row and read across to find the number closest to 7.50, which is 7.46. Read up to find the APR, which is 13.5%.

3. Find the finance charge per $100 of the amount financed.

$$\frac{1070}{6600} \times 100 = \$16.21$$

In Table 3, find the "30 payments" row and read across to find the number closest to 16.21, which is 16.24. Read up to find the APR, which is 12.0%.

5. First find the amount financed by subtracting the down payment from the purchase price.

$$\$3000 - \$500 = \$2500$$

Now find the total payments by adding the amount financed to the finance charge.

$$\$2500 + \$150 = \$2650$$

The monthly payment is

$$\frac{\$2650}{24} = \$110.42.$$

7. First find the amount financed by subtracting the down payment from the purchase price.

$$\$3950 - \$300 = \$3650$$

Now find the total payments by adding the amount financed to the finance charge.

$$\$3650 + \$800 = \$4450$$

The monthly payment is

$$\frac{\$4450}{48} = \$92.71.$$

9. First find the amount financed by subtracting the down payment from the purchase price.

$$\$4190 - \$390 = \$3800$$

Now find the finance charge. The interest rate of 6% will be charged on the amount financed for 1 year (12 payments).

$$I = Prt$$
$$= (3800)(.06)(1)$$
$$= \$228$$

Next find the finance charge per $100 of the amount financed.

$$\frac{228}{3800} \times 100 = \$6.00$$

In Table 3, the number closest to 6.00 in the "12 payments" row is 6.06. Read up to find the APR, which is 11.0%.

11. First find the amount financed by subtracting the down payment from the purchase price.

$$\$7480 - \$2200 = \$5280$$

Now find the finance charge. The interest rate of 8% will be charged on the amount financed for 1.5 years (18 payments).

$$\begin{aligned} I &= Prt \\ &= (5280)(.08)(1.5) \\ &= \$633.60 \end{aligned}$$

Next find the finance charge per $100 of the amount financed.

$$\frac{633.60}{5280} \times 100 = \$12.00$$

In Table 3, the number closest to 12.00 in the "18 payments" row is 11.87. Read up to find the APR, which is 14.5%.

13. (a) The amount of the total payments is

$$(24)(\$94.50) = \$2268.$$

The finance charge is the difference between this amount and the purchase price.

$$\$2268 - \$1990 = \$278$$

(b) Find the finance charge per $100 of the amount financed.

$$\frac{278}{1990} \times 100 = \$13.97$$

In Table 3, the number closest to 13.97 in the "24 payments" row is 14.10. Read up to find the APR, which is 13.0%.

15. (a) To find the finance charge use the simple interest formula. The interest rate of 6% will be charged on the amount financed for 1.5 years.

$$\begin{aligned} I &= Prt \\ &= (\$2000)(.06)(1.5) \\ &= \$180 \end{aligned}$$

(b) Find the finance charge per $100 of the amount financed.

$$\frac{180}{2000} \times 100 = \$9.00$$

In Table 3, the number closest to 9.00 in the "18 payments" row is 8.93. Read up to find the APR, which is 11.0%.

17. (a) Find the intersection of the row for 18 payments and the column for 11.0% to find the value of h: $8.93.

(b) Use the formula $U = kR\left(\frac{h}{100+h}\right)$ to find unearned interest, with $k = $ remaining number of payments, $R = $ regular monthly payment, and $h = $ finance charge per $100.

$$\begin{aligned} U &= (18)(346.70)\left(\frac{8.93}{100+8.93}\right) \\ &= (18)(346.70)\left(\frac{8.93}{108.93}\right) \\ &= \$511.60 \end{aligned}$$

(c) The payoff amount is equal to the current payment plus the sum of the scheduled remaining payments minus the unearned interest.

$$\$346.70 + (18)(\$346.70) - 511.60 = \$6075.70$$

19. (a) Find the intersection of the row for 6 payments and the column for 13.5% to find the value of h: $3.97.

(b) Use the formula $U = kR\left(\frac{h}{100+h}\right)$ to find unearned interest, with $k = $ remaining number of payments, $R = $ regular monthly payment, and $h = $ finance charge per $100.

$$\begin{aligned} U &= (6)(\$595.80)\left(\frac{\$3.97}{\$100+\$3.97}\right) \\ &= (6)(\$595.80)\left(\frac{\$3.97}{\$103.97}\right) \\ &= \$136.50 \end{aligned}$$

(c) The payoff amount is equal to the current payment plus the sum of the scheduled remaining payments minus the unearned interest.

$$\$595.80 + (6)(\$595.80) - \$136.50 = \$4034.10.$$

21. (a) Using the actuarial method, first find the APR. If the loan were not paid off early, the total payments would be

$$(18)(\$201.85) = \$3633.30,$$

and the finance charge would be

$$\$3633.30 - \$3310 = \$323.30.$$

The finance charge per $100 of the amount financed would be

$$\frac{\$323.30}{\$3310} \times 100 = \$9.77.$$

In Table 3, the number 9.77 appears in the "18 payments" row and the 12.0% column, so the APR is 12%. Because the total number of payments remaining is 6, move up to the "6 payments" row to find $h = \$3.53$, the value of h needed in the actuarial method.

Use the formula $U = kR\left(\dfrac{h}{100+h}\right)$ to find unearned interest, with $k =$ remaining number of payments, $R =$ regular monthly payment, and $h =$ finance charge per \$100.

$$U = (6)(\$201.85)\left(\frac{\$3.53}{\$100 + \$3.53}\right)$$
$$= (6)(\$201.85)\left(\frac{\$3.53}{\$103.53}\right)$$
$$= \$41.29$$

(b) From part (a), the original finance charge is \$323.30. Use the rule of 78 to find unearned interest, with $k = 6$, $n = 18$, and $F = \$323.30$.

$$U = \frac{k(k+1)}{n(n+1)} \times F$$
$$= \frac{6(6+1)}{18(18+1)} \times \$323.30$$
$$= \$39.70$$

23. (a) Using the actuarial method, first find the APR. If the loan were not paid off early, the total payments would be

$$(60)(\$641.58) = \$38,494.80,$$

and the finance charge would be

$$\$38494.80 - \$29850 = \$8644.80.$$

The finance charge per \$100 of the amount financed would be

$$\frac{\$8644.80}{\$29850} \times 100 = \$28.96.$$

In Table 3, the number 28.96 appears in the "60 payments" row and the 10.5% column, so the APR is 10.5%. Because the total number of payments remaining is 12, move up to the "12 payments" row to find $h = \$5.78$, the value of h needed in the actuarial method.

Use the formula $U = kR\left(\dfrac{h}{100+h}\right)$ to find unearned interest, with $k =$ remaining number of payments, $R =$ regular monthly payment, and $h =$ finance charge per \$100.

$$U = (12)(\$641.58)\left(\frac{\$5.78}{\$100 + \$5.78}\right)$$
$$= (12)(\$641.58)\left(\frac{\$5.78}{\$105.78}\right)$$
$$= \$420.68$$

(b) From part (a), the original finance charge is \$8644.80. Use the rule of 78 to find unearned interest, with $k = 12$, $n = 60$, and $F = \$8644.80$.

$$U = \frac{k(k+1)}{n(n+1)} \times F$$
$$= \frac{12(12+1)}{60(60+1)} \times \$8644.80$$
$$= \$368.47$$

25. (a) Use the finance charge formula with $n = 4$ and APR $= .086$.

$$h = \frac{n \times \frac{APR}{12} \times \$100}{1 - \left(1 + \frac{APR}{12}\right)^{-n}} - \$100$$
$$= \frac{4 \times \frac{.086}{12} \times \$100}{1 - \left(1 + \frac{.086}{12}\right)^{-4}} - \$100$$
$$= \$1.80$$

(b) Use the actuarial formula for unearned interest with $k = 4$, $R = \$212$, and $h = \$1.80$.

$$U = kR\left(\frac{h}{\$100 + h}\right)$$
$$= 4(\$212)\left(\frac{\$1.80}{\$100 + \$1.80}\right)$$
$$= \$14.99$$

(c) The payoff amount is equal to the current payment added to the sum of the remaining payments minus the unearned interest.

$$\$212 + 4(\$212) - \$14.99 = \$1045.01$$

27. *Finance Company*
Find the sum of the amount borrowed and the interest that will be charged.

$$\$5000 + (\$5000)(.065)(3) = \$5975$$

Then, the finance charge is

$$\$5975 - \$5000 = \$975.$$

The finance charge per \$100 of the amount financed would be

$$\frac{\$975}{\$5000} \times 100 = \$19.50.$$

In Table 3, the number closest to 19.50 in the "36 payments" row is 19.57. Read up to find the APR, which is 12.0%.

Credit Union
Find the total amount she would pay.

$$36(\$164.50) = \$5922.00$$

Then, the finance charge is

$$\$5922 - \$5000 = \$922.$$

The finance charge per $100 of the amount financed would be

$$\frac{\$922}{\$5000} \times 100 = \$18.44.$$

In Table 3, the number closest to 18.44 in the "36 payments" row is 18.71. Read up to find the APR, which is 11.5%.

The credit union offers the better choice.

29.   Use the finance charge formula with $n = 6$ and $APR = .115$. (See Exercise 27.)

$$h = \frac{n \times \frac{APR}{12} \times \$100}{1 - \left(1 + \frac{APR}{12}\right)^{-n}} - \$100$$

$$= \frac{6 \times \frac{.115}{12} \times \$100}{1 - \left(1 + \frac{.115}{12}\right)^{-6}} - \$100$$

$$= \$3.38$$

Then, use the actuarial formula for unearned interest with $k = 6$, $R = \$164.50$, and $h = \$3.38$.

$$U = kR\left(\frac{h}{\$100 + h}\right)$$

$$= 6(\$164.50)\left(\frac{\$3.38}{\$100 + \$3.38}\right)$$

$$= \$32.27$$

31.   Writing exercise

33.   Writing exercise

35.   Use the rule of 78 to set up an inequality with $n = 36$.

$$\frac{k(k+1)}{n(n+1)} \times F \geq .10F$$

$$\frac{k(k+1)}{n(n+1)} \geq .10$$

$$k(k+1) \geq .10 \times n(n+1)$$

$$k(k+1) \geq .10 \times 36(36+1)$$

$$k^2 + k \geq 133.2$$

This could be solved as a quadratic inequality; however, trial and error could also be applied. Because k must be an integer, try $k = 11$. Test the inequality:

$$11^2 + 11 \geq 133.2. \quad \text{Not true.}$$

Try $k = 12$. Test the inequality:

$$12^2 + 12 \geq 133.2. \quad \text{True.}$$

The value of k must be at least 12.

37.   Writing exercise

39.   Writing exercise

41.   Writing exercise

13.4 EXERCISES

1.   In Table 4, find the 10.0% row and read across to the column for 20 years to find entry 9.65022. Since this is the monthly payment amount needed to amortize a loan of $1000, and this loan is for $70,000, the required monthly payment is

$$70 \times 9.65022 = \$675.52.$$

3.   Because the interest rate of 8.7 is not in Table 4, use the formula for regular monthly payment with $P = \$57,300$, $r = .087$, and $t = 25$.

$$R = \frac{P\left(\frac{r}{12}\right)\left(1 + \frac{r}{12}\right)^{12t}}{\left(1 + \frac{r}{12}\right)^{12t} - 1}$$

$$= \frac{\$57300\left(\frac{.087}{12}\right)\left(1 + \frac{.087}{12}\right)^{12(25)}}{\left(1 + \frac{.087}{12}\right)^{12(25)} - 1}$$

$$= \$469.14$$

5.   In Table 4, find the 12.5% row and read across to the column for 25 years to find entry 10.90354. Since this is the monthly payment amount needed to amortize a loan of $1000, and this loan is for $27,750, the required monthly payment is

$$\frac{27750}{1000} \times 10.90354 = \$302.57.$$

7. Because the interest rate of 7.6 and the term of the loan, 22 years, are not in Table 4, use the formula for regular monthly payment with $P = \$132,500$, $r = .076$, and $t = 22$.

$$R = \frac{P\left(\frac{r}{12}\right)\left(1 + \frac{r}{12}\right)^{12t}}{\left(1 + \frac{r}{12}\right)^{12t} - 1}$$

$$= \frac{\$132,500\left(\frac{.076}{12}\right)\left(1 + \frac{.076}{12}\right)^{12(22)}}{\left(1 + \frac{.076}{12}\right)^{12(22)} - 1}$$

$$= \$1034.56$$

9. (a) To find the total payment, use Table 4. Find the row for 10.0% interest; read over to the 30 year column to find 8.77572. Since this is the monthly payment amount needed to amortize a loan of $1000, and this loan is for $58,500, the required monthly payment is

$$\frac{58500}{1000} \times 8.77572 = \$513.38.$$

(b) This total payment includes both principal and interest. For the first month, interest is charged on the full amount of the mortgage, so use the formula $I = Prt$, with $P = 58500$, $r = .10$, and $t = 1/12$.

$$(\$58500)(.10)\left(\frac{1}{12}\right) = \$487.50$$

(c) The remainder of the total payment is applied to the principal, so the principal payment is

$$\$513.38 - \$487.50 = \$25.88.$$

(d) The balance of the principal is

$$\$58500 - \$25.88 = \$58,474.12.$$

11. (a) To find the total payment, use Table 4. Find the row for 7.5% interest; read over to the 15 year column to find 9.27012. Since this is the monthly payment amount needed to amortize a loan of $1000, and this loan is for $93,200, the required monthly payment is

$$\frac{93200}{1000} \times 9.27012 = \$863.98.$$

(b) This total payment includes both principal and interest. For the first month, interest is charged on the full amount of the mortgage, so use the formula $I = Prt$, with $P = \$93,200$, $r = .075$, and $t = 1/12$.

$$(\$93200)(.075)\left(\frac{1}{12}\right) = \$582.50.$$

(c) The remainder of the total payment is applied to the principal, so the principal payment is

$$\$863.98 - \$582.50 = \$281.48.$$

(d) The balance of the principal is

$$\$93200 - \$281.48 = \$92,918.52.$$

(e) Every monthly payment is the same, so the second monthly payment is $863.98.

(f) The interest payment for the second month is

$$(\$92,918.52)(.075)\left(\frac{1}{12}\right) = \$580.74.$$

(g) The principal payment for the second month is

$$\$863.98 - \$580.74 = \$283.24.$$

(h) The balance of principal after the second month is

$$\$92918.52 - \$283.24 = \$92,635.28$$

13. (a) Because the interest rate of 8.2 is not in Table 4, use the formula for regular monthly payment with $P = \$113,650$, $r = .082$, and $t = 10$.

$$R = \frac{P\left(\frac{r}{12}\right)\left(1 + \frac{r}{12}\right)^{12t}}{\left(1 + \frac{r}{12}\right)^{12t} - 1}$$

$$= \frac{\$113650\left(\frac{.082}{12}\right)\left(1 + \frac{.082}{12}\right)^{12(10)}}{\left(1 + \frac{.082}{12}\right)^{12(10)} - 1}$$

$$= \$1390.93$$

(b) This total payment includes both principal and interest. For the first month, interest is charged on the full amount of the mortgage, so use the formula $I = Prt$, with $P = \$113,650$, $r = .082$, and $t = 1/12$.

$$(\$113650)(.082)\left(\frac{1}{12}\right) = \$776.61$$

(c) The remainder of the total payment is applied to the principal, so the principal payment is

$$\$1390.93 - \$776.61 = \$614.32.$$

(d) The balance of the principal is

$$\$113650 - \$614.32 = \$113,035.68.$$

(e) Every monthly payment is the same, so the second monthly payment is $1390.93.

(f) The interest payment for the second month is

$$(\$113035.68)(.082)\left(\frac{1}{12}\right) = \$772.41.$$

(g) The principal payment for the second month is

$$\$1390.93 - \$772.41 = \$618.52.$$

(h) The balance of principal after the second month is

$$\$113,035.68 - \$618.52 = \$112,417.16.$$

15. Use Table 4 to find the monthly amortization payment (principal and interest).

$$\frac{\$62300}{\$1000} \times 12.43521 = \$774.71$$

The monthly tax and insurance payment is

$$\frac{\$610 + \$220}{12} = \$69.17.$$

The total monthly payment, including taxes and insurance, is

$$\$774.71 + \$69.17 = \$843.88.$$

17. Use Table 4 to find the monthly amortization payment (principal and interest).

$$\frac{\$89560}{\$1000} \times 11.35480 = \$1016.94$$

The monthly tax and insurance payment is

$$\frac{\$915 + \$309}{12} = \$102.$$

The total monthly payment, including taxes and insurance, is

$$\$1016.94 + \$102 = \$1118.94.$$

19. Because the interest rate is not in Table 4, use the formula for regular monthly payment with $P = \$115,400$, $r = .088$, and $t = 20$.

$$R = \frac{P\left(\frac{r}{12}\right)\left(1 + \frac{r}{12}\right)^{12t}}{\left(1 + \frac{r}{12}\right)^{12t} - 1}$$

$$= \frac{\$115400\left(\frac{.088}{12}\right)\left(1 + \frac{.088}{12}\right)^{12(20)}}{\left(1 + \frac{.088}{12}\right)^{12(20)} - 1}$$

$$= \$1023.49$$

The monthly tax and insurance payment is

$$\frac{\$1295.16 + \$444.22}{12} = \$144.95.$$

The total monthly payment, including taxes and insurance, is

$$\$1023.49 + \$144.95 = \$1168.44.$$

21. $12 \times 30 = 360$

23. The total interest is

$$(360 \times \$672.68) - \$80000 = \$162,164.80.$$

25. (a) In Table 5 read the heading of the column on the left to see that the monthly payment is $304.01.

(b) Read the heading of the column on the right to see that the monthly payment is $$734.73.

27. Payment number 12 is the last payment for the year.

(a) Balance of principal is $59,032.06.

(b) Balance of principal is $59,875.11.

29. Compare the Interest Payment column with the Principal Payment column.

(a) The first payment in which the principal payment is higher is number 176.

(b) The first payment in which the principal payment is higher is number 304

31. Use Table 4 and an annual rate of 7.5% interest. Read across to the 10-year column to find 11.87018; this is the monthly payment for a $1000 mortgage. Multiply by 60 to obtain the monthly payment for a $60,000 mortgage:

$$60 \times \$11.87018 = \$712.21$$

There would be $10 \times 12 = 120$ monthly payments:

$$120 \times \$712.21 = \$85,465.20.$$

Then, the interest is the difference between this amount and the loan amount.

$$\$85,465.20 - \$60000 = \$25,465.20$$

33. Use Table 4 and an annual rate of 7.5% interest. Read across to the 30-year column to find 6.99215; this is the monthly payment for a $1000 mortgage. Multiply by 60 to obtain the monthly payment for a $60,000 mortgage:

$$60 \times \$6.99215 = \$419.53.$$

There would be $30 \times 12 = 360$ monthly payments:

$$360 \times \$419.53 = \$151,030.80.$$

Then, the interest is the difference between this amount and the loan amount is

$$\$151030.80 - \$60000 = \$91,030.80.$$

35. (a) Add the initial index rate and the margin to obtain the ARM interest rate.

$$6.5 + 2.5 = 9.0\%$$

In Table 4 find 9.0% in the annual rate column and read across to the column for a 20-year mortgage to find 8.99726. Multiply this figure by 75 to obtain the initial monthly payment.

$$75 \times \$8.99726 = \$674.79$$

(b) The interest rate for the second adjustment period is given by the ARM interest rate.

$$8.0 + 2.5 = 10.5\%$$

Use the formula for Regular monthly payment with $P = \$73,595.52$ (the Adjusted Balance), $r = .105$, and $t = 19$.

$$R = \frac{P\left(\frac{r}{12}\right)\left(1 + \frac{r}{12}\right)^{12t}}{\left(1 + \frac{r}{12}\right)^{12t} - 1}$$

$$= \frac{\$73595.52\left(\frac{.105}{12}\right)\left(1 + \frac{.105}{12}\right)^{12(19)}}{\left(1 + \frac{.105}{12}\right)^{12(19)} - 1}$$

$$= \$746.36$$

(c) The change in monthly payment is the difference in the two amounts from parts (a) and (b).

$$\$746.36 - \$674.79 = \$71.57$$

37. (a) The ARM interest rate is 2% plus 7.5% or 9.5%. Then,

$$\frac{.095 \times \$50000}{12} = \$395.83.$$

(b) To find the first monthly payment first add 2% and 7.5% to obtain 9.5% as the ARM interest rate. In Table 4 find 9.5% in the annual rate column and read across to the 20-year mortgage column to find 9.32131. Multiply this figure by 50 to obtain the monthly payment for the $50,000 mortgage.

$$50 \times \$9.32131 = \$466.07$$

39. From Exercise 38, the monthly payment for the first month of the second year is $531.09. The monthly adjustment at the end of the second year is the difference between this amount and and the monthly payment amount at the end of the first year.

$$\$531.09 - \$466.07 = \$65.02$$

41. The down payment is 20% of the purchase price of the house.

$$.20 \times \$95000 = \$19,000$$

Then, the mortgage amount is the difference between the purchase price of the house and this figure.

$$\$95000 - \$19000 = \$76,000$$

43. From Exercise 42, the Loan fee is $.02 \times \$76000 = \$1520$. Add this figure to the other closing costs listed in the text to obtain the total closing costs of $2705.

45. Writing exercise

47. Writing exercise

49. Writing exercise

51. Writing exercise

53. Writing exercise

13.5 EXERCISES

1. (a) The total purchase price is

$$(\$20 \text{ per share}) \times (40 \text{ shares}) = \$800.$$

(b) The total dividend amount is

$$(\$2 \text{ per share}) \times (40 \text{ shares}) = \$80.$$

(c) The capital gain is found by multiplying the change in price per share times the number of shares.

$$(\$44 - \$20) \times (40 \text{ shares}) = \$960$$

(d) The total return is the sum of the dividends and the capital gain.

$$\$80 + \$960 = \$1040$$

(e) The percentage return is the quotient of the total return and the total cost as a percent.

$$\frac{\$1040}{\$800} \times 100 = 130\%$$

3. (a) The total purchase price is

$$(\$12.50 \text{ per share}) \times (100 \text{ shares}) = \$1250.$$

(b) The total dividend amount is

$$(\$1.28 \text{ per share}) \times (100 \text{ shares}) = \$128.$$

(c) The capital gain is found by multiplying the change in price per share times the number of shares.

$$(\$15.10 - \$12.50) \times (100 \text{ shares}) = \$260$$

(d) The total return is the sum of the dividends and the capital gain.

$$\$128 + \$260 = \$388$$

(e) The percentage return is the quotient of the total return and the total cost as a percent.

$$\frac{\$388}{\$1250} \times 100 = 31\%$$

5. Using the formula for simple interest with $P = \$1000$, $r = .055$, and $t = 5$ the total return is:

$$I = Prt$$
$$= \$1000 \times .055 \times 5$$
$$= \$275.$$

7. Using the formula for simple interest with $P = \$10,000$, $r = .0711$, and $t = 3/12$ the total return is:

$$I = Prt$$
$$= \$10000 \times .0711 \times \frac{3}{12}$$
$$= \$177.75.$$

9. (a) Use the formula for net asset value with $A = \$875$ million, $L = \$36$ million, and $N = 80$ million.

$$NAV = \frac{A - L}{N}$$
$$= \frac{875 - 36}{80}$$
$$= \$10.49$$

(b) Find the number of shares purchased by dividing the amount invested by the net asset value. Round to the nearest share.

$$\frac{\$3500}{\$10.49} = 334 \text{ shares}$$

11. (a) Use the formula for net asset value with $A = \$2.31$ billion ($\$2,310$ million), $L = \$135$ million, and $N = 263$ million.

$$NAV = \frac{A - L}{N}$$
$$= \frac{2310 - 135}{263}$$
$$= \$8.27$$

(b) Find the number of shares purchased by dividing the amount invested by the net asset value. Round to the nearest share.

$$\frac{\$25470}{\$8.27} = 3080 \text{ shares}$$

13. (a) Find the monthly return by multiplying the monthly percentage return, as a decimal, times the amount invested.

$$.013 \times \$645 = \$8.39$$

(b) Find the annual return by multiplying the monthly percentage return by 12, and then multiplying this annual percentage times the amount invested.

$$12 \times .013 \times \$645 = \$100.62$$

(c) The annual percentage return is the ratio of the annual return to the amount invested, expressed as a percent.

$$\frac{\$100.62}{\$645} \times 100 = 15.6\%$$

15. (a) Find the monthly return by multiplying the monthly percentage return, as a decimal, times the amount invested.

$$.023 \times \$2498 = \$57.45$$

(b) Find the annual return by multiplying the monthly percentage return by 12, and then multiplying this annual percentage times the amount invested.

$$12 \times .023 \times \$2498 = \$689.45$$

(c) The annual percentage return is the ratio of the annual return to the amount invested, expressed as a percent.

$$\frac{\$689.45}{\$2498} \times 100 = 27.6\%$$

17. (a) The beginning value of the investment is the product of the beginning net asset value and the number of shares purchased.

$$\$9.63 \times 125 = \$1203.75$$

(b) To find the first monthly return, multiply the monthly percentage return, in decimal form, times the beginning value from part (a).

$$.015 \times \$1203.75 = \$18.06$$

(c) Using Example 5 from the text as a guide, find the effective annual rate of return as follows. First find the return relative.

$$1 + 1.5\% = 1 + .015 = 1.015$$

Raise this value to the 12th power to get the annual return relative and subtract 1 to get the percentage rate.

$$(1.015)^{12} - 1 \approx .1956, \text{ which is } 19.56\%$$

19. (a) The beginning value of the investment is the product of the beginning net asset value and the number of shares purchased.

$$\$11.94 \times 350 = \$4179$$

(b) To find the first monthly return, multiply the monthly percentage return, in decimal form, times the beginning value from part (a).

$$.0183 \times \$4179 = \$76.48$$

(c) Using Example 5 from the text as a guide, find the effective annual rate of return as follows. First find the return relative.

$$1 + 1.83\% = 1 + .0183 = 1.0183$$

Raise this value to the 12th power to get the annual return relative and subtract 1 to get the percentage rate.

$$(1.0183)^{12} - 1 \approx .2431, \text{ which is } 24.31\%$$

21. *Aggressive Growth*

$$7\% \text{ of } \$20,000 = .07 \times \$20000 = \$1400$$

Growth

$$43\% \text{ of } \$20,000 = .43 \times \$20000 = \$8600$$

Growth & Income

$$31\% \text{ of } \$20,000 = .31 \times \$20000 = \$6200$$

Income

$$14\% \text{ of } \$20,000 = .14 \times \$20000 = \$2800$$

Cash

$$5\% \text{ of } \$20,000 = .05 \times \$20000 = \$1000$$

23. *Aggressive Growth*

$$2\% \text{ of } \$400,000 = .02 \times \$400000 = \$8000$$

Growth

$$28\% \text{ of } \$400,000 = .28 \times \$400000 = \$112,000$$

Growth & Income

$$36\% \text{ of } \$400,000 = .36 \times \$400000 = \$144,000$$

Income

$$29\% \text{ of } \$400,000 = .29 \times \$400000 = \$116,000$$

Cash

$$5\% \text{ of } \$400,000 = .05 \times \$400000 = \$20,000$$

25. First find the difference of 100% and 25%.

$$100\% - 25\% = 75\%$$

Then find 75% of 5%.

$$.75 \times .05 = .0375$$

The tax-exempt rate of return is 3.75%.

27. First find the difference of 100% and 35%.

$$100\% - 35\% = 65\%$$

Then find 65% of 8%.

$$.65 \times .08 = .052$$

The tax-exempt rate of return is 5.2%.

29. Writing exercise

31. Writing exercise

33. Writing exercise

CHAPTER 13 TEST

1. Use the formula $A = P(1 + rt)$, with $P = \$100$; $r = .06$; and $t = 5$.

$$\begin{aligned}
A &= P(1 + rt) \\
&= 100(1 + .06 \cdot 5) \\
&= 100(1 + .30) \\
&= 100(1.30) \\
&= \$130
\end{aligned}$$

2. To find the final amount, use the formula for future value $A = P\left(1 + \frac{r}{m}\right)^n$, with $P = \$50$, $r = .08$, $m = 4$ and $n = 4 \cdot 2 = 8$ (compounded 4 times per year for 2 years).

$$\begin{aligned}
A &= P\left(1 + \frac{r}{m}\right)^n \\
&= 50\left(1 + \frac{.08}{4}\right)^8 \\
&= 50(1.02)^8 \\
&= 50(1.171659) \\
&= \$58.58
\end{aligned}$$

3. Use the formula $Y = \left(1 + \frac{r}{m}\right)^m - 1$, with $r = .06$ and $m = 12$.

$$\begin{aligned}
Y &= \left(1 + \frac{.06}{12}\right)^{12} - 1 \\
&= (1.005)^{12} - 1 \\
&= .0617
\end{aligned}$$

This is 6.17%, to the nearest hundredth of a percent.

4. Years to double $= \dfrac{70}{\text{Annual inflation rate}}$

Years to double $= \dfrac{70}{12}$

Years to double ≈ 6 years

5. Use the present value formula for compound interest with $A = \$100,000$, $r = .10$, $m = 2$, and $n = 2 \cdot 10 = 20$.

$$P = \frac{A}{\left(1 + \frac{r}{m}\right)^n}$$

$$= \frac{100000}{\left(1 + \frac{.10}{2}\right)^{20}}$$

$$= \frac{100000}{(1.05)^{20}}$$

$$= \$37688.95$$

6. The interest due is the 1.6% of $680.

$$.016 \times \$680 = \$10.88$$

7. Use the simple interest formula with $P = \$3000$, $r = .08$, and $t = 2$. Because he made a down payment of $1000, he will pay interest on only $3000. In this formula, t must be expressed in years, so 24 months is equivalent to 2 years.

$$I = Prt$$

$$= (3000)(.08)(2)$$

$$= \$480$$

8. The total amount owed is

$$P + I = \$3000 + \$480$$

$$= \$3480.$$

There are 24 monthly payments, so the amount of each monthly payment is

$$\frac{\$3480}{24} = \$145.$$

9. Find the finance charge per $100 of the amount financed.

$$\frac{\$480}{\$3000} \times 100 = \$16$$

In Table 3, find the "24 payments" row and read across to find the number closest to 16, which is 15.80. Read up to find the APR, which is 14.5%.

10. Use the formula $U = kR\left(\dfrac{h}{100 + h}\right)$ to find unearned interest, with $k =$ remaining number of payments, $R =$ regular monthly payment, and $h =$ finance charge per $100. In Table 3, the number 15.80, which is closest to 16, appears in the "24 payments" row. Because the total number of payments remaining is 6, move up to the "6 payments" row to find $h = \$4.27$, the value of h needed in the actuarial method.

$$U = (6)(\$145)\left(\frac{\$4.27}{100 + \$4.27}\right)$$

$$= (6)(\$145)\left(\frac{\$4.27}{\$104.27}\right)$$

$$= \$35.63$$

11. First find the finance charge by multiplying $7 by 6.

$$6 \times \$7 = \$42$$

Find the finance charge per $100 of the amount financed.

$$\frac{\$7}{\$150} \times \$100 \approx \$4.67$$

In Table 3, the number closest to 4.67 in the "6 payments" row is 4.72. Read up to find the APR, which is 16.0%.

12. Writing exercise

13. In Table 4, find the 12.0% row and read across to the column for 20 years to find entry 11.01086. Since this is the monthly payment amount needed to amortize a loan of $1000, and this loan is for $50,000, the required monthly payment is

$$\frac{50000}{1000} \times 11.01086 = \$550.54.$$

14. Subtract the down payment from the purchase price of the house to find the principal.

$$\$88000 - (.20 \times \$88000) = \$70,400$$

In Table 4 find the 10.5% row and read across to the column for 30 years to find entry 9.14739. Since this is the monthly payment amount needed to amortize a loan of $1000, and this loan is for $70,400, the required monthly payment is

$$\frac{\$70400}{\$1000} \times \$9.14739 = \$643.98.$$

The monthly tax and insurance payment is

$$\frac{\$900 + \$600}{12} = \$125.$$

The total monthly payment, including taxes and insurance, is

$$\$643.98 + \$125 = \$768.98.$$

15. The "two points" charged by the lender mean 2%. This additional cost is the percentage taken of the amount borrowed, the principal from Exercise 14.

$$.02 \times \$70400 = \$1408$$

16. Writing exercise

17. If the index started at 7.85%, adding the margin of 2.25% gives

$$7.85\% + 2.25\% = 10.1\%.$$

Because the periodic rate cap is 2%, the interest rate during the second year cannot exceed

$$10.1\% + 2.00\% = 12.1\%.$$

18. First find how much she paid for the stock by multiplying the price per share by 1000.

$$\$12.75 \times 1000 = \$12,750$$

Calculate the amount of money she received from both dividends.

$$\$1.38 \times 1000 + \$1.02 \times 1000 = \$2400$$

Subtract this amount from the total price she paid.

$$\$12750 - \$2400 = \$10350$$

Finally subtract this amount from the price for which she sold the stock.

$$(\$10.36 \times 1000) - \$10350 = \$10$$

19. Compare her amount of return to how much she originally paid for the stock.

$$\frac{\$10}{\$12750} = .00078$$

This is .078%.

20. This percentage is not the annual rate of return, because the time period was for more than one year.

14 | GRAPH THEORY

14.1 Basic Concepts
14.2 Euler Circuits
14.3 Hamilton Circuits
14.4 Trees and Minimum Spanning Trees
Chapter 14 Test

Chapter Goals

The major goal of this chapter is to provide an overview of graph theory topics and their applications, both current and historical in nature. Upon completion of this chapter, the student should be able to

- Distinguish between coordinate "graphs" of the relations and functions found in Chapter 8 of the text and the "graphs" of this chapter (convenient diagrams that show connections and relations between objects).
- Understand the meaning of simple graphs, isomorphic graphs, connected graphs, and complete graphs.
- Determine the relationship between number of vertices and edges of a graph.
- Distinguish between a "walk," "path," and a "circuit."
- Apply information about graphs to problem solving.
- Understand and identify Euler circuits
- Identify problems that can be modeled and solved by the use of Euler circuits.
- Understand and identify Hamilton circuits
- Identify problems that can be modeled and solved by the use of Hamilton circuits.
- Understand and identify Spanning trees
- Identify problems that can be modeled and solved by the use of Spanning trees.

Chapter Summary

Graph theory is a field of mathematics that looks at the relations or connections between objects in some collection. The graphs are just convenient diagrams of these connections.

A **graph** is a collection of vertices (at least one) and edges. Each edge goes from a vertex to a vertex. The following is an example of a connected graph. Each dot–named in this example by a letter–represents a vertex, and each line segment, an edge.

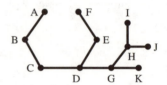

A graph is **connected** if we can get from each vertex of the graph to every other vertex of the graph *along edges of the graph*.

Graphs are used in a variety of fields of interest to diagram, analyze, and solve problems. These fields are as diverse as Chemistry, where graphs are used to designate chemical compounds, to the analysis of subway services, telephone networks, the world wide web, and even garden mazes and the clarification of rhythm schemes in poetry.

Most graphs have subgraphs associated with them.

A **subgraph** of a graph is a graph consisting of some of the vertices and some of the original edges between these vertices. ABCDEF of the graph above is an example of a subgraph.

The following represent useful special cases of subgraphs.

1.   A **walk** in a graph is a sequence of vertices, each linked to the next vertex by a specified edge of the graph. In the graph above F→E→D→G→K would represent a walk. A→F→E would not be a walk.

A special "walk" of interest is called a "path."

2.   A **path** in a graph is a walk that uses no edge more than once. The same walk, F→E→D→G→K, above represents a path. Do you see others?

A special "path" of interest is called a "circuit."

3. A **circuit** in a graph is a path that begins and ends at the same vertex.

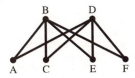

In this connected graph, an example of a circuit would be: B→C→D→A→B. Do you see others?

Graphs can look different yet show the same relationships between vertices. We call such graphs "isomorphic."

Two graphs are **isomorphic** if there is a one-to-one correspondence between vertices of the two graphs with the property that whenever there is an edge between two vertices of either one of the graphs, there is an edge between the corresponding vertices of the other graph.

An important characteristic of most graphs is the associated degree for each vertex. The **degree** of a vertex is the number of edges joined to that vertex. In the above graph, for example, the degree of vertex A is 2 and the degree of vertex D is 4.

The following theorem indicates the relationship between the sum of degrees of the vertices of a graph and the number of edges.

Theorem: In any graph, the sum of the degrees of the vertices equals twice the number of edges.

EXAMPLE A   Relationship of edges to vertices

In the above graph count the degree of each vertex and add these values. From A to F the sum of the degrees is given by

$$2 + 4 + 2 + 4 + 2 + 2 = 16.$$

Counting the number of edges, we see there are a total of 8. This is predicted by the theorem.

EXAMPLE B   Application

There are 7 people at a business meeting. One of these shakes hands with 4 people, 4 shake hands with 2 people, and 2 shake hands with 3 people. How many handshakes occur?

Solution

Each of the 7 people represents a vertex. Vertex 1 (first person) has degree 4, since he/she shakes hands with 4. Vertices 2–5 have degree 2. Vertices 6 and 7 have degree 3. Thus, the sum of the degrees is

$$4 + 2 + 2 + 2 + 2 + 3 + 3 = 18.$$

Since there are one half as many edges as the sum of the degrees, we arrive at 9 edges, or handshakes.

Useful in application work are weighted graphs.

A **weighted graph** is a graph with numbers indicated along edges. The numbers are called **weights**. The following is an example of a weighted graph.

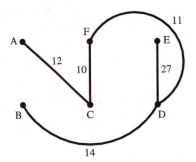

The total weight for this graph is 74.

Another useful graph is called a complete graph.

A **complete graph** is a graph in which there is exactly one edge going from each vertex to each other vertex in the graph. The following is an example of a complete graph.

The two prominent and very useful circuits studied in this chapter are "Euler circuits" and "Hamilton circuits." The study of Euler Circuits are effective in planning efficient routes for snowplows, mail delivery and many other practical problems.

An **Euler circuit** in a graph is a circuit that uses every edge of the graph exactly once.

Probably the most famous historical problem which involves Euler circuits is the Köenigsberg bridge problem, finding a path through the historical city which crosses all seven bridges only once.

The following is a summary of important or useful theorems related to Euler Circuits.

Euler's theorem provides a simple way to determine if any given graph has an Euler circuit.

Euler's theorem:
Suppose we have a connected graph.

1. If the graph has an Euler circuit then each vertex of the graph has even degree.

2. If each vertex of the graph has even degree, then the graph has an Euler circuit.

EXAMPLE C   Application of an Euler Circuit

Examine the floor plan given. Is it possible to start outside, walk through each door exactly once, and end up back outside?

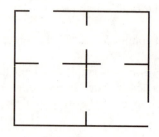

Solution
Treat outside and each room as a vertex, the doorways as edges and draw connections to each vertex.

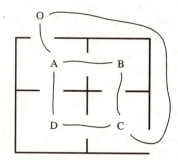

Since several vertices have odd degree (e.g. A and C), an Euler circuit does not exist. Thus, it will not be possible to walk through each door exactly once and end up back outside.

One common method for finding an Euler circuit for a graph is called Fleury's algorithm. The algorithm makes use of the definition of "cut edge."

A **cut edge** in a graph is an edge whose removal disconnects a component of the graph.

Fleury's Algorithm

1. Start at any vertex. Go along any edge from this vertex to another vertex. Remove this edge from the graph.
2. You are now on a vertex of the revised graph. Choose any edge from this vertex, subject to only one condition: do not use a cut edge (of the revised graph) unless you have no other option!
3. Keep going like this until you have used all the edges and got back to where you started.

See text or *Study Guide and Solutions Manual* (EXERCISES 14.2) for examples related to finding Euler circuits.

The study of Hamilton circuits often involves very large numbers and, as the text points out, gives rise to many unsolved problems in both mathematics and computer science.

A **Hamilton circuit** in a graph is a circuit that visits each vertex exactly once (returning to the starting vertex to complete the circuit).

It is often hard to decide whether or not a particular graph has a Hamilton circuit. But remember that, by the following theorem, a "complete graph" with three or more vertices will always have a Hamilton circuit.

Theorem: Any complete graph with three or more vertices has a Hamilton circuit.

Theorem: A complete graph with n vertices has $(n-1)!$ Hamilton circuits.

Factorial products such as $(n-1)!$ indicated in the above theorem become large numbers very quickly. For example, a complete graph with 26 vertices will have

$$(26 - 1)! = 25 \cdot 24 \cdot 23 \cdots 3 \cdot 2 \cdot 1$$
$$= 15,511,310,043,330,985,984,000,000$$
$$= 1.6 \times 10^{25}.$$

In other words, there are more than a trillion trillions Hamilton circuits for such a graph.

Many applications require finding a Hamilton circuit with a minimum weight associated with the circuit. In a weighted graph, a **minimum Hamilton circuit** is a Hamilton circuit with smallest possible total weight. One method of finding a minimum Hamilton circuit for a graph is called the "Brute force algorithm."

Brute force algorithm

1. Choose a starting point
2. List all the Hamilton circuits with that starting point
3. Find the total weight of each circuit.
4. Choose a Hamilton circuit with smallest total weight. Such an approach soon becomes impractical. A complete graph with just 6 vertices will have $5! = 120$ Hamilton circuits to analyze (7 vertices will have 720). As we get to important applications with much larger numbers of vertices, even computers can soon be overwhelmed using this approach.

A much more efficient, but approximate, approach often used is the "Nearest Neighbor Algorithm." The underlying idea of this algorithm is that from each vertex we proceed to a "nearest" available vertex.

Nearest Neighbor Algorithm

1. Choose a starting point for the circuit. (e.g. vertex A)
2. Check all the edges joined to A, and choose one that has smallest weight; proceed along this edge to the next vertex.
3. At each vertex you reach, check the edges from there to vertices not yet visited. Choose one with smallest weight. Proceed along this edge to the next vertex.
4. Continue until you have visited all the vertices.
5. Return to starting vertex.

See text or *Study Guide and Solutions Manual* (EXERCISES 14.3) for examples related to finding a minimum Hamilton circuit.

In the initial sections one is studying graphs of special kinds of circuits. In the last section (14.4) graphs, called **trees**, that have no circuits are the focus. The following graphs are examples of trees.

Although "connected graph" was defined early, an alternative definition indicated by the text is useful in the discussion of "trees."

A **connected graph** is one in which there is *at least one path* between each pair of vertices. Thus, "**tree**" is the special name for graphs that *are connected* and *have no circuits*.

Since trees are connected there is always <u>at least one path</u> between each pair of vertices. Also, if the graph is a tree there <u>can not be two different paths</u> between a pair of vertices. This property, which falls directly from the definition of a "tree," can be stated as follows:

In a tree there is always **exactly one path** from each vertex in the graph to any other vertex in the graph.

It is this property that makes trees so useful in real world applications (see Example D).

A subgraph is called a **spanning tree** for a graph if it "spans" the graph in the sense that it connects all the vertices of the original graph into a subgraph.

We can find a spanning tree for any connected graph and if our original graph has at least one circuit, it will have a number of different spanning trees.

A spanning tree that has minimum total weight is called a **minimum spanning tree** for the graph.

Minimum spanning trees turn out to be useful in a wide variety of fields from the creation of binary codes to cellular biology.

Kruskal's algorithm is a useful algorithm for finding a minimum spanning tree for any connected weighted graph.

Kruskal's algorithm:

Choose edges for the spanning tree as follows:

1. Choose any edge with minimum weight.
2. Find an edge with minimum weight amongst those not yet selected, and choose this as the next edge. (It does not matter if your subgraph at this stage looks disconnected!)
3. Continue to choose edges of minimum weight from those not yet selected, except do not select any edge that creates a circuit in the subgraph.
4. Keep going until you have a tree that connects all vertices of your original graph.

See text or *Study Guide and Solutions Manual* (EXERCISES 14.4) for examples related to finding a minimum spanning tree for any connected weighted graph.

There is a simple relationship between the number of vertices and the number of edges in a tree.

Theorem: If a graph is a tree, then the number of edges in the graph is one less than the number of vertices. So a tree with n vertices has $n - 1$ edges.

If the graph is connected, then the converse for this theorem is also true.

Theorem: For a connected graph, if the number of edges is one less than the number of vertices, then the graph is a tree.

EXAMPLE D Application of a Spanning Tree

Maria Jimenez has 12 vegetable and flower beds in her garden and wants to build flagstone paths between the beds so that she can get from each bed to every other bed along a flagstone path. She also wants a path linking her front door to one of the beds. Determine the minimum number of paths she must build to achieve this.

Solution
Treating each of the 12 vegetable and flower beds along with her front door as a vertex in a tree graph, there would be one less edge than the resulting 13 vertices. Thus, it will take 12 edges to complete the tree and she will have to build a minimum of 12 paths.

14.1 EXERCISES

1. By counting, there are 7 vertices and 7 edges.

3. There are 10 vertices and 9 edges.

5. There are 6 vertices and 9 edges.

7. Two vertices have degree 3, three have degree 2, and two have degree 1. The sum of the degrees is

$$3 + 3 + 2 + 2 + 2 + 1 + 1 = 14.$$

This is twice the the number of edges which is 7.

9. Six vertices have degree 1. Four vertices have degree 3. The sum of the degrees is

$$1 + 1 + 1 + 1 + 1 + 1 + 3 + 3 + 3 + 3 = 18.$$

This is twice the the number of edges which is 9.

11. No, the two graphs are not isomorphic. There is one more vertex in the (b) graph.

13. Yes, the graphs are isomorphic.

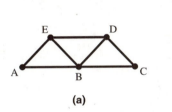   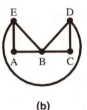

    (a)   (b)

15. No, the two graphs are not isomorphic. Two vertices in graph (b) have degree 1 while only one vertex in graph (a) has a degree of 1.

17. The graph is connected with only 1 component.

19. The graph is disconnected with 3 components.

21. The graph is disconnected with 2 components.

23. Since the sum of the degrees is given by

$$4 \times 3 = 12,$$

there are

$$\frac{1}{2} \times 12 = 6 \text{ edges.}$$

25. Since the sum of the degrees is given by

$$1 + 1 + 1 + 2 + 3 = 8,$$

there are

$$\frac{1}{2} \times 8 = 4 \text{ edges.}$$

27. (a) Yes, A→B→C is a walk.

(b) Yes, B→A→D is not a walk since there is no edge from A to D.

(c) No, E→F→A→E is not a walk since there is no edge from A to E.

(d) Yes, B→D→F→B→D is a walk.

(e) Yes, D→E is a walk.

(f) Yes, C→B→C→B is a walk.

29. (a) No, A→B→C→D→E→F is not a circuit since the path does not return to the starting vertex.

(b) Yes, A→B→D→E→F→A is a circuit.

(c) No, C→F→E→D→C is not a circuit since there is no edge from C to F.

(d) No, G→F→D→E→F is not a circuit since the path does not return to the starting vertex.

(e) No, F→D→F→E→D→F is not a circuit since the edge from F to D is used more than once.

31. (a) No, A→B→C is not a path since there is no edge from B to C.

(b) No, J→G→I→G→F is not a path since the edge from I to G is used more than once.

(c) No, D→E→I→G→F is not a path since it is not a walk (there is no edge from E to I).

(d) Yes, C→A is a path.

(e) Yes, C→A→D→E is a path.

(f) No, C→A→D→E→D→A→B is not a path since the edges A to D and D to E are used more than once.

33. A→B→C is a walk and also a path (no edges are repeated) but not a circuit (since the path does not return to the starting vertex.)

35. A→B→A→C→D→A is a walk, not a path (since the edge A to B is used twice), and hence, not a circuit.

37. C→A→B→C→D→A→E is a walk and a path but not a circuit (since it doesn't end at C).

39. No, this is not a complete graph since there is no edge going from A to C (nor B to D).

41. No, this is not a complete graph since there is no edge going from A to F, for example.

43. Yes, this is a complete graph since there is exactly one edge going from each vertex to each other vertex in the graph.

45.

Counting the edges, there are 6 games to be played in this competition.

47. One can draw a graph and count the edges or one may reason that each of the members of one team must shake the hand of each of the 6 members of the second team. Thus, there are $6 \times 6 = 36$ handshakes in total.

49. Each of the 8 people represents a vertex. Two of these people had 4 conversations. Thus, the degree for each of these vertices is 4. Another person had 3 conversations. Thus, the degree for this vertex is 3. Four people had 2 conversations each. Therefore, each of these 4 vertices have degree 2. One person had 1 conversation so the degree of this vertex is 1. Thus, the sum of the degrees is

$$4 + 4 + 3 + 2 + 2 + 2 + 2 + 1 = 20.$$

Since there are one half as many edges as the sum of the degrees, we arrive at 10 edges, or telephone conversations.

51.

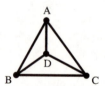

One of several circuits could be A→B→D→A. This corresponds to tracing round the edges of a single face.

53. The sum of degrees is twice the number of edges, so it must be even. But if there were an odd number of vertices with odd degree, the sum of degrees would be odd.

55.

The degree of each vertex is 4, which shows that each face has common boundaries with 4 other faces.

57. Writing exercise

59. Writing exercise

14.2 EXERCISES

1. (a) No, the sequence of vertices, A→B→C→D→A→B→C→D→A, does not represent an Euler circuit since it is not a path. Remember that a path can traverse an edge only once.

   (b) Yes, the sequence of vertices, C→B→A→D→C, is an Euler circuit since it is a circuit and each edge is traversed only once.

   (c) No, the sequence, A→C→D→B→A, is not an Euler circuit since it is not a path.

   (d) No, the sequence, A→B→C→D, is not an Euler circuit since it is not a circuit.

3. (a) No, the sequence of vertices, A→B→C→D→E→F→A, does not represent an Euler circuit since it doesn't use all of the edges (e.g. B→F).

   (b) Yes, the sequence of vertices, A→B→C→D→E→G→C→E→F→G→B→F→A, is an Euler circuit since it is a circuit and each edge is traversed only once.

   (c) No, the sequence of vertices, A→B→C→D→E→C→G→E→F→G→E→F→A, is not an Euler circuit since it is not a path (the edge E to F is traversed twice).

   (d) Yes, the sequence of vertices, A→B→G→E→D→C→G→F→B→C→E→F→A, is an Euler circuit since it is a circuit and each edge is traversed only once.

5. Yes, the graph will have an Euler circuit since all vertices have even degree.

7. No, the graph will not have an Euler circuit since some vertices (e.g. G) have odd degree.

9. If we assume that a vertex occurs at each intersection of curves and check the degree of each such vertex, we see that all vertices have an even degree. Thus, by Euler's theorem, all edges are traversed exactly one time —that is, we can find an Euler circuit. The answer is <u>yes</u> since this is equivalent to tracing the pattern without lifting your pencil nor going over any line more than one time.

11. Since all vertices have even degree, the graph has an Euler circuit. Because one must pass through vertex I more than once to complete any circuit, no circuit will visit each vertex exactly once.

13. Since all vertices have even degree, the graph has an Euler circuit. The sequence, A→B→H→C→G→D→F→E→A, visits each vertex exactly once.

15. Some vertices (e.g. B) have odd degree. Therefore, no Euler circuit exists. No circuit exists that will pass through each vertex exactly once.

Exercises 16–19 correspond to deciding if an Euler circuit exists when we assume that each intersection of line segments forms a vertex.

17. The upper left (and lower right) corner is of odd degree. Hence, there is no Euler circuit (or continuous path to apply the grout).

19. Since all vertices are of even degree, there exists an Euler circuit and thus, a continuous path to apply the grout without retracing.

21. There are no cut edges since no break along an edge will disconnect the graph.

23. The student has a choice of B→E or B→D . Choosing B→F, for example, is not allowed since each is a cut edge.

25. The student has a choice of B→C or B→H. Choosing B→G, for example, is not allowed since it is a cut edge.

There are many different correct answers for the following exercises.

27. Beginning at A, one Euler circuit that results is A→C→B→F→E→D→C→F→D→A.

29. Since each vertex is of even degree, there is an Euler circuit. Beginning at A, one such circuit is A→G→H→J→I→L→J→K→I→H→F→G→E→F→D→E→C→D→B→C→A→E→B→A.

31. Some vertices (e.g. C) have odd degree. Therefore, no Euler circuit exists.

33. Such a route exists. Since all vertices are of even degree, an Euler circuit exists for the graph. Using Fleury's algorithm, one such route is given by A→D→B→C→A→H→D→E→B→H→G→E→F→H→J→L→C→M→A→K→M→L→K→J→A.

35. Draw a map such as the following.

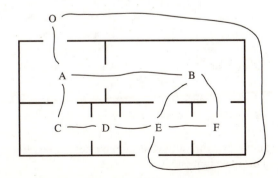

Since several vertices have odd degree (e.g. A), an Euler circuit does not exist. Thus, it will not be possible to walk through each door exactly once and end up back outside.

37. There is no Euler path since there are more than two vertices with odd degree.

39. Yes, there is an Euler path since there are exactly two vertices with odd degree. The path must begin or end at B or G, the two vertices with odd degree.

Exercises 41–43 are solved by thinking of the corresponding graphs for Exercises 34–36. Remember that the rooms represent vertices and the doors are edges. To answer each question in the affirmative, an Euler path must exist.

41. Yes, it is possible because there are exactly two rooms with an odd number of doors (odd degree).

43. No, there is no Euler path due to the fact that all rooms have an even number of doors (or degree).

45. One possible circuit is A→B→D→C→A. There are 4 edges in any such circuit.

47. One possible circuit is A→B→C→D→E→F→G→H→C→A→H→I→A. There are 12 edges in any such circuit.

49. Only those *complete graphs* for which the number of vertices is an odd number greater than or equal to 3 have an Euler circuit. In a complete graph with n vertices, the degree of each vertex is $n-1$. And $n-1$ is even if, and only if, n is odd.

51. Writing exercise

14.3 EXERCISES

1. (a) No, A→E→C→D→E→B→A, is not a Hamilton circuit since it visits vertex E twice.

   (b) Yes, A→E→C→D→B→A, is a Hamilton circuit since it visits all vertices (except the first) only once.

   (c) No, D→B→E→A→B, is not a Hamilton circuit since it does not visit C.

   (d) No, E→D→C→B→E, is not a Hamilton circuit since it does not visit A.

3. (a) The path, A→B→C→D→E→A, is not a circuit (since BC is not an edge), and hence, is not an Euler circuit nor a Hamilton circuit.

   (b) The path, B→E→C→D→A→B, is a circuit, is an Euler circuit (travels each edge only once), and is a Hamilton circuit (travels through each vertex only once).

(c) The path, E→B→A→D→A→D→C→E, is not a circuit, is not an Euler circuit (travels edge AD twice), and is not a Hamilton circuit (travels through several vertices more than once).

5. This graph has a Hamilton circuit. One example is A→B→D→E→F→C→A.

7. This graph has a Hamilton circuit. One example is G→H→J→I→G.

9. This graph has a Hamilton circuit. One example is X→T→U→W→V→X.

11.

A→B→C→D→A is a Hamilton circuit. There is no Euler circuit since at least one of the vertices has odd degree. (In fact, all have odd degree.)

13.

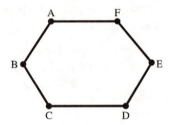

A→B→C→D→E→F→A is both a Hamilton and an Euler circuit.

15. This exercise could be solved by a Hamilton circuit since each vertex represents a bandstand.

17. This exercise could be solved by an Euler circuit since each path corresponds to an edge.

19. This exercise could be solved by a Hamilton circuit since each vertex represents a country.

21. $4! = 4 \cdot 3 \cdot 2 \cdot 1 = 24$.

23. $9! = 9 \cdot 8 \cdot 7 \cdot 6 \cdot 5 \cdot 4 \cdot 3 \cdot 2 \cdot 1 = 362,880$

25. There are

$$(10 - 1)! = 9!$$

Hamilton circuits.

27. There are

$$(18 - 1)! = 17!$$

Hamilton circuits.

29. Since this is a complete graph with 4 vertices, there are

$$(4 - 1)! = 3! = 3 \cdot 2 \cdot 1 = 6$$

Hamilton circuits. Choosing P as a beginning vertex, they are:

P→Q→R→S→P; P→Q→S→R→P; P→R→Q→S→P;
P→R→S→Q→P; P→S→Q→R→P; P→S→R→Q→P.

Note: A tree diagram might be helpful here. P would represent the first node; Q, R and S, the second node; the third node would represent each of the remaining two vertices.

31. Hamilton circuits starting with E→H→I would include:

E→H→I→F→G→E and E→H→I →G→F→E.

33. Hamilton circuits starting with E→F would include:

E→F→G →H→I→E;   E→F→G →I→H→E;
E→F→H→G→I→E;   E→F →H→I→G→E;
E→F→I→G→H→E;   E→F→I→H→G→E.

35. Hamilton circuits starting with E→G would include:

E→G→F→H→I→E;   E→G→F→I→H→E;
E→G→H→F→I→E;   E→G →H→I→F→E;
E→G→I→F→H→E;   E→G→I→H→F→E.

37. Hamilton circuits starting with E would include:

A→B→C→D→E→A;   A→C→B→D→E→A;
A→B→C→E→D→A;   A→C→B→E→D→A;
A→B→D→C→E→A;   A→C→D→B→E→A;
A→B→D→E→C→A;   A→C→D→E→B→A;
A→B→E→C→D→A;   A→C→E→B→D→A;
A→B→E→D→C→A;   A→C→E→D→B→A;
A→D→B→C→E→A;   A→E→B→C→D→A;
A→D→B→E→C→A;   A→E→B→D→C→A;
A→D→C→B→E→A;   A→E→C→B→D→A;
A→D→C→E→B→A;   A→E→C→D→B→A;
A→D→E→B→C→A;   A→E→D→B→C→A;
A→D→E→C→B→A;   A→E→D→C→B→A.

39. Using the Brute Force Algorithm:

Circuit:	Total weight of circuit:
1. P→Q→R→S→P	$550 + 640 + 500 + 510 = 2200$
2. P→Q→S→R→P	$550 + 790 + 500 + 600 = 2440$
3. P→S→Q→R→P	$510 + 790 + 640 + 600 = 2540$
4. P→R→S→Q→P (opposite of 2)	2440
5. P→S→R→Q→P (opposite of 1)	2200
6. P→R→Q→S→P (opposite of 3)	2540

Thus, the Minimum Hamilton circuit is P→Q→R→S→P and the weight is 2200.

41. Observe that for a complete graph with 5 vertices we will have $(5-1)=4!=24$ Hamilton circuits.

Circuit:	Total weight of circuit:
1. C→D→E→F→G→C	$12+10+17+15+10=64$
2. C→D→E→G→F→C	$12+10+13+15+15=65$
3. C→F→D→E→G→C	$15+21+10+13+10=69$
4. C→D→F→G→E→C	$12+21+17+13+10=73$
5. C→E→D→F→G→C	$15+10+21+15+10=71$
6. C→D→G→E→F→C	$12+13+13+17+15=70$
7. C→E→F→D→G→C	$15+17+21+13+10=76$
8. C→E→D→G→F→C	$15+10+13+15+15=68$
9. C→F→E→D→G→C	$15+17+10+13+10=65$
10. C→F→G→D→E→C	$15+15+13+10+15=68$
11. C→D→F→E→G→C	$12+21+17+13+10=73$
12. C→D→F→G→E→C	$12+21+15+13+10=71$

13. − 24. (opposites of above)

Thus, the Minimum Hamilton circuit is C→D→E→F→G→C and the weight is 64.

43. (a) Using the nearest neighbor algorithm, choose the first edge with the minimum weight, A$\xrightarrow{1}$C. Keep track of weight by noting its value over the arrow. Choose the second edge, C$\xrightarrow{2}$E, which has the minimum weight. Choose the third edge, E$\xrightarrow{7}$D, which has the minimum weight. Choose the fourth edge, D$\xrightarrow{6}$B, which has the minimum weight. Finally, choose the remaining edge, B$\xrightarrow{4}$A. Note that this is your only choice for the last remaining edge. The resulting approximate minimum Hamilton circuit is, therefore,

$$A→C→E→D→B→A.$$

The circuit has a total weight of

$$1+2+7+6+4=20.$$

(b) Using the nearest neighbor algorithm, choose the edge which has the minimum weight, C$\xrightarrow{1}$A. Keep track of weight by noting its value over the arrow. Choose the second edge with corresponding minimum weight, A$\xrightarrow{4}$B. Choose the third edge, which has the minimum weight, B$\xrightarrow{6}$D. Choose the fourth edge, which has the minimum weight, D$\xrightarrow{7}$E. Finally, choose the remaining edge, E$\xrightarrow{2}$C. The resulting approximate minimum Hamilton circuit is, therefore,

$$C→A→B→D→E→C.$$

The circuit has a total weight of

$$1+4+6+7+2=20.$$

(c) Using the nearest neighbor algorithm, choose the edge which has the minimum weight, D$\xrightarrow{3}$C. Choose the second edge with corresponding minimum weight, C$\xrightarrow{1}$A. Choose the third edge, which has the minimum weight, A$\xrightarrow{4}$B. Choose the fourth edge, which has the minimum weight, B$\xrightarrow{8}$E. Finally, choose the remaining edge, E$\xrightarrow{7}$D.

The resulting approximate minimum Hamilton circuit is, therefore,

$$D→C→A→B→E→D.$$

The circuit has a total weight of

$$3+1+4+8+7=23.$$

(d) Using the nearest neighbor algorithm, choose the edge which has the minimum weight, E$\xrightarrow{2}$C. Choose the second edge with corresponding minimum weight, C$\xrightarrow{1}$A. Choose the third edge, which has the minimum weight, A$\xrightarrow{4}$B. Choose the fourth edge, which has the minimum weight, B$\xrightarrow{6}$D. Finally, choose the remaining edge, D$\xrightarrow{7}$E. The resulting approximate minimum Hamilton circuit is, therefore,

$$E→C→A→B→D→E.$$

The circuit has a total weight of

$$2+1+4+6+7=20.$$

45. (a) Beginning with vertex A, choose the edge with the minimum weight, A$\xrightarrow{5}$C. Choose the second edge, C$\xrightarrow{7}$D. Choose the third edge, D$\xrightarrow{10}$E. Choose the fourth edge, E$\xrightarrow{50}$B. Finally, to the last vertex, A, we must choose B$\xrightarrow{15}$A. The resulting approximate minimum Hamilton circuit is

$$A→C→D→E→B→A.$$

The circuit has a total weight of

$$5+7+10+50+15=87.$$

Beginning with vertex B, choose the edge with the minimum weight, B$\xrightarrow{11}$C. Choose the second edge, C$\xrightarrow{5}$A. Choose the third edge, A$\xrightarrow{9}$E. Choose the fourth edge, E$\xrightarrow{10}$D. Finally, to the last vertex, B, we must choose D$\xrightarrow{60}$B. The resulting approximate minimum Hamilton circuit is

$$B→C→A→E→D→B.$$

The circuit has a total weight of

$$11+5+9+10+60=95.$$

Beginning with vertex C, choose the edge with the minimum weight, C$\xrightarrow{5}$A. Choose the second edge, A$\xrightarrow{9}$E. Choose the third edge, E$\xrightarrow{10}$D. Choose the fourth edge, D$\xrightarrow{60}$B. Finally, to the last vertex, C, we must choose B$\xrightarrow{11}$C. The resulting approximate minimum Hamilton circuit is

$$C→A→E→D→B→C.$$

The circuit has a total weight of

$$5+9+10+60+11=95.$$

Beginning with vertex D, choose the edge with the minimum weight, D$\xrightarrow{7}$C. Choose the second edge, C$\xrightarrow{5}$A. Choose the third edge, A$\xrightarrow{9}$E. Choose the fourth edge, E$\xrightarrow{50}$B. Finally, to the last vertex, D, we must choose B$\xrightarrow{60}$D. The resulting approximate minimum Hamilton circuit is

$$D \rightarrow C \rightarrow A \rightarrow E \rightarrow B \rightarrow D.$$

The circuit has a total weight of

$$7 + 5 + 9 + 50 + 60 = 131.$$

Beginning with vertex E, choose the edge with the minimum weight, E$\xrightarrow{8}$C. Choose the second edge, C$\xrightarrow{5}$A. Choose the third edge, A$\xrightarrow{14}$D. Choose the fourth edge, D$\xrightarrow{60}$B. Finally, to the last vertex, D, we must choose B$\xrightarrow{50}$E. The resulting approximate minimum Hamilton circuit is

$$E \rightarrow C \rightarrow A \rightarrow D \rightarrow B \rightarrow E.$$

The circuit has a total weight of

$$8 + 5 + 14 + 60 + 50 = 137.$$

(b) The best solution is the circuit

$$A \rightarrow C \rightarrow D \rightarrow E \rightarrow B \rightarrow A.$$

since its total weight, 87, is the smallest.

(c) One example would be

$$A \rightarrow B \rightarrow C \rightarrow D \rightarrow E \rightarrow A$$

with a total weight of

$$15 + 11 + 7 + 10 + 9 = 52.$$

47. All Hamilton circuits include:

A→B→C→D→E→F→A; A→B→C→F→E→D→A;
A→B→E→D→C→F→A; A→B→E→F→C→D→A;
A→D→E→F→C→B→A; A→D→E→B→C→F→A;
A→D→C→B→E→F→A; A→D→C→F→E→B→A;
A→F→E→B→C→D→A; A→F→E→D→C→B→A;
A→F→C→D→E→B→A; A→F→C→B→E→D→A.

A tree diagram, such as below, can be a help in creating all of the Hamilton circuits.

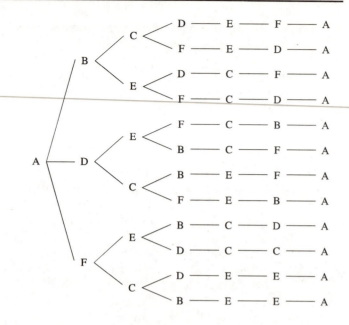

49. All Hamilton circuits include:

$$A \rightarrow B \rightarrow C \rightarrow D \rightarrow E \rightarrow F \rightarrow A;$$
$$A \rightarrow B \rightarrow C \rightarrow E \rightarrow D \rightarrow F \rightarrow A;$$
$$A \rightarrow B \rightarrow C \rightarrow E \rightarrow F \rightarrow D \rightarrow A;$$
$$A \rightarrow B \rightarrow C \rightarrow F \rightarrow E \rightarrow D \rightarrow A;$$
$$A \rightarrow D \rightarrow E \rightarrow F \rightarrow C \rightarrow B \rightarrow A;$$
$$A \rightarrow D \rightarrow F \rightarrow E \rightarrow C \rightarrow B \rightarrow A;$$
$$A \rightarrow F \rightarrow D \rightarrow E \rightarrow C \rightarrow B \rightarrow A;$$
$$A \rightarrow F \rightarrow E \rightarrow D \rightarrow C \rightarrow B \rightarrow A.$$

51. (a) For graph (a), $n/2$: $6/2 = 3$. Condition not satisfied. For example, vertex A is degree 2
For graph (b), $n/2$: $6/2 = 3$. Condition is satisfied since all vertices are 3 or larger in degree.
For graph (c), $n/2$: $5/2 = 2.5$. Condition not satisfied. For example, vertex G is degree 2.
For graph (d), $n/2$: $5/2 = 2.5$. Condition is satisfied since all vertices are 3 or larger in degree.
For graph (e), $n/2$: $7/2 = 3.5$. Condition not satisfied. For example, vertex B is degree 3.

(b) By Dirac's theorem, graphs (b) and (d) are predicted to have Hamilton circuits.

(c) We can not be sure that a graph which doesn't satisfy Dirac's theorem will not have a Hamilton circuit. For example, graph (a), which doesn't satisfy Dirac's theorem, has a Hamilton circuit

$$A \rightarrow B \rightarrow C \rightarrow D \rightarrow E \rightarrow F \rightarrow A.$$

(d) Dirac's theorem is not true for $n < 3$ since such a graph will not have any Hamilton circuit at all.

(e) The degree of each vertex in a complete graph with n vertices is $(n-1)$. If $n \geq 3$, then $(n-1) \geq n/2$. Thus, we can conclude that the graph has a Hamilton circuit.

Note that there are different possible answers for question 53.

53. A→F→G→R→S→T→U→Q→P→N→M→L→K→J→I→H→ B→C→D→E→A is a Hamilton circuit.

55. Writing exercise

57. Writing exercise

14.4 EXERCISES

1. The graph is a tree since it is connected and has no circuits.

3. The graph is a not a tree since it is not connected.

5. The graph is a tree since it is connected and has no circuits.

7. Such a graph is a not a tree since it has a circuit. Remember that Euler's theorem tells us that all connected graphs with all vertices of even degree will have an Euler circuit.

9. It is not possible to form a tree by adding an extra edge since the graph has a circuit.

11. Yes, such a graph is a tree since it is connected and has no circuits.

13. No, such a graph would not necessarily be a tree since circuits may be formed.

15. The statement "Every connected graph in which each edge is a cut edge is a tree" is true. Since all edges are cut edges, there can be no circuits.

17. The statement "Every graph in which each edge is a cut edge is a tree" is false. For example, disconnected graphs as the following

are not trees.

19.

21.

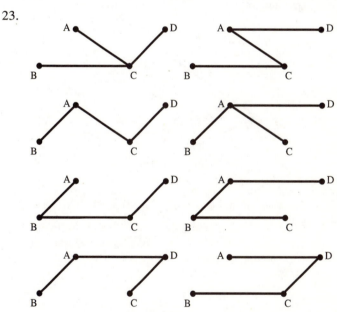

23.

25. There are two circuits in this graph. We can drop any one of 4 edges from the first circuit and any one of 5 edges from the second circuit to form a spanning tree. Thus, there would be a total of $4 \times 5 = 20$ different spanning trees.

27. In a connected graph has circuits, none of which have common edges, the number of spanning trees for the graph is the product of the number of degrees in each circuit.

29. Choose the edge with minimum weight GF (2). Choose the next edge with minimum weight, FC (3). The next edge with minimum weight is FD (8). The next edge of choice would be DE (10). Note here that, although DC represents a remaining edge with minimum weight (9), if it were chosen, we would then have a circuit, FDC, which is not allowed. Choose next edge with minimum weight FB (12) followed by BA (16). Note BG (15) is not allowed because the circuit, BGF, is then formed. We now have the following minimum spanning tree:

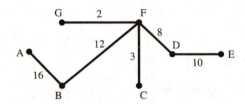

The total weight is $2 + 3 + 8 + 10 + 12 + 16 = 51$.

31. Choose initial minimum edge AD (6), followed by BG (8), then EF (9), then DC (10), and then AB (13). The last remaining minimum edge to choose, without completing a circuit, is AE (20). Thus, the minimum spanning tree is:

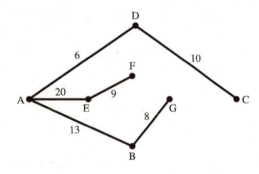

The total weight is $6 + 8 + 9 + 10 + 13 + 20 = 66$.

33. Choose the initial minimum edge, BF (23), followed by either AE (25) or EF (25). Then choose the other edge with weight 25. Finally, choose AD (32) followed by DC (35).

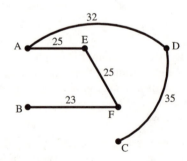

The total length of the minimum pathway is

$$23 + 25 + 25 + 32 + 35 = 140 \text{ ft.}$$

35. A tree with 34 vertices will have one less, or 33 edges.

37. A spanning tree for a complete graph with 63 vertices will have one less, or 62 edges.

39. Different spanning trees must have the same number of edges since the number of vertices in the tree is the number of vertices in the original graph, and the number of edges has to be one less than the number of vertices.

41. Consider a tree with 10 vertices.
    (a) There will be $10 - 1 = 9$ edges.

    (b) The sum of the degrees for all vertices will be twice the number of edges, or $2 \times 9 = 18$.

    (c) The smallest number of vertices of degree four on this graph will be 0. For example, a tree may be drawn with the first and last vertex of degree 1, and the remaining 8 vertices of degree 2.

    (d) The largest number of vertices of degree four in this graph is 2. Let us use the strategy of trial and error. If we use 4 vertices with degree 4, this will contribute 16 to the total degree and would leave $18 - 16 = 2$ as the degree sum of the remaining 6 vertices. But this means that some of those vertices would have no edges joined to them, so that our graph would not be connected, and would not be a tree. Similarly, using 3 vertices with degree 4 would leave $18 - 12 = 6$ as the sum of (the remaining vertex) degrees. Thus, of the last 7 vertices, at least one would have no degree (or connecting edge). This would not be a problem if we choose two vertices with degree 4. The following tree, for example, would satisfy our conditions.

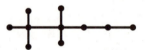

43. Treating each of the 23 employee's computers as a vertex and the network as a tree, there will be one less edge than the number of vertices. Hence, there will be 22 cables that will have to be run between the computers.

45. It is possible to draw in the same number of vertices as edges. The graph must be a tree since we still have one less edge than vertex.

47. It is possible to draw in more edges than vertices. But the graph can not be a tree since we would now have at least as many edges as vertices.

49. Using Cayley's theorem with $n = 3$, "A complete graph with n vertices has n^{n-2} spanning trees" we arrive at $3^{3-2} = 3^1 = 3$ spanning trees. A complete graph is as follows.

The three spanning trees are as follows.

51. For a graph with 5 vertices, we will have $5^{5-2} = 5^3 = 125$ spanning trees.

53. There are just 3 non-isomorphic trees with 5 vertices. They are:

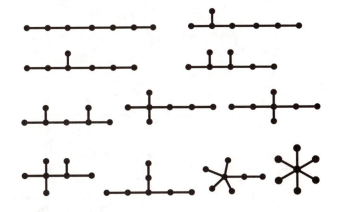

55. There are 11 non-isomorphic trees with 7 vertices. They are:

57. Writing exercise

CHAPTER 14 TEST

1. By counting, there are 7 vertices.

2. The sum of the degrees of the vertices are:
$$4 + 2 + 2 + 2 + 6 + 2 + 2 = 20.$$

3. By counting, there are 10 edges.

4. (a) No, B→A→C→E→B→A is not a path since edge AB is used twice.

   (b) Yes, A→B→E→A is a path.

   (c) No, A→C→D→E is not a path since there is no edge from C to D.

5. (a) Yes, A→B→E→D→A is a circuit.

   (b) No, A→B→C→D→E→F→G→A is not a circuit since, for example, there is no edge from B to C.

   (c) Yes, A→B→E→F→G→E→D→A→E→C→A is a circuit.

6. A graph with 2 components, for example, is:

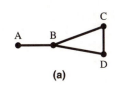

7. The sum of degrees of the vertices is:
$$4 + 4 + 4 + 2 + 2 + 2 + 2 + 2 + 2 + 2 = 26.$$

Thus, there are $26 \div 2 = 13$ edges.

8. The graphs are isomorphic.

(a)   (b)

9. The graph is:

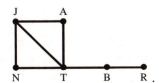

The graph is connected. Tina knows the largest number of other guests.

10. Let each of the 8 contestants represent a vertex of a complete graph (graph with each vertex connected by an edge to all other vertices). The degree of each vertex is 7. Thus, the sum of the degrees is $8 \cdot 7 = 56$. Since the sum of the degrees is twice the number of edges, there must by $56 \div 2 = 28$ edges. Since each edge represents a game to be played, there are 28 games in the competition.

11. Yes, the graph is a complete graph since there is an edge from each vertex to each of the remaining 6 vertices.

12. (a) No, A→B→E→D→A is not an Euler circuit since it does not use all of the edges.

(b) No, A→B→C→D→E→F→G→A is not a circuit because, for example, there is no edge from B to C.

(c) Yes, A→B→E→F→G→E→D→A→E→C→A is an Euler circuit.

13. No, the graph will not have an Euler circuit since some of the vertices have odd degree.

14. Yes, the graph will have an Euler circuit since all vertices have even degree.

15. No, since two of the rooms have an odd number of doors. Note that we are considering each room to be a vertex and asking the question "Can an Euler circuit be formed?"

16. A resulting Euler circuit is F→B→E→D→B→C→D→K→ B→A→H→G→F→A→G→J→F. Note, BF is the only cut edge after F→B, thus you may choose any vertex in the right subgraph after B.

17. (a) No, A→B→E→D→A, is not a Hamilton circuit since it does not visit all vertices.

(b) No, A→B→C→D→E→F→G→A is not a circuit because, for example, there is no edge from B to C.

(c) No, A→B→E→F→G→E→D→A→E→C→A is not an Hamilton circuit since it visits some vertices twice before returning to starting vertex.

18. F→G→H→I→E→F;   F→G→H→E→I→F;
    F→G→I→H→E→F;   F→G→I→E→H→F;
    F→G→E→H→I→F;   F→G→E→I→H→F.

There are 6 such Hamilton circuits.

19. Using the Brute force algorithm and P as the starting vertex we get the following circuits. Use a tree diagram as an aid.

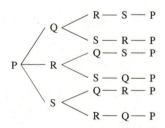

Circuit:		Total weight of circuit:
1. P→Q→R→S→P		$7 + 8 + 7 + 7 = 29$
2. P→Q→S→R→P		$7 + 9 + 7 + 4 = 27$
3. P→R→Q→S→P		$4 + 8 + 9 + 7 = 28$
4. P→R→S→Q→P (opposite of 2)		27
5. P→S→Q→R→P (opposite of 3)		28
6. P→S→R→Q→P (opposite of 1).		29.

Thus, P→Q→S→R→P is the minimum Hamilton circuit with a weight of 27.

20. Using the nearest neighbor algorithm, choose the first edge with the minimum weight, A$\overset{1.6}{\to}$E. Keep track of weight by noting its value over the arrow. Choose the second edge, E$\overset{1.95}{\to}$D, which has the minimum weight. Choose the third edge, D$\overset{1.8}{\to}$C, which has the minimum weight. Choose C$\overset{2.3}{\to}$F. Choose F$\overset{1.7}{\to}$B. Finally, choose B$\overset{2.5}{\to}$A. Note that this is your only choice for the last remaining edge. The resulting approximate minimum Hamilton circuit is, therefore,

$$A→E→D→C→F→B→A.$$

The circuit has a total weight of

$$1.6 + 1.95 + 1.8 + 2.3 + 1.7 + 2.5 = 11.85.$$

21. For a complete graph with 25 vertices, there will be $(25 - 1)! = 24!$ Hamilton circuits.

22. This problem calls for a Hamilton circuit since the band wants to visit each city (vertex) only once.

23. Any three of the following, for example, would satisfy the stated conditions:

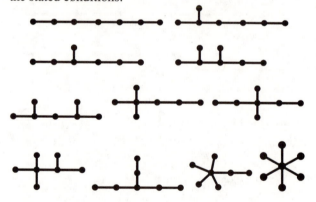

24. The statement "Every tree has a Hamilton circuit" is false since in many trees, you will have to visit the same vertex more than once.

25. The statement "In a tree each edge is a cut edge" is true.

26. The statement "Every tree is connected" is true since one can always move from each vertex of the graph along edges to every other vertex.

27. The following represent the different spanning trees for the accompanying graph.

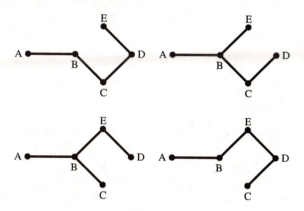

There are 4 spanning trees.

28. Using Kruskal's algorithm, choose the edge with minimum weight, B$\overset{3}{\to}$E. Choose the next edge with minimum weight, E$\overset{5}{\to}$D. Continuing in this fashion, we choose C$\overset{7}{\to}$D followed by D$\overset{9}{\to}$A.

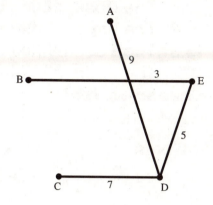

The weight is

$$3 + 5 + 7 + 9 = 24.$$

29. The number of edges in a tree is one less than the number of vertices. Thus, there are $50 - 1 = 49$ edges.

15   VOTING AND APPORTIONMENT

15.1   The Possibilities of Voting
15.2   The Impossibilities of Voting
15.3   The Possibilities of Apportionment
15.4   The Impossibilities of Apportionment
Chapter 15 Test

Chapter Goals

After completing this chapter the student should be able to

- Understand and use several different voting methods.
- Show that voting methods can violate fairness properties.
- Understand and use different methods of apportionment.
- Identify the flaws and paradoxes of apportionment methods.

Chapter Summary

There are several different methods of voting that can be used. The Plurality method selects a winner by the greatest number of first place votes.

EXAMPLE A   Plurality Method

Here is a voter profile for a contest among four candidates.

Votes	Ranking
3	a > b > c > d
1	b > c > a > d
2	d > b > c > a
1	c > b > a > d
2	a > d > c > b
1	b > c > d > a
1	c > b > d > a
2	a > b > d > c

By the plurality method the candidate with the most votes wins. Examine the table above to see the total number of first place votes for each candidate.

Candidate	1st place votes
a	$3 + 2 + 2 = 7$
b	$1 + 1 = 2$
c	$1 + 1 = 2$
d	$2 = 2$

Candidate a has the most first place votes and is the winner.

Another common voting method is the pairwise comparison method in which pairs of candidates are compared and assigned points for the winning the comparison. Here is an example and explanation of this method.

EXAMPLE B   Pairwise Comparison Method

Given the following voter profile, here is an explanation of the pairwise comparison method.

Votes	Ranking	Row
4	a > b > c	1
2	b > c > a	2
4	b > a > c	3
3	c > a > b	4

First each pair of candidates must be compared. Make a table to compute the votes in comparisons.

Comparison	Votes	Rows
a > b	$4 + 3 = 7$	1, 4
b > a	$2 + 4 = 6$	2, 3
a > c	$4 + 4 = 8$	1, 3
c > a	$2 + 3 = 5$	2, 4
b > c	$4 + 2 + 4 = 10$	1, 2, 3
c > b	$3 = 3$	4

Next make another table to award pairwise points. The winner of each comparison is awarded one point; if there is a tie, each receives 1/2 point.

Pairs	Votes	Pairwise Points
a : b	7 : 6	a, 1
a : c	8 : 5	a, 1
b : c	10 : 3	b, 1

The winner is candidate a with 2 pairwise points. Candidate b received only 1 point, and candidate c received 0 points.

A third voting method is the Borda method, which assigns points to each candidate according to the position in the ranking. The points are then summed, and the winner is the candidate with the highest point total. Here is an example and explanation of the Borda method.

EXAMPLE C Borda Method

Using the same voter profile from Example B, make a table to examine the points that are assigned according to the ranking of the candidate. Because there are three choices, a first place vote awards 2 points, a second place vote awards 1 point, and a last place vote awards 0 points.

Votes	Ranking	Points 2	1	0
4	a > b > c	a	b	c
2	b > c > a	b	c	a
4	b > a > c	b	a	c
3	c > a > b	c	a	b

Examine the last three columns and compute the weighted sum for each candidate. That is, notice that four first-place votes have been given to candidate a for a total of $4 \cdot 2 = 8$ points; four second-place votes have been given to candidate b for a total of $4 \cdot 1 = 4$ votes; and four third-place votes have been given to candidate c for a total of $4 \cdot 0 = 0$ votes.

Candidate	Borda Points
a	$4 \cdot 2 + 2 \cdot 0 + 4 \cdot 1 + 3 \cdot 1 = 15$
b	$4 \cdot 1 + 2 \cdot 2 + 4 \cdot 2 + 3 \cdot 0 = 16$
c	$4 \cdot 0 + 2 \cdot 1 + 4 \cdot 0 + 3 \cdot 2 = 8$

The winner is candidate b with 16 Borda points.

The Hare method looks for the candidate with the majority of first-place votes. If no candidate has a majority, the candidate(s) with the least number of votes are eliminated, and another vote is taken. This process continues until one candidate receives a majority of the votes.

EXAMPLE D Hare Method

Using the same voter profile from Examples B and C, there are 13 voters so a candidate must receive a majority

of the votes or 7 votes to win by the Hare Method. No candidate received a majority; candidates a, b, and c received 4, 6, and 3 votes respectively. By this method, candidate c is eliminated, and another vote is taken. Now compare only candidates a and b.

Comparison	Votes	Rows
a > b	$4 + 3 = 7$	1, 4
b > a	$2 + 4 = 6$	2, 3

Candidate a now receives a majority and wins.

The sequential pairwise method is a variation of the plurality method compares the entire field of candidates two at a time in a predetermined order. For each comparison, the candidate with the greater number of votes wins and goes on to challenge the next candidate.

EXAMPLE E Sequential Pairwise Method

Using the same voter profile from Examples B, C and D and the predetermined order c, b, a, begin by comparing c and b in the original voting.

Comparison	Votes	Row
b > c	$4 + 2 + 4 = 10$	1, 2, 3
c > b	$3 = 3$	4

Candidate b wins this competition, 10 to 3. Now compare b to a.

Comparison	Votes	Row
b > a	$2 + 4 = 6$	2, 3
a > b	$4 + 3 = 7$	1, 4

Candidate a wins the next competition, 7 to 6, and also wins the entire competition in the sequential pairwise comparison method.

Finally, the approval voting method is also another form of the plurality method. Each of the voters gives approval to as many candidates as he wishes, and the candidate with the most approval votes wins the election.

Unfortunately each of these voting methods often has flaws and problems associated with them. These flaws and problems are identified by the fact that they violate one or more of the following criteria for fairness: the majority criterion, the Condorcet criterion, the monotonicity criterion, and the independence of irrelevant alternatives criterion. The majority criterion is straightforward: If a candidate wins a majority of the votes in a voter profile, that candidate should be the winner of the election. The following example will show, however, that this criterion can be violated, for example, by the Borda method of voting.

EXAMPLE F   **Violation of Majority Criterion by the Borda Method**

Here is a profile for 36 voters.

Votes	Ranking
4	$b > c > a > d$
9	$c > a > d > b$
4	$a > c > b > d$
19	$d > c > b > a$

Notice that d has the majority of first place votes. Because there are 36 voters, at least 19 votes constitutes a majority. Alternative d received 19 votes. According to the majority criterion, alternative d should win the election.

Now use the information make a table to examine the points in the Borda method.

Votes	Ranking	Points			
		3	2	1	0
4	$b > c > a > d$	b	c	a	d
9	$c > a > d > b$	c	a	d	b
4	$a > c > b > d$	a	c	b	d
19	$d > c > b > a$	d	c	b	a

Examine the last four columns and compute the weighted sum for each alternative.

Alternative	Borda Points
a	$4 \cdot 1 + 9 \cdot 2 + 4 \cdot 3 + 19 \cdot 0 = 34$
b	$4 \cdot 3 + 9 \cdot 0 + 4 \cdot 1 + 19 \cdot 1 = 35$
c	$4 \cdot 2 + 9 \cdot 3 + 4 \cdot 2 + 19 \cdot 2 = 81$
d	$4 \cdot 0 + 9 \cdot 1 + 4 \cdot 0 + 19 \cdot 3 = 66$

The winner is alternative c with 81 Borda points. The Borda method violates the majority criterion because it fails to select the majority candidate.

The Condorcet candidate is the candidate that wins in a pairwise comparison with every other candidate. Certain voting techniques violate the Condorcet criterion, as in the following example.

EXAMPLE G   **The Condorcet Criterion**

Here is a profile for 13 voters.

Votes	Ranking
3	$a > c > b$
4	$c > b > a$
2	$b > a > c$
4	$b > c > a$

First make a table to compute the votes in comparisons:

Comparison	Votes	Rows
$a > b$	$3 = 3$	1
$b > a$	$4 + 2 + 4 = 10$	2, 3, 4
$a > c$	$3 + 2 = 5$	1, 3
$c > a$	$4 + 4 = 8$	2, 4
$b > c$	$2 + 4 = 6$	3, 4
$c > b$	$3 + 4 = 7$	1, 2

Make another table to award pairwise points.

Pairs	Votes	Pairwise Points
$a : b$	$3 : 10$	b, 1
$a : c$	$5 : 8$	c, 1
$b : c$	$6 : 7$	c, 1

Using the pairwise comparison method, the winner is candidate c with 2 pairwise points and is the Condorcet candidate.

Now examine the voter profile to see that candidate a receives 3 first place votes; c receives 4 first place votes; b receives $2 + 4 = 6$ first place votes to win by the plurality method. Because c is not the winner, the plurality method violates the Condorcet criterion.

Next test the Hare method of voting. There are 13 votes, which makes 7 votes a majority. No candidate wins a majority on this vote. Candidate a has the fewest votes, so compare only b and c.

Comparison	Votes	Rows
$b > c$	$2 + 4 = 6$	3, 4
$c > b$	$3 + 4 = 7$	1, 2

Candidate c now has a majority and is selected by the Hare method. This agrees with the Condorcet candidate.

Finally, test the Borda method of voting. Use the voter profile to make a table to examine the points in the Borda method.

Votes	Ranking	Points		
		2	1	0
3	$a > c > b$	a	c	b
4	$c > b > a$	c	b	a
2	$b > a > c$	b	a	c
4	$b > c > a$	b	c	a

Examine the last three columns and compute the weighted sum for each alternative.

Alternative	Borda Points
a	$3 \cdot 2 + 4 \cdot 0 + 2 \cdot 1 + 4 \cdot 0 = 8$
b	$3 \cdot 0 + 4 \cdot 1 + 2 \cdot 2 + 4 \cdot 2 = 16$
c	$3 \cdot 1 + 4 \cdot 2 + 2 \cdot 0 + 4 \cdot 1 = 15$

Candidate b wins with 16 Borda points.

In conclusion, the plurality and Borda methods violate the Condorcet criterion in this example, because neither method selects candidate c, the Condorcet candidate. The Hare Method does not violate the criterion because the Condorcet candidate wins.

The monotonicity criterion states that the winner of an election should also be the winner of a second election, if voters rearrange their rankings and move that candidate to first place in a second vote. Again, this criterion is sometimes violated by some of the voting methods.

EXAMPLE H   The Monotonicity Criterion

Here is a profile for 19 voters.

Votes	Ranking
7	$c > d > s > b$
5	$d > b > s > c$
3	$b > c > s > d$
4	$s > c > b > d$

Suppose a preliminary non-binding vote is taken by the Borda method. First make a table to examine the points.

Votes	Ranking	3	2	1	0
7	$c > d > s > b$	c	d	s	b
5	$d > b > s > c$	d	b	s	c
3	$b > c > s > d$	b	c	s	d
4	$s > c > b > d$	s	c	b	d

Examine the last four columns and compute the weighted sum for each activity.

Candidate	Borda Points
c	$7 \cdot 3 + 5 \cdot 0 + 3 \cdot 2 + 4 \cdot 2 = 35$
d	$7 \cdot 2 + 5 \cdot 3 + 3 \cdot 0 + 4 \cdot 0 = 29$
s	$7 \cdot 1 + 5 \cdot 1 + 3 \cdot 1 + 4 \cdot 3 = 27$
b	$7 \cdot 0 + 5 \cdot 2 + 3 \cdot 3 + 4 \cdot 1 = 23$

The winner is c with 35 Borda points in the preliminary non-binding decision.

Now the voters in the last two rows change their ranking for the official vote. Here is the new voter profile showing that change.

Votes	Ranking
7	$c > d > s > b$
5	$d > b > s > c$
3	$c > d > b > s$
4	$c > d > s > b$

Make another table to examine the Borda points.

Votes	Ranking	3	2	1	0
7	$c > d > s > b$	c	d	s	b
5	$d > b > s > c$	d	b	s	c
3	$c > d > b > s$	c	d	b	s
4	$c > d > s > b$	c	d	s	b

Examine the last four columns and compute the weighted sum for each activity.

Candidate	Borda Points
c	$7 \cdot 3 + 5 \cdot 0 + 3 \cdot 3 + 4 \cdot 3 = 42$
d	$7 \cdot 2 + 5 \cdot 3 + 3 \cdot 2 + 4 \cdot 2 = 43$
s	$7 \cdot 1 + 5 \cdot 1 + 3 \cdot 0 + 4 \cdot 1 = 16$
b	$7 \cdot 0 + 5 \cdot 2 + 3 \cdot 1 + 4 \cdot 0 = 13$

This time the winner is d with 43 Borda points. The monotonicity criterion is violated because candidate c does not win the second election.

Finally the independence of irrelevant alternatives criterion says that if a candidate wins a first election, he should also win a second election if one or more of the losing candidates withdraws. Here is an example of that situation.

EXAMPLE I   The Independence of Irrelevant Alternatives Criterion

Four candidates, a, b, c, and d are ranked by 95 voters according to the following profile.

Votes	Ranking
40	$a > c > d > b$
30	$c > a > b > d$
15	$b > c > d > a$
10	$d > b > c > a$

Calculate how many points each candidate obtained from the 95 first place votes from the voter profile:

Candidate	1st place votes
a	40
b	15
c	30
d	10

Candidate a receives the most votes and wins by the plurality method.

Now compare the rankings if candidate b drops out.

Votes	Ranking	Row
40	a > c > d	1
30	c > a > d	2
15	c > d > a	3
10	d > c > a	4

$$
\begin{aligned}
a & \quad 40 = 40 \\
c & \quad 30 + 15 = 45 \\
d & \quad 10 = 10
\end{aligned}
$$

Candidate c wins with 45 votes. The Independence of Irrelevant Alternatives criterion is violated because candidate a does not win the second election.

Several different methods of apportionment have been developed over the course of history: Hamilton, Jefferson, Webster, Adams, and Hill-Huntington. Each method attempts to apportion seats in the House of Representatives in a fair way. Each method calculates a divisor and then a quota, followed by rounding rules leading toward final apportionment. The Hamilton method follows the steps in the next example.

EXAMPLE J   The Hamilton Method

Here is an enrollment profile for four classes at a university. The math/science department has nine teaching assistant positions that must be assigned to each course.

Course	Bio	Chem	Physics	Math	Total
Enrollment	35	40	28	95	198

Using the Hamilton method:

1. Compute a standard divisor, d, by dividing the total enrollment by the number of teaching assistants.

$$d = \frac{198}{9} = 22$$

2. Compute a standard quota, Q, for each course by dividing the number of students enrolled by the standard divisor.

$$Q = \frac{\text{enrollment}}{d}$$

3. Round each standard quota down to the nearest integer.

Course	Enr.	Q	Rounded Q
Bio	35	$\frac{35}{22} \approx 1.591$	1
Chem	40	$\frac{40}{22} \approx 1.818$	1
Physics	28	$\frac{28}{22} \approx 1.273$	1
Math	95	$\frac{95}{22} \approx 4.318$	4
Totals	198		7

4. Give any additional assistants, one at a time, to the course(s) with the largest fractional part of their standard quota.

The remaining two teaching assistants will be assigned to chemistry and biology. Notice the fractional part of chemistry's quota is .818 and the fractional part of biology's quota is .591. These are the courses with the largest fractional portion of their quotas. The final apportionment is

Course	Enr.	Q	Number of TA's
Bio	35	1.591	$1 + 1 = 2$
Chem	40	1.818	$1 + 1 = 2$
Physics	28	1.273	1
Math	95	4.318	4
Totals	198		9

The Jefferson method of apportionment is similar, except that a modified divisor is found so that when the modified quotas are rounded down, the integers sum to exactly the number to be apportioned. The next example shows the apportionment of Example J using the Jefferson method.

EXAMPLE K   The Jefferson Method

1. Compute md, the modified divisor. This modified divisor is found by slowly increasing or decreasing the standard divisor, a process of trial and error that is possibly best accomplished by using the power of a spreadsheet. The standard divisor is 22. A modified divisor, md, of 18 works for this problem.

2. Compute mQ, the modified quota for each course

$$mQ = \frac{\text{enrollment}}{md}$$

3. Round each quota down to the nearest integer.

4. Give each course this integer number of assistants.

Using a modified divisor of 18 to apportion the teaching assistants of Example J, here is the apportionment.

Course	Enr.	Q	Rounded Q
Bio	35	$\frac{35}{18} \approx 1.944$	1
Chem	40	$\frac{40}{18} \approx 2.222$	2
Physics	28	$\frac{28}{18} \approx 1.556$	1
Math	95	$\frac{95}{18} \approx 5.278$	5
Totals	198		9

The Webster method of apportionment follows the same steps as the Jefferson method, using a modified divisor. However, in this method, rounding is done according to the usual rules of rounding. Here is an example of the method.

EXAMPLE L   The Webster Method

1. Compute md, the modified divisor. Again, this modified divisor is found by slowly increasing or decreasing the standard divisor, a process of trial and error that is possibly best accomplished by using the power of a spreadsheet. The standard divisor is 22. A modified divisor, md, of 21.5 works for this problem.

2. Compute mQ, the modified quota for each course

$$mQ = \frac{enrollment}{md}$$

3. Round each quota up to the nearest integer if its fractional part is greater than or equal to .5; round down to the nearest integer if its fractional part is less than .5.
4. Give each course this integer number of assistants.

Using a modified divisor of 21.5 to apportion the teaching assistants of Example J, here is the apportionment.

Course	Enr.	Q	Rounded Q
Bio	35	$\frac{35}{21.5} \approx 1.628$	2
Chem	40	$\frac{40}{21.5} \approx 1.860$	2
Physics	28	$\frac{28}{21.5} \approx 1.302$	1
Math	95	$\frac{95}{21.5} \approx 4.419$	4
Totals	198		9

The Adams method follows the same basic four steps using a modified divisor; however, all modified quotas are rounded up to give the exact number of items to be apportioned. The Hill-Huntington method uses a more involved rounding scheme, as described in the text.

Just as the various voting methods have flaws, so the apportionment methods also have problems associated with them. The quota rule dictates that the number of items apportioned must be either the standard quota rounded down or rounded up. It is not unusual for this rule to be violated by some of the apportionment methods.

EXAMPLE M   Quota Rule Violation

Here is a population profile for five states. The apportionment of 150 legislative seats follows.

State	a	b	c	d	e
Population	1720	3363	6960	24223	8800

First the sum of the populations is

$$1720 + 3363 + 6960 + 24223 + 8800 = 45066.$$

The standard divisor is

$$d = \frac{45066}{150} = 300.44.$$

Using the standard quota, the population of each state is found by dividing by d. Each quota is rounded down to the nearest integer.

State	Population	Q	Rounded Q
a	1720	$\frac{1720}{300.44} \approx 5.725$	5
b	3363	$\frac{3363}{300.44} \approx 11.194$	11
c	6960	$\frac{6960}{300.44} \approx 23.166$	23
d	24223	$\frac{24223}{300.44} \approx 80.625$	**80**
e	8800	$\frac{8800}{300.44} \approx 29.290$	29
Totals	45066		148

Now using the Jefferson Method with $md = 295$

State	Population	Q	Rounded Q
a	1720	$\dfrac{1720}{295} \approx 5.831$	5
b	3363	$\dfrac{3363}{295} = 11.4$	11
c	6960	$\dfrac{6960}{295} \approx 23.593$	23
d	24223	$\dfrac{24223}{295} \approx 82.146$	**82**
e	8800	$\dfrac{8800}{295} \approx 29.831$	29
Totals	45066		150

The Jefferson method violates the quota rule because state d receives two more seats than its apportionment from the standard quota.

The Alabama paradox occurs when an increase in the number of objects to be apportioned actually forces a state to lose one of its legislative seats.

EXAMPLE N   The Alabama Paradox

Here is a population profile for five states. Initially, there are 45 legislative seats to be apportioned. Notice the number of seats apportioned to State b when the number of seats is increased to 46.

State	a	b	c	d	e
Population	309	289	333	615	465

First calculate the apportionment with $n = 45$. The total population is

$$309 + 289 + 333 + 615 + 465 = 2011.$$

The standard divisor for 45 seats is

$$d = \frac{2011}{45} \approx 44.68889.$$

Using the Hamilton Method

State	Population	Q	Rounded Q
a	309	$\dfrac{309}{44.68889} \approx 6.914$	6
b	289	$\dfrac{289}{44.68889} \approx 6.467$	6
c	333	$\dfrac{333}{44.68889} \approx 7.452$	7
d	615	$\dfrac{615}{44.68889} \approx 13.762$	13
e	465	$\dfrac{465}{44.68889} \approx 10.405$	10
Totals	2011		42

The three remaining seats will be apportioned to States a, d, and b, because the fractional parts of their quotas are the largest. The final apportionment is

State	Q	Number of seats
a	6.914	$6 + 1 = 7$
b	6.467	$6 + 1 = 7$
c	7.452	7
d	13.762	$13 + 1 = 14$
e	10.405	10
Totals		45

Now increase the number of seats to 46. The standard divisor is

$$d = \frac{2011}{46} \approx 43.71739.$$

Using the Hamilton Method with the new d

State	Population	Q	Rounded Q
a	309	$\dfrac{309}{43.71739} \approx 7.068$	**7**
b	289	$\dfrac{289}{43.71739} \approx 6.611$	6
c	333	$\dfrac{333}{43.71739} \approx 7.617$	7
d	615	$\dfrac{615}{43.71739} \approx 14.068$	14
e	465	$\dfrac{465}{43.71739} \approx 10.636$	10
Totals	3459		44

The two remaining seats will be apportioned to States e and c, because the fractional parts of their quotas are the largest. The final apportionment is

State	Q	Number of seats
a	7.068	7
b	6.611	**6**
c	7.617	$7 + 1 = 8$
d	14.068	14
e	10.636	$10 + 1 = 11$
Totals		46

State b is a victim of the Alabama Paradox, because it has lost a seat despite the fact that the overall number of seats has increased.

The population paradox occurs when there is a transfer of seats between two states based upon updated population figures, but the faster-growing population actually loses a seat.

EXAMPLE O   The Population Paradox

Here is a population profile for three states showing the initial population and the revised population figures.

State	a	b	c
Initial Population	930	738	415
Revised Population	975	752	422

First calculate the apportionment for the initial populations. The total population is

$$930 + 738 + 415 = 2083.$$

There are 13 legislative seats to be apportioned, so the standard divisor is

$$d = \frac{2083}{13} = 160.2308.$$

Using the Hamilton Method of apportionment

State	Population	Q	Rounded Q
a	930	$\frac{930}{160.2308} \approx 5.804$	5
b	738	$\frac{738}{160.2308} \approx 4.606$	4
c	415	$\frac{415}{160.2308} \approx 2.590$	2
Totals	2083		11

The two remaining seats will be apportioned to States a and b, because the fractional parts of their quotas are the largest. The final apportionment is

State	Q	Number of seats
a	5.804	$5 + 1 = 6$
b	4.606	$4 + 1 = 5$
c	2.590	2
Totals		13

Now apportion the seats for the revised populations.

$$975 + 752 + 422 = 2149.$$

The new standard divisor is

$$d = \frac{2149}{13} \approx 165.3077.$$

Using the Hamilton Method with the new values

State	Population	Q	Rounded Q
a	975	$\frac{975}{165.3077} \approx 5.898$	5
b	752	$\frac{752}{165.3077} \approx 4.549$	4
c	422	$\frac{422}{165.3077} \approx 2.553$	2
Totals	2149		11

The two remaining seats will be apportioned to States a and c, because the fractional parts of their quotas are the largest. The final apportionment is

State	Q	Number of seats
a	5.898	$5 + 1 = 6$
b	4.549	4
c	2.553	$2 + 1 = 3$
Totals		13

Here is a final summary:

State	Old Pop	New Pop	% Inc.	Old No. of Seats	New No. of Seats
a	930	975	4.84	6	6
b	738	752	**1.90**	**5**	**4**
c	415	422	**1.69**	**2**	**3**
Totals	2083	2150		13	13

Notice from the table that there was a greater percent increase in growth for State b than for State c. Yet State c gained a seat and State b lost a seat. This is an example of the population paradox.

The new states paradox occurs when there is a shift in the apportionment of the original states that is caused by the addition of seats because a new state has been added.

EXAMPLE P   The New States Paradox

Here is a population profile for two original states and the new state that is added.

State	Original State a	Original State b	New State c
Population	134	52	39

First calculate the apportionment for the initial populations. The total population is

$$134 + 52 = 186.$$

There are 16 seats to be apportioned, so the standard divisor is

$$d = \frac{186}{16} = 11.625.$$

Using the Hamilton method

State	Population	Q	Rounded Q
a	134	$\frac{134}{11.625} \approx 11.527$	11
b	52	$\frac{52}{11.625} \approx 4.473$	4
Totals	186		15

The one remaining seat will be apportioned to State a because the fractional part of its quota is larger. The final apportionment is

State	Q	Number of seats
a	11.527	$11 + 1 = 12$
b	4.473	4
Totals		16

Now apportion the seats to include the new state. The standard quota of the new state is

$$Q = \frac{39}{11.625} \approx 3.3548.$$

Rounded down to 3, add 3 new seats to the original 16 to obtain 19 seats to be apportioned. Now the new population is

$$134 + 52 + 39 = 225.$$

The new standard divisor is

$$d = \frac{225}{19} \approx 11.8421.$$

State	Population	Q	Rounded Q
a	134	$\dfrac{134}{11.8421} \approx 11.316$	11
b	52	$\dfrac{52}{11.8421} \approx 4.391$	4
c	39	$\dfrac{39}{11.8421} \approx 3.293$	3
Totals	225		18

The one remaining seat will be apportioned to State b because the fractional part of its quota is the largest. The final apportionment is

State	Q	Number of seats
a	11.316	11
b	4.391	$4 + 1 = 5$
c	3.293	3
Totals		19

The New State Paradox has occurred because the addition of the new state has caused a shift in the apportionment of the original states. States a and b originally had 12 and 4 seats, respectively; now they have 11 and 5, respectively.

Many of the exercises in this chapter are done most efficiently by using a spreadsheet.

15.1 EXERCISES

1. (a) Because there are four breeds the number of ways that a staff member can complete her ballot is

$$4! = 4 \cdot 3 \cdot 2 \cdot 1 = 24.$$

(b) The voter profile is

Votes	Ranking	Voters
3	$b > c > a > d$	1, 2, 9
2	$a > d > c > b$	3, 11
1	$c > d > b > a$	4
1	$d > c > b > a$	5
1	$d > a > b > c$	6
2	$c > a > d > b$	7, 8
1	$a > c > d > b$	10
2	$a > b > c > d$	12, 13

(c) By the plurality method the breed with the most votes wins. The Australian Shepherd wins.

Breed	1st place votes
a	5
b	3
c	3
d	2

3. For $n = 5$, the number of rankings is

$$5! = 5 \cdot 4 \cdot 3 \cdot 2 \cdot 1 = 120.$$

For $n = 7$, the number of rankings is

$$7! = 7 \cdot 6 \cdot 5 \cdot 4 \cdot 3 \cdot 2 \cdot 1 = 5040.$$

5. Writing exercise

7. The number of pairwise comparisons needed to learn the outcome of an election involving $n = 6$ candidates is the number of combinations of 6, taken 2 at a time. Mathematically this means

$$_6C_2 = \frac{6!}{2!(6-2)!} = \frac{6 \cdot 5 \cdot 4 \cdot 3 \cdot 2 \cdot 1}{2 \cdot 1 \cdot 4 \cdot 3 \cdot 2 \cdot 1} = 15.$$

For $n = 8$ candidates, find the combinations of 8, taken 2 at a time.

$$_8C_2 = \frac{8!}{2!(8-2)!} = \frac{8 \cdot 7 \cdot 6 \cdot 5 \cdot 4 \cdot 3 \cdot 2 \cdot 1}{2 \cdot 1 \cdot 6 \cdot 5 \cdot 4 \cdot 3 \cdot 2 \cdot 1} = 28.$$

9. Writing exercise

11. Use the voter profile to answer the questions.

Votes	Ranking	Row
3	$a > c > b$	1
4	$c > b > a$	2
2	$b > a > c$	3
4	$b > c > a$	4

(a) Using the plurality method to determine the chairperson, the winner is candidate b with 6 first place votes in rows 3 and 4. Notice from the table that candidate a received 3 votes in row 1, and candidate c received 4 votes in row 2.

(b) First, the number of comparisons is

$$_3C_2 = \frac{3!}{2!(3-2)!} = \frac{3 \cdot 2 \cdot 1}{2 \cdot 1 \cdot 1} = 3.$$

Make a table to compute the votes in comparisons.

Comparison	Votes	Rows
$a > b$	$3 = 3$	1
$b > a$	$4 + 2 + 4 = 10$	2, 3, 4
$a > c$	$3 + 2 = 5$	1, 3
$c > a$	$4 + 4 = 8$	2, 4
$b > c$	$2 + 4 = 6$	3, 4
$c > b$	$3 + 4 = 7$	1, 2

Make another table to award pairwise points.

Pairs	Votes	Pairwise Points
$a : b$	$3 : 10$	b, 1
$a : c$	$5 : 8$	c, 1
$b : c$	$6 : 7$	c, 1

Using the pairwise comparison method , the winner is candidate c with 2 pairwise points. Candidate b received only 1 point, and candidate a received 0 points.

(c) Use the information from part (a) to make a table to examine the points in the Borda method.

Votes	Ranking	Points 2	1	0
3	$a > c > b$	a	c	b
4	$c > b > a$	c	b	a
2	$b > a > c$	b	a	c
4	$b > c > a$	b	c	a

Examine the last three columns and compute the weighted sum for each candidate.

Candidate	Borda Points
a	$3\cdot2+4\cdot0+2\cdot1+4\cdot0=8$
b	$3\cdot0+4\cdot1+2\cdot2+4\cdot2=16$
c	$3\cdot1+4\cdot2+2\cdot0+4\cdot1=15$

The winner is candidate b with 16 Borda points.

(d) There are 13 voters so a candidate must receive a majority of the votes or 7 votes to win by the Hare method. No candidate received a majority; candidates a, b, and c received 3, 6, and 4 votes respectively. (See part a.) By this method, candidate a is eliminated, and another vote is taken. Now compare only candidates b and c.

Comparison	Votes	Rows
$b > c$	$2+4=6$	3, 4
$c > b$	$3+4=7$	1, 2

Candidate c now receives a majority and wins.

(e) Using the predetermined order c, b, a, begin by comparing c and b in the original voting.

Comparison	Votes	Row
$b > c$	$2+4=6$	3, 4
$c > b$	$3+4=7$	1, 2

Candidate c wins this competition, 7 to 6. Now compare c to a.

Comparison	Votes	Row
$c > a$	$4+4=8$	2, 4
$a > c$	$3+2=5$	1, 3

Candidate c wins this competition, 8 to 5, and also wins the entire competition in the sequential pairwise comparison method.

13. Use the voter profile to answer the questions.

Votes	Ranking	Row
6	$a > b > c$	1
1	$b > c > a$	2
3	$b > a > c$	3
3	$c > a > b$	4

(a) Using the plurality method to determine the logo, the winner is logo a with 6 first place votes in row 1. Notice from the table that logo b received a total of 4 votes in rows 2 and 3, and logo c received 3 votes in row 4.

(b) First, the number of comparisons is

$$_3C_2 = \frac{3!}{2!(3-2)!} = \frac{3\cdot2\cdot1}{2\cdot1\cdot1} = 3.$$

Make a table to compute the votes in comparisons.

Comparison	Votes	Rows
$a > b$	$6+3=9$	1, 4
$b > a$	$1+3=4$	2, 3
$a > c$	$6+3=9$	1, 3
$c > a$	$1+3=4$	2, 4
$b > c$	$6+1+3=10$	1, 2, 3
$c > b$	$3=3$	4

Make another table to award pairwise points.

Pairs	Votes	Pairwise Points
$a:b$	9: 4	a, 1
$a:c$	9: 4	a, 1
$b:c$	10: 3	b, 1

Using the pairwise comparison method , the winner is logo a with 2 pairwise points. Logo b received only 1 point, and logo c received 0 points.

(c) Use the information from part (a) to make a table to examine the points in the Borda method.

		Points		
Votes	Ranking	2	1	0
6	$a > b > c$	a	b	c
1	$b > c > a$	b	c	a
3	$b > a > c$	b	a	c
3	$c > a > b$	c	a	b

Examine the last three columns and compute the weighted sum for each logo.

Logo	Borda Points
a	$6\cdot2+1\cdot0+3\cdot1+3\cdot1=18$
b	$6\cdot1+1\cdot2+3\cdot2+3\cdot0=14$
c	$6\cdot0+1\cdot1+3\cdot0+3\cdot2=7$

The winner is logo a with 18 Borda points.

(d) There are 13 voters so a logo must receive a majority of the votes or at least 7 votes to win by the Hare method. No logo received a majority. Logo a received 6 votes; logo b received 4 votes; logo c received 3 votes. Eliminate logo c and compare the remaining logos.

Votes	Ranking	Row
6	$a > b$	1
1	$b > a$	2
3	$b > a$	3
3	$a > b$	4

Now logo a receives $6 + 3 = 9$ votes; logo b receives $1 + 3 = 4$ votes. Logo a now receives a majority and wins.

(e) Using the predetermined order b, c, a, begin by comparing b and c in the original voting.

Comparison	Votes	Row
$b > c$	$6 + 1 + 3 = 10$	1, 2, 3
$c > b$	$3 = 3$	4

Logo b wins this competition, 10 to 3. Now compare b to a.

Comparison	Votes	Row
$a > b$	$6 + 3 = 9$	1, 4
$b > a$	$1 + 3 = 4$	2, 3

Logo a wins this competition, 9 to 4, and also wins the entire competition in the sequential pairwise comparison method.

15. Use the voter profile to answer the questions.

Votes	Ranking	Row
6	$h > j > g > e$	1
5	$e > g > j > h$	2
4	$g > j > h > e$	3
3	$j > h > g > e$	4
3	$e > j > h > g$	5

(a) Using the plurality method to determine the highest priority issue, the winner is issue e with a total of 8 first place votes in rows 2 and 5. Notice from the table that the remaining issues received the following number of first place votes: h, 6 votes; g, 4 votes; j, 3 votes.

(b) First, the number of comparisons is

$$_4C_2 = \frac{4!}{2!(4-2)!} = \frac{4 \cdot 3 \cdot 2 \cdot 1}{2 \cdot 1 \cdot 2 \cdot 1} = 6.$$

Make a table to compute the votes in comparisons.

Comparison	Votes	Rows
$e > h$	$5 + 3 = 8$	2, 5
$h > e$	$6 + 4 + 3 = 13$	1, 3, 4
$e > g$	$5 + 3 = 8$	2, 5
$g > e$	$6 + 4 + 3 = 13$	1, 3, 4
$e > j$	$5 + 3 = 8$	2, 5
$j > e$	$6 + 4 + 3 = 13$	1, 3, 4
$h > g$	$6 + 3 + 3 = 12$	1, 4, 5
$g > h$	$5 + 4 = 9$	2, 3
$h > j$	$6 = 6$	1
$j > h$	$5 + 4 + 3 + 3 = 15$	2, 3, 4, 5
$j > g$	$6 + 3 + 3 = 12$	1, 4, 5
$g > j$	$5 + 4 = 9$	2, 3

Make another table to award pairwise points.

Pairs	Votes	Pairwise Points
e: h	8: 13	h, 1
e: g	8: 13	g, 1
e: j	8: 13	j, 1
h: g	12:9	h, 1
h: j	6:15	j, 1
j: g	12:9	j, 1

Using the pairwise comparison method, the winner is issue j with 3 pairwise points. Issue h received 2 points; issue g received 1 point; issue e received no points.

(c) Use the information from part (a) to make a table to examine the points in the Borda method.

Votes	Ranking	Points 3	2	1	0
6	$h > j > g > e$	h	j	g	e
5	$e > g > j > h$	e	g	j	h
4	$g > j > h > e$	g	j	h	e
3	$j > h > g > e$	j	h	g	e
3	$e > j > h > g$	e	j	h	g

Examine the last four columns and compute the weighted sum for each issue.

Issue	Borda Points
h	$6 \cdot 3 + 5 \cdot 0 + 4 \cdot 1 + 3 \cdot 2 + 3 \cdot 1 = 31$
j	$6 \cdot 2 + 5 \cdot 1 + 4 \cdot 2 + 3 \cdot 3 + 3 \cdot 2 = 40$
g	$6 \cdot 1 + 5 \cdot 2 + 4 \cdot 3 + 3 \cdot 1 + 3 \cdot 0 = 31$
e	$6 \cdot 0 + 5 \cdot 3 + 4 \cdot 0 + 3 \cdot 0 + 3 \cdot 3 = 24$

The winner is issue j with 40 Borda points.

(d) There are 21 voters so an issue must receive a majority of the votes or at least 11 votes to win by the Hare method. No issue received a majority. See part (a) that issue e received 8 first place votes; issue h received 6 votes; issue g received 4 votes; and issue j received 3 votes. Eliminate issue j because it has the least number of votes, and compare only h, g, and e.

Votes	Ranking	Row
6	$h > g > e$	1
5	$e > g > h$	2
4	$g > h > e$	3
3	$h > g > e$	4
3	$e > h > g$	5

Issue h receives $6 + 3 = 9$ votes.
Issue g receives 4 votes.
Issue e receives $5 + 3 = 8$ votes.
No issue has received a majority so another vote is taken with issue g eliminated.

Votes	Ranking	Row
6	$h > e$	1
5	$e > h$	2
4	$h > e$	3
3	$h > e$	4
3	$e > h$	5

Issue h receives $6 + 4 + 3 = 13$ votes.
Issue e receives $5 + 3 = 8$ votes.
Now issue h has a majority and wins.

(e) Using the predetermined order h, j, e, g begin by comparing h and j in the original voting.

Comparison	Votes	Row
$h > j$	$6 = 6$	1
$j > h$	$5 + 4 + 3 + 3 = 15$	2, 3, 4, 5

Issue j wins this competition, 15 to 6. Now compare j to e.

Comparison	Votes	Row
$j > e$	$6 + 4 + 3 = 13$	1, 3, 4
$e > j$	$5 + 3 = 8$	2, 5

Issue j wins this competition, 13 to 8. Now compare j and g.

Comparison	Votes	Row
$j > g$	$6 + 3 + 3 = 12$	1, 4, 5
$g > j$	$5 + 4 = 9$	2, 3

Finally issue j wins this competition, 12 to 9, and also wins the entire competition in the sequential pairwise comparison method.

17. The voter profile from Exercise 1 (b) is

Votes	Ranking	Voter	Row
3	$b > c > a > d$	1, 2, 9	1
2	$a > d > c > b$	3, 11	2
1	$c > d > b > a$	4	3
1	$d > c > b > a$	5	4
1	$d > a > b > c$	6	5
2	$c > a > d > b$	7, 8	6
1	$a > c > d > b$	10	7
2	$a > b > c > d$	12, 13	8

(a) Using the pairwise comparison method, first make a table to compute the votes in comparison.

Compare	Votes	Rows
$a > b$	$2 + 1 + 2 + 1 + 2 = 8$	2, 5, 6, 7, 8
$b > a$	$3 + 1 + 1 = 5$	1, 3, 4
$a > c$	$2 + 1 + 1 + 2 = 6$	2, 5, 7, 8
$c > a$	$3 + 1 + 1 + 2 = 7$	1, 3, 4, 6
$a > d$	$3 + 2 + 2 + 1 + 2 = 10$	1, 2, 6, 7, 8
$d > a$	$1 + 1 + 1 = 3$	3, 4, 5
$b > c$	$3 + 1 + 2 = 6$	1, 5, 8
$c > b$	$2 + 1 + 1 + 2 + 1 = 7$	2, 3, 4, 6, 7
$b > d$	$3 + 2 = 5$	1, 8
$d > b$	$2 + 1 + 1 + 1 + 2 + 1 = 8$	2, 3, 4, 5, 6, 7
$c > d$	$3 + 1 + 2 + 1 + 2 = 9$	1, 3, 6, 7, 8
$d > c$	$2 + 1 + 1 = 4$	2, 4, 5

Make another table to award pairwise points.

Pairs	Votes	Pairwise Points
$a : b$	$8 : 5$	a, 1
$a : c$	$6 : 7$	c, 1
$a : d$	$10 : 3$	a, 1
$b : c$	$6 : 7$	c, 1
$b : d$	$5 : 8$	d, 1
$c : d$	$9 : 4$	c, 1

Using the pairwise comparison method, the winner is c with 3 pairwise points. Breed a received 2 points; breed d received 1 point; breed b received no points.

(b) Use the information from part (a) to make a table to examine the points in the Borda method.

Votes	Ranking	Points			
		3	2	1	0
3	b > c > a > d	b	c	a	d
2	a > d > c > b	a	d	c	b
1	c > d > b > a	c	d	b	a
1	d > c > b > a	d	c	b	a
1	d > a > b > c	d	a	b	c
2	c > a > d > b	c	a	d	b
1	a > c > d > b	a	c	d	b
2	a > b > c > d	a	b	c	d

Examine the last four columns and compute the weighted sum for each dog breed.

Breed	Borda Points
a	$3 \cdot 1 + 2 \cdot 3 + 1 \cdot 0 + 1 \cdot 0$ $+ 1 \cdot 2 + 2 \cdot 2 + 1 \cdot 3 + 2 \cdot 3 = 24$
b	$3 \cdot 3 + 2 \cdot 0 + 1 \cdot 1 + 1 \cdot 1$ $+ 1 \cdot 1 + 2 \cdot 0 + 1 \cdot 0 + 2 \cdot 2 = 16$
c	$3 \cdot 2 + 2 \cdot 1 + 1 \cdot 3 + 1 \cdot 2$ $+ 1 \cdot 0 + 2 \cdot 3 + 1 \cdot 2 + 2 \cdot 1 = 23$
d	$3 \cdot 0 + 2 \cdot 2 + 1 \cdot 2 + 1 \cdot 3$ $+ 1 \cdot 3 + 2 \cdot 1 + 1 \cdot 1 + 2 \cdot 0 = 15$

The winner is breed a with 24 Borda points.

(c) Calculate how many points each breed obtained from the thirteen first place votes from the voter profile.

Votes	Ranking	Voter	Row
3	b > c > a > d	1, 2, 9	1
2	a > d > c > b	3, 11	2
1	c > d > b > a	4	3
1	d > c > b > a	5	4
1	d > a > b > c	6	5
2	c > a > d > b	7, 8	6
1	a > c > d > b	10	7
2	a > b > c > d	12, 13	8

Breed	1st place votes
a	5
b	3
c	3
d	2

None of the dog breeds received a majority, 7, of the votes. Therefore, eliminate breed d, the breed that received the least number of votes. Compare the other three.

Votes	Ranking	Voter	Row
3	b > c > a	1, 2, 9	1
2	a > c > b	3, 11	2
1	c > b > a	4	3
1	c > b > a	5	4
1	a > b > c	6	5
2	c > a > b	7, 8	6
1	a > c > b	10	7
2	a > b > c	12, 13	8

Breed a: $2 + 1 + 1 + 2 = 6$ points.
Breed b: 3 points.
Breed c: $1 + 1 + 2 = 4$ points.
None of the breeds has received a majority; therefore, eliminate b and compare first place points again for a and c.

Votes	Ranking	Voter	Row
3	c > a	1, 2, 9	1
2	a > c	3, 11	2
1	c > a	4	3
1	c > a	5	4
1	a > c	6	5
2	c > a	7, 8	6
1	a > c	10	7
2	a > c	12, 13	8

Breed a: $2 + 1 + 1 + 2 = 6$
Breed c: $3 + 1 + 1 + 2 = 7$
Breed c now has a majority and wins.

19. The voter profile is

Votes	Ranking	Row
18	t > b > h > k > c	1
12	c > h > b > k > t	2
10	k > c > h > b > t	3
9	b > k > h > c > t	4
4	h > c > b > k > t	5
2	h > k > b > c > t	6

(a) If the plurality method is used, activity t is selected with 18 first place votes.

(b) Using the pairwise comparison method, first make a table to compute the votes in comparison.

Compare	Votes	Rows
$t > b$	$18 = 18$	1
$b > t$	$12 + 10 + 9 + 4 + 2 = 37$	2, 3, 4, 5, 6
$t > h$	$18 = 18$	1
$h > t$	$12 + 10 + 9 + 4 + 2 = 37$	2, 3, 4, 5, 6
$t > k$	$18 = 18$	1
$k > t$	$12 + 10 + 9 + 4 + 2 = 37$	2, 3, 4, 5, 6
$t > c$	$18 = 18$	1
$c > t$	$12 + 10 + 9 + 4 + 2 = 37$	2, 3, 4, 5, 6
$b > h$	$18 + 9 = 27$	1, 4
$h > b$	$12 + 10 + 4 + 2 = 28$	2, 3, 5, 6
$b > k$	$18 + 12 + 9 + 4 = 43$	1, 2, 4, 5
$k > b$	$10 + 2 = 12$	3, 6
$b > c$	$18 + 9 + 2 = 29$	1, 4, 6
$c > b$	$12 + 10 + 4 = 26$	2, 3, 5
$h > k$	$18 + 12 + 4 + 2 = 36$	1, 2, 5, 6
$k > h$	$10 + 9 = 19$	3, 4
$h > c$	$18 + 9 + 4 + 2 = 33$	1, 4, 5, 6
$c > h$	$12 + 10 = 22$	2, 3
$k > c$	$18 + 10 + 9 + 2 = 39$	1, 3, 4, 6
$c > k$	$12 + 4 = 16$	2, 5

Make another table to award pairwise points.

Pairs	Votes	Pairwise Points
$t : b$	$18 : 37$	b, 1
$t : h$	$18 : 37$	h, 1
$t : k$	$18 : 37$	k, 1
$t : c$	$18 : 37$	c, 1
$b : h$	$27 : 28$	h, 1
$b : k$	$43 : 12$	b, 1
$b : c$	$29 : 26$	b, 1
$h : k$	$36 : 19$	h, 1
$h : c$	$33 : 22$	h, 1
$k : c$	$39 : 16$	k, 1

Using the pairwise comparison method, the winner is h with 4 pairwise points. Activity b received 3 points; activity k received 2 points; activity c received 1 point; activity t received no points.

(c) Use the information from part (a) to make a table to examine the points in the Borda method,

Votes	Ranking	Points				
		4	3	2	1	0
18	$t > b > h > k > c$	t	b	h	k	c
12	$c > h > b > k > t$	c	h	b	k	t
10	$k > c > h > b > t$	k	c	h	b	t
9	$b > k > h > c > t$	b	k	h	c	t
4	$h > c > b > k > t$	h	c	b	k	t
2	$h > k > b > c > t$	h	k	b	c	t

Examine the last five columns and compute the weighted sum for each activity.

Breed	Borda Points
t	$18 \cdot 4 + 12 \cdot 0 + 10 \cdot 0 + 9 \cdot 0$ $+ 4 \cdot 0 + 2 \cdot 0 = 72$
b	$18 \cdot 3 + 12 \cdot 2 + 10 \cdot 1 + 9 \cdot 4$ $+ 4 \cdot 2 + 2 \cdot 2 = 136$
h	$18 \cdot 2 + 12 \cdot 3 + 10 \cdot 2 + 9 \cdot 2$ $+ 4 \cdot 4 + 2 \cdot 4 = 134$
k	$18 \cdot 1 + 12 \cdot 1 + 10 \cdot 4 + 9 \cdot 3$ $+ 4 \cdot 1 + 2 \cdot 3 = 107$
c	$18 \cdot 0 + 12 \cdot 4 + 10 \cdot 3 + 9 \cdot 1$ $+ 4 \cdot 3 + 2 \cdot 1 = 101$

The winner is b with 136 Borda points.

(d) Calculate how many points each activity obtained from the 55 first place votes from the voter profile.

Votes	Ranking	Row
18	$t > b > h > k > c$	1
12	$c > h > b > k > t$	2
10	$k > c > h > b > t$	3
9	$b > k > h > c > t$	4
4	$h > c > b > k > t$	5
2	$h > k > b > c > t$	6

Activity	1st place votes
t	18
c	12
k	10
b	9
h	6

None of the activities received a majority, 28, of the votes. Therefore, eliminate activity h, the activity that received the least number of votes. Compare the first place votes of the others.

Votes	Ranking	Row
18	$t > b > k > c$	1
12	$c > b > k > t$	2
10	$k > c > b > t$	3
9	$b > k > c > t$	4
4	$c > b > k > t$	5
2	$k > b > c > t$	6

Activity t: 18 points.
Activity c: $12 + 4 = 16$ points.
Activity k: $10 + 2 = 12$ points.
Activity b: 9 points.
There is no activity that has received a majority; therefore, eliminate b with only 9 points, and compare the first place votes of the others.

Votes	Ranking	Row
18	$t > k > c$	1
12	$c > k > t$	2
10	$k > c > t$	3
9	$k > c > t$	4
4	$c > k > t$	5
2	$k > c > t$	6

Activity t: 18 points.
Activity c: $12 + 4 = 16$ points.
Activity k: $10 + 9 + 2 = 21$ points.
Again there is not a majority; therefore, eliminate c and compare the first place votes of t and k.

Votes	Ranking	Row
18	$t > k$	1
12	$k > t$	2
10	$k > t$	3
9	$k > t$	4
4	$k > t$	5
2	$k > t$	6

Activity t: 18 points.
Activity k: $12 + 10 + 9 + 4 + 2 = 37$ points.
Finally k wins with 37 points.

(e) Using the predetermined order b, t, c, k, h, begin by comparing b and t in the original voting.

Comparison	Votes	Row
$b > t$	$12 + 10 + 9 + 4 + 2 = 37$	2, 3, 4, 5, 6
$t > b$	$18 = 18$	1

Activity b wins this competition, 37 to 18. Now compare b to c.

Comparison	Votes	Row
$b > c$	$18 + 9 + 2 = 29$	1, 4, 6
$c > b$	$12 + 10 + 4 = 26$	2, 3, 5

Activity b wins this competition, 29 to 26. Now compare b and k.

Comparison	Votes	Row
$b > k$	$18 + 12 + 9 + 4 = 43$	1, 2, 4, 5
$k > b$	$10 + 2 = 12$	3, 6

Activity b wins this competition, 43 to 12. Now compare b and h.

Comparison	Votes	Row
$b > h$	$18 + 9 = 27$	1, 4
$h > b$	$12 + 10 + 4 + 2 = 28$	2, 3, 5, 6

Finally activity h wins this competition, 28 to 27, and also wins the entire competition in the sequential pairwise comparison method.

21. The voter profile is

Votes	Ranking	Row
18	$t > b > h > k > c$	1
12	$c > h > b > k > t$	2
10	$k > c > h > b > t$	3
9	$b > k > h > c > t$	4
4	$h > c > b > k > t$	5
2	$h > k > b > c > t$	6

The activities with the most votes are t and c. Eliminate the remaining candidates to examine a second runoff election between these two candidates.

Votes	Ranking	Row
18	$t > c$	1
12	$c > t$	2
10	$c > t$	3
9	$c > t$	4
4	$c > t$	5
2	$c > t$	6

Activity t has 18 votes.
Activity c: $12 + 10 + 9 + 4 + 2 = 37$.
Activity c wins.

23. The voter profile is:

Votes	Ranking	Row
6	$h > j > g > e$	1
5	$e > g > j > h$	2
4	$g > j > h > e$	3
3	$j > h > g > e$	4
3	$e > j > h > g$	5

The issues with the most votes are h and e. Eliminate the remaining candidates to examine a second runoff election between these two issue.

Votes	Ranking	Row
6	h > e	1
5	e > h	2
4	h > e	3
3	h > e	4
3	e > h	5

Issue h: $6 + 4 + 3 = 13$
Issue e: $5 + 3 = 8$.
Issue h wins.

25. The voter profile is

Votes	Ranking	Row
18	t > b > h > k > c	1
12	c > h > b > k > t	2
10	k > c > h > b > t	3
9	b > k > h > c > t	4
4	h > c > b > k > t	5
2	h > k > b > c > t	6

The activities that rank second and third in first place votes are c and k, respectively. Here is a run off election between these two activities:

Votes	Ranking	Row
18	k > c	1
12	c > k	2
10	k > c	3
9	k > c	4
4	c > k	5
2	k > c	6

Activity k: $18 + 10 + 9 + 2 = 39$
Activity c: $12 + 4 = 16$
Now k faces the first place candidate, t:

Votes	Ranking	Row
18	t > k	1
12	k > t	2
10	k > t	3
9	k > t	4
4	k > t	5
2	k > t	6

Activity t has 18 votes.
Candidate k: $12 + 10 + 9 + 4 + 2 = 37$
Activity k wins with 37 votes.

27. The voter profile is:

Votes	Ranking	Row
6	a > b > c	1
1	b > c > a	2
3	b > a > c	3
3	c > a > b	4

The logos that rank second and third in first place votes are b and c, respectively. Here is a run off election between these two logos.

Votes	Ranking	Row
6	b > c	1
1	b > c	2
3	b > c	3
3	c > b	4

Logo b: $6 + 1 + 3 = 10$
Logo c: 3
Now b faces the first place logo, a:

Votes	Ranking	Row
6	a > b	1
1	b > a	2
3	b > a	3
3	a > b	4

Logo a: $6 + 3 = 9$
Logo b: $1 + 3 = 4$
Logo a wins.

29. (a) Count the votes in the table shown in the text to see that Joan receives 8 votes; Lori receives 7; Mary receives 10; and Alison receives 9. Mary wins by the approval method.

(b) Mary and Alison win with 10 and 9 votes, respectively.

31. (a) For $n = 7$ candidates, find the combinations of 7, taken 2 at a time:

$$_7C_2 = \frac{7!}{2!(7-2)!} = \frac{7 \cdot 6 \cdot 5 \cdot 4 \cdot 3 \cdot 2 \cdot 1}{2 \cdot 1 \cdot 5 \cdot 4 \cdot 3 \cdot 2 \cdot 1} = 21.$$

The sum of the number of comparisons listed in the table in the text is:

$$3 + 5 + 7 + 1 + 2 + 1 = 19$$

Only two comparisons remain, so f wins two points.

(b) Examine the table in the text to see that c wins 7 pairwise points.

33. (a) For $n = 8$ candidates, find the combinations of 8, taken 2 at a time:

$$_8C_2 = \frac{8!}{2!(8-2)!} = \frac{8 \cdot 7 \cdot 6 \cdot 5 \cdot 4 \cdot 3 \cdot 2 \cdot 1}{2 \cdot 1 \cdot 6 \cdot 5 \cdot 4 \cdot 3 \cdot 2 \cdot 1} = 28.$$

The sum of the number of comparisons listed in the table in the text is:

$$2 + 6 + 3 + 4 + 2 + 2 + 2 = 21$$

$28 - 21 = 7$ comparisons remain, so e wins seven points.

(b) Candidate e wins with 7 pairwise points.

35. Notice that in the Borda method, the sum of all the points must equal the product of the number of voters and the number of possible points for each voter's selection. For example, in this exercise there are 15 voters choosing among 3 candidates. Each voter has a total of 3 points to assign: 2 points for first place and 1 point for second place. Therefore, the total number of points is

$$15 \cdot 3 = 45.$$

If Candidate a receives 15 points and Candidate b receives 14 points, then Candidate c must receive

$$45 - (15 + 14) = 16 \text{ points}.$$

Candidate c wins the Borda election.

37. Notice that in the Borda method, the sum of all the points must equal the product of the number of voters and the number of possible points for each voter's selection. For example, in this exercise there are 20 voters choosing among 5 candidates. Each voter has a total of 10 points to assign: 4 points for first place, 3 points for second place, 2 points for third place, and 1 point for fourth place. Therefore, the total number of points is

$$20 \cdot 10 = 200.$$

Candidate c must receive the difference between 200 and the sum of the points of the other candidates.

$$200 - (35 + 40 + 40 + 30) = 55 \text{ points}.$$

Candidate c wins the Borda election with 55 points.

39. The voter profile is:

Votes	Ranking	Row
6	$h > j > g > e$	1
5	$e > g > j > h$	2
4	$g > j > h > e$	3
3	$j > h > g > e$	4
3	$e > j > h > g$	5

Using the Coombs method eliminate the issue with the most last place votes, which is e, with 13 last place votes.

Compare the remaining issues.

Votes	Ranking	Row
6	$h > j > g$	1
5	$g > j > h$	2
4	$g > j > h$	3
3	$j > h > g$	4
3	$j > h > g$	5

Again compare totals of last place votes.
Issue g: $6 + 3 + 3 = 12$
Issue h: $5 + 4 = 9$
Eliminate g because it has the most last place votes.
Now h and j remain.

Votes	Ranking	Row
6	$h > j$	1
5	$j > h$	2
4	$j > h$	3
3	$j > h$	4
3	$j > h$	5

Again compare totals of last place votes.
Issue j: 6
Issue h: $5 + 4 + 3 + 3 = 15$
Issue j is the winner.

41. Using the predetermined order, n, a_x, e, examine the table to see that:

$$n \text{ beats } a_x$$
$$n \text{ beats } e$$

The new bill, n wins.

43. Using the predetermined order, n, a_x, a_y, e, examine the table to see that:

$$n \text{ beats } a_x$$
$$a_y \text{ beats } n$$
$$a_y \text{ beats } e$$

The amended bill, a_y wins.

45. Here is one possible arrangement.

Votes	Ranking	Row
2	$a > b > d > c$	1
4	$b > c > d > a$	2
5	$c > d > a > b$	3
7	$d > a > b > c$	4
3	$a > d > c > b$	5

Plurality method
Candidate a receives $2 + 3 = 5$ votes.
Candidate b receives 4 votes.
Candidate c receives 5 votes.
Candidate d receives 7 votes.

Pairwise Comparison method
First make a table to compute the votes in comparison.

Compare	Votes	Rows
$a > b$	$2 + 5 + 7 + 3 = 17$	$1, 3, 4, 5$
$b > a$	$4 = 4$	2
$a > c$	$2 + 7 + 3 = 12$	$1, 4, 5$
$c > a$	$4 + 5 = 9$	$2, 3$
$a > d$	$2 + 3 = 5$	$1, 5$
$d > a$	$4 + 5 + 7 = 16$	$2, 3, 4$
$b > c$	$2 + 4 + 7 = 13$	$1, 2, 4$
$c > b$	$5 + 3 = 8$	$3, 5$
$b > d$	$2 + 4 = 6$	$1, 2$
$d > b$	$5 + 7 + 3 = 15$	$3, 4, 5$
$c > d$	$4 + 5 = 9$	$2, 3$
$d > c$	$2 + 7 + 3 = 12$	$1, 4, 5$

Make another table to award pairwise points.

Pairs	Votes	Pairwise Points
$a : b$	$17 : 4$	$a, 1$
$a : c$	$12 : 9$	$a, 1$
$a : d$	$5 : 16$	$d, 1$
$b : c$	$13 : 8$	$b, 1$
$b : d$	$6 : 15$	$d, 1$
$c : d$	$9 : 12$	$d, 1$

Using the pairwise comparison method , the winner is d with 3 pairwise points. Candidate a received 2 points; candidate b received 1 point; candidate c received no points.

Borda method
Make a table to examine the points in the Borda method.

Votes	Ranking	Points 3	2	1	0
2	$a > b > d > c$	a	b	d	c
4	$b > c > d > a$	b	c	d	a
5	$c > d > a > b$	c	d	a	b
7	$d > a > b > c$	d	a	b	c
3	$a > d > c > b$	a	d	c	b

Examine the last four columns and compute the weighted sum for each candidate.

Issue	Borda Points
a	$2 \cdot 3 + 4 \cdot 0 + 5 \cdot 1 + 7 \cdot 2 + 3 \cdot 3 = 34$
b	$2 \cdot 2 + 4 \cdot 3 + 5 \cdot 0 + 7 \cdot 1 + 3 \cdot 0 = 23$
c	$2 \cdot 0 + 4 \cdot 2 + 5 \cdot 3 + 7 \cdot 0 + 3 \cdot 1 = 26$
d	$2 \cdot 1 + 4 \cdot 1 + 5 \cdot 2 + 7 \cdot 3 + 3 \cdot 2 = 43$

The winner is d with 43 Borda points.

Hare method
There are 21 voters so a candidate must receive a majority of the votes or at least 11 votes to win by the Hare method. Examine the original ranking to see that no candidate received a majority.   Eliminate b because it has the least number of votes, and compare only a, c, and d.

Votes	Ranking	Row
2	$a > d > c$	1
4	$c > d > a$	2
5	$c > d > a$	3
7	$d > a > c$	4
3	$a > d > c$	5

Candidate a receives $2 + 3 = 5$ votes.
Candidate c receives $4 + 5 = 9$ votes.
Candidate d receives 7 votes.
Again no candidate has received a majority so another vote is taken with a eliminated.

Votes	Ranking	Row
2	$d > c$	1
4	$c > d$	2
5	$c > d$	3
7	$d > c$	4
3	$d > c$	5

Candidate c receives $4 + 5 = 9$ votes.
Candidate d receives $2 + 7 + 3 = 12$ votes.
Finally d has a majority and wins.

47. Writing exercise

49. Writing exercise

51. Writing exercise

15.2 EXERCISES

1. (a) Read the table in the text to see that a has the majority of first place votes. Because there are 11 voters, 6 votes constitutes a majority.

(b) Examine the following table to calculate the Borda points.

Votes	Ranking	Points 2	1	0
6	$a > b > c$	a	b	c
3	$b > c > a$	b	c	a
2	$c > b > a$	c	b	a

Alternative a: $6 \cdot 2 + 3 \cdot 0 + 2 \cdot 0 = 12$
Alternative b: $6 \cdot 1 + 3 \cdot 2 + 2 \cdot 1 = 14$
Alternative c: $6 \cdot 0 + 3 \cdot 1 + 2 \cdot 2 = 7$
Alternative b wins with 14 Borda points.

(c) The Borda method violates the majority criterion because it fails to select the majority candidate.

3. (a) Read the table in the text to see that a has the majority of first place votes. Because there are 36 voters, at least 19 votes constitutes a majority. Alternative a received 20 votes.

(b) Use the information from the text to make a table to examine the points in the Borda method.

Votes	Ranking	Points			
		3	2	1	0
20	$a > b > c > d$	a	b	c	d
6	$b > c > d > a$	b	c	d	a
5	$c > b > d > a$	c	b	d	a
5	$d > b > a > c$	d	b	a	c

Examine the last four columns and compute the weighted sum for each alternative.

Alternative	Borda Points
a	$20 \cdot 3 + 6 \cdot 0 + 5 \cdot 0 + 5 \cdot 1 = 65$
b	$20 \cdot 2 + 6 \cdot 3 + 5 \cdot 2 + 5 \cdot 2 = 78$
c	$20 \cdot 1 + 6 \cdot 2 + 5 \cdot 3 + 5 \cdot 0 = 47$
d	$20 \cdot 0 + 6 \cdot 1 + 5 \cdot 1 + 5 \cdot 3 = 26$

The winner is alternative b with 78 Borda points.

(c) The Borda method violates the majority criterion because it fails to select the majority candidate.

5. (a) Read the table in the text to see that a has the majority of first place votes. Because there are 30 voters, at least 16 votes constitutes a majority. Alternative a received 16 votes.

(b) Use the information from the text to make a table to examine the points in the Borda method.

Votes	Ranking	Points				
		4	3	2	1	0
16	$a > b > c > d > e$	a	b	c	d	e
3	$b > c > d > e > a$	b	c	d	e	a
5	$c > d > b > e > a$	c	d	b	e	a
3	$d > b > c > a > e$	d	b	c	a	e
3	$e > c > d > a > b$	e	c	d	a	b

Examine the last five columns and compute the weighted sum for each alternative.

Alternative	Borda Points
a	$16 \cdot 4 + 3 \cdot 0 + 5 \cdot 0 + 3 \cdot 1 + 3 \cdot 1 = 70$
b	$16 \cdot 3 + 3 \cdot 4 + 5 \cdot 2 + 3 \cdot 3 + 3 \cdot 0 = 79$
c	$16 \cdot 2 + 3 \cdot 3 + 5 \cdot 4 + 3 \cdot 2 + 3 \cdot 3 = 76$
d	$16 \cdot 1 + 3 \cdot 2 + 5 \cdot 3 + 3 \cdot 4 + 3 \cdot 2 = 55$
e	$16 \cdot 0 + 3 \cdot 1 + 5 \cdot 1 + 3 \cdot 0 + 3 \cdot 4 = 20$

The winner is alternative b with 79 Borda points.

(c) The Borda method violates the majority criterion because it fails to select the majority candidate.

7. (a) Make a table to compute the votes in comparisons.

Comparison	Votes	Rows
$a > b$	$4 + 3 = 7$	1, 4
$b > a$	$2 + 4 = 6$	2, 3
$a > c$	$4 + 4 = 8$	1, 3
$c > a$	$2 + 3 = 5$	2, 4
$b > c$	$4 + 2 + 4 = 10$	1, 2, 3
$c > b$	$3 = 3$	4

Make another table to award pairwise points.

Pairs	Votes	Pairwise Points
$a : b$	$7 : 6$	a, 1
$a : c$	$8 : 5$	a, 1
$b : c$	$10 : 3$	b, 1

Using the pairwise comparison method, the winner is candidate a with 2 pairwise points and is the Condorcet candidate.

(b) Examine the table in the text to see that a receives 4 first place votes; c receives 3 first place votes; b receives $2 + 4 = 6$ first place votes to win by the plurality method.

(c) There are 13 votes, which makes 7 votes a majority. No candidate wins a majority on this vote. Candidate c has the fewest votes, so compare only a and b.

Comparison	Votes	Rows
$a > b$	$4 + 3 = 7$	1, 4
$b > a$	$2 + 4 = 6$	2, 3

Candidate a now has a majority and is selected by the Hare method.

(d) Use the information from the text to make a table to examine the points in the Borda method.

Votes	Ranking	Points		
		2	1	0
4	$a > b > c$	a	b	c
2	$b > c > a$	b	c	a
4	$b > a > c$	b	a	c
3	$c > a > b$	c	a	b

Examine the last three columns and compute the weighted sum for each alternative.

Alternative	Borda Points
a	$4 \cdot 2 + 2 \cdot 0 + 4 \cdot 1 + 3 \cdot 1 = 15$
b	$4 \cdot 1 + 2 \cdot 2 + 4 \cdot 2 + 3 \cdot 0 = 16$
c	$4 \cdot 0 + 2 \cdot 1 + 4 \cdot 0 + 3 \cdot 2 = 8$

Candidate b wins with 16 Borda points.

(e) The plurality and the Borda methods violate the Condorcet criterion because neither method selects candidate a, the Condorcet candidate. The Hare method does not violate the criterion because the Condorcet candidate wins.

9. (a) Make a table to compute the votes in comparisons.

Comparison	Votes	Rows
$e > h$	$3 + 5 + 4 + 3 = 15$	1, 3, 4, 5
$h > e$	$6 = 6$	2
$e > g$	$3 + 6 + 3 = 12$	1, 2, 5
$g > e$	$5 + 4 = 9$	3, 4
$e > j$	$3 + 6 + 4 = 13$	1, 2, 4
$j > e$	$5 + 3 = 8$	3, 5
$h > g$	$3 + 6 + 3 = 12$	1, 2, 5
$g > h$	$5 + 4 = 9$	3, 4
$h > j$	$3 + 6 + 4 = 13$	1, 2, 4
$j > h$	$5 + 3 = 8$	3, 5
$j > g$	$5 + 3 = 8$	3, 5
$g > j$	$3 + 6 + 4 = 13$	1, 2, 4

Make another table to award pairwise points.

Pairs	Votes	Pairwise Points
e: h	15: 6	e, 1
e: g	12: 9	e, 1
e: j	13: 8	e, 1
h: g	12:9	h, 1
h: j	13:8	h, 1
j: g	8:13	g, 1

Using the pairwise comparison method , the winner is candidate e with 3 pairwise points and is the Condorcet candidate.

(b) Using the plurality method to determine the highest priority issue, the winner is issue j with a total of 8 first place votes in rows 3 and 5.

(c) There are 21 votes, which makes 11votes a majority. No candidate wins a majority on this vote. Eliminate issue e because it has the least number of votes, and compare only h, g, and j.

Votes	Ranking	Row
3	$h > g > j$	1
6	$h > g > j$	2
5	$j > g > h$	3
4	$g > h > j$	4
3	$j > h > g$	5

Issue h receives $3 + 6 = 9$ votes.
Issue g receives 4 votes.
Issue j receives $5 + 3 = 8$ votes.
Again no issue has received a majority. Therefore, eliminate g with the least number of votes, and compare only h and j.

Votes	Ranking	Row
3	$h > j$	1
6	$h > j$	2
5	$j > h$	3
4	$h > j$	4
3	$j > h$	5

Issue h receives $3 + 6 + 4 = 13$ votes.
Issue j receives $5 + 3 = 8$ votes.
Finally issue h is the winner.

(d) Make a table to examine the points in the Borda method.

Votes	Ranking	Points 3	2	1	0
3	$e > h > g > j$	e	h	g	j
6	$h > e > g > j$	h	e	g	j
5	$j > g > e > h$	j	g	e	h
4	$g > e > h > j$	g	e	h	j
3	$j > e > h > g$	j	e	h	g

Examine the last four columns and compute the weighted sum for each logo.

Issue	Borda Points
e	$3 \cdot 3 + 6 \cdot 2 + 5 \cdot 1 + 4 \cdot 2 + 3 \cdot 2 = 40$
h	$3 \cdot 2 + 6 \cdot 3 + 5 \cdot 0 + 4 \cdot 1 + 3 \cdot 1 = 31$
g	$3 \cdot 1 + 6 \cdot 1 + 5 \cdot 2 + 4 \cdot 3 + 3 \cdot 0 = 31$
j	$3 \cdot 0 + 6 \cdot 0 + 5 \cdot 3 + 4 \cdot 0 + 3 \cdot 3 = 24$

The winner is issue e with 40 Borda points.

(e) The plurality and Hare methods violate the Condorcet criterion because issue e, the Condorcet candidate does not win. The Borda method does not violate the criterion.

11. (a) Using the pairwise comparison method, first make a table to compute the votes in comparison.

Compare	Votes	Rows
t > k	$18 = 18$	1
k > t	$12 + 10 + 9 + 4 + 2 = 37$	2, 3, 4, 5, 6
t > h	$18 = 18$	1
h > t	$12 + 10 + 9 + 4 + 2 = 37$	2, 3, 4, 5, 6
t > b	$18 = 18$	1
b > t	$12 + 10 + 9 + 4 + 2 = 37$	2, 3, 4, 5, 6
t > c	$18 = 18$	1
c > t	$12 + 10 + 9 + 4 + 2 = 37$	2, 3, 4, 5, 6
k > h	$18 + 9 = 27$	1, 4
h > k	$12 + 10 + 4 + 2 = 28$	2, 3, 5, 6
k > b	$18 + 12 + 9 + 4 = 43$	1, 2, 4, 5
b > k	$10 + 2 = 12$	3, 6
k > c	$18 + 9 + 2 = 29$	1, 4, 6
c > k	$12 + 10 + 4 = 26$	2, 3, 5
h > b	$18 + 12 + 4 + 2 = 36$	1, 2, 5, 6
b > h	$10 + 9 = 19$	3, 4
h > c	$18 + 9 + 4 + 2 = 33$	1, 4, 5, 6
c > h	$12 + 10 = 22$	2, 3
b > c	$18 + 10 + 9 + 2 = 39$	1, 3, 4, 6
c > b	$12 + 4 = 16$	2, 5

Make another table to award pairwise points.

Pairs	Votes	Pairwise Points
t : k	18 : 37	k, 1
t : h	18 : 37	h, 1
t : b	18 : 37	b, 1
t : c	18 : 37	c, 1
k : h	27 : 28	h, 1
k : b	43 : 12	k, 1
k : c	29 : 26	k, 1
h : b	36 : 19	h, 1
h : c	33 : 22	h, 1
b : c	39 : 16	b, 1

Using the pairwise comparison method, the winner is h with 4 pairwise points. This is the Condorcet candidate.

(b) If the plurality method is used, activity t is selected with 18 first place votes.

(c) Calculate how many points each activity obtained from the 55 first place votes from the voter profile.

Votes	Ranking	Row
18	t > k > h > b > c	1
12	c > h > k > b > t	2
10	b > c > h > k > t	3
9	k > b > h > c > t	4
4	h > c > k > b > t	5
2	h > b > k > c > t	6

Activity	1st place votes
t	18
c	12
b	10
k	9
h	6

None of the activities received a majority, 28, of the votes. Therefore, eliminate activity h, the activity that received the least number of votes. Compare the first place votes of the others.

Votes	Ranking	Row
18	t > k > b > c	1
12	c > k > b > t	2
10	b > c > k > t	3
9	k > b > c > t	4
4	c > k > b > t	5
2	b > k > c > t	6

Activity t: 18 points.
Activity c: $12 + 4 = 16$ points.
Activity b: $10 + 2 = 12$ points.
Activity k: 9 points.
There is no activity that has received a majority; therefore, eliminate k with only 9 points, and compare the first place votes of the others.

Votes	Ranking	Row
18	t > b > c	1
12	c > b > t	2
10	b > c > t	3
9	b > c > t	4
4	c > b > t	5
2	b > c > t	6

Activity t: 18 points.
Activity c: $12 + 4 = 16$ points.
Activity b: $10 + 9 + 2 = 21$ points.
Again there is not a majority; therefore, eliminate c and compare the first place votes of t and b.

Votes	Ranking	Row
18	t > b	1
12	b > t	2
10	b > t	3
9	b > t	4
4	b > t	5
2	b > t	6

Activity t: 18 points.
Activity b: $12 + 10 + 9 + 4 + 2 = 37$ points.
Finally b wins with 37 points in the Hare method.

(d) Make a table to examine the points in the Borda method.

Votes	Ranking	Points 4	3	2	1	0
18	t > k > h > b > c	t	k	h	b	c
12	c > h > k > b > t	c	h	k	b	t
10	b > c > h > k > t	b	c	h	k	t
9	k > b > h > c > t	k	b	h	c	t
4	h > c > k > b > t	h	c	k	b	t
2	h > b > k > c > t	h	b	k	c	t

Examine the last five columns and compute the weighted sum for each activity.

Activity	Borda Points
t	$18 \cdot 4 + 12 \cdot 0 + 10 \cdot 0 + 9 \cdot 0$ $+ 4 \cdot 0 + 2 \cdot 0 = 72$
k	$18 \cdot 3 + 12 \cdot 2 + 10 \cdot 1 + 9 \cdot 4$ $+ 4 \cdot 2 + 2 \cdot 2 = 136$
h	$18 \cdot 2 + 12 \cdot 3 + 10 \cdot 2 + 9 \cdot 2$ $+ 4 \cdot 4 + 2 \cdot 4 = 134$
b	$18 \cdot 1 + 12 \cdot 1 + 10 \cdot 4 + 9 \cdot 3$ $+ 4 \cdot 1 + 2 \cdot 3 = 107$
c	$18 \cdot 0 + 12 \cdot 4 + 10 \cdot 3 + 9 \cdot 1$ $+ 4 \cdot 3 + 2 \cdot 1 = 101$

The winner is k with 136 Borda points.

(e) All three methods violate the Condorcet criterion because none of them select the Condorcet candidate.

13. (a) Using the pairwise comparison method, first make a table to compute the votes in comparison.

Compare	Votes	Rows
m > c	$5 + 3 = 8$	1, 3
c > m	$4 + 2 = 6$	2, 4
m > s	$5 + 2 = 7$	1, 4
s > m	$4 + 3 = 7$	2, 3
m > b	$5 + 2 = 7$	1, 4
b > m	$4 + 3 = 7$	2, 3
c > s	$5 + 2 = 7$	1, 4
s > c	$4 + 3 = 7$	2, 3
c > b	$5 + 2 = 7$	1, 4
b > c	$4 + 3 = 7$	2, 3
s > b	$5 + 2 = 7$	1, 4
b > s	$4 + 3 = 7$	2, 3

Make another table to award pairwise points.

Pairs	Votes	Pairwise Points
m: c	8 : 6	m, 1
m: s	7 : 7	m, $\frac{1}{2}$; s, $\frac{1}{2}$
m: b	7 : 7	m, $\frac{1}{2}$; b, $\frac{1}{2}$
c: s	7 : 7	c, $\frac{1}{2}$; s, $\frac{1}{2}$
c: b	7 : 7	c, $\frac{1}{2}$; b, $\frac{1}{2}$
s: b	7 : 7	s, $\frac{1}{2}$; b, $\frac{1}{2}$

Using the pairwise comparison method, here are the total points each city has received.

$$m: \quad 1 + \frac{1}{2} + \frac{1}{2} = 2$$

$$s: \quad \frac{1}{2} + \frac{1}{2} + \frac{1}{2} = 1\frac{1}{2}$$

$$b: \quad \frac{1}{2} + \frac{1}{2} + \frac{1}{2} = 1\frac{1}{2}$$

$$c: \quad \frac{1}{2} + \frac{1}{2} = 1$$

Montreal, m, is selected.

(b) Make a table to compute the votes in comparison.

Compare	Votes	Rows
m > c	$5 + 3 + 2 = 10$	1, 3, 4
c > m	$4 = 4$	2
m > s	$5 + 2 = 7$	1, 4
s > m	$4 + 3 = 7$	2, 3
m > b	$5 + 2 = 7$	1, 4
b > m	$4 + 3 = 7$	2, 3
c > s	$5 + 2 = 7$	1, 4
s > c	$4 + 3 = 7$	2, 3
c > b	$5 = 5$	1
b > c	$4 + 3 + 2 = 9$	2, 3, 4
s > b	$5 = 5$	1
b > s	$4 + 3 + 2 = 9$	2, 3, 4

Make another table to award pairwise points.

Pairs	Votes	Pairwise Points
m: c	10: 4	m, 1
m: s	7: 7	m, $\frac{1}{2}$; s, $\frac{1}{2}$
m: b	7: 7	m, $\frac{1}{2}$; b, $\frac{1}{2}$
c: s	7: 7	c, $\frac{1}{2}$; s, $\frac{1}{2}$
c: b	5: 9	b, 1
s: b	5: 9	b, 1

Using the pairwise comparison method, here are the total points each city has received for this vote.

$$m: \quad 1 + \frac{1}{2} + \frac{1}{2} = 2$$

$$s: \quad \frac{1}{2} + \frac{1}{2} = 1$$

$$b: \quad \frac{1}{2} + 1 + 1 = 2\frac{1}{2}$$

$$c: \quad \frac{1}{2}$$

Boston, b, is selected.

(c) The Monotonicity criterion is violated because candidate m does not win the second election.

15. (a) Make a table to examine the points in the Borda method.

		Points			
Votes	Ranking	3	2	1	0
7	s > b > c > d	s	b	c	d
5	b > d > c > s	b	d	c	s
3	d > s > c > b	d	s	c	b
4	c > s > d > b	c	s	d	b

Examine the last four columns and compute the weighted sum for each activity.

City	Borda Points
s	$7 \cdot 3 + 5 \cdot 0 + 3 \cdot 2 + 4 \cdot 2 = 35$
b	$7 \cdot 2 + 5 \cdot 3 + 3 \cdot 0 + 4 \cdot 0 = 29$
c	$7 \cdot 1 + 5 \cdot 1 + 3 \cdot 1 + 4 \cdot 3 = 27$
d	$7 \cdot 0 + 5 \cdot 2 + 3 \cdot 3 + 4 \cdot 1 = 23$

The winner is s with 35 Borda points in the preliminary non-binding decision.

(b) Make a table to examine the points in the Borda method.

		Points			
Votes	Ranking	3	2	1	0
7	s > b > c > d	s	b	c	d
5	b > d > c > s	b	d	c	s
3	s > b > d > c	s	b	d	c
4	s > b > c > d	s	b	c	d

Examine the last four columns and compute the weighted sum for each activity.

City	Borda Points
s	$7 \cdot 3 + 5 \cdot 0 + 3 \cdot 3 + 4 \cdot 3 = 42$
b	$7 \cdot 2 + 5 \cdot 3 + 3 \cdot 2 + 4 \cdot 2 = 43$
c	$7 \cdot 1 + 5 \cdot 1 + 3 \cdot 0 + 4 \cdot 1 = 16$
d	$7 \cdot 0 + 5 \cdot 2 + 3 \cdot 1 + 4 \cdot 0 = 13$

The winner is b with 43 Borda points.

(c) The Monotonicity criterion is violated because candidate s does not win the second election.

17. (a) Make a table to examine the points in the Borda method.

		Points			
Votes	Ranking	3	2	1	0
6	a > c > d > b	a	c	d	b
5	c > b > d > a	c	b	d	a
4	b > a > d > c	b	a	d	c
3	d > a > b > c	d	a	b	c

Examine the last four columns and compute the weighted sum for each activity.

City	Borda Points
a	$6 \cdot 3 + 5 \cdot 0 + 4 \cdot 2 + 3 \cdot 2 = 32$
c	$6 \cdot 2 + 5 \cdot 3 + 4 \cdot 0 + 3 \cdot 0 = 27$
d	$6 \cdot 1 + 5 \cdot 1 + 4 \cdot 1 + 3 \cdot 3 = 24$
b	$6 \cdot 0 + 5 \cdot 2 + 4 \cdot 3 + 3 \cdot 1 = 25$

The winner is a with 32 Borda points in the preliminary non-binding decision.

(b) Make a table to examine the points in the Borda method.

		Points			
Votes	Ranking	3	2	1	0
6	a > c > d > b	a	c	d	b
5	c > b > d > a	c	b	d	a
4	a > c > b > d	a	c	b	d
3	a > c > d > b	a	c	d	b

Examine the last four columns and compute the weighted sum for each activity.

City	Borda Points
a	$6 \cdot 3 + 5 \cdot 0 + 4 \cdot 3 + 3 \cdot 3 = 39$
c	$6 \cdot 2 + 5 \cdot 3 + 4 \cdot 2 + 3 \cdot 2 = 41$
d	$6 \cdot 1 + 5 \cdot 1 + 4 \cdot 0 + 3 \cdot 1 = 14$
b	$6 \cdot 0 + 5 \cdot 2 + 4 \cdot 1 + 3 \cdot 0 = 14$

The winner is c with 41 Borda points.

(c) The monotonicity criterion is violated because candidate a does not win the second election.

19. (a) Calculate how many points each city obtained from the 17 first place votes from the voter profile.

City	1st place votes
a	6
c	$4 + 2 = 6$
b	5
d	0

None of the cities received a majority, 9, of the votes. Therefore, eliminate city d, the city that received the least number of votes. Compare the first place votes of the others.

Votes	Ranking	Row
6	$a > c > b$	1
5	$b > a > c$	2
4	$c > b > a$	3
2	$c > a > b$	4

$$a: \quad\quad 6 = 6$$
$$b: \quad\quad 5 = 5$$
$$c: \quad 4 + 2 = 6$$

Again no city has received a majority; therefore, eliminate b with only 5 points and compare the first place votes of a and c.

Votes	Ranking	Row
6	$a > c$	1
5	$a > c$	2
4	$c > a$	3
2	$c > a$	4

$$a: \quad 6 + 5 = 11$$
$$c: \quad 4 + 2 = 6$$

Finally a wins with 11 points in the Hare method.

(b) Calculate how many points each city obtained from the 17 first place votes from the voter profile.

City	1st place votes
a	$6 + 2 = 8$
b	5
c	4
d	0

None of the cities received a majority, 9, of the votes. Therefore, eliminate city d, the city that received the least number of votes. Compare the first place votes of the others.

Votes	Ranking	Row
6	$a > c > b$	1
5	$b > a > c$	2
4	$c > b > a$	3
2	$a > c > b$	4

$$a: \quad 6 + 2 = 8$$
$$b: \quad\quad 5 = 5$$
$$c: \quad\quad 4 = 4$$

Again no city has received a majority; therefore, eliminate c with only 4 points and compare the first place votes of the a and b.

Votes	Ranking	Row
6	$a > b$	1
5	$b > a$	2
4	$b > a$	3
2	$a > b$	4

$$a: \quad 6 + 2 = 8$$
$$b: \quad 5 + 4 = 9$$

Finally b wins with 9 points in the Hare method.

(c) The monotonicity criterion is violated because candidate a does not win the second election.

21. (a) Calculate how many points each city obtained from the 24 first place votes from the voter profile.

City	1st place votes
a	8
b	$3 + 4 = 7$
c	$6 + 2 = 8$
d	1

None of the cities received a majority, 13, of the votes. Therefore, eliminate city d, the city that received the least number of votes. Compare the first place votes of the others.

Votes	Ranking	Row
8	a > b > c	1
3	b > c > a	2
4	b > a > c	3
6	c > b > a	4
2	c > a > b	5
1	a > b > c	6

a: $8 + 1 = 9$
b: $3 + 4 = 7$
c: $6 + 2 = 8$

Again no city has received a majority; therefore, eliminate b with only 7 points and compare the first place votes of a and c.

Votes	Ranking	Row
8	a > c	1
3	c > a	2
4	a > c	3
6	c > a	4
2	c > a	5
1	a > c	6

a: $8 + 4 + 1 = 13$
c: $3 + 6 + 2 = 11$

Finally a wins with 13 points in the Hare method.

(b) Calculate how many points each city obtained from the 24 first place votes from the voter profile.

City	1st place votes
a	$8 + 2 = 10$
b	$3 + 4 = 7$
c	6
d	1

None of the cities received a majority, 13, of the votes. Therefore, eliminate city d, the city that received the least number of votes. Compare the first place votes of the others.

Votes	Ranking	Row
8	a > b > c	1
3	b > c > a	2
4	b > a > c	3
6	c > b > a	4
2	a > c > b	5
1	a > b > c	6

a: $8 + 2 + 1 = 11$
b: $3 + 4 = 7$
c: $6 = 6$

Again no city has received a majority; therefore, eliminate c with only 6 points and compare the first place votes of the a and b.

Votes	Ranking	Row
8	a > b	1
3	b > a	2
4	b > a	3
6	b > a	4
2	a > b	5
1	a > b	6

a: $8 + 2 + 1 = 11$
b: $3 + 4 + 6 = 13$

Finally b wins with 13 points in the Hare method.

(c) The monotonicity criterion is violated because candidate a does not win the second election.

23. (a) Calculate how many points each candidate obtained from the 13 first place votes from the voter profile.

City	1st place votes
a	6
b	5
c	2

Candidate a receives the most votes and wins by the plurality method.

(b) Compare the rankings if candidate c drops out.

Votes	Ranking	Row
6	a > b	1
5	b > a	2
2	b > a	3

Now candidate a has 6 first place votes and candidate b has 7 first place votes; b is the winner.

(c) The Independence of Irrelevant Alternatives criterion is violated because candidate a does not win the second election.

25. (a) Calculate how many points each candidate obtained from the 175 first place votes from the voter profile.

City	1st place votes
a	75
b	30
c	50
d	20

Candidate a receives the most votes and wins by the plurality method.

(b) Compare the rankings if candidate b drops out.

Votes	Ranking	Row
75	$a > c > d$	1
50	$c > a > d$	2
30	$c > d > a$	3
20	$d > c > a$	4

$$a \qquad 75 = 75 .$$
$$c \quad 50 + 30 = 80$$
$$d \qquad 20 = 20$$

Now c wins with 80 votes.

(c) The Independence of Irrelevant Alternatives criterion is violated because candidate a does not win the second election.

27. (a) Using the pairwise comparison method, first make a table to compute the votes in comparison.

Compare	Votes	Rows
$e > j$	$5 + 2 + 2 = 9$	1, 3, 4
$j > e$	$2 = 2$	2
$e > h$	$5 + 2 + 2 = 9$	1, 2, 4
$h > e$	$2 = 2$	3
$e > g$	$5 + 2 + 2 = 9$	1, 2, 3
$g > e$	$2 = 2$	4
$e > m$	$5 = 5$	1
$m > e$	$2 + 2 + 2 = 6$	2, 3, 4
$j > h$	$5 + 2 + 2 = 9$	1, 2, 4
$h > j$	$2 = 2$	3
$j > g$	$5 + 2 = 7$	1, 2
$g > j$	$2 + 2 = 4$	3, 4
$j > m$	$5 = 5$	1
$m > j$	$2 + 2 + 2 = 6$	2, 3, 4
$h > g$	$5 + 2 + 2 = 9$	1, 2, 3
$g > h$	$2 = 2$	4
$h > m$	$5 + 2 = 7$	1, 3
$m > h$	$2 + 2 = 4$	2, 4
$g > m$	$5 + 2 = 7$	1, 4
$m > g$	$2 + 2 = 4$	2, 3

Make another table to award pairwise points.

Pairs	Votes	Pairwise Points
e: j	9 : 2	e, 1
e: h	9 : 2	e, 1
e: g	9 : 2	e, 1
e: m	5 : 6	m, 1
j: h	9 : 2	j, 1
j: g	7 : 4	j, 1
j: m	5 : 6	m, 1
h: g	9 : 2	h, 1
h: m	7 : 4	h, 1
g: m	7 : 4	g, 1

Using the pairwise comparison method, here are the total points each issue has received.

$$e: \quad 1 + 1 + 1 = 3$$
$$m: \quad 1 + 1 = 2$$
$$j: \quad 1 + 1 = 2$$
$$h: \quad 1 + 1 = 2$$
$$g: \quad 1 = 1$$

The education issue, e, was voted most important with 3 pairwise points.

(b) Now omitting h and g, make a table to compute the votes in comparison.

Compare	Votes	Rows
$e > j$	$5 + 2 + 2 = 9$	1, 3, 4
$j > e$	$2 = 2$	2
$e > m$	$5 = 5$	1
$m > e$	$2 + 2 + 2 = 6$	2, 3, 4
$j > m$	$5 = 5$	1
$m > j$	$2 + 2 + 2 = 6$	2, 3, 4

Make another table to award pairwise points.

Pairs	Votes	Pairwise Points
e: j	9 : 2	e, 1
e: m	5 : 6	m, 1
j: m	5 : 6	m, 1

Using the pairwise comparison method, m has two pairwise points and e has only one pairwise point.

(c) The Independence of Irrelevant Alternatives criterion is violated because candidate e fails to receive the highest priority in the second comparison.

29. In Example 4 percussionist x wins the most comparisons in the pairwise selection. If w had dropped out instead of y, here are the comparisons

Pairs	Pairwise Points
v: x	v, 1
y: v	y, 1
v: z	v, 1
x: y	x, 1
x: z	x, 1
z: y	z, 1

Using the pairwise comparison method, here are the total points each percussionist has received.

$$v: \quad 1+1 = 2$$
$$x: \quad 1+1 = 2$$
$$y: \qquad 1 = 1$$
$$z: \qquad 1 = 1$$

The Independence of Irrelevant Alternatives criterion is not violated; this second pairwise vote has resulted in a tie between v and x.

31. (a) Make a table to examine the points in the Borda method.

Votes	Ranking	Points 2	1	0
13	$c > b > a$	c	b	a
8	$b > a > c$	b	a	c
4	$b > c > a$	b	c	a

Examine the last three columns and compute the weighted sum for each candidate.

Candidate	Borda Points
a	$13 \cdot 0 + 8 \cdot 1 + 4 \cdot 0 = 8$
b	$13 \cdot 1 + 8 \cdot 2 + 4 \cdot 2 = 37$
c	$13 \cdot 2 + 8 \cdot 0 + 4 \cdot 1 = 30$

The winner is candidate b with 37 Borda points.

(b) If a drops out:

Votes	Ranking	Points 1	0
13	$c > b$	c	b
8	$b > c$	b	c
4	$b > c$	b	c

Examine the last two columns and compute the weighted sum for each candidate.

Candidate	Borda Points
b	$13 \cdot 0 + 8 \cdot 1 + 4 \cdot 1 = 12$
c	$13 \cdot 1 + 8 \cdot 0 + 4 \cdot 0 = 13$

Candidate c wins this election with 13 Borda points.

(c) According to the Independence of Irrelevant Alternatives criterion, candidate b should win the second election; therefore, the Borda method violates the criterion.

33. (a) Make a table to examine the points in the Borda method with k eliminated..

Votes	Ranking	Points 3	2	1	0
18	$t > b > h > c$	t	b	h	c
12	$c > h > b > t$	c	h	b	t
10	$c > h > b > t$	c	h	b	t
9	$b > h > c > t$	b	h	c	t
4	$h > c > b > t$	h	c	b	t
2	$h > b > c > t$	h	b	c	t

Examine the last four columns and compute the weighted sum for each activity.

Activity	Borda Points
t	$18 \cdot 3 + 12 \cdot 0 + 10 \cdot 0 + 9 \cdot 0$ $+ 4 \cdot 0 + 2 \cdot 0 = 54$
b	$18 \cdot 2 + 12 \cdot 1 + 10 \cdot 1 + 9 \cdot 3$ $+ 4 \cdot 1 + 2 \cdot 2 = 93$
h	$18 \cdot 1 + 12 \cdot 2 + 10 \cdot 2 + 9 \cdot 2$ $+ 4 \cdot 3 + 2 \cdot 3 = 98$
c	$18 \cdot 0 + 12 \cdot 3 + 10 \cdot 3 + 9 \cdot 1$ $+ 4 \cdot 2 + 2 \cdot 1 = 85$

The winner is h with 98 Borda points.

(b) According to the Independence of Irrelevant Alternatives criterion, candidate b should win the second election; therefore, the Borda method violates the criterion.

35. (a) Calculate how many points each candidate obtained from the 34 first place votes from the voter profile.

City	1st place votes
a	12
b	10
c	$8 + 4 = 12$

None of the candidates received a majority, 18, of the votes. Therefore, eliminate candidate b, the candidate that received the least number of votes. Compare the first place votes of a and c.

Votes	Ranking	Row
12	a > c	1
10	a > c	2
8	c > a	3
4	c > a	4

a: $12 + 10 = 22$

c: $8 + 4 = 12$

Now a wins with 22 votes.

(b) If candidate c drops out, the rankings are:

Votes	Ranking	Row
12	a > b	1
10	b > a	2
8	b > a	3
4	a > b	4

a: $12 + 4 = 16$

b: $10 + 8 = 18$

Now b wins with 18 votes.

(c) According to the Independence of Irrelevant Alternatives criterion, candidate a should win the second election; therefore, the Hare method violates the criterion.

37. (a) This is the original non-binding vote.

Voters	Ranking
6	a > c > b > d
5	b > d > a > c
4	c > d > b > a
2	c > a > b > d

If Chicago withdraws from the vote, compare the first place votes of the others.

Votes	Ranking	Row
6	a > b > d	1
5	b > d > a	2
4	d > b > a	3
2	a > b > d	4

a: $6 + 2 = 8$

b: $5 = 5$

d: $4 = 4$

No city has received a majority; therefore, eliminate d with only 4 points and compare the first place votes of a and b.

Votes	Ranking	Row
6	a > b	1
5	b > a	2
4	b > a	3
2	a > b	4

a: $6 + 2 = 8$

b: $5 + 4 = 9$

Finally b wins with 9 points in the Hare method.

(b) According to the Independence of Irrelevant Alternatives criterion, candidate a should win the second election; therefore, the Hare method violates the criterion.

39. Writing exercise

41. One possibility is

Votes	Ranking
10	a > b > c > d > e > f
9	b > f > e > c > d > a

Candidate a wins the majority of the 19 votes. If a pairwise comparison is made, candidate a beats each of the others in turn to earn 5 votes, from the 10 votes in the top row. From the information in the second row, notice that b beats f, e, c, and d and will receive 4 votes in the pairwise comparison.

43. One possibility is

Votes	Ranking	3	2	1	0
21	a > b > c > d	a	b	c	d
5	b > c > a > d	b	c	a	d
9	c > b > d > a	c	b	d	a
5	d > c > b > a	d	c	b	a

Examine the last four columns and compute the weighted sum for each activity.

Candidate	Borda Points
a	$21 \cdot 3 + 5 \cdot 1 + 9 \cdot 0 + 5 \cdot 0 = 68$
b	$21 \cdot 2 + 5 \cdot 3 + 9 \cdot 2 + 5 \cdot 1 = 80$
c	$21 \cdot 1 + 5 \cdot 2 + 9 \cdot 3 + 5 \cdot 2 = 68$
d	$21 \cdot 0 + 5 \cdot 0 + 9 \cdot 1 + 5 \cdot 3 = 24$

Although candidate a has the majority of the votes with a total of 21, the winner of the Borda election is b with 80 Borda points.

45. One possible profile is from Exercise 17 in 15.1.

Votes	Ranking	Voter	Row
3	b > c > a > d	1, 2, 9	1
2	a > d > c > b	3, 11	2
1	c > d > b > a	4	3
1	d > c > b > a	5	4
1	d > a > b > c	6	5
2	c > a > b > d	7, 8	6
1	a > c > d > b	10	7
2	a > b > c > d	12, 13	8

Candidate c is the Condorcet candidate because c is preferred over each of the other candidates. (See Exercise 17.)

Borda method

Here is the table from Exercise 17 to examine the points in the Borda method.

Votes	Ranking	Points 3	2	1	0
3	b > c > a > d	b	c	a	d
2	a > d > c > b	a	d	c	b
1	c > d > b > a	c	d	b	a
1	d > c > b > a	d	c	b	a
1	d > a > b > c	d	a	b	c
2	c > a > b > d	c	a	b	d
1	a > c > d > b	a	c	d	b
2	a > b > c > d	a	b	c	d

Examine the last four columns and compute the weighted sum for each.

Candidate	Borda Points
a	$3 \cdot 1 + 2 \cdot 3 + 1 \cdot 0 + 1 \cdot 0$ $+ 1 \cdot 2 + 2 \cdot 2 + 1 \cdot 3 + 2 \cdot 3 = 24$
b	$3 \cdot 3 + 2 \cdot 0 + 1 \cdot 1 + 1 \cdot 1$ $+ 1 \cdot 1 + 2 \cdot 1 + 1 \cdot 0 + 2 \cdot 2 = 18$
c	$3 \cdot 2 + 2 \cdot 1 + 1 \cdot 3 + 1 \cdot 2$ $+ 1 \cdot 0 + 2 \cdot 3 + 1 \cdot 2 + 2 \cdot 1 = 23$
d	$3 \cdot 0 + 2 \cdot 2 + 1 \cdot 2 + 1 \cdot 3$ $+ 1 \cdot 3 + 2 \cdot 0 + 1 \cdot 1 + 2 \cdot 0 = 13$

The winner is Candidate a with 24 Borda points. The Condorcet candidate has not won.

Hare method

Here is the information from Exercise 17.
Calculate how many points each candidate obtained from the thirteen first place votes from the voter profile.

Votes	Ranking	Voter	Row
3	b > c > a > d	1, 2, 9	1
2	a > d > c > b	3, 11	2
1	c > d > b > a	4	3
1	d > c > b > a	5	4
1	d > a > b > c	6	5
2	c > a > b > d	7, 8	6
1	a > c > d > b	10	7
2	a > b > c > d	12, 13	8

Candidate	1st place votes
a	5
b	3
c	3
d	2

None of the candidates received a majority, 7, of the votes. Therefore, eliminate d, the candidate that received the least number of votes. Compare the other three.

Votes	Ranking	Voter	Row
3	b > c > a	1, 2, 9	1
2	a > c > b	3, 11	2
1	c > b > a	4	3
1	c > b > a	5	4
1	a > b > c	6	5
2	c > a > b	7, 8	6
1	a > c > b	10	7
2	a > b > c	12, 13	8

Candidate a: $2 + 1 + 1 + 2 = 6$ points.
Candidate b: 3 points.
Candidate c: $1 + 1 + 2 = 4$ points.
None has received a majority; therefore, eliminate b and compare first place points again for a and c.

Votes	Ranking	Voter	Row
3	c > a	1, 2, 9	1
2	a > c	3, 11	2
1	c > a	4	3
1	c > a	5	4
1	a > c	6	5
2	c > a	7, 8	6
1	a > c	10	7
2	a > c	12, 13	8

Candidate a: $2 + 1 + 1 + 2 = 6$
Candidate c: $3 + 1 + 1 + 2 = 7$
Candidate c now has a majority and wins.

Plurality method

See from the Hare method above that a has the most first place votes to win.

47. (a) Use the information from the text to make a table to examine the points in the Borda method.

Votes	Ranking	Points			
		3	2	1	0
5	$m > b > s > c$	m	b	s	c
4	$b > s > c > m$	b	s	c	m
3	$s > m > c > b$	s	m	c	b
2	$c > m > s > b$	c	m	s	b

Examine the last three columns and compute the weighted sum for each candidate.

Candidate	Borda Points
m	$5 \cdot 3 + 4 \cdot 0 + 3 \cdot 2 + 2 \cdot 2 = 25$
b	$5 \cdot 2 + 4 \cdot 3 + 3 \cdot 0 + 2 \cdot 0 = 22$
s	$5 \cdot 1 + 4 \cdot 2 + 3 \cdot 3 + 2 \cdot 1 = 24$
c	$5 \cdot 0 + 4 \cdot 1 + 3 \cdot 1 + 2 \cdot 3 = 13$

Candidate m wins with 25 Borda points.

(b) Make a table to examine the points in the Borda method with the bottom two rows changed as follows:

Votes	Ranking	Points			
		3	2	1	0
5	$m > b > s > c$	m	b	s	c
4	$b > s > c > m$	b	s	c	m
3	$m > b > s > c$	m	b	s	c
2	$m > b > c > s$	m	b	c	s

Examine the last three columns and compute the weighted sum for each candidate.

Candidate	Borda Points
m	$5 \cdot 3 + 4 \cdot 0 + 3 \cdot 3 + 2 \cdot 3 = 30$
b	$5 \cdot 2 + 4 \cdot 3 + 3 \cdot 2 + 2 \cdot 2 = 32$
s	$5 \cdot 1 + 4 \cdot 2 + 3 \cdot 1 + 2 \cdot 0 = 16$
c	$5 \cdot 0 + 4 \cdot 1 + 3 \cdot 0 + 2 \cdot 1 = 6$

Candidate b wins with 32 Borda points.

49. Here is one possibility. If candidate c is deleted, the voter profile is:

Votes	Ranking
15	$a > b > d$
8	$b > a > d$
9	$b > a > d$
6	$d > b > a$

Now candidate b wins with $8 + 9 = 17$ votes. Because candidate a wins in the original profile, the plurality

method violates the Independence of Irrelevant Alternatives criterion in the second election.

51. Here is one possible profile:

Votes	Ranking	Points		
		2	1	0
21	$g > j > e$	g	j	e
12	$j > e > g$	j	e	g
8	$j > g > e$	j	g	e

Examine the last three columns and compute the weighted sum for each candidate.

Candidate	Borda Points
g	$21 \cdot 2 + 12 \cdot 0 + 8 \cdot 1 = 50$
j	$21 \cdot 1 + 12 \cdot 2 + 8 \cdot 2 = 61$
e	$21 \cdot 0 + 12 \cdot 1 + 8 \cdot 0 = 12$

The winner is candidate j with 61 Borda points.

For the second election, candidate e drops out.

Votes	Ranking	Points	
		1	0
21	$g > j$	g	j
12	$j > g$	j	g
8	$j > g$	j	g

Examine the last two columns and compute the weighted sum for each candidate.

Candidate	Borda Points
g	$21 \cdot 1 + 12 \cdot 0 + 8 \cdot 0 = 21$
j	$21 \cdot 0 + 12 \cdot 1 + 8 \cdot 1 = 20$

Candidate g wins with 21 Borda points, which violates the Independence of Irrelevant Alternatives criterion.

53. This is the original non-binding vote.

Voters	Ranking
8	$d > c > b > a$
3	$c > b > a > d$
4	$c > d > b > a$
6	$b > a > c > d$
2	$b > d > a > c$
1	$a > d > c > b$

Candidate d wins by the Hare method with the most first place votes. Here are the rankings with a and b deleted from the original voter profile:

Votes	Ranking	Row
8	d > c	1
3	c > d	2
4	c > d	3
6	c > d	4
2	d > c	5
1	d > c	6

$$d: \quad 8 + 2 + 1 = 11$$
$$c: \quad 3 + 4 + 6 = 13$$

Now c wins with 13 votes, which violates the Independence of Irrelevant Alternatives criterion.

55. Writing exercise

57. Writing exercise

15.3 EXERCISES

1. (a)

State Park	Acres
a	1429
b	8639
c	7608
d	6660
e	5157
Totals	29,493

The standard divisor, d, is found by dividing the total number of acres by the number of trees.

$$d = \frac{29493}{239} \approx 123.4017$$

(b) Using the Hamilton method to apportion the trees, set up a table as seen below. Remember that the standard quota, Q, is found by dividing the number of acres of land by the standard divisor, d, from part (a).

Park	Acres	Q	Rounded Q
a	1429	$\frac{1429}{123.4017} \approx 11.580$	11
b	8639	$\frac{8639}{123.4017} \approx 70.007$	70
c	7608	$\frac{7608}{123.4017} \approx 61.652$	61
d	6660	$\frac{6660}{123.4017} \approx 53.970$	53
e	5157	$\frac{5157}{123.4017} \approx 41.790$	41
Totals	29,493		236

There are 3 trees remaining to be apportioned to those parks that have the largest fractional parts of the standard quota: c, d, and e. Here are the final numbers:

Park	Rounded Q	Trees Apportioned
a	11	11
b	70	70
c	61	61 + 1 = 62
d	53	53 + 1 = 54
e	41	41 + 1 = 42
Totals	236	239

(c) Using the Jefferson method to apportion the trees, first calculate the standard divisor, md, by dividing the total number of acres by the number of trees as in part (a).

$$d = \frac{29493}{239} \approx 123.4017$$

Set up a table as seen below by using the table in part (a). Remember that the standard quota, Q, is found by dividing the number of acres of land by the standard divisor, d, from part (a). If d is decreased to 123, the modified quotas add up to approximately 239.8, which is too high. An md of 122 works.

Park	Acres	Q	mQ
a	1429	$\frac{1429}{122} \approx 11.713$	11
b	8639	$\frac{8639}{122} \approx 70.811$	70
c	7608	$\frac{7608}{122} \approx 62.361$	62
d	6660	$\frac{6660}{122} \approx 54.590$	54
e	5157	$\frac{5157}{122} \approx 42.270$	42
Totals	29,493		239

(d)

Standard Quota	Rounded Traditionally
11.580	12
70.007	70
61.652	62
53.970	54
41.790	42
	240

This sum is greater than the number of trees to be apportioned.

(e) The value of md for the Webster method should be greater than the standard divisor, d, because larger divisors create smaller quotas with a smaller total sum.

(f) Using the Webster method, again build a table using a modified divisor of 124.

Park	Acres	Q	mQ
a	1429	$\dfrac{1429}{124} \approx 11.524$	12
b	8639	$\dfrac{8639}{124} \approx 69.669$	70
c	7608	$\dfrac{7608}{124} \approx 61.355$	61
d	6660	$\dfrac{6660}{124} \approx 53.710$	54
e	5157	$\dfrac{5157}{124} \approx 41.589$	42
Totals	29,493		239

(g) The apportionment for the Hamilton and Jefferson methods are the same; the apportionment for the Webster method is different.

3. (a) The total enrollment is

$$56 + 35 + 78 + 100 = 269.$$

The standard divisor is

$$d = \frac{269}{11} = 24.\overline{45}.$$

(b) Using the Hamilton method,

Course	Enrollment	Q	mQ
Fiction	56	$\dfrac{56}{24.45455} \approx 2.290$	2
Poetry	35	$\dfrac{35}{24.45455} \approx 1.431$	1
Short Story	78	$\dfrac{78}{24.45455} \approx 3.190$	3
Multicultural	100	$\dfrac{100}{24.45455} \approx 4.089$	4
Totals	269		10

There is 1 section remaining to be apportioned to the course that has the largest fractional parts of the standard quota, Poetry. Here are the final numbers:

Course	mQ	Sections Apportioned
Fiction	2	2
Poetry	1	$1 + 1 = 2$
Short Story	3	3
Multicultural	4	4
Totals	10	11

(c) Using the Jefferson method to apportion the standard divisor must be modified. Set up a table as seen below. An md of 20 works.

Course	Enrollment	Q	mQ
Fiction	56	$\dfrac{56}{20} = 2.8$	2
Poetry	35	$\dfrac{35}{20} = 1.75$	1
Short Story	78	$\dfrac{78}{20} = 3.9$	3
Multicultural	100	$\dfrac{100}{20} = 5$	5
Totals	269		11

(d)

Standard Quota	Rounded Traditionally
2.290	2
1.431	1
3.190	3
4.089	4
	10

This sum is less the number of sections to be apportioned.

(e) The value of md for the Webster method should be less than the standard divisor, d, because smaller divisors create larger quotas with a larger total sum.

(f) Using the Webster method with an md of 23,

Course	Enrollment	Q	mQ
Fiction	56	$\dfrac{56}{23} \approx 2.4354$	2
Poetry	35	$\dfrac{35}{23} \approx 1.522$	2
Short Story	78	$\dfrac{78}{23} \approx 3.391$	3
Multicultural	100	$\dfrac{100}{23} \approx 4.348$	4
Totals	269		11

(g) The Hamilton and Webster apportionments are the same; the Jefferson apportionment differs.

(h) A Poetry student would hope that either the Hamilton or Webster method would be used, because both apportion 2 sections of the class rather than 1. This would create a smaller class size.

(i) A Multicultural student would hope that the Jefferson method be used, because this apportions 5 sections rather than 4.

5. (a) Using the Hamilton method to apportion the 131 seats, the standard divisor is found by dividing the entire population by the number of seats

$$d = \frac{47841}{131} \approx 365.1985$$

State	Population	Q	Rounded Q
Abo	5672	$\frac{5672}{365.1985} \approx 15.531$	15
Boa	8008	$\frac{8008}{365.1985} \approx 21.928$	21
Cio	2400	$\frac{2400}{365.1985} \approx 6.572$	6
Dao	6789	$\frac{6789}{365.1985} \approx 18.590$	18
Effo	4972	$\frac{4972}{365.1985} \approx 13.615$	13
Foti	20,000	$\frac{20,000}{365.1985} \approx 54.765$	54
Totals	47,841		127

There are $131 - 127 = 4$ seats remaining to be apportioned to those states that have the largest fractional parts of the standard quota: Boa, Foti, Effo, and Dao. Here are the final numbers:

State	Rounded Q	Seats Apportioned
Abo	15	15
Boa	21	$21 + 1 = 22$
Cio	6	6
Dao	18	$18 + 1 = 19$
Effo	13	$13 + 1 = 14$
Foti	54	$54 + 1 = 55$
Totals	127	131

(b) The Hamilton, Jefferson, and Webster apportionments are all different.

7. Using the Webster method with an md of 366.9730:

State	Population	mQ	Rounded Q
Abo	5672	$\frac{5672}{366.9730} \approx 15.456178$	15
Boa	8008	$\frac{8008}{366.9730} \approx 21.821769$	22
Cio	2400	$\frac{2400}{366.9730} \approx 6.539991$	7
Dao	6789	$\frac{6789}{366.9730} \approx 18.4999999$	18
Effo	4972	$\frac{4972}{366.9730} \approx 13.548681$	14
Foti	20,000	$\frac{20,000}{366.9730} \approx 54.499922$	54
Totals	47,841		**130**

Notice that the total of the last column is 130 seats.

9. (a) The total number of beds is

$$137 + 237 + 337 + 455 + 555 = 1721.$$

The standard divisor for the apportionment of nurses is

$$d = \frac{1721}{40} = 43.025.$$

(b) Using the Hamilton method to apportion the nurses:

Hosp.	No. of Beds	Q	Rounded Q
A	137	$\frac{137}{43.025} \approx 3.184$	3
B	237	$\frac{237}{43.025} \approx 5.508$	5
C	337	$\frac{337}{43.025} \approx 7.833$	7
D	455	$\frac{455}{43.025} \approx 10.575$	10
E	555	$\frac{555}{43.025} \approx 12.899$	12
Totals	1721		37

There are 3 more nurses to apportion to those hospitals with the greatest fractional part remaining in the Q column. The final apportionment is

Hospital	Q	mQ
A	$\frac{137}{43.025} \approx 3.184$	3
B	$\frac{237}{43.025} \approx 5.508$	5
C	$\frac{337}{43.025} \approx 7.833$	$7 + 1 = 8$
D	$\frac{455}{43.025} \approx 10.575$	$10 + 1 = 11$
E	$\frac{555}{43.025} \approx 12.899$	$12 + 1 = 13$
Totals		40

(c) Using the Jefferson method with an md of 40

Hospital	No. of Beds	Q	Rounded Q
A	137	$\frac{137}{40} = 3.425$	3
B	237	$\frac{237}{40} = 5.925$	5
C	337	$\frac{337}{40} = 8.425$	8
D	455	$\frac{455}{40} = 11.375$	11
E	555	$\frac{555}{40} = 13.875$	13
Totals	1721		40

(d)

Standard Quota	Rounded Traditionally
3.184	3
5.508	6
7.833	8
10.575	11
12.899	13
	41

The traditionally rounded values add to 41, which is greater than the number of nurses to be apportioned.

(e) The value of md for the Webster method should be greater than the standard divisor, because larger divisors create smaller modified quotas with a smaller total sum.

(f) Using the Webster method with an md of 43.1

Hospital	No. of Beds	Q	Rounded Q
A	137	$\frac{137}{43.1} \approx 3.179$	3
B	237	$\frac{237}{43.1} \approx 5.499$	5
C	337	$\frac{337}{43.1} \approx 7.819$	8
D	455	$\frac{455}{43.1} \approx 10.557$	11
E	555	$\frac{555}{43.1} \approx 12.877$	13
Totals	1721		40

(g) All three apportionments are the same.

11. Here is one possible population profile.

State	a	b	c	d	e	Total
Population	50	230	280	320	120	1000

Hamilton method
The standard divisor is

$$d = \frac{1000}{100} = 10.$$

State	Pop.	Q
a	50	$\frac{50}{10} = 5$
b	230	$\frac{230}{10} = 23$
c	280	$\frac{280}{10} = 28$
d	320	$\frac{320}{10} = 32$
e	120	$\frac{120}{10} = 12$
Totals	1000	100

It is unnecessary to find a modified divisor for the Jefferson and Webster methods, because the value of d divides into each population evenly. No rounding is necessary. That is, the modified divisor is 10 for both methods.

13. Here is one possible ridership profile.

Bus Route	a	b	c	d	e	Total
No. of Riders	131	140	303	178	197	949

Hamilton method
The standard divisor is

$$d = \frac{949}{16} = 59.3125.$$

Bus Route	Riders	Q	Rounded Q
a	131	$\frac{131}{59.3125} \approx 2.209$	2
b	140	$\frac{140}{59.3125} \approx 2.360$	2
c	303	$\frac{303}{59.3125} \approx 5.109$	5
d	178	$\frac{178}{59.3125} \approx 3.001$	3
e	197	$\frac{197}{59.3125} \approx 3.321$	3
Totals	949		15

The remaining bus will be assigned to route b because it has the largest fractional portion in its quotient. The final apportionment is

Bus Route	Riders	Q	Rounded Q
a	131	2.209	2
b	140	2.360	2 + 1 = 3
c	303	5.109	5
d	178	3.001	3
e	197	3.321	3
Totals	949		16

Jefferson method

Using a modified divisor of 50, here is the apportionment.

Bus Route	Riders	Q	Rounded Q
a	131	$\frac{131}{50} = 2.62$	2
b	140	$\frac{140}{50} = 2.8$	2
c	303	$\frac{303}{50} = 6.06$	6
d	178	$\frac{178}{50} = 3.56$	3
e	197	$\frac{197}{50} = 3.94$	3
Totals	949		16

Webster method

Using a modified divisor of 56.2, round each value of Q according to normal rules of rounding.

Bus Route	Riders	Q	Rounded Q
a	131	$\frac{131}{56.2} \approx 2.331$	2
b	140	$\frac{140}{56.2} \approx 2.491$	2
c	303	$\frac{303}{56.2} \approx 5.391$	5
d	178	$\frac{178}{56.2} \approx 3.167$	3
e	197	$\frac{197}{56.2} \approx 3.505$	4
Totals	949		16

15. (a) Using the Adams method with an md of 29

Course	Enrollment	Q	mQ
Fiction	56	$\frac{56}{29} \approx 1.931$	2
Poetry	35	$\frac{35}{29} \approx 1.207$	2
Short Story	78	$\frac{78}{29} \approx 2.690$	3
Multicultural	100	$\frac{100}{29} \approx 3.448$	4
Totals	269		11

(b) The Adams apportionment is the same as the Hamilton and Webster apportionments. It is different from the Jefferson apportionment.

17. (a) Using the Adams method with an md of 377.3

State	Population	Q	Rounded Q
Abo	5672	$\frac{5672}{377.3} \approx 15.033$	16
Boa	8008	$\frac{8008}{377.3} \approx 21.224$	22
Cio	2400	$\frac{2400}{377.3} \approx 6.361$	7
Dao	6789	$\frac{6789}{377.3} \approx 17.994$	18
Effo	4972	$\frac{4972}{377.3} \approx 13.177$	14
Foti	20,000	$\frac{20,000}{377.3} \approx 53.008$	54
Totals	47,841		131

(b) Each method produces a different apportionment.

19. The cutoff point for rounding the modified quota of 56.498 up to 57 is calculated by finding the geometric mean of 56 (the integer part of 56.498) and 57.

$$\sqrt{56 \cdot 57} = \sqrt{3192} \approx 56.498.$$

21. The cutoff point for rounding the modified quota of 32.497 up to 33 is calculated by finding the geometric mean of 32 (the integer part of 32.497) and 33.

$$\sqrt{32 \cdot 33} = \sqrt{1056} \approx 32.496.$$

23. If the sum of the traditionally rounded Q values is greater than the number of objects being apportioned, then the modified divisor is found by slowly increasing the value of d. A larger divisor produces smaller modified quotas with a smaller sum.

25. (a) Using the Hill-Huntington method with an md of 24

Course	Enrollment	Q	mQ
Fiction	56	$\frac{56}{24} = 2.\overline{3}$	2
Poetry	35	$\frac{35}{24} \approx 1.458$	2
Short Story	78	$\frac{78}{24} = 3.25$	3
Multicultural	100	$\frac{100}{24} = 4.1\overline{6}$	4
Totals	269		11

(b) The Hill-Huntington apportionment is the same as the Hamilton, Webster, and Adams apportionments. It is different from the Jefferson apportionment.

27. (a) Using the Hill-Huntington method with an md of 367, each value of mQ is found by dividing the population figure by 367. See the explanation in the text for computation of the geometric mean.

State	Population	mQ	Geo. Mean	Rounded Q
Abo	5672	15.4550	15.4919	15
Boa	8008	21.8202	21.4942	22
Cio	2400	6.5395	6.4807	7
Dao	6789	18.4986	18.4932	19
Effo	4972	13.5477	13.4907	14
Foti	20,000	54.4959	54.4977	54
Totals	47,841			131

(b) The Hill-Huntington apportionment is the same as the Webster apportionment. The other three are all different.

29. Writing exercise

31. Writing exercise

15.4 EXERCISES

1. The sum of the populations is

$$17179 + 7500 + 49400 + 5824 = 79,903.$$

The standard divisor is

$$d = \frac{79903}{132} = 605.326.$$

Using the standard quota

State	Population	Q	Rounded Q
a	17179	$\frac{17179}{605.326} \approx 28.380$	28
b	7500	$\frac{7500}{605.326} \approx 12.390$	12
c	49400	$\frac{49400}{605.326} \approx 81.609$	**81**
d	5824	$\frac{5824}{605.326} \approx 9.621$	9
Totals	79,903		130

Using the Jefferson method with $md = 595$

State	Population	Q	Rounded Q
a	17179	$\frac{17179}{595} \approx 28.872$	28
b	7500	$\frac{7500}{595} \approx 12.605$	12
c	49400	$\frac{49400}{595} \approx 83.025$	**83**
d	5824	$\frac{5824}{595} \approx 9.788$	9
Totals	79,903		132

The Jefferson method violates the Quota Rule because state c receives receives two more seats than its apportionment from the standard quota.

3. From Exercise 5 in Section 15.3, the standard divisor is found by dividing the entire population by the number of seats

$$d = \frac{47841}{131} \approx 365.1985$$

State	Population	Q	Rounded Q
Abo	5672	$\frac{5672}{365.1985} \approx 15.531$	15
Boa	8008	$\frac{8008}{365.1985} \approx 21.928$	21
Cio	2400	$\frac{2400}{365.1985} \approx 6.572$	6
Dao	6789	$\frac{6789}{365.1985} \approx 18.590$	18
Effo	4972	$\frac{4972}{365.1985} \approx 13.615$	13
Foti	20000	$\frac{20,000}{365.1985} \approx 54.765$	**54**
Totals	47841		127

Using the Jefferson method with $md = 356$

State	Population	Q	Rounded Q
Abo	5672	$\frac{5672}{356} \approx 15.933$	15
Boa	8008	$\frac{8008}{356} \approx 22.494$	22
Cio	2400	$\frac{2400}{356} \approx 6.742$	6
Dao	6789	$\frac{6789}{356} \approx 19.070$	19
Effo	4972	$\frac{4972}{356} \approx 13.966$	13
Foti	20,000	$\frac{20,000}{356} \approx 56.180$	**56**
Totals	47,841		131

The Jefferson method violates the Quota Rule because Foti receives two more seats than its apportionment from the standard quota.

5. The sum of the populations is

$$2567 + 1500 + 8045 + 950 + 1099 = 14161.$$

The standard divisor is

$$d = \frac{14161}{290} \approx 48.8310.$$

Using the standard quota

State	Population	Q	Rounded Q
a	2567	$\frac{2567}{48.8310} \approx 52.569$	52
b	1500	$\frac{1500}{48.8310} \approx 30.718$	30
c	8045	$\frac{8045}{48.8310} \approx 164.752$	**164**
d	950	$\frac{950}{48.8310} \approx 19.455$	19
e	1099	$\frac{1099}{48.8310} \approx 22.506$	22
Totals	14, 161		287

Using the Jefferson method with $md = 48.4$

State	Population	Q	Rounded Q
a	2567	$\frac{2567}{48.4} \approx 53.037$	53
b	1500	$\frac{1500}{48.4} \approx 30.992$	30
c	8045	$\frac{8045}{48.4} \approx 166.219$	**166**
d	950	$\frac{950}{48.4} \approx 19.628$	19
e	1099	$\frac{1099}{48.4} \approx 22.707$	22
Totals	14, 161		290

The Jefferson method violates the Quota Rule because state c receives two more seats than its apportionment from the standard quota.

7. First calculate the apportionment with $n = 204$. The total population is

$$3462 + 7470 + 4265 + 5300 = 20,497.$$

The standard divisor for 204 seats is

$$d = \frac{20497}{204} \approx 100.4755.$$

Using the Hamilton method

State	Population	Q	Rounded Q
a	3462	$\frac{3462}{100.4755} \approx 34.456$	34
b	7470	$\frac{7470}{100.4755} \approx 74.346$	74
c	4265	$\frac{4265}{100.4755} \approx 42.448$	42
d	5300	$\frac{5300}{100.4755} \approx 52.749$	52
Totals	20497		202

The two remaining seats will be apportioned to State d and State a, because the fractional parts of their quotas are the largest. The final apportionment is

State	Population	Q	Number of seats
a	3462	34.456	$34 + 1 = $ **35**
b	7470	74.346	74
c	4265	42.448	42
d	5300	52.749	$52 + 1 = 53$
Totals	20497		204

Now increase the number of seats to 205. The standard divisor is

$$d = \frac{20497}{205} \approx 99.98537.$$

Using the Hamilton method with the new d

State	Population	Q	Rounded Q
a	3462	$\frac{3462}{99.98537} \approx 34.623$	**34**
b	7470	$\frac{7470}{99.98537} \approx 74.711$	74
c	4265	$\frac{4265}{99.98537} \approx 42.656$	42
d	5300	$\frac{5300}{99.98537} \approx 53.008$	53
Totals	20497		203

The two remaining seats will be apportioned to State b and State c, because the fractional parts of their quotas are the largest. The final apportionment is

State	Q	Number of seats
a	34.623	**34**
b	74.711	$74 + 1 = 75$
c	42.656	$42 + 1 = 43$
d	53.008	53
Totals		205

State a is a victim of the Alabama Paradox, because it has lost a seat despite the fact that the overall number of seats has increased.

9. First calculate the apportionment with $n = 126$. The total population is

$$263 + 808 + 931 + 781 + 676 = 3459.$$

The standard divisor for 126 seats is

$$d = \frac{3459}{126} \approx 27.4524.$$

Using the Hamilton method

State	Population	Q	Rounded Q
a	263	$\frac{263}{27.4524} \approx 9.580$	9
b	808	$\frac{808}{27.4524} \approx 29.433$	29
c	931	$\frac{931}{27.4524} \approx 33.913$	33
d	781	$\frac{781}{27.4524} \approx 28.449$	28
e	676	$\frac{676}{27.4524} \approx 24.624$	24
Totals	3459		123

The three remaining seats will be apportioned to States c, e, and a, because the fractional parts of their quotas are the largest. The final apportionment is

State	Q	Number of seats
a	9.580	$9 + 1 = \mathbf{10}$
b	29.433	29
c	33.913	$33 + 1 = 34$
d	28.449	28
e	24.624	$24 + 1 = 25$
Totals		126

Now increase the number of seats to 127. The standard divisor is

$$d = \frac{3459}{127} \approx 27.2362.$$

Using the Hamilton method with the new d

State	Population	Q	Rounded Q
a	263	$\frac{263}{27.2362} \approx 9.656$	**9**
b	808	$\frac{808}{27.2362} \approx 29.666$	29
c	931	$\frac{931}{27.2362} \approx 34.182$	34
d	781	$\frac{781}{27.2362} \approx 28.675$	28
e	676	$\frac{676}{27.2362} \approx 24.820$	24
Totals	3459		124

The three remaining seats will be apportioned to States e,

d and b, because the fractional parts of their quotas are the largest. The final apportionment is

State	Q	Number of seats
a	9.656	**9**
b	29.666	$29 + 1 = 30$
c	34.182	34
d	28.675	$28 + 1 = 29$
e	24.820	$24 + 1 = 25$
Totals		127

State a is a victim of the Alabama Paradox, because it has lost a seat despite the fact that the overall number of seats has increased.

11. First calculate the apportionment with $n = 149$. The total population is

$$5552 + 8260 + 5968 + 6256 + 5150 = 31186.$$

The standard divisor for 149 seats is

$$d = \frac{31186}{149} \approx 209.302.$$

Using the Hamilton method

State	Population	Q	Rounded Q
a	5552	$\frac{5552}{209.302} \approx 26.526$	26
b	8260	$\frac{8260}{209.302} \approx 39.465$	39
c	5968	$\frac{5968}{209.302} \approx 28.514$	28
d	6256	$\frac{6256}{209.302} \approx 29.890$	29
e	5150	$\frac{5150}{209.302} \approx 24.606$	24
Totals	31, 186		146

The three remaining seats will be apportioned to States d, e, and a, because the fractional parts of their quotas are the largest. The final apportionment is

State	Q	Number of seats
a	26.526	$26 + 1 = \mathbf{27}$
b	39.465	39
c	28.514	28
d	29.890	$29 + 1 = 30$
e	24.606	$24 + 1 = 25$
Totals		149

Now increase the number of seats to 150. The standard divisor is

$$d = \frac{31186}{150} \approx 207.9067.$$

Using the Hamilton method with the new d

State	Population	Q	Rounded Q
a	5552	$\dfrac{5552}{207.9067} \approx 26.704$	26
b	8260	$\dfrac{8260}{207.9067} \approx 39.729$	39
c	5968	$\dfrac{5968}{207.9067} \approx 28.705$	28
d	6256	$\dfrac{6256}{207.9067} \approx 30.090$	30
e	5150	$\dfrac{5150}{207.9067} \approx 24.771$	24
Totals	31186		147

The three remaining seats will be apportioned to States e, b, and c, because the fractional parts of their quotas are the largest. The final apportionment is

State	Q	Number of seats
a	26.704	**26**
b	39.729	$39 + 1 = 40$
c	28.705	$28 + 1 = 29$
d	30.090	30
e	24.771	$24 + 1 = 25$
Totals		150

State a is a victim of the Alabama Paradox, because it has lost a seat despite the fact that the overall number of seats has increased.

13. First calculate the apportionment for the initial populations. The total population is

$$55 + 125 + 190 = 370.$$

The standard divisor for 11 seats is

$$d = \frac{370}{11} = 33.6364.$$

Using the Hamilton method

State	Population	Q	Rounded Q
a	55	$\dfrac{55}{33.6364} \approx 1.635$	1
b	125	$\dfrac{125}{33.6364} \approx 3.716$	3
c	190	$\dfrac{190}{33.6364} \approx 5.649$	5
Totals	370		9

The two remaining seats will be apportioned to States b and c, because the fractional parts of their quotas are the largest. The final apportionment is

State	Q	Number of seats
a	1.635	1
b	3.716	$3 + 1 = 4$
c	5.649	$5 + 1 = 6$
Totals		11

Now apportion the seats for the revised populations.

$$61 + 148 + 215 = 424$$

The new standard divisor is

$$d = \frac{424}{11} \approx 38.5455.$$

Using the Hamilton method with the new d

State	Population	Q	Rounded Q
a	61	$\dfrac{61}{38.5455} \approx 1.583$	1
b	148	$\dfrac{148}{38.5455} \approx 3.840$	3
c	215	$\dfrac{215}{38.5455} \approx 5.578$	5
Totals	424		9

The two remaining seats will be apportioned to States a and b, because the fractional parts of their quotas are the largest. The final apportionment is

State	Q	Number of seats
a	1.583	$1 + 1 = 2$
b	3.840	$3 + 1 = 4$
c	5.578	5
Totals		11

Here is a final summary:

State	Old Pop	New Pop	% Inc.	Old No. of Seats	New No. of Seats
a	55	61	**10.9**	1	**2**
b	125	148	18.4	4	4
c	190	215	**13.2**	6	5
Totals	370	424		11	11

Notice from the table that there was a greater percent increase in growth for State c than for State a. Yet State a gained a seat and State c lost a seat. This is an example of the Population Paradox.

15. First calculate the apportionment for the initial populations. The total population is

$$930 + 738 + 415 = 2083.$$

The standard divisor for 13 seats is

$$d = \frac{2083}{13} = 160.2308.$$

Using the Hamilton method

State	Population	Q	Rounded Q
a	930	$\frac{930}{160.2308} \approx 5.804$	5
b	738	$\frac{738}{160.2308} \approx 4.606$	4
c	415	$\frac{415}{160.2308} \approx 2.590$	2
Totals	2083		11

The two remaining seats will be apportioned to States a and b, because the fractional parts of their quotas are the largest. The final apportionment is

State	Q	Number of seats
a	5.804	$5 + 1 = 6$
b	4.606	$4 + 1 = 5$
c	2.590	2
Totals		13

Now apportion the seats for the revised populations.

$$975 + 750 + 421 = 2146.$$

The new standard divisor is

$$d = \frac{2146}{13} \approx 165.0769.$$

Using the Hamilton method with the new values

State	Population	Q	Rounded Q
a	975	$\frac{975}{165.0769} \approx 5.906$	5
b	750	$\frac{750}{165.0769} \approx 4.543$	4
c	421	$\frac{421}{165.0769} \approx 2.550$	2
Totals	2146		11

The two remaining seats will be apportioned to States a and c, because the fractional parts of their quotas are the largest. The final apportionment is

State	Q	Number of seats
a	5.906	$5 + 1 = 6$
b	4.543	4
c	2.550	$2 + 1 = 3$
Totals		13

Here is a final summary:

State	Old Pop	New Pop	% Inc.	Old No. of Seats	New No. of Seats
a	930	975	4.84	6	6
b	738	750	**1.63**	**5**	**4**
c	415	421	**1.45**	**2**	**3**
Totals	2083	2146		13	13

Notice from the table that there was a greater percent increase in growth for State b than for State c. Yet State c gained a seat and State b lost a seat. This is an example of the Population Paradox.

17. First calculate the apportionment for the initial populations. The total population is

$$89 + 125 + 225 = 439.$$

The standard divisor for 13 seats is

$$d = \frac{439}{13} \approx 33.7692.$$

Using the Hamilton method

State	Population	Q	Rounded Q
a	89	$\frac{89}{33.7692} \approx 2.636$	2
b	125	$\frac{125}{33.7692} \approx 3.702$	3
c	225	$\frac{225}{33.7692} \approx 6.663$	6
Totals	439		11

The two remaining seats will be apportioned to States b and c, because the fractional parts of their quotas are the largest. The final apportionment is

State	Q	Number of seats
a	2.636	2
b	3.702	$3 + 1 = 4$
c	6.663	$6 + 1 = 7$
Totals		13

Now apportion the seats for the revised populations.

$$97 + 145 + 247 = 489.$$

The new standard divisor is

$$d = \frac{489}{13} \approx 37.6154.$$

Using the Hamilton method with the new values

State	Population	Q	Rounded Q
a	97	$\frac{97}{37.6154} \approx 2.579$	2
b	145	$\frac{145}{37.6154} \approx 3.855$	3
c	247	$\frac{247}{37.6154} \approx 6.566$	6
Totals	489		11

The two remaining seats will be apportioned to States a and b, because the fractional parts of their quotas are the largest. The final apportionment is

State	Q	Number of seats
a	2.579	$2 + 1 = 3$
b	3.855	$3 + 1 = 4$
c	6.566	6
Totals		13

Here is a final summary:

State	Old Pop	New Pop	% Inc.	Old No. of Seats	New No. of Seats
a	89	97	**8.99**	**2**	**3**
b	125	145	16.00	4	4
c	225	247	**9.78**	**7**	**6**
Totals	439	489		13	13

Notice from the table that there was a greater percent increase in growth for State c than for State a. Yet State a gained a seat and State c lost a seat. This is an example of the Population Paradox.

19. First calculate the apportionment for the initial populations. The total population is

$$134 + 52 = 186.$$

The standard divisor for 16 seats is

$$d = \frac{186}{16} = 11.625.$$

Using the Hamilton method

State	Population	Q	Rounded Q
a	134	$\frac{134}{11.625} \approx 11.527$	11
b	52	$\frac{52}{11.625} \approx 4.473$	4
Totals	186		15

The one remaining seat will be apportioned to State a because the fractional parts of its quota is larger. The final apportionment is

State	Q	Number of seats
a	11.527	$11 + 1 = 12$
b	4.473	4
Totals		16

Now apportion the seats to include the new state. The standard quota of the new state is:

$$Q = \frac{38}{11.625} \approx 3.269.$$

Rounded down to 3, add 3 new seats to the original 16 to obtain 19 seats to be apportioned. Now the new population is

$$134 + 52 + 38 = 224.$$

The new standard divisor is

$$d = \frac{224}{19} \approx 11.7895.$$

State	Population	Q	Rounded Q
a	134	$\frac{134}{11.7895} \approx 11.366$	11
b	52	$\frac{52}{11.7895} \approx 4.411$	4
c	38	$\frac{38}{11.7895} \approx 3.223$	3
Totals	224		18

The one remaining seat will be apportioned to State b because the fractional part of its quota is the largest. The final apportionment is

State	Q	Number of seats
a	11.366	11
b	4.411	$4 + 1 = 5$
c	3.223	3
Totals		19

The New State Paradox has occurred because the addition of the new state has caused a shift in the apportionment of the original states. States a and b originally had 12 and 4 seats, respectively; now they have 11 and 5, respectively.

21. First calculate the apportionment for the initial populations. The total population is

$$3184 + 8475 = 11,659.$$

The standard divisor for 75 seats is

$$d = \frac{11659}{75} \approx 155.4533.$$

Using the Hamilton method

State	Population	Q	Rounded Q
a	3184	$\frac{3184}{155.4533} \approx 20.482$	20
b	8475	$\frac{8475}{155.4533} \approx 54.518$	54
Totals	11,659		74

The one remaining seat will be apportioned to State b because the fractional parts of its quota is larger. The final apportionment is

State	Q	Number of seats
a	20.482	20
b	54.518	$54 + 1 = 55$
Totals		75

Now apportion the seats to include the new state. The standard quota of the new state is:

$$Q = \frac{330}{155.4533} \approx 2.123.$$

Rounded down to 2, add 2 new seats to the original 75 to obtain 77 seats to be apportioned. Now the new population is

$$3184 + 8475 + 330 = 11,989.$$

The new standard divisor is

$$d = \frac{11989}{77} \approx 155.7013.$$

State	Population	Q	Rounded Q
a	3184	$\frac{3184}{155.7013} \approx 20.449$	20
b	8475	$\frac{8475}{155.7013} \approx 54.431$	54
c	330	$\frac{330}{155.7013} \approx 2.119$	2
Totals	11,989		76

The one remaining seat will be apportioned to State a because the fractional part of its quota is the largest. The final apportionment is

State	Q	Number of seats
a	20.449	$20 + 1 = 21$
b	54.431	54
c	2.119	2
Totals		77

The New State Paradox has occurred because the addition of the new state has caused a shift in the apportionment of the original states. States a and b originally had 20 and 55 seats, respectively; now they have 21 and 54, respectively.

23. First calculate the apportionment for the initial populations. The total population is

$$7500 + 9560 = 17,060.$$

The standard divisor for 83 seats is

$$d = \frac{17060}{83} \approx 205.5422.$$

Using the Hamilton method

State	Population	Q	Rounded Q
a	7500	$\frac{7500}{205.5422} \approx 36.489$	36
b	9560	$\frac{9560}{205.5422} \approx 46.511$	46
Totals	17,060		82

The one remaining seat will be apportioned to State b because the fractional parts of its quota is larger. The final apportionment is

State	Q	Number of seats
a	36.489	36
b	46.511	$46 + 1 = 47$
Totals		83

Now apportion the seats to include the new state. The standard quota of the new state is:

$$Q = \frac{1500}{205.5422} \approx 7.298.$$

Rounded down to 7, add 7 new seats to the original 83 to obtain 90 seats to be apportioned. Now the new population is

$$7500 + 9560 + 1500 = 18,560.$$

The new standard divisor is

$$d = \frac{18560}{90} \approx 206.2222.$$

State	Population	Q	Rounded Q
a	7500	$\dfrac{7500}{206.2222} \approx 36.369$	36
b	9560	$\dfrac{9560}{206.2222} \approx 46.358$	46
c	1500	$\dfrac{1500}{206.2222} \approx 7.274$	7
Totals	18,560		89

The one remaining seat will be apportioned to State a because the fractional part of its quota is the largest. The final apportionment is

State	Q	Number of seats
a	36.369	$36 + 1 = 37$
b	46.358	46
c	7.274	7
Totals		90

The New State Paradox has occurred because the addition of the new state has caused a shift in the apportionment of the original states. States a and b originally had 36 and 47 seats, respectively; now they have 37 and 46, respectively.

25. Use the Jefferson method to apportion the teaching assistants for the following enrollment profile.

Class	Enrollment
a	225
b	45
c	35
d	30
Total	335

first calculate the standard divisor, d, by dividing the total enrollment by the number of teaching assistants, 25.

$$d = \frac{335}{25} = 13.4.$$

Calculate the standard quota, Q, for each class by dividing the enrollment by the standard divisor, d. If d is used, the modified quotas total to only 23. If d is slowly decreased to 12.4, the modified quotas add up to 25.

Class	Enrollment	Standard Q	Rounded Q	New Q	mQ
a	225	$\dfrac{225}{13.4} \approx 16.8$	16	$\dfrac{225}{12.4} \approx 18.1$	18
b	45	$\dfrac{45}{13.4} \approx 3.4$	3	$\dfrac{45}{12.4} \approx 3.6$	3
c	35	$\dfrac{35}{13.4} \approx 2.6$	2	$\dfrac{35}{12.4} \approx 2.8$	2
d	30	$\dfrac{30}{13.4} \approx 2.2$	2	$\dfrac{30}{12.4} \approx 2.4$	2
Totals	335		23		25

For class a, rounding 16.8 up to 17 indicates that this class could indeed receive one more teaching assistant without violating the quota rule. However, class a receives two additional teaching assistants by the Jefferson method.

27. Here is the profile from Exercise 5, using 200 seats to be apportioned. The standard divisor is

$$d = \frac{14161}{200} = 70.805.$$

State	Population	Q	Rounded Q
a	2567	$\dfrac{2567}{70.805} \approx 36.255$	36
b	1500	$\dfrac{1500}{70.805} \approx 21.185$	21
c	8045	$\dfrac{8045}{70.805} \approx 113.622$	**113**
d	950	$\dfrac{950}{70.805} \approx 13.417$	13
e	1099	$\dfrac{1099}{70.805} \approx 15.522$	15
Totals	14,161		198

Using the Jefferson method with $md = 69.5$.

State	Population	Q	Rounded Q
a	2567	$\dfrac{2567}{69.5} \approx 36.94$	36
b	1500	$\dfrac{1500}{69.5} \approx 21.58$	21
c	8045	$\dfrac{8045}{69.5} \approx 115.76$	**115**
d	950	$\dfrac{950}{69.5} \approx 13.67$	13
e	1099	$\dfrac{1099}{69.5} \approx 15.81$	15
Totals	14,161		200

The Jefferson method violates the Quota Rule because state c receives two more seats than its apportionment from the standard quota.

29. First calculate the apportionment with $n = 49$. The total population is

$$465 + 552 + 385 + 251 = 1653.$$

The standard divisor for 49 seats is

$$d = \frac{1653}{49} \approx 33.7347.$$

Using the Hamilton method

State	Population	Q	Rounded Q
a	465	$\frac{465}{33.7347} \approx 13.784$	13
b	552	$\frac{552}{33.7347} \approx 16.363$	16
c	385	$\frac{385}{33.7347} \approx 11.413$	11
d	251	$\frac{251}{33.7347} \approx 7.440$	7
Totals	1653		47

The two remaining seats will be apportioned to State a and State d, because the fractional parts of their quotas are the largest. The final apportionment is

State	Population	Q	Number of seats
a	465	13.784	$13 + 1 = 14$
b	552	16.363	16
c	385	11.413	11
d	251	7.440	$7 + 1 = 8$
Totals	1653		49

Now increase the number of seats to 50. The standard divisor is

$$d = \frac{1653}{50} = 33.06.$$

Using the Hamilton method with the new d

State	Population	Q	Rounded Q
a	465	$\frac{465}{33.06} \approx 14.065$	14
b	552	$\frac{552}{33.06} \approx 16.697$	16
c	385	$\frac{385}{33.06} \approx 11.645$	11
d	251	$\frac{251}{33.06} \approx 7.592$	**7**
Totals	1653		48

The two remaining seats will be apportioned to State b and State c, because the fractional parts of their quotas are the largest. The final apportionment is

State	Q	Number of seats
a	14.065	14
b	16.697	$16 + 1 = 17$
c	11.645	$11 + 1 = 12$
d	7.592	**7**
Totals		50

State d is a victim of the Alabama Paradox, because it has lost a seat despite the fact that the overall number of seats has increased.

31. From Exercise 13, calculate the apportionment for the preliminary enrollments. The total enrollment is

$$55 + 125 + 190 = 370.$$

The standard divisor for 11 teaching assistants is

$$d = \frac{370}{11} = 33.6364.$$

Using the Hamilton method

Course	Enrollment	Q	Rounded Q
a	55	$\frac{55}{33.6364} \approx 1.635$	1
b	125	$\frac{125}{33.6364} \approx 3.716$	3
c	190	$\frac{190}{33.6364} \approx 5.649$	5
Totals	370		9

The two remaining teaching assistants will be apportioned to Courses b and c, because the fractional parts of their quotas are the largest. The final apportionment is

Course	Q	Number of TA's
a	1.635	1
b	3.716	$3 + 1 = 4$
c	5.649	$5 + 1 = 6$
Totals		11

Now try the following actual enrollments.

$$61 + 145 + 213 = 419.$$

The new standard divisor is

$$d = \frac{419}{11} \approx 38.0909.$$

Using the Hamilton method with the new d

Course	Enrollment	Q	Rounded Q
a	61	$\frac{61}{38.0909} \approx 1.601$	1
b	145	$\frac{145}{38.0909} \approx 3.807$	3
c	213	$\frac{213}{38.0909} \approx 5.592$	5
Totals	419		9

The two remaining teaching assistants will be apportioned to Courses a and b, because the fractional parts of their quotas are the largest. The final apportionment is

Course	Q	Number of TA's
a	1.601	$1 + 1 = 2$
b	3.807	$3 + 1 = 4$
c	5.592	5
Totals		11

Here is a final summary:

Course	Pre. Enr.	Actual Enr.	% Inc.	Old No. of TA's	New No. of TA's
a	55	61	**10.9**	1	**2**
b	125	145	16.0	4	4
c	190	213	**12.1**	**6**	**5**
Totals	370	419		11	11

Notice from the table that there was a greater percent increase in growth for Course c than for Course a. Yet Course a gained a teaching assistant and Course c lost one. This is an example of the Population Paradox.

33. Here is the apportionment from Example 4, with a population of 531 for the second subdivision. The total population would be

$$8500 + 1671 + 531 = 10,702.$$

The standard divisor would be

$$d = \frac{10702}{105} \approx 101.9238$$

Community	Pop.	Standard Quota	Rounded down Q
Original	8500	$\frac{8500}{101.9238} \approx 83.396$	83
1st Annexed	1671	$\frac{1671}{101.9238} \approx 16.395$	16
2nd Annexed	531	$\frac{531}{101.9238} \approx 5.210$	5
Totals	10,702		104

The one remaining seat would be apportioned to the original community because the fractional parts of its quota is the largest. The final apportionment is

Community	Q	Number of seats
a	83.396	$83 + 1 = 84$
b	16.395	16
c	5.210	5
Totals		105

Now increasing the population of the second annexed subdivision to 532, here are the calculations. The total population would be

$$8500 + 1671 + 532 = 10,703.$$

The standard divisor would be

$$d = \frac{10703}{105} \approx 101.9333.$$

Community	Pop.	Standard Quota	Rounded down Q
Original	8500	$\frac{8500}{101.9333} \approx 83.388$	83
1st Annexed	1671	$\frac{1671}{101.9333} \approx 16.393$	16
2nd Annexed	532	$\frac{532}{101.9333} \approx 5.219$	5
Totals	10,703		104

Now the one remaining seat would be apportioned to the 1st annexed community because the fractional parts of its quota is the largest. The final apportionment is

Community	Q	Number of seats
a	83.388	83
b	16.396	$16 + 1 = 17$
c	5.219	5
Totals		105

The New State Paradox occurs when the population of the 2nd annexed subdivision increases to 532.

35. The sum of the populations is

$$1720 + 3363 + 6960 + 24223 + 8800 = 45066.$$

The standard divisor is

$$d = \frac{45066}{220} \approx 204.8455.$$

State	Population	Q	Rounded Q
a	1720	$\frac{1720}{204.8455} \approx 8.397$	8
b	3363	$\frac{3363}{204.8455} \approx 16.417$	16
c	6960	$\frac{6960}{204.8455} \approx 33.977$	33
d	24223	$\frac{24223}{204.8455} \approx 118.250$	**118**
e	8800	$\frac{8800}{204.8455} \approx 42.959$	42
Totals	45066		217

Using the Adams method with a modified divisor of 208

State	Population	Q	Rounded Q
a	1720	$\frac{1720}{208} \approx 8.269$	9
b	3363	$\frac{3363}{208} \approx 16.168$	17
c	6960	$\frac{6960}{208} \approx 33.462$	34
d	24223	$\frac{24223}{208} \approx 116.457$	**117**
e	8800	$\frac{8800}{208} \approx 42.308$	43
Totals	45066		220

This is a violation of the Quota Rule because State d receives only 117 seats, although it should receive at least 118.

37. Writing exercise

39. Writing exercise

41. Writing exercise

CHAPTER 15 TEST

1. The voter profile is

Votes	Ranking
5	$a > b > d > c$
6	$b > c > a > d$
5	$c > d > b > a$
7	$d > a > b > c$
4	$c > a > d > b$

By the plurality method the destination with the most votes wins. Cancun receives $5 + 4 = 9$ first place votes to win.

2. Use the ranking information from Exercise 1 to make a table to examine the points in the Borda method.

Votes	Ranking	Points 3	2	1	0
5	$a > b > d > c$	a	b	d	c
6	$b > c > a > d$	b	c	a	d
5	$c > d > b > a$	c	d	b	a
7	$d > a > b > c$	d	a	b	c
4	$c > a > d > b$	c	a	d	b

Examine the last four columns and compute the weighted sum for each candidate.

Candidate	Borda Points
a	$5 \cdot 3 + 6 \cdot 1 + 5 \cdot 0 + 7 \cdot 2 + 4 \cdot 2 = 43$
b	$5 \cdot 2 + 6 \cdot 3 + 5 \cdot 1 + 7 \cdot 1 + 4 \cdot 0 = 40$
c	$5 \cdot 0 + 6 \cdot 2 + 5 \cdot 3 + 7 \cdot 0 + 4 \cdot 3 = 39$
d	$5 \cdot 1 + 6 \cdot 0 + 5 \cdot 2 + 7 \cdot 3 + 4 \cdot 1 = 40$

The winner is Aruba with 43 Borda points.

3. There are 27 sorority sisters voting, so a destination must receive a majority of the votes or at least 14 votes to win by the Hare method. No destination received a majority. See from Exercise 1 that Aruba received 5 first place votes; the Bahamas received 6 votes; Cancun received 9 votes; and the Dominican Republic received 7 votes. Eliminate Aruba because it has the least number of votes, and compare only b, c, and d.

Votes	Ranking	Row
5	$b > d > c$	1
6	$b > c > d$	2
5	$c > d > b$	3
7	$d > b > c$	4
4	$c > d > b$	5

The Bahamas received $5 + 6 = 11$ votes.
Cancun receives $5 + 4 = 9$ votes.
The Dominican Republic receives 7 votes.
Again no destination has received a majority. Therefore, eliminate d with the least number of votes, and compare only b and c.

Votes	Ranking	Row
5	$b > c$	1
6	$b > c$	2
5	$c > b$	3
7	$b > c$	4
4	$c > b$	5

The Bahamas receives $5 + 6 + 7 = 18$ votes.
Cancun receives $5 + 4 = 9$ votes.
Finally the Bahamas is the winner.

4. Using the predetermined order a, c, b, d begin by comparing a and c in the original voting.

Comparison	Votes	Row
$a > c$	$5 + 7 = 12$	1, 4
$c > a$	$6 + 5 + 4 = 15$	2, 3, 5

Cancun wins this competition, 15 to 12. Now compare c to b.

Comparison	Votes	Row
c > b	$5 + 4 = 9$	3, 5
b > c	$5 + 6 + 7 = 18$	1, 2, 4

The Bahamas wins this competition, 18 to 9. Now compare b and d.

Comparison	Votes	Row
b > d	$5 + 6 = 11$	1, 2
d > b	$5 + 7 + 4 = 16$	3, 4, 5

Finally Dominican Republic wins this competition, 16 to 11, and also wins the entire competition in the sequential pairwise comparison method.

5.   Writing exercise

6.   Using the approval voting method with each sister voting for her first and second choice, here are the number of votes for each destination:

Aruba	$5 + 7 + 4 = 16$
Bahamas	$5 + 6 = 11$
Cancun	$6 + 5 + 4 = 15$
Dominican Republic	$5 + 7 = 12$

Aruba wins.

7.   Using the approval voting method with the first 5 sisters giving approval to all destinations except Cancun, the approvals look like this:

Number of Voters	Ranking
5	a, b, d
6	b, c
5	c, d
7	d, a
4	c, a

Here are the number of votes for each destination:

Aruba	$5 + 7 + 4 = 16$
Bahamas	$5 + 6 = 11$
Cancun	$6 + 5 + 4 = 15$
Dominican Republic	$5 + 5 + 7 = 17$

The Dominican Republic wins.

8.   Using the pairwise comparison method make a table to compute the votes in comparisons.

Comparison	Votes	Rows
a > b	$5 + 7 + 4 = 16$	1, 4, 5
b > a	$6 + 5 = 11$	2, 3
a > c	$5 + 7 = 12$	1, 4
c > a	$6 + 5 + 4 = 15$	2, 3, 5
a > d	$5 + 6 + 4 = 15$	1, 2, 5
d > a	$5 + 7 = 12$	3, 4
b > c	$5 + 6 + 7 = 18$	1, 2, 4
c > b	$5 + 4 = 9$	3, 5
b > d	$5 + 6 = 11$	1, 2
d > b	$5 + 7 + 4 = 16$	3, 4, 5
c > d	$6 + 5 + 4 = 15$	2, 3, 5
d > c	$5 + 7 = 12$	1, 4

Make another table to award pairwise points.

Pairs	Votes	Pairwise Points
a : b	16 : 11	a, 1
a : c	12 : 15	c, 1
a : d	15 : 12	a, 1
b : c	18 : 9	b, 1
b : d	11 : 16	d, 1
c : d	15 : 12	c, 1

Using the pairwise comparison method, here are the final point totals.

$$a: \quad 1 + 1 = 2$$
$$b: \quad \quad 1 = 1$$
$$c: \quad 1 + 1 = 2$$
$$d: \quad \quad 1 = 1$$

There is a tie between Aruba and Cancun.

9.   Using the predetermined order a, b, c, d begin by comparing a and b in the original voting.

Comparison	Votes	Row
a > b	$5 + 7 + 4 = 16$	1, 4, 5
b > a	$6 + 5 = 11$	2, 3

Aruba wins this competition, 16 to 11. Now compare a to c.

Comparison	Votes	Row
a > c	$5 + 7 = 12$	1, 4
c > a	$6 + 5 + 4 = 15$	2, 3, 5

Cancun wins this competition, 15 to 12. Now compare c and d.

Comparison	Votes	Row
$c > d$	$6 + 5 + 4 = 15$	$2, 3, 5$
$d > c$	$5 + 7 = 12$	$1, 4$

Finally Cancun wins this competition, 15 to 12, and also wins the entire competition in the sequential pairwise comparison method.

10. Writing exercise

11. Writing exercise

12. Writing exercise

13. Writing exercise

14. Use the ranking information from the text to make a table to examine the points in the Borda method.

Votes	Ranking	Points 2	1	0
16	$a > b > c$	a	b	c
8	$b > c > a$	b	c	a
7	$c > b > a$	c	b	a

Examine the last three columns and compute the weighted sum for each candidate.

Candidate	Borda Points
a	$16 \cdot 2 + 8 \cdot 0 + 7 \cdot 0 = 32$
b	$16 \cdot 1 + 8 \cdot 2 + 7 \cdot 1 = 39$
c	$16 \cdot 0 + 8 \cdot 1 + 7 \cdot 2 = 22$

Although a has the majority of first place votes in the ranking, candidate b wins by the Borda method. This violates the majority criterion.

15. Make a table to compute the votes in comparisons.

Comparison	Votes	Rows
$a > b$	$5 = 5$	1
$b > a$	$6 + 4 + 6 = 16$	$2, 3, 4$
$a > c$	$5 + 4 = 9$	$1, 3$
$c > a$	$6 + 6 = 12$	$2, 4$
$b > c$	$4 + 6 = 10$	$3, 4$
$c > b$	$5 + 6 = 11$	$1, 2$

Make another table to award pairwise points.

Pairs	Votes	Pairwise Points
$a : b$	$5 : 16$	b, 1
$a : c$	$9 : 12$	c, 1
$b : c$	$10 : 11$	c, 1

Using the pairwise comparison method, the winner is c with 2 pairwise points. Alternative b received only 1 point, and a received 0 points.

16. Alternative c is the Condorcet candidate.

Plurality method

$$
\begin{array}{ll}
a & 5 = 5 \\
b & 4 + 6 = 10 \\
c & 6 = 6
\end{array}
$$

Alternative b is selected, which violates the Condorcet criterion.

Borda method
Use the ranking information from the text to make a table to examine the points in the Borda method.

Votes	Ranking	Points 2	1	0
5	$a > c > b$	a	c	b
6	$c > b > a$	c	b	a
4	$b > a > c$	b	a	c
6	$b > c > a$	b	c	a

Examine the last three columns and compute the weighted sum for each candidate.

Candidate	Borda Points
a	$5 \cdot 2 + 6 \cdot 0 + 4 \cdot 1 + 6 \cdot 0 = 14$
b	$5 \cdot 0 + 6 \cdot 1 + 4 \cdot 2 + 6 \cdot 2 = 26$
c	$5 \cdot 1 + 6 \cdot 2 + 4 \cdot 0 + 6 \cdot 1 = 23$

Alternative b wins by the Borda method. This violates the Condorcet criterion.

Hare method
None of the candidates has a majority of first place votes, which is at least 11. Eliminate a because it has the least number of votes at 5, and compare only b and c.

Votes	Ranking
5	$c > b$
6	$c > b$
4	$b > c$
6	$b > c$

Now alternative c receives $5 + 6 = 11$ votes and wins. This does not violate the Condorcet criterion.

17. Make a table to compute the votes in comparisons.

Comparison	Votes	Rows
c > m	8 + 6 = 14	1, 3
m > c	7 + 5 = 12	2, 4
c > b	8 + 5 = 13	1, 4
b > c	7 + 6 = 13	2, 3
c > s	8 + 5 = 13	1, 4
s > c	7 + 6 = 13	2, 3
m > b	8 + 5 = 13	1, 4
b > m	7 + 6 = 13	2, 3
m > s	8 + 5 = 13	1, 4
s > m	7 + 6 = 13	2, 3
b > s	8 + 5 = 13	1, 4
s > b	7 + 6 = 13	2, 3

Make another table to award pairwise points.

Pairs	Votes	Pairwise Points
c:m	14:12	c, 1
c:b	13:13	c, $\frac{1}{2}$; b, $\frac{1}{2}$
c:s	13:13	c, $\frac{1}{2}$; s, $\frac{1}{2}$
m:b	13:13	m, $\frac{1}{2}$; b, $\frac{1}{2}$
m:s	13:13	m, $\frac{1}{2}$; s, $\frac{1}{2}$
b:s	13:13	b, $\frac{1}{2}$; s, $\frac{1}{2}$

Using the pairwise comparison method, here are the final point totals.

$$c: \quad 1 + \frac{1}{2} + \frac{1}{2} = 2$$
$$b: \quad \frac{1}{2} + \frac{1}{2} + \frac{1}{2} = 1\frac{1}{2}$$
$$m: \quad \frac{1}{2} + \frac{1}{2} = 1$$
$$s: \quad \frac{1}{2} + \frac{1}{2} + \frac{1}{2} = 1\frac{1}{2}$$

Using the pairwise comparison method, the winner is c with 2 pairwise points.

Now change the ranking of the last 5 voters as shown in the text and make a table to compute the votes in comparisons.

Comparison	Votes	Rows
c > m	8 + 6 + 5 = 19	1, 3, 4
m > c	7 = 7	2
c > b	8 + 5 = 13	1, 4
b > c	7 + 6 = 13	2, 3
c > s	8 + 5 = 13	1, 4
s > c	7 + 6 = 13	2, 3
m > b	8 + 5 = 13	1, 4
b > m	7 + 6 = 13	2, 3
m > s	8 = 8	1
s > m	7 + 6 + 5 = 18	2, 3, 4
b > s	8 = 8	1
s > b	7 + 6 + 5 = 18	2, 3, 4

Make another table to award pairwise points.

Pairs	Votes	Pairwise Points
c:m	19:7	c, 1
c:b	13:13	c, $\frac{1}{2}$; b, $\frac{1}{2}$
c:s	13:13	c, $\frac{1}{2}$; s, $\frac{1}{2}$
m:b	13:13	m, $\frac{1}{2}$; b, $\frac{1}{2}$
m:s	8:18	s, 1
b:s	8:18	s, 1

Using the pairwise comparison method, here are the final point totals.

$$c: \quad 1 + \frac{1}{2} + \frac{1}{2} = 2$$
$$b: \quad \frac{1}{2} + \frac{1}{2} = 1$$
$$m: \quad \frac{1}{2} = \frac{1}{2}$$
$$s: \quad \frac{1}{2} + 1 + 1 = 2\frac{1}{2}$$

Using the pairwise comparison method, the winner is s with $2\frac{1}{2}$ pairwise points. Although c was moved to the top of the ranking when the 5 voters changed their votes, c did not win. The pairwise comparison method has violated the monotonicity criterion.

18. Count the number of first place votes to see that choice a has 9; choice b has 8, choice c has 12, and choice d has 0. None of the candidates has a majority of first place votes, which is at least 15. The elimination of d does not change the number of first place votes of a, b, and c. Therefore, eliminate b and compare only a and c.

Votes	Ranking
9	$a > c$
8	$a > c$
7	$c > a$
5	$c > a$

Now a receives $9 + 8 = 17$ votes and wins.

Now change the ranking of the last 5 voters as shown in the text and apply the Hare method again. Count the number of first place votes to see that choice a now has 14, choice b still has 8, and choice c now has 7. Because none of the candidates has a majority, eliminate c and compare only a and b.

Votes	Ranking
9	$a > b$
8	$b > a$
7	$b > a$
5	$a > b$

This time b received $8 + 7 = 15$ votes and wins. Although a was moved to the top of the ranking when the 5 voters changed their votes, a did not win. The Hare method has violated the monotonicity criterion.

19. Examine the voter profile in the text to see that candidate a receives the most votes at 10 and is the winner by the plurality method. If alternative c is dropped from the selection process the rankings are:

Votes	Ranking
10	$a > b$
7	$b > a$
5	$b > a$

Now alternative b has 12 votes and wins. Although losing alternative c is dropped, candidate a does not win the second election. This is a violation of the Independence of Irrelevant Alternatives criterion.

20. Use the ranking information from the text to make a table to examine the points in the Borda method.

Votes	Ranking	Points			
		3	2	1	0
7	$a > b > c > d$	a	b	c	d
5	$b > c > d > a$	b	c	d	a
4	$d > c > a > b$	d	c	a	b

Examine the last four columns and compute the weighted sum for each candidate.

Candidate	Borda Points
a	$7 \cdot 3 + 5 \cdot 0 + 4 \cdot 1 = 25$
b	$7 \cdot 2 + 5 \cdot 3 + 4 \cdot 0 = 2$
c	$7 \cdot 1 + 5 \cdot 2 + 4 \cdot 2 = 25$
d	$7 \cdot 0 + 5 \cdot 1 + 4 \cdot 3 = 17$

Alternative b wins by the Borda method.

Now repeat the Borda method after eliminating alternative d. Make a table to examine the points in the Borda method.

Votes	Ranking	Points		
		2	1	0
7	$a > b > c$	a	b	c
5	$b > c > a$	b	c	a
4	$c > a > b$	c	a	b

Examine the last three columns and compute the weighted sum for each candidate.

Candidate	Borda Points
a	$7 \cdot 2 + 5 \cdot 0 + 4 \cdot 1 = 18$
b	$7 \cdot 1 + 5 \cdot 2 + 4 \cdot 0 = 17$
c	$7 \cdot 0 + 5 \cdot 1 + 4 \cdot 2 = 13$

This time alternative a wins by the Borda method. This is a violation of the Independence of Irrelevant Alternatives criterion.

21. Count the number of first place votes to see that choice a has 7; choice b has 5, choice d has 4, and choice c has 0. None of the candidates has a majority of first place votes, which is at least 9. The elimination of c does not change the number of first place votes of a, b, and d. Therefore, eliminate d and compare only a and b.

Votes	Ranking
7	a > b
5	b > a
4	a > b

Now a receives $7 + 4 = 11$ votes and wins.

Now eliminate candidate b apply the Hare method again.

Here is a new voter profile:

Votes	Ranking
7	a > c > d
5	c > d > a
4	d > c > a

Count the number of first place votes to see that choice a now has 7, choice c has 5, and choice d now has 4. Because none of the candidates has a majority, eliminate d and compare only a and c.

Votes	Ranking
7	a > c
5	c > a
4	c > a

This time c received $5 + 4 = 9$ votes and wins. Although b was dropped from the ranking, the original preferred candidate a did not win. This is a violation of the Independence of Irrelevnt Alternative criterion.

22. Writing exercise

23. Using the Hamilton method, the standard divisor is

$$1429 + 8639 + 7608 + 6660 + 1671 = 26007.$$

The standard divisor is

$$d = \frac{26007}{195} \approx 133.369.$$

Ward	Population	Q	Rounded Q
1st	1429	$\frac{1429}{133.369} \approx 10.715$	10
2nd	8639	$\frac{8639}{133.369} \approx 64.775$	64
3rd	7608	$\frac{7608}{133.369} \approx 57.045$	57
4th	6660	$\frac{6660}{133.369} \approx 49.937$	49
5th	1671	$\frac{1671}{133.369} \approx 12.529$	12
Totals	26007		192

The three remaining seats will be apportioned to the 1st, 2nd and 4th wards, because the fractional parts of their quotas are the largest. The final apportionment is

Ward	Q	Number of seats
1st	10.715	$10 + 1 = 11$
2nd	64.775	$64 + 1 = 65$
3rd	57.045	57
4th	49.937	$49 + 1 = 50$
5th	12.529	12
Totals		195

24. Using the Jefferson method with $md = 131$

Ward	Population	Q	Rounded Q
1st	1429	$\frac{1429}{131} \approx 10.908$	10
2nd	8639	$\frac{8639}{131} \approx 65.947$	65
3rd	7608	$\frac{7608}{131} \approx 58.076$	58
4th	6660	$\frac{6660}{131} \approx 50.840$	50
5th	1671	$\frac{1671}{131} \approx 12.756$	12
Totals	26007		195

25. Using the Webster method with $md = 133.7$

Ward	Population	Q	Rounded Q
1st	1429	$\frac{1429}{133.7} \approx 10.688$	11
2nd	8639	$\frac{8639}{133.7} \approx 64.615$	65
3rd	7608	$\frac{7608}{133.7} \approx 56.904$	57
4th	6660	$\frac{6660}{133.7} \approx 49.813$	50
5th	1671	$\frac{1671}{133.7} \approx 12.498$	12
Totals	26007		195

26. Writing exercise

27. Writing exercise

28. Writing exercise

29. Writing exercise

30. The sum of the populations is

$$2354 + 4500 + 5598 + 23000 = 35452.$$

The standard divisor is

$$d = \frac{35452}{100} = 354.52.$$

Using the standard quota

State	Population	Q	Rounded Q
a	2354	$\frac{2354}{354.52} \approx 6.640$	6
b	4500	$\frac{4500}{354.52} \approx 12.693$	12
c	5598	$\frac{5598}{354.52} \approx 15.790$	15
d	23000	$\frac{23000}{354.52} \approx 64.876$	**64**
Totals	35,452		97

Using the Jefferson method with $md = 347$

State	Population	Q	Rounded Q
a	2354	$\frac{2354}{347} \approx 6.784$	6
b	4500	$\frac{4500}{347} \approx 12.968$	12
c	5598	$\frac{5598}{347} \approx 16.133$	16
d	23000	$\frac{23000}{347} \approx 66.282$	**66**
Totals	35,452		100

The Jefferson method violates the Quota Rule because state d receives two more seats than its apportionment from the standard quota.

31. First calculate the apportionment with $n = 126$. The total population is

$$263 + 809 + 931 + 781 + 676 = 3460.$$

The standard divisor for 126 seats is

$$d = \frac{3460}{126} \approx 27.4603.$$

Using the Hamilton method

State	Population	Q	Rounded Q
a	263	$\frac{263}{27.4603} \approx 9.577$	9
b	809	$\frac{809}{27.4603} \approx 29.461$	29
c	931	$\frac{931}{27.4603} \approx 33.903$	33
d	781	$\frac{781}{27.4603} \approx 28.441$	28
e	676	$\frac{676}{27.4603} \approx 24.617$	24
Totals	3460		123

The three remaining seats will be apportioned to States c, e, and a, because the fractional parts of their quotas are the largest. The final apportionment is

State	Q	Number of seats
a	9.577	$9 + 1 = $ **10**
b	29.461	29
c	33.903	$33 + 1 = 34$
d	28.441	28
e	24.617	$24 + 1 = 25$
Totals		126

Now increase the number of seats to 127. The standard divisor is

$$d = \frac{3460}{127} \approx 27.2441.$$

Using the Hamilton method with the new d

State	Population	Q	Rounded Q
a	263	$\frac{263}{27.2441} \approx 9.653$	9
b	809	$\frac{809}{27.2441} \approx 29.695$	29
c	931	$\frac{931}{27.2441} \approx 34.173$	34
d	781	$\frac{781}{27.2441} \approx 28.667$	28
e	676	$\frac{676}{27.2441} \approx 24.813$	24
Totals	3460		124

The three remaining seats will be apportioned to States e, b, and d, because the fractional parts of their quotas are the largest. The final apportionment is

State	Q	Number of seats
a	9.653	**9**
b	29.695	$29 + 1 = 30$
c	34.173	34
d	28.667	$28 + 1 = 29$
e	24.813	$24 + 1 = 25$
Totals		127

State a is a victim of the Alabama Paradox, because it has lost a seat despite the fact that the overall number of seats has increased.

32. First calculate the apportionment for the initial populations. The total population is

$$55 + 125 + 190 = 370.$$

The standard divisor for 11 seats is

$$d = \frac{370}{11} = 33.6364.$$

Using the Hamilton method

State	Population	Q	Rounded Q
a	55	$\frac{55}{33.6364} \approx 1.635$	1
b	125	$\frac{125}{33.6364} \approx 3.716$	3
c	190	$\frac{190}{33.6364} \approx 5.649$	5
Totals	370		9

The two remaining seats will be apportioned to States b and c, because the fractional parts of their quotas are the largest. The final apportionment is

State	Q	Number of seats
a	1.635	1
b	3.716	$3 + 1 = 4$
c	5.649	$5 + 1 = 6$
Totals		11

Now apportion the seats for the revised populations.

$$63 + 150 + 220 = 433.$$

The new standard divisor is

$$d = \frac{433}{11} \approx 39.3636.$$

Using the Hamilton method with the new values

State	Population	Q	Rounded Q
a	63	$\frac{63}{39.3636} \approx 1.600$	1
b	150	$\frac{150}{39.3636} \approx 3.811$	3
c	220	$\frac{220}{39.3636} \approx 5.589$	5
Totals	433		9

The two remaining seats will be apportioned to States a and b, because the fractional parts of their quotas are the largest. The final apportionment is

State	Q	Number of seats
a	1.600	$1 + 1 = 2$
b	3.811	$3 + 1 = 4$
c	5.589	5
Totals		11

Here is a final summary:

State	Old Pop	New Pop	% Inc.	Old No. of Seats	New No. of Seats
a	55	63	**14.55**	1	**2**
b	125	150	20.00	4	4
c	190	220	**15.79**	**6**	**5**
Totals				11	11

Notice from the table that there was a greater percent increase in growth for State c than for State a. Yet State a gained a seat and State c lost a seat. This is an example of the Population Paradox.

33. First calculate the apportionment for the initial populations. The total population is

$$49 + 160 = 209.$$

The standard divisor for 100 seats is

$$d = \frac{209}{100} = 2.09.$$

Using the Hamilton method

State	Population	Q	Rounded Q
a	49	$\frac{49}{2.09} \approx 23.445$	23
b	160	$\frac{160}{2.09} \approx 76.555$	76
Totals	209		99

The one remaining seat will be apportioned to State b because the fractional parts of its quota is larger. The final apportionment is

State	Q	Number of seats
a	23.445	23
b	76.555	$76 + 1 = 77$
Totals		100

Now apportion the seats to include the new state. The standard quota of the new state is:

$$Q = \frac{32}{2.09} \approx 15.311.$$

Rounded down to 15, add 15 new seats to the original 100 to obtain 115 seats to be apportioned. Now the new population is

$$49 + 160 + 32 = 241.$$

The new standard divisor is

$$d = \frac{241}{115} \approx 2.0957.$$

State	Population	Q	Rounded Q
a	49	$\dfrac{49}{2.0957} \approx 23.381$	23
b	160	$\dfrac{160}{2.0957} \approx 76.347$	76
c	32	$\dfrac{32}{2.0957} \approx 15.269$	15
Totals	241		114

The one remaining seat will be apportioned to State a because the fractional part of its quota is the largest. The final apportionment is

State	Q	Number of seats
a	23.381	$23 + 1 = 24$
b	76.347	76
c	15.269	15
Totals		115

The New State Paradox has occurred because the addition of the new state has caused a shift in the apportionment of the original states. States a and b originally had 23 and 77 seats, respectively; now they have 24 and 76, respectively.

34. Writing exercise

THE METRIC SYSTEM

1. $\dfrac{8\text{ m}}{1} \cdot \dfrac{1000\text{ mm}}{1\text{ m}} = 8000\text{ mm}$

3. $\dfrac{8500\text{ cm}}{1} \cdot \dfrac{1\text{ m}}{100\text{ cm}} = 85\text{ m}$

5. $\dfrac{68.9\text{ cm}}{1} \cdot \dfrac{10\text{ mm}}{1\text{ cm}} = 689\text{ mm}$

7. $\dfrac{59.8\text{ mm}}{1} \cdot \dfrac{1\text{ cm}}{10\text{ mm}} = 5.98\text{ cm}$

9. $\dfrac{5.3\text{ km}}{1} \cdot \dfrac{1000\text{ m}}{1\text{ km}} = 5300\text{ m}$

11. $\dfrac{27,500\text{ m}}{1} \cdot \dfrac{1\text{ km}}{1000\text{ m}} = 27.5\text{ km}$

13. 2.54 cm; 25.4 mm

15. 5 cm; 50 mm

17. $\dfrac{6\text{ L}}{1} \cdot \dfrac{10^2\text{ cl}}{1\text{ L}} = 6 \times 100 = 600\text{ cl}$

19. $\dfrac{8.7\text{ L}}{1} \cdot \dfrac{10^3\text{ ml}}{1\text{ L}} = 8.7 \times 1000 = 8700\text{ ml}$

21. $\dfrac{925\text{ cl}}{1} \cdot \dfrac{1\text{ L}}{10^2\text{ cl}} = \dfrac{925}{100} = 9.25\text{ L}$

23. $\dfrac{8974\text{ ml}}{1} \cdot \dfrac{1\text{ L}}{10^3\text{ ml}} = \dfrac{8974}{1000} = 8.974\text{ L}$

25. $\dfrac{8000\text{ g}}{1} \cdot \dfrac{1\text{ kg}}{10^3\text{ g}} = \dfrac{8000}{1000} = 8\text{ kg}$

27. $\dfrac{5.2\text{ kg}}{1} \cdot \dfrac{10^3\text{ g}}{1\text{ kg}} = 5.2 \times 1000 = 5200\text{ g}$

29. $\dfrac{4.2\text{ g}}{1} \cdot \dfrac{10^3\text{ mg}}{1\text{ g}} = 4.2 \times 1000 = 4200\text{ mg}$

31. $\dfrac{598\text{ mg}}{1} \cdot \dfrac{1\text{ g}}{10^3\text{ mg}} = \dfrac{598}{1000} = .598\text{ g}$

33. $C = \dfrac{5(F-32)}{9} = \dfrac{5(86-32)}{9} = \dfrac{5(54)}{9} = 30°$

35. $C = \dfrac{5(F-32)}{9} = \dfrac{5(-114-32)}{9} = \dfrac{5(-146)}{9} = -81°$

37. $F = \dfrac{9}{5}C + 32 = \dfrac{9}{5} \cdot 10 + 32 = 18 + 32 = 50°$

39. $F = \dfrac{9}{5}C + 32 = \dfrac{9}{5} \cdot -40 + 32 = -72 + 32 = -40°$

41. $\dfrac{1\text{ kg}}{1} \cdot \dfrac{1000\text{ g}}{1\text{ kg}} \cdot \dfrac{1\text{ nickel}}{5\text{ g}} = 200\text{ nickels}$

43. $\dfrac{1\text{ L}}{1} \cdot \dfrac{1000\text{ ml}}{1\text{ L}} \cdot \dfrac{.0002\text{ g}}{1\text{ ml}} = .2\text{ g}$

45. $\dfrac{7\text{ strips}}{1} \cdot \dfrac{67\text{ cm}}{1\text{ strip}} \cdot \dfrac{1\text{ m}}{100\text{ cm}} \cdot \dfrac{\$8.74}{1\text{ m}} = \$40.99$

47. $A = 128\text{ cm} \cdot 174\text{ cm} = 22,272\text{ cm}^2$

$\dfrac{22,272\text{ cm}^2}{1} \cdot \dfrac{(1\text{ m})^2}{(100\text{ cm})^2} \cdot \dfrac{\$174.20}{1\text{ m}^2} = \$387.98$

49. $\dfrac{82\text{ cm}}{1} \cdot \dfrac{1\text{ m}}{100\text{ cm}} = .82\text{ m}$

$V = .82\text{ m} \cdot 1.1\text{ m} \cdot 1.2\text{ m} = 1.0824\text{ m}^3$

$\dfrac{1.0824\text{ m}^3}{1} \cdot \dfrac{(100\text{ cm})^3}{(1\text{ m})^3} = 1,082,400\text{ cm}^3$

51. $\dfrac{160\text{ L}}{1} \cdot \dfrac{1000\text{ ml}}{1\text{ L}} \cdot \dfrac{1\text{ bottle}}{800\text{ ml}} = 200\text{ bottles}$

53. $\dfrac{982\text{ yd}}{1} \cdot \dfrac{.9144\text{ m}}{1\text{ yd}} = 897.9\text{ m}$

55. $\dfrac{125\text{ mi}}{1} \cdot \dfrac{1.609\text{ km}}{1\text{ mi}} = 201.1\text{ km}$

57. $\dfrac{1816\text{ g}}{1} \cdot \dfrac{.0022\text{ lb}}{1\text{ g}} = 3.995\text{ lb}$

59. $\dfrac{47.2\text{ lb}}{1} \cdot \dfrac{454\text{ g}}{1\text{ lb}} = 21,428.8\text{ g}$

61. $\dfrac{28.6\text{ L}}{1} \cdot \dfrac{1.0567\text{ qt}}{1\text{ L}} = 30.22\text{ qt}$

63. $\dfrac{28.2\text{ gal}}{1} \cdot \dfrac{3.785\text{ L}}{1\text{ gal}} = 106.7\text{ L}$

65. Unreasonable; $\dfrac{2\text{ kg}}{1} \cdot \dfrac{2.2\text{ lb}}{1\text{ kg}} = 4.4\text{ lb}$

67. Reasonable;

$\dfrac{25\text{ ml}}{1} \cdot \dfrac{1\text{ L}}{1000\text{ ml}} \cdot \dfrac{1.0567\text{ qt}}{1\text{ L}} \cdot \dfrac{32\text{ oz}}{1\text{ qt}} \cdot \dfrac{2\text{ T}}{1\text{ oz}} \approx 1.7\text{ T}$

69. Unreasonable; $\dfrac{.5\text{ L}}{1} \cdot \dfrac{1.0567\text{ qt}}{1\text{ L}} \approx .5\text{ qt}$

71. B; $\dfrac{3\text{ m}}{1} \cdot \dfrac{3.2808\text{ ft}}{1\text{ m}} \approx 9.8\text{ ft}$

73. B; $\dfrac{5000\text{ km}}{1} \cdot \dfrac{.6214\text{ mi}}{1\text{ km}} \approx 3100\text{ mi}$

75. C; $\dfrac{198\text{ mm}}{1} \cdot \dfrac{1\text{ in}}{25.4\text{ cm}} \approx 7.8\text{ in}$

77. A; $\dfrac{1300\text{ kg}}{1} \cdot \dfrac{2.20\text{ lb}}{1\text{ kg}} \approx 2900\text{ lb}$

79. A; $\dfrac{180\text{ cm}}{1} \cdot \dfrac{1\text{ m}}{100\text{ cm}} \cdot \dfrac{3.2808\text{ ft}}{1\text{ m}} \approx 5.9\text{ ft}$

81. C; $\dfrac{800 \text{ m}}{1} \cdot \dfrac{1 \text{ km}}{1000 \text{ m}} \cdot \dfrac{.6214 \text{ mi}}{1 \text{ km}} \approx .5 \text{ mi}$

83. A; $\dfrac{70 \text{ cm}}{1} \cdot \dfrac{1 \text{ m}}{100 \text{ cm}} \cdot \dfrac{39.37 \text{ in}}{1 \text{ m}} \approx 28 \text{ in}$

85. B; $\dfrac{50 \text{ cm}}{1} \cdot \dfrac{1 \text{ m}}{100 \text{ cm}} \cdot \dfrac{39.37 \text{ in}}{1 \text{ m}} \approx 20 \text{ in}$

87. B; $\dfrac{9 \text{ mm}}{1} \cdot \dfrac{1 \text{ in}}{25.4 \text{ mm}} \approx .35 \text{ in}$

89. A; This is the freezing temperature of water in Celsius.

91. B; $F = \dfrac{9}{5} \cdot 60 + 32 = 108 + 32 = 140°F$

93. C; $F = \dfrac{9}{5} \cdot 10 + 32 = 18 + 32 = 50°F$

95. B; $F = \dfrac{9}{5} \cdot 170 + 32 = 306 + 32 \approx 340°F$